Ventilation of Buildings

D0420672

Essential reading

Building Services Design Methodology, a practical guide
D. Bownass   pbk: 0-419-25280-0
                                                    Spon Press

Building Services Engineering, 3rd edition
D. Chadderton   hbp: 0-419-25730-6
                                                    pbk: 0-419-25740-3
                                                    Spon Press

**Naturally Ventilated Buildings, building for the senses,
the economy and society**
Edited by D. Clements-Croome   hbk: 0-419-21520-4
                                                    Spon Press

Indoor Air Quality Issues
D. L. Hansen   hbk: 1-56032-866-5
                                                    Spon Press

Information and ordering details
For price availability and ordering visit our website www.sponpress.com
Alternatively our books are available from all good bookshops.

Ventilation of Buildings
Second edition

Hazim B. Awbi

Spon Press
Taylor & Francis Group

LONDON AND NEW YORK

First published 2003
by Spon Press
2 Park Square, MiltonPark, Abingdon, Oxon OX144RN

Simultaneously published in the USA and Canada
by Spon Press
270 Madison Ave, NewYork, NY 10016

Transferred to Digital Printing 2008

Spon Press is an imprint of the Taylor & Francis Group, an informa businesss

© 2003 Hazim B. Awbi

Typeset in Sabon by
Newgen Imaging Systems (P) Ltd, Chennai, India
Printed and bound in Great Britain by
TJI Digital, Padstow, Cornwall

British Library Cataloguing in Publication Data
A catalogue record for this book is available
from the British Library

Library of Congress Cataloging in Publication Data
A catalog record for this book has been requested

ISBN 10: 0-415-27056-1 (pbk)
ISBN 10: 0-415-27055-3 (hbk)
ISBN 13: 978-0-415-27056-4 (pbk)
ISBN 13: 978-0-415-27055-7 (hbk)

Contents

Preface to the second edition

During the last three decades ventilation philosophy has been experiencing major changes. In the first decade of this period, considerable efforts were made towards understanding the mechanisms of air infiltration in buildings in order to control and often reduce the fortuitous ventilation and conserve energy. In some cases, the reduction in air infiltration created problems associated with the air quality in buildings and the generic term 'sick building syndrome' came into being. The second decade of the same period experienced concerted efforts to understand the causes of sick buildings, which resulted in the introduction of new ventilation concepts, such as the age of air, and new air quality units, and a consensus for increased outdoor air flow rates. In the third decade, the emphasis on reducing energy consumption and environmental consciousness has focused the minds of researchers and designers alike on the potential of natural ventilation and user control of the local environment. As a result of these changes, new ventilation standards and guidelines have been written to reflect the importance of ventilation on the quality of indoor environment.

Almost 12 years have passed since the publication of the first edition of *Ventilation of Buildings*. During this time ventilation has truly been established as a discipline for scientists to research and engineers to apply. Previously, ventilation was considered to be on the fringe of scientific research; this is no more to be. The subject now attracts respected researchers from different backgrounds all round the world, who are engaged in developing new theories and applications into more effective and sustainable ventilation systems. This shift on emphasis did not happen accidentally but as a result of man's needs for better indoor environment.

In this second edition, almost every chapter from the first edition has been thoroughly updated and a new chapter on natural ventilation has been added. Recent developments in ventilation concepts and room air distribution methods are included within the updated chapters. It was the intention that this edition provides the reader with recent developments in the subject but, at the same time, emphasizes the practical aspects that are needed for modern ventilation system design.

With the recognition of ventilation as a science, much good quality research has been done. I tried to capture some of this in the new edition, but in a text of this kind it is impractical to include all of the good quality research that has been done and hope that I have not omitted the obvious!

Finally, my thanks and appreciations to those colleagues who provided their research data and illustrations to include in this edition. My thanks also go to Michela Marchetti for producing the CAD drawings for some of the new illustrations.

Hazim B. Awbi
Reading, UK, 2002

Preface to the first edition

The term air-conditioning is generally understood to mean the heating, cooling and control of moisture in buildings. Consequently, this involves heating and cooling load calculations in addition to the design of plant components, ductwork and control systems. Ventilation, on the other hand, refers to the provision of sufficient quantities of outside air in the building for the occupants to breathe and to dilute the concentration of pollution generated by people, equipment and materials inside the building. This is necessary for both air-conditioned and non-air-conditioned buildings. The dilution of indoor contaminants is influenced by the quantity and quality of the outside air supplied to the building as well as the way this air is distributed around the space. The method of air distribution is also a vital part of any air-conditioning system. In this book, the emphasis is on the last two themes, namely quantifying the outdoor air flow rate to a building and distributing this ventilation air around the space.

During the last 15 years ventilation philosophy has been experiencing major changes. In the first part of this period, considerable efforts were made to understand the mechanisms of air infiltration in buildings in order to control, and often to reduce, the fortuitous ventilation and conserve energy. In some cases, the reduction in air infiltration created problems associated with the air quality in the building and the generic term 'sick building syndrome' came into being. The second half of the same period experienced concerted efforts to understand the causes of sick buildings, which resulted in the introduction of two new air quality units by Professor P. O. Fanger, namely the olf and the decipol, and a consensus for increased outdoor air flow rates. As a result of these changes, some ventilation standards, which initially recommended a reduction in outdoor air requirement for occupancy, had to increase these rates beyond those recommended prior to this period. Reflecting the current attitude, the new ASHRAE Standard 62-1989 has increased the minimum outdoor air per person from 2.5 to $7.5\,\mathrm{l\,s^{-1}}$, i.e. a threefold increase.

This book has been designed to complement rather than replicate the HVAC handbooks such as those by ASHRAE and CIBSE. Where appropriate the theory of a design problem is given to broaden the readers' horizon of the subject. Recent developments in ventilation requirements, thermal comfort, indoor air quality and room air distribution are also included. The text is intended for the practitioner in the building services industry, the architect, the postgraduate student taking courses or researching in HVAC or general building services and the undergraduate studying building services as a major subject. The book assumes that readers are familiar with the basic principles of fluid flow and heat transfer and knowledgeable with regard to the thermal characteristics of building fabric. However, Chapter 7 requires more advanced knowledge of partial differential equations, which describe the turbulent flow and heat transfer processes of fluids.

Chapter 1 presents the theory and practice of thermal comfort and indoor air quality (AIQ) in some detail, including recent advances in these areas based largely on the work of Fanger and his associates. Chapters 2 and 3 describe the procedures used in determining ventilation rate requirements for different occupancy and the methods of determining air infiltration rates and the design of passive and smoke ventilation techniques. Chapters 4 and 5 outline the theory of different types of air jets, including the influence of buoyancy, Coanda force and obstructions and the aerodynamic characteristics of various air terminal devices used in practice. Chapter 6 presents, in some detail, the theory of physical modelling as applied to room air movement and discusses the influence of various fluid and heat flow parameters on the modelling process. This chapter also contains case studies involving reduced-scale and prototype measurements. Design procedures for different room air distribution methods are also presented here. Chapter 7 is devoted to the theory of computational fluid dynamics (CFD) as applied to ventilation and room air movement and presents the results of recent publications to illustrate the state of the art in this rapidly expanding field of ventilation research. Finally, Chapter 8 deals with the principles of measuring air temperature, radiant temperature, humidity, pressure, air velocity, air flow rate, thermal comfort, indoor air contamination and air flow visualization.

This book provides the reader with recent developments in the subject which are largely missing from other titles currently available. The ultimate aim is, of course, better design of ventilation systems to reduce the frequency of public complaints about HVAC systems which have increased in recent years. Finally, with the subject of intelligent buildings of the future frequently being mentioned, it is hoped that this book will provide the designer of ventilation systems for these buildings with some of the necessary information.

Hazim B. Awbi
1991

1 Human comfort and ventilation

1.1 Introduction

The purpose of a ventilation system is to provide acceptable microclimate in the space being ventilated. In this context, microclimate refers to thermal environment as well as air quality. These two factors must be considered in the design of a ventilation system for a room or a building, as they are fundamental to the comfort and well-being of the human occupants or the performance of industrial processes within these spaces. In a modern technological society, people spend more than 90% of their time in an artificial environment (a dwelling, a workplace or a transport vehicle). As a result of the energy-saving measures which started in the early 1970s these artificially created 'internal' or 'indoor' environments have undergone radical changes, some being positive and others negative. On the positive side, increased levels of thermal comfort have become *fait accompli* through improved thermal insulation and more advanced air-conditioning or heating system design. On the negative side, a deterioration of the indoor air quality has been experienced particularly among air-conditioned buildings [1]. Indeed, the term 'sick building syndrome' is synonymous with the energy-saving era. These indoor air quality problems have been associated with poor plant maintenance, high concentrations of internally generated pollutants and low outdoor air supply rates.

The designers and operators of ventilation systems should be familiar with the comfort requirements and the quality of air necessary to achieve acceptable indoor climate. These require knowledge of the heat balance between the human body and the internal environment, the factors that influence thermal comfort and discomfort as well as the indoor pollution concentrations that can be tolerated by the occupants. In recent years, there has been a flux of published information, notably from Scandinavia and the USA, on these important issues of the built environment. The fundamental work that has been done on thermal comfort and air quality will be discussed in the first part of this chapter and more recent developments will be presented in the second part. This chapter will then become the fulcrum for the remaining chapters of this book.

1.2 Heat balance equations

1.2.1 Body thermoregulation

The primary function of thermoregulation is to maintain the body core, which contains the vital organs, within the rather narrow range of temperature which is vital for their proper functioning. The temperature control centre is the hypothalamus, a part of the

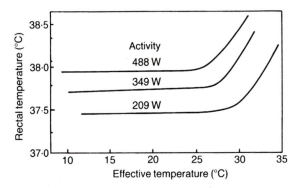

Figure 1.1 Rectal temperature as a function of ambient temperature for three activities.

brain that is linked to thermoreceptors in the brain, the skin and other parts of the body such as the muscles. The hypothalamus receives nerve pulses from the temperature sensors and coordinates information to different body organs to maintain a constant body core temperature. The thermoreceptors are particularly sensitive to changes in temperature and temperature change rates as small as $+0.001$ and $-0.004\,\mathrm{Ks}^{-1}$ can be detected. Temperature regulation is carried out by controlling metabolic heat production rate, control of blood flow, sweating, muscle contraction and shivering in extreme cold situations. Under normal conditions the body core temperature, t_c, is approximately $37\,°\mathrm{C}$ and this is maintained at a constant value despite changes in the ambient temperature, as shown in Figure 1.1 for three levels of activity [2]. The figure shows the ability of the temperature-regulating mechanisms to maintain a constant core temperature up to a certain ambient temperature beyond which core temperature cannot be maintained because evaporative cooling of sweat becomes ineffective. It is also shown in Figure 1.1 that the core temperature is not always constant but depends on activity, i.e. increases with increase in metabolism, and may be as high as $39.5\,°\mathrm{C}$ in extreme activities.

While the body core temperature remains almost constant over a wide range of ambient temperatures, the skin temperature changes in response to changes in the environment and is usually different for different parts of the body. However, the variation in skin temperature over the body is reduced when the body is in a state of thermal equilibrium and comfort. The variation of mean skin temperature, t_s, for a nude person with ambient temperature is shown in Figure 1.2, which also shows the rectal temperature for the purpose of comparison [3].

1.2.2 *Heat transfer between body and environment*

The heat balance equation for the human body is obtained by equating the rate of heat production in the body by metabolism and performance of external work to the heat loss from the body to the environment by the processes of evaporation, respiration, radiation, convection and conduction from the surface of clothing. Thus:

$$S = M + W + R + C + K - E - RES \qquad (1.1)$$

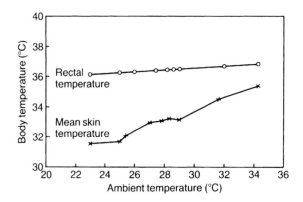

Figure 1.2 Variation of mean skin and rectal temperatures for a nude body with ambient temperature.

where S = heat storage in body ($\propto \Delta t_b/\theta$ where θ is time), W; M = metabolic rate, W; W = mechanical work, W; R = heat exchange by radiation, W; C = heat exchange by convection, W; K = heat exchange by conduction, W; E = evaporative heat loss, W and RES = heat loss by respiration, W. A positive value of S indicates a rising body temperature, t_b, a negative value suggests a falling t_b, and when $S = 0$ the body is in thermal equilibrium. The expressions that are used to calculate the terms on the right-hand side are given below, but for a more detailed assessment of each term readers should consult the books by Fanger [4] and McIntyre [5].

Metabolism (M)   This is the body heat production rate resulting from the oxidation of food. Its value for every person depends upon their diet and level of activity and may be estimated using the equation below:

$$M = 2.06 \times 10^4 \, \dot{V}(F_{oi} - F_{oe}) \quad \text{W} \tag{1.2}$$

where $\dot{V}$ = air breathing rate, $1\,s^{-1}$; and F_{oi} and F_{oe} are the fraction of oxygen in the inhaled and exhaled air, respectively. The value of F_{oi} is normally 0.209 but F_{oe} varies with the composition of food used in the metabolism; for fat diet $F_{oe} \approx 0.159$ and for carbohydrates $F_{oe} \approx 0.163$.

The rate of metabolism per square metre of body surface area can be obtained using equation (1.2) and Du Bois' area given as:

$$A_D = 0.202 \, m^{0.425} H^{0.725} \quad \text{m}^2 \tag{1.3}$$

where m = body mass, kg and H = body height, m. For an average man of mass 70 kg and height 1.73 m, $A_D = 1.83\,\text{m}^2$. Metabolism is often given the unit 'met' which corresponds to the metabolism of a relaxed, seated person, i.e. 1 met = 58.15 W m^{-2}. Typical activity values are given in Table 1.1.

Work (W)   If work is produced by the body, the metabolism increases to provide extra energy to perform this work. Because the thermal efficiency (W/M) of a human is poor, i.e. less than 20%, for every watt of work power produced an increase in

Table 1.1 Metabolic rates of different activities [6]

Activity	Metabolic rate	
	(W m^{-2})	(met)
Reclining	46	0.8
Seated, relaxed	58	1.0
Standing, relaxed	70	1.2
Sedentary activity (office, dwelling, school, laboratory)	70	1.2
Standing activity (shopping, laboratory, light industry)	93	1.6
Standing activity (shop assistant, domestic work, machine work)	116	2.0
Medium activity (heavy machine work, garage work)	165	2.8

metabolism of 5 W will be needed. Most of the work produced by a person is positive; however, in some cases negative work can occur such as a person walking down a steep hill, in which case some of the potential energy will be converted to heat in the muscles to keep a constant walking speed.

Radiation (R)   The heat exchange by radiation occurs between the surface of the body (clothing and skin) and the surrounding surfaces such as internal room surfaces and heat sources or sinks. It is calculated using the Stefan–Boltzmann equation:

$$R = f_{eff}f_{cl}\varepsilon\sigma\left(t_{cl}^4 - \bar{t}_r^4\right) \quad \text{W m}^{-2} \tag{1.4}$$

where f_{eff} = factor of effective radiation area, i.e. ratio of the effective radiation area to the total surface area of clothed body; f_{cl} = factor of clothing area, i.e. ratio of surface area of clothed body to surface area of nude body; ε = emissivity of clothed body; σ = Stefan–Boltzmann constant, i.e. 5.67×10^{-8} W m^{-2} K^{-4}; t_{cl} = surface temperature of clothing, K and $\bar{t}_r$ = mean radiant temperature, i.e. effective temperature of room surfaces, K.

The factor f_{cl} is used in equation (1.4) because the heat transfer is based on the surface area of the nude body, i.e. Du Bois' area, A_D. Because some parts of the human body act as a shield to other parts, the factor f_{eff} has a value of 0.696 for a seated person and 0.725 for a standing person; a mean value of 0.71 is usually used. The combined emissivity of skin and clothing, ε, is approximately 0.97 and, for long-wave radiation, this is independent of the skin or clothing colour. Substituting the values for f_{eff}, f_{cl}, ε and σ, equation (1.4) reduces to:

$$R = 3.9 \times 10^{-8}f_{cl}\left(t_{cl}^4 - \bar{t}_r^4\right) \quad \text{W m}^{-2} \tag{1.5}$$

The mean radiant temperature, $\bar{t}_r$, can be either calculated from areas, view factors and temperatures of room surfaces (e.g. see Fanger [4]) or measured directly (see Chapter 9). The clothing temperature, $\bar{t}_{cl}$, depends on the metabolic rate, thermal resistance of clothing and air temperature (see later).

The range of temperatures in the indoor environment is usually small (typically 10–30 °C) and in this case the fourth power law of equation (1.5) can be adequately replaced by the linear equation:

$$R = f_{cl}h_r\left(t_{cl} - \bar{t}_r\right) \tag{1.6}$$

where the radiant heat transfer coefficient, h_r, can be approximated by:

$$h_r = 4.6(1 + 0.01\bar{t}_r) \tag{1.7}$$

For normal room conditions, $h_r \approx 5.7\,\mathrm{W\,m^{-2}\,K^{-1}}$.

Convection (C) The heat transfer between a body and the surrounding air is primarily by convection which can be either free (natural) caused by buoyancy or forced (mechanical) caused by a relative movement between the body and air. The general heat convection equation is:

$$C = f_{cl}h_c(t_{cl} - t_a) \quad \mathrm{W\,m^{-2}} \tag{1.8}$$

where h_c = convective heat transfer coefficient, $\mathrm{W\,m^{-2}\,K^{-1}}$; t_a = air temperature, °C. The convective heat transfer coefficient, h_c, depends on the mode of heat transfer. For free convection it is given by:

$$h_c = 2.38(t_{cl} - t_a)^{0.25} \tag{1.9}$$

For forced convection it is:

$$h_c = 12.1\sqrt{(v)} \tag{1.10}$$

where v is the relative velocity between the body and air, $\mathrm{m\,s^{-1}}$.

Combined radiation and convection (R + C) The radiation and convection heat transfer may be combined into a single equation to give the sensible heat transfer from the body to the surroundings, i.e. using equations (1.6) and (1.8):

$$R + C = f_{cl}[h_r(t_{cl} - \bar{t}_r) + h_c(t_{cl} - t_a)]$$

This equation may be simplified to:

$$R + C = f_{cl}h(t_{cl} - t_o) \tag{1.11}$$

where h is a combined radiation and convection heat transfer coefficient, $\mathrm{W\,m^{-2}\,K^{-1}}$, and t_o is the operative temperature, °C. These are given by:

$$h = h_r + h_c \tag{1.12}$$

$$t_o = (h_r\bar{t}_r + h_c t_a)/(h_r + h_c) \tag{1.13}$$

The operative temperature can be defined as the average of the mean radiant and air temperatures weighted by their respective heat transfer coefficients. At air speeds $v \leq 0.4\,\mathrm{m\,s^{-1}}$ and $\bar{t}_r < 50\,°C$, t_o is approximately the simple average of the air and mean radiant temperatures. This is an environmental index that is described later.

Conduction through clothing (K)   The heat conduction through clothing using the normal heat conduction equation is:

$$K = h_{cl}(t_s - t_{cl}) \quad \text{W m}^{-2} \tag{1.14}$$

where h_{cl} = heat conductive coefficient of clothing, $\text{W m}^{-2}\text{K}^{-1}$; t_s = average skin temperature, °C.

The conductive coefficient is often replaced by the reciprocal of the thermal resistance of the clothing, I_{cl}. The clothing thermal resistance is usually given the unit 'clo' which is given by:

$$1 \text{ clo} = 0.155 \text{ m}^2 \text{ K W}^{-1}$$

Hence:

$$h_{cl} = 1/(0.1555 I_{cl}) = 6.45/I_{cl} \tag{1.15}$$

where I_{cl} has the unit clo.

Typical values of I_{cl} for different clothing ensembles are given in Table 1.2. The clothing area factor f_{cl} is directly related to the thermal resistance of clothing and is calculated from the following relations:

$$f_{cl} = \begin{cases} 1.00 + 0.2 I_{cl} & \text{for } I_{cl} < 0.5 \text{ clo} \\ 1.05 + 0.1 I_{cl} & \text{for } I_{cl} < 0.5 \text{ clo} \end{cases} \tag{1.16}$$

The mean skin temperature, t_s, may be estimated from an empirical relation derived by Fanger [4] which relates the mean skin temperature to activity as follows:

$$t_s = 35.7 - 0.0275(M - W) \quad °C \tag{1.17}$$

This formula has been derived for $4 > M > 1$.

Table 1.2 Thermal resistance of clothing ensembles [6]

Clothing ensemble	I_{cl}	
	$(\text{m}^2 \text{ K W}^{-1})$	(clo)
Nude	0	0
Shorts	0.015	0.1
Typical tropical clothing ensemble: briefs, shorts, open-neck shirt with short sleeves, light socks and sandals	0.045	0.3
Light summer clothing: briefs, long lightweight trousers, open-neck shirt with short sleeves, light socks and shoes	0.08	0.5
Light working ensemble: light underwear, cotton work shirt with long sleeves, work trousers, woollen socks and shoes	0.11	0.7
Typical indoor winter clothing ensemble: underwear, shirt with long sleeves, trousers, jacket or sweater with long sleeves, heavy socks and shoes	0.16	1.0
Heavy traditional European business suit: cotton underwear with long legs and sleeves, shirt, suit including trousers, jacket and waistcoat, woollen socks and heavy shoes	0.23	1.5

Evaporative heat loss (E)   Heat loss by evaporation is partly due to diffusion of water vapour through the skin tissues, E_d, and partly due to evaporation of sweat from the skin surface, E_{sw}. In both cases heat is absorbed from the skin and this process controls the rise in body temperature. The water diffusion is a continuous process that occurs even in a cool environment but the sweat evaporation only occurs in a hot environment and when the body activity is higher than normal.

The diffusion heat loss depends on the difference between the saturated vapour pressure at skin temperature, p_{ss}, and the vapour pressure of the surrounding air, p_a. Olesen [3] gives the following equation for E_d:

$$E_d = 3.05 \times 10^{-3}(p_{ss} - p_a) \text{ W m}^{-2} \tag{1.18}$$

where p_{ss} and p_a are in pascals (Pa).

The saturated vapour pressure may be obtained using an equation such as that given in the CIBSE Guide [7] which is:

$$\log_{10} p_{ss} = 30.59051 - 8.2 \log_{10} t_s + 2.4804 \times 10^{-3} t_s - 3.14231 \times 10^3 / t_s \tag{1.19}$$

Here, p_{ss} is in kPa and t_s is absolute skin temperature, K.

Over the skin temperature range $27\,°C < t_s < 37\,°C$ a linear expression for p_{ss} can be approximated to within 3% error using steam table data. Thus:

$$p_{ss} = 256 t_s - 3373 \text{ Pa} \tag{1.20}$$

where t_s is in °C.

Equations (1.18) and (1.20) may be combined to give:

$$E_d = 3.05 \times 10^{-3}(256 t_s - 3373 - p_a) \text{ W m}^{-2} \tag{1.21}$$

p_a can be calculated from a knowledge of the air temperature and relative humidity as shown in Section 2.2.9. The value of E_d represents the minimum heat loss by evaporation.

McIntyre [5] gives a different expression for E_d as follows:

$$E_d = 4.0 + 1.2 \times 10^{-3}(p_{ss} - p_a) \text{ W m}^{-2} \tag{1.22}$$

where p_{ss} and p_a are in Pa.

The evaporation of sweat from skin is the most effective way of maintaining a constant body temperature in a hot environment and at a high metabolic rate. Consequently, the rate of heat loss by sweat is influenced by the ambient temperature and metabolism. The maximum heat loss by sweat occurs when the skin is completely wet and is given by:

$$(E_{sw})_{max} = f_{pcl} h_e (p_{ss} - p_a) \text{ W m}^{-2} \tag{1.23}$$

where h_e is the evaporative heat transfer coefficient, $\text{W m}^{-2}\,\text{Pa}^{-1}$, and f_{pcl} is a permeation factor of clothing for water vapour which, for porous clothing, may be represented by [8]:

$$f_{pcl} = 1/(1 + 0.143 h_c / h_{cl}) \tag{1.24}$$

where h_c is the convective transfer coefficient and h_{cl} is the heat conductance of clothing, both in $W\,m^{-2}\,K^{-1}$. The coefficient h_e in equation (1.23) may be represented by the convective coefficient h_c through the Lewis relation, i.e.:

$$h_e = 16.7h_c$$

Hence

$$(E_{sw})_{max} = 16.7f_{pcl}h_c(p_{ss} - p_a) \quad W\,m^{-2} \tag{1.25}$$

The effect of clothing absorbing sweat, which is then transmitted to the surface by the capillary action of the clothing fabric, is not considered in the above equations. When this happens the evaporation of sweat takes place from within the clothing and not from the skin. This reduces the efficiency of sweat in removing excess body heat and more sweat will be required to produce the same heat loss from the skin surface. A sweating efficiency may be defined by:

$$\eta_{sw} = 1/(1 + h/h_{cl}) \tag{1.26}$$

where h is the combined radiative and convective heat transfer coefficient. The actual heat removed by sweating then becomes:

$$E_{sw} = \eta_{sw}(E_{sw})_{max} \tag{1.27}$$

Fanger [4] has produced a formula based on experimental measurements correlating E_{sw} to metabolism as follows:

$$E_{sw} = 0.42(M - W - 58.15) \quad W\,m^{-2} \tag{1.28}$$

This equation applies to $4 > M > 1$. The diffusion loss, E_d, is used to calculate evaporative losses when the skin is not wet with sweat. However, E_d is ignored when the skin is completely wet and in this case equation (1.25) or (1.28) is used to calculate E_{sw}.

Respiration heat loss (RES) Inspired air is both warmed and humidified by its passage through the respiratory system. The sensible and latent heat losses are proportional to the volume flow rate of air to the lungs that in turn is proportional to the metabolic rate. The sensible heat loss is given by:

$$S_{res} = 0.0014\,M(34 - t_a) \quad W\,m^{-2} \tag{1.29}$$

where t_a is the ambient air temperature, °C. S_{res} is a small quantity in comparison with the latent heat loss which is given as:

$$L_{res} = 1.72 \times 10^{-5}\,M(5867 - p_a) \quad W\,m^{-2} \tag{1.30}$$

where p_a is the ambient water vapour pressure, Pa. Respiration heat loss is only significant at high activity and under normal sedentary activity it is less than $6\,W\,m^{-2}$ and can be neglected.

1.3 Environmental indices

In the preceding section it was shown that there are four environmental parameters – air temperature, t_a, mean radiant temperature, $\bar{t}_r$, air velocity, v, and water vapour pressure in the air, p_a – and three personal parameters – metabolism, M, work, W and thermal insulation of clothing, I_{cl}. The heat storage, S, is not included in the personal parameters when heat balance between the body and the environment is present. During the last 80 years a number of studies have been carried out to define environmental indices in terms of these parameters.

Current environmental indices may be divided into three categories: direct; rational and empirical. Direct indices are based on the measurement of a simple instrument that responds to the environmental variables in a manner similar to that of a human. An example is a globe thermometer which responds to changes in air temperature, air velocity and radiant temperature. Rational indices are based upon models of human responses to the thermal environment taking into consideration thermoregulation and heat exchange between the body and the environment. These models can be used to predict human responses to certain environmental conditions. The operative temperature, t_o, is a rational index which combines the heat exchange between the body and the environment by radiation and convection which forms the basis of ASHRAE Standard 55-1992 [9].

An empirical environmental index is based on a model of the energy exchange between the human body and the environment which is validated by subjecting a large population sample of known activity and clothing to a range of environmental conditions and recording their thermal sensation. The body is represented by a single node, two nodes or multi-node system for the purpose of evaluating the energy exchange. The data is then analysed to assess the effect of each variable and from this an environmental index is established. Some of the earliest works in this area using a single node system was that by Houghton and Yaglou [10], which introduced the effective temperature index that was later adopted by ASHRAE, and the well-known work carried out by Fanger and his collaborators in Denmark which introduced the predicted percentage of dissatisfied, *PPD* [4]. These and other more recent models are described in Section 1.4 but in the following sections environmental indices that are based on a single 'equivalent' temperature are first introduced.

1.3.1 *Effective temperature* (*ET**)

The effective temperature index (*ET*), derived by Houghton and Yaglou [10], combines the effect of dry-bulb and wet-bulb temperatures with air movement to yield equal sensations of warmth or cold. This scale, which was adopted by ASHRAE, was used by HVAC designers for almost 50 years until it was replaced by the new (or revised) effective temperature scale (*ET**). Criticisms of the old *ET* scale were that it overemphasized the effect of humidity in cooler and neutral conditions, underemphasized its effect in warm conditions, and failed to account for air velocity under hot and humid conditions. This scale is no longer in use now.

Results of tests carried out by Gagge [11] in environmental chambers have shown that skin wettedness is an excellent predictor of discomfort under transient environments. He then developed the new index (*ET**) which is defined in terms of the operative temperature, t_o, and combines the effects of air and radiant temperatures as well as water vapour pressure. Gagge correlated his measurements with a two-node

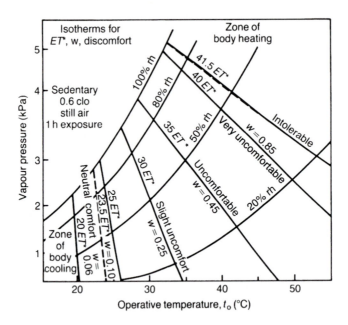

Figure 1.3 The new effective temperature scale-lines of constant ET^* [12].

model that was used to evaluate skin temperature, skin wettedness and evaporation rate within the zone of operation. A single-node model, such as the PMV/PPD model, is only valid under steady-state thermal conditions and to study transient effects or non-uniform thermal environments, two- or multi-node models are required. In the ET^* scale, skin temperature, t_s, skin wettedness, w, and clothing permeability index are used to define the thermal state of a person. The skin wettedness, w, is the ratio of the actual evaporative loss at the skin surface to the maximum loss that could occur in the same environment, i.e. when the skin is completely wet. This scale has been adopted by ASHRAE [12] to replace the old ET scale.

The ASHRAE effective temperature (ET^*) is the dry-bulb temperature of uniform enclosure at 50% relative humidity (rh) in which people have the same net heat exchange by radiation, convection and evaporation as they do in varying humidities of the test environment. The ET^* scale is based on equal air and operative temperatures, a clothing resistance of 0.6 clo, an air speed of $0.2 \, \mathrm{m \, s^{-1}}$, a sedentary activity ($\approx$ 1 met) and an exposure time of 1 h. As shown in Figure 1.3 the new effective temperature for thermal comfort at these conditions and an rh of 50% is 23.5 °C for $w = 0.06$, i.e. no sweating. The figure also gives the values of ET^* for different skin wettedness ratios. The principle used to develop ET^* has been extended to higher levels of activities, representing higher levels of sweating. These are discussed in more detail in the ASHRAE Handbook of Fundamentals [12].

1.3.2   Operative temperature (t_o)

The ASHRAE Standard 55-1992 [9] uses the operative temperature as the environmental variable for evaluating thermal comfort at different activity and clothing insulation.

It is defined as the uniform temperature of a radiantly black enclosure in which an occupant would exchange the same amount of heat by radiation plus convection as in the actual non-uniform environment. It is expressed by equation (1.13), which defines the average of the mean radiant and air temperatures, weighted by their respective heat transfer coefficients. The ASHRAE Standard specifies environmental conditions that are acceptable to 80% or more of the occupants. It is mainly applicable to sedentary activity ($\approx$1.2 met) with normal winter or summer clothing ensembles, i.e. 0.8–1.2 clo winter clothing or 0.6–0.8 clo summer clothing. The acceptable ranges of operative temperature and humidity for winter and summer are defined by the shaded areas in the psychrometric chart of Figure 1.4 which is based on 10% dissatisfaction [9]. The figure shows an overlap of the winter and summer zone in the range of $t_0 = 23$–24 °C because in this region people in summer clothing would tend to feel slightly cool while those in winter garments would be near the slightly warm sensation.

The maximum average air speed in the occupied zone is specified by the Standard as 0.15 m s^{-1} in winter and 0.25 m s^{-1} in summer environments. However, in summer, the comfort zone (see Figure 1.4) may be extended beyond 26 °C operative temperature

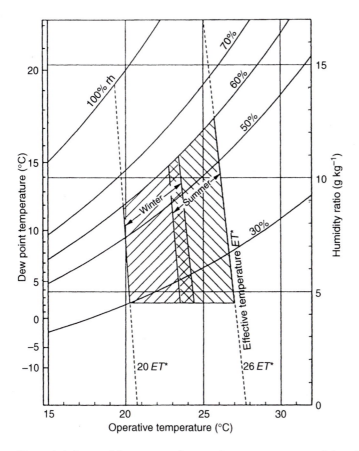

Figure 1.4 Acceptable ranges of operative temperature and humidity for winter and summer clothing and sedentary activity, 1.2 met [9].

if the average air speed is increased by $0.27\,\mathrm{m\,s^{-1}\,K^{-1}}$ of increased temperature to a maximum of $0.8\,\mathrm{m\,s^{-1}}$. Air speeds of $0.8\,\mathrm{m\,s^{-1}}$ are unacceptably high in normal occupancy as loose paper, hair and other objects may be blown about at this speed.

The comfort zone temperature in Figure 1.4 should be decreased when the activity level is higher than sedentary, i.e. $M > 1.2$ met. The operative temperature for activity range $1.2 < M < 3$ is obtained using [9]:

$$t_\mathrm{o}\ \text{active} = t_\mathrm{o}\ \text{sedentary} - 3(1 + \mathrm{clo})(\mathrm{met} - 1.2)$$

However, the minimum allowable temperature is $15\,°\mathrm{C}$.

1.3.3   Dry resultant temperature (t_res)

Missenard (see McIntyre [5]) was the first to define the dry resultant temperature which was the equilibrium temperature recorded by a 90-mm-diameter globe thermometer. He also defined the wet resultant temperature as the temperature recorded by a globe thermometer that is partly kept moist by wet muslin. This temperature is rarely used nowadays. Missenard selected a 90-mm-diameter sphere so that the ratio of radiant to convective heat exchange with the environment is the same as a human body, i.e. $1:0.9$. In still air the dry resultant temperature measured by a 90-mm-diameter globe is:

$$t_\mathrm{res} = 0.47t_\mathrm{a} + 0.53\bar{t}_\mathrm{r} \tag{1.31}$$

However, Missenard gave no correction for air movement and t_res was undefined in moving air.

CIBSE [13] has adopted the dry resultant temperature as a recommended environmental index for design but allowance has been made for air speed. It is defined as the temperature recorded by a thermometer at the centre of a blackened globe of 100 mm diameter. The original globe thermometer of Missenard was increased from 90 to 100 mm so that, at low air speeds, t_res becomes the mean of t_a and $\bar{t}_\mathrm{r}$, i.e.:

$$t_\mathrm{res} = 0.5(t_\mathrm{a} + \bar{t}_\mathrm{r}) \tag{1.32}$$

Taking the effect of air speed into account, CIBSE defines t_res by:

$$t_\mathrm{res} = \left[\bar{t}_\mathrm{r} + t_\mathrm{a}\sqrt{(10\,v)}\right] \Big/ \left[1 + \sqrt{(10\,v)}\right] \tag{1.33}$$

where v is the air speed, in $\mathrm{m\,s^{-1}}$. Equation (1.33) reduces to equation (1.32) when $v = 0.1\,\mathrm{m\,s^{-1}}$, i.e. 'still' air.

The CIBSE Guide [13] gives a table of recommended t_res values in still air environment and a relative humidity range of 40–70%. These values range from $13\,°\mathrm{C}$ for heavy activity in factories to $26\,°\mathrm{C}$ in swimming pools. They implicitly take into account the activity and clothing insulation. The recommended value for offices is $20\,°\mathrm{C}$. Where air speeds depart from the still air value (i.e. $v > 0.1\,\mathrm{m\,s^{-1}}$), CIBSE recommends an elevation of t_res according to Figure 1.5, but speeds greater than $0.3\,\mathrm{m\,s^{-1}}$ are not recommended, except in summer.

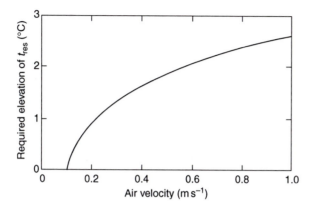

Figure 1.5 Correction to dry resultant temperature to allow for air movement.

1.4 Thermal comfort models

1.4.1 Fanger's model

The environmental indices discussed so far were based on statistical analyses of extensive experimental data obtained in the laboratory or on site. Therefore, each index strictly applies only over the range of physical conditions covered by the experiments. Because of the large tasks involved in conducting such experiments, most of the measurements were restricted to lightly clothed sedentary subjects and it would be imprudent to extend the range of application of these indices beyond that covered experimentally. Fanger [4] has developed a more fundamental approach that is based on physical assessment of the body heat exchange with the environment and supported by extensive tests on human subjects placed under strictly controlled environments.

Thermal comfort equation Fanger approached the problem of arriving at a universal environmental index from the principle that a person senses his/her own temperature and not those of the environment. He stipulated three requirements for achieving thermal comfort:

1 The body must be in thermal equilibrium with the environment, i.e. the rate of heat loss to the environment balances the rate of heat production. This implies a steady-state condition, i.e. a single-node model is used.
2 Thermal sensation is related to skin temperature and therefore the mean skin temperature, t_s, should be at an appropriate level. Measurements have shown that t_s decreases with increasing metabolic rate (equation (1.17)).
3 There should be a preferred rate of sweating, e.g. at sedentary activity people prefer not to sweat. The rate of sweating increases with metabolic rate as given by equation (1.28).

Applying the three thermal comfort requirements and using the heat exchange expressions presented in Section 1.2, Fanger reduced the heat balance equation

(equation (1.1)) to the functional relationship:

$$f(M, W, I_{cl}, t_a, \bar{t}_r, v, p_a) = 0 \tag{1.34}$$

As a result, the function f represents a thermal comfort expression which includes the three personal parameters M, W and I_{cl} and the four environmental parameters t_a, $\bar{t}_r$, v and p_a. Theoretically, 'any' combination of the four environmental parameters may be chosen to satisfy the expression, hence achieving thermal comfort, for a person with a given clothing (I_{cl}), metabolism (M) and work (W).

On substituting the appropriate formulae for heat exchange between the body and environment in the heat balance equation (1.1) as well as the expressions for skin temperature (equation (1.17)) and the evaporative losses (equations (1.21) and (1.28)), Fanger arrived at the now well-known thermal comfort equation given below:

$$(M - W) - 3.05 \times 10^{-3}[5733 - 6.99(M - W) - p_a] - 0.42[(M - W) - 58.15]$$
$$- 1.7 \times 10^{-5}M(5867 - p_a) - 0.0014M(34 - t_a)$$
$$= 3.96 \times 10^{-8}f_{cl}[(t_{cl} + 273)^4 - (\bar{t}_r + 273)^4] - f_{cl}h_c(t_{cl} - t_a) \tag{1.35}$$

where the clothing surface temperature is given as:

$$t_{cl} = 35.7 - 0.028(M - W) - 0.155I_{cl}\{(M - W) - 3.05 \times 10^{-3}$$
$$\times [5733 - 6.99(M - W) - p_a] - 0.42[(M - W) - 58.15]$$
$$- 1.7 \times 10^{-5}M(5867 - p_a) - 0.0014M(34 - t_a)\} \tag{1.36}$$

The convective heat transfer coefficient, h_c, in equation (1.35) is calculated using equations (1.9) and (1.10) and taking the greater of the two values. The clothing area factor, f_{cl}, is obtained from equation (1.16).

The thermal comfort equation (1.35), which includes the personal and environmental parameters, is too complex to solve except by computer. For this reason, comfort charts representing solutions to the comfort equation are often used by various combinations of two variables to represent a line of a third variable as a constant. Figure 1.6 is a typical comfort chart with the mean radiant temperature as the ordinate, the air temperature as the abscissa and a family of lines of constant air speed. Many other similar charts may be drawn to provide a quick reference for designers of environmental systems, see [4, 12].

Many experiments were carried out in the 1970s, mainly on lightly clothed sedentary people in environmental chambers, to validate Fanger's comfort equation. In particular the influence of such variables as age, race, seasonal variations, adaptation and background colour and noise on the preferred temperature were tested. The variation in the preferred temperature by the people tested was within 1.5 K. These experiments confirmed the predictions of the comfort equation under mainly sedentary activity but there have been few studies at high activity levels. However, under these conditions people are less sensitive to the environmental temperature and any errors that may arise from applying the comfort equation are not expected to be significant in assessing the comfort at a high level of activity.

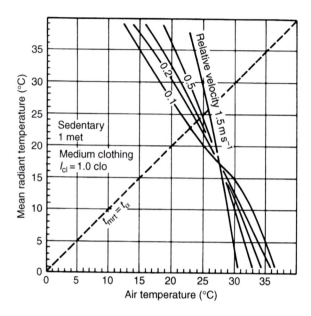

Figure 1.6 Comfort chart for 1 met activity, 1 clo and 50% rh ($w = 0$).

Predicted mean vote (PMV) The thermal comfort equation is only applicable to a person in thermal equilibrium with the environment. If not in equilibrium, the body will be under physiological strain to activate the effector mechanism that is necessary to change the skin temperature to achieve a new heat balance. Fanger assumed that the thermal sensation at a given activity level is related to this strain. He used the heat balance equation (1.35) to predict a value for the degree of sensation using his own experimental data and other published data for different activity levels. The thermal sensation index that has been adopted by Fanger is based on the seven-point psychophysical scale:

$$\text{cold} - \text{cool} - \text{slightly cool} - \text{neutral} - \text{slightly warm} - \text{warm} - \text{hot}$$
$$-3 \qquad -2 \qquad -1 \qquad 0 \qquad +1 \qquad +2 \qquad +3$$

This index is termed the predicted mean vote (PMV) which is the mean vote one would expect to get by averaging the thermal sensation vote of a large group of people in a given environment. The *PMV* is a complex mathematical expression involving activity, clothing and the four environmental parameters. It is expressed by:

$$PMV = (0.303e^{-0.036M} + 0.028)\{(M - W)$$
$$- 3.05 \times 10^{-3}[5733 - 6.99(M - W) - p_a]$$
$$- 0.42[(M - W) - 58.15] - 1.7 \times 10^{-5}M(5867 - p_a)$$
$$- 0.0014M(34 - t_a) - 3.96 \times 10^{-8}f_{cl}[(t_{cl} + 273)^4$$
$$- (\bar{t}_r + 273)^4] - f_{cl}h_c(t_{cl} - t_a)\} \tag{1.37}$$

where

$$t_{cl} = 35.7 - 0.028(M - W) - 0.155\,I_{cl}\{3.96 \times 10^{-8} f_{cl}[(t_{cl} + 273)^4$$
$$- (\bar{t}_r + 273)^4] + f_{cl}h_c(t_{cl} - t_a)\} \tag{1.38}$$

The convective heat transfer coefficient, h_c, again takes the greater of the values given by equations (1.9) and (1.10). Equation (1.38) is a transcendental equation which can only be solved by an iterative process. Equations (1.37) and (1.38) are more conveniently solved using a computer. A computer program written in FORTRAN is given in [6].

Predicted percentage of dissatisfied (PPD)   From the experimental data available to him, Fanger correlated the percentage ratio of the people who were dissatisfied with the thermal environment with the predicted mean vote. This ratio was called the predicted percentage dissatisfied and is shown plotted against *PMV* in Figure 1.7. The figure shows a symmetrical curve with a minimum value of 5% corresponding to the lowest PD subjects, i.e. in an ideal environment. ISO Standard 7730 [6] recommends a *PPD* limit of 10% corresponding to $-0.5 \le PMV \le +0.5$. Figure 1.7 may be represented by the expression [6]:

$$PPD = 100 - 95 \exp -\{0.03353(PMV)^4 + 0.2179(PMV)^2\} \tag{1.39}$$

The simplified expression below [14] may be used giving good accuracy for $|PMV| \le 2$:

$$PPD = 5 + 20.97|PMV|^{1.79} \tag{1.40}$$

Although Fanger's thermal comfort criterion is the most comprehensive one derived to date it still has some deficiencies. The criterion produces good results for the standard conditions of sedentary activity and light clothing (representing the conditions at which experimental data were obtained), but it is less satisfactory at more extreme conditions of activity and heavy clothing and in less well-controlled environments, such as free-running buildings. Further work will be required to test the criterion at these extreme conditions. However, measurements at high activity and in free-running buildings where transient thermal conditions exist become less certain as it becomes

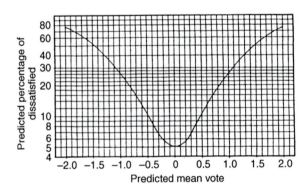

Figure 1.7 Predicted percentage of dissatisfied as a function of predicted mean vote.

difficult to maintain a constant activity or constant clothing ensembles for sufficiently long periods.

Application of PMV/PPD   For a room with a uniform thermal environment, a single value of *PMV* and *PPD* can be obtained using equations (1.37) and (1.38) and Figure 1.7 or equation (1.39) from the values of t_a, t_r, v and p_a (or relative humidity) at a single point in the room and known values of M and I_{cl}. If the calculated *PMV* and *PPD* are outside the recommended range, then a simple adjustment to room temperature can restore them to acceptable values.

In most rooms, however, complete thermal uniformity does not exist, owing to the presence of hot or cold surfaces which produce temperature gradients and varying air velocity in the room. Hence, to analyse the comfort level in a room, measurement of the environmental variables should be carried out at various points in the occupied zone and these are then used to calculate the distribution of *PMV* throughout the zone. Normally, the occupied zone is divided into 1 or 0.5 m square grids and measurements of t_a, t_r and v are made at each grid point for three or four heights above the floor. Typical heights normally chosen are 0.15 m (ankle level), 0.6 m (back level, seated), 1.2 m (head level, seated) and 1.8 m (head level, standing). The mean radiant temperature at each point can be calculated using equation (1.33) from a measurement of the dry resultant temperature, t_{res}, using a 100 mm globe thermometer. Thus:

$$\bar{t}_r = \left[1 + \sqrt{(10\,v)}\right] t_{res} - \sqrt{(10\,v)} t_a \qquad (1.41)$$

The water vapour pressure usually has the same value throughout the room and only one measurement of the relative humidity in the room is needed to obtain p_a. The *PMV* values at each grid point are volume averaged to obtain the mean *PMV* for the whole of the occupied zone, from which the *PPD* for the zone is calculated. Typical *PMV* contours obtained for a test room with a side-wall air supply (heating) having one window and two outside walls are shown in Figure 1.8 [14]. It can be seen that the distribution of *PMV* for most of the occupied zone is below −0.5 which corresponds to *PPD* > 10%. The average *PMV* in the occupied zone is −0.69 which corresponds to a *PPD* of 15.7%. This is partly due to a low room temperature and partly due to

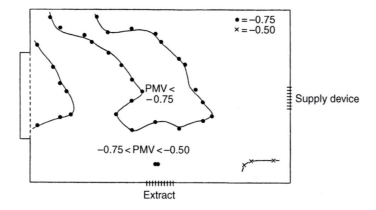

Figure 1.8 *PMV* contours for a room heated by a side-wall air supply.

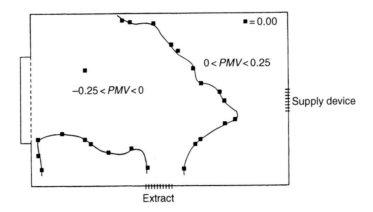

Figure 1.9 Corrected *PMV* contours for the room in Figure 1.8.

thermal non-uniformity of the room (i.e. presence of a cold window and two walls). Therefore, this environment would not be suitable for long-term occupancy.

By altering the air temperature level in the room to give a zero average *PMV* value, it would be possible to examine the thermal uniformity of the occupied zone more closely. The effect of moderate changes in the average room air temperature on the velocity field and on the temperature gradients in the occupied zone will be negligible. This can be achieved by adjusting the supply temperature to produce the necessary change in the mean temperature of the occupied zone. The change in room temperature will produce an average *PMV* for the occupied zone of zero. It can be calculated either from equations (1.37) and (1.38) by setting *PMV* = 0 or from *PMV* tables given by Fanger [4]. To examine the thermal uniformity of the occupied zone the mean *PMV* value is subtracted from the *PMV* value at each grid point in the occupied zone and new or corrected *PMV* contours are then plotted. From the corrected *PMV*s the new *PPD*s are calculated to give what Fanger called the lowest possible percentage of dissatisfied (*LPPD*). The average of *LPPD*s in the occupied zone should be less than 10% for an acceptable environment. If *LPPD* > 10% the causes should be examined closely to suggest improvements to the air distribution system or thermal insulation of the building or both.

The *PMV* values in Figure 1.8 were increased by +0.69 to give an average value for the occupied zone of zero. The corrected *PMV* distribution is shown in Figure 1.9 which is in the range −0.25 < *PMV* < +0.25 and this is quite acceptable for long-term occupancy. The supply air temperature in this case was increased by 2.45 K and the average *LPPD* was 5.24%.

1.4.2 Multi-node thermal comfort models

The human body is not a simple thermal system that can be assumed homogenous under different environmental conditions. Different parts of the body react to the environment in different ways and are normally at different temperatures. It has been recognized that treating the human body as a homogeneous system does only produce accurate predictions under transient environmental conditions or in non-uniform environments. Consequently, models were developed such that the human body is represented by two or more nodes to allow for inhomogeneous effects in the environment. Such models of the human body are called multi-node models.

The earliest such model was that developed by Gagge [11] in the 1970s which represents the human body by two nodes, an internal core and the skin (which also included a layer of clothing) with heat transfer taking place between the two nodes by blood flow. Under transient conditions sweating from the skin was also considered in the total energy balance. This model was adequate for predicting the overall response of the body at a moderate activity in a homogeneous environment that can also be transient. Since then a number of models involving, in some cases, a large number of nodes have been developed for dealing with transient and inhomogeneous environmental conditions. Thellier *et al.* [15] have developed a 25-node model by dividing the body into six segments (head, trunk, arms, hands, legs and feet) with each segment made up of four layers (core, muscle, fat and skin) and the 25th node is the blood forming a thermal link with the other nodes. The model was validated using experimental measurements under transient and inhomogeneous thermal environments with good correlations.

Conceição [16] has developed a model by dividing the human body into 35 cylindrical and spherical elements with each element made up of three layers (internal, central and external), i.e. using a total of 105 nodes. The internal layer was represented by a vein and an artery except for an element in an extremity (e.g. head, fingers and foot) when the artery was not considered. The results from this model were used to compare the local skin temperature at four locations of nude people in an environmental chamber under transient thermal environments with good correlations. The number of layers was later extended [17] to 12 with one layer representing the core, two layers for the muscle, two layers for the fat and seven layers for the skin with the possibility of clothing protection for each element. The extended model was again evaluated by simulating people exposed to inhomogeneous environments and predicting the percentage of dissatisfied in steady- and unsteady-state conditions.

1.4.3 Models-based on field measurements

The temperature indices described in Section 1.3 have largely been derived from data obtained in environmental chambers and laboratory measurements where consistent control of the environmental parameters is maintained during the experiments. Although such tests provide accurate evaluation of the influence of each parameter or a group of parameters on the thermal sensation of people, the subjects involved are not in their familiar surroundings nor are they engaged in their usual work activities throughout the test periods. The subjects in the laboratory tests may therefore lack the opportunity to adjust to the thermal environment as people do in normal life. This anomaly can be overcome by conducting field studies where people's thermal response is investigated in their usual work environments. The results from these measurements could then be used to develop thermal comfort models for the occupants of actual buildings. A number of field investigations are reported in the literature, e.g. [18–25].

Humphreys [18, 19] analysed the results of more than 30 field studies of thermal comfort conducted over a period of 40 years in about 12 countries of different climatic conditions. All the investigations were conducted indoors, the majority of which were

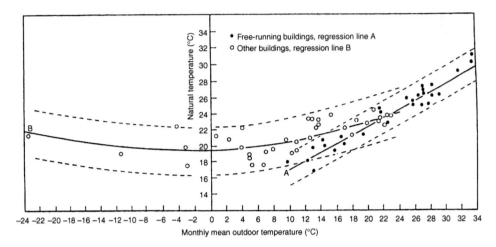

Figure 1.10 Relationship between the comfort temperature and monthly mean outdoor temperature [18].

in offices, i.e. light activity. The mean temperatures for these studies range from 17 °C for English homes in winter to 37 °C for offices in Iraq in the summer. In the latter study, the temperature was adjusted to 34 °C to allow for the abnormally high air velocities present. The temperatures recorded in these investigations were air and/or globe temperatures which differed little in most cases. From the temperature measurements and the survey results it was possible to establish a comfort temperature, t_c, for the subjects involved. This is the temperature at which the respondent to the survey is thermally neutral. Allowing for the high air movement in a hot environment, Humphreys obtained a range of comfort temperatures from 17 to 30 °C (see Figure 1.10).

Other, but less wide-ranging, surveys in offices were carried out by Auliciems [20] in Australia, Fishman and Pimbert [21] in England and Schiller *et al.* [22] in USA. These studies produced comfort temperatures of 20.5–23.1 °C; 22.0; 22.0 °C (winter) and 22.6 °C (summer), respectively. In the study by Schiller *et al.* the winter comfort temperature compares well with the comfort temperature recommended by ASHRAE Standard 55-1992 [9], however the comfort temperature in the summer was lower than that recommended by the Standard (24.5 °C). This indicates that people tend to prefer cooler conditions in the summer than those suggested by the Standard, which is largely based on laboratory test data.

The results from these and other field studies [22–25] suggest that people tend to adjust to their environment. However, one must remember that field data normally have lower degrees of correlation than laboratory-based measurements and further research is needed for an accurate assessment of the thermal acceptability of daily environments in different types of buildings, particularly those that are naturally ventilated.

1.4.4 Adaptive models

Based on the believe that people naturally adapt to their thermal environment by making various adjustments to themselves and their surroundings in order to reduce discomfort and physiological strain and the fact that a single-node model such as the *PMV/PPD* model does not always predict conditions in free-running buildings, new models called *Adaptive Models* have been developed during the last two decades. The single-node model may be acceptable to buildings with environmental control systems but would not apply in the case of unconditioned, e.g. naturally ventilated buildings. In such cases, there will invariably be interaction between occupants' behaviour and the environment to influence their thermal sensation. Thermal interaction with the environment can take different forms, e.g.

- Clothing changes
- Changes of posture and activity
- Movement between different thermal zones in the building
- Adapting the available thermal control devices to change the environment.

These interactions are achieved under conscious control which will also affect the human physiological regulatory mechanisms. They are also time-dependent and a different type of adaptation may occur at different times, e.g.

- Short-term adaptation: changing of clothing, posture, environmental control.
- Long-term adaptation: seasonal changes in clothing, seasonal changes in activity, changes in furnishing and changes in building control devices.

The adaptive models do not actually predict comfort responses of people but rather the thermal conditions under which people are expected to be comfortable in buildings. Based on field tests, such as those mentioned earlier, it has been found that an acceptable degree of comfort in residential and office buildings could be accomplished over a very wide temperature range. Another significant parameter which is inherent in most adaptive models is the external weather condition and its influence on both people's perception and on the climate inside. Mean monthly external temperature is used as a parameter in most adaptive models developed thus far to predict an acceptable or a comfort temperature, t_c, inside a building. Extensive studies in unconditioned (free-running) buildings [18–29] have found a correlation between the mean outdoor temperature and the indoor 'comfort' temperature. Most of these correlations are between the comfort (neutral) temperature t_c and the monthly mean outdoor temperature t_m. Some of these correlations are given below:

Humphreys' relation for free-running buildings [18]:

$$t_c = 12.1 + 0.53 t_m \tag{1.42}$$

Humphreys' relation for climate-controlled buildings [19]:

$$t_c = 24.2 + 0.43(t_m - 22) \exp\left\{ (t_m - 22)/24\sqrt{2} \right\} \tag{1.43}$$

Auliciems' relation [20]: $\quad\quad\quad\quad t_c = 17.6 + 0.31t_m$ $\quad\quad\quad$ (1.44)

Williamson *et al.*'s relation [27]: $\quad t_c = 10.5 + 0.58t_m$ $\quad\quad\quad$ (1.45)

Nicol's relation [28]: $\quad\quad\quad\quad t_c = 17.0 + 0.38t_m$ $\quad\quad\quad$ (1.46)

De Dear and Brager's relation [29]: $\quad t_c = 17.8 + 0.31t_m$ $\quad\quad\quad$ (1.47)

A mean expression for free-running buildings based on the relations (1.42) to (1.47) (not including equation (1.43)) is:

$$t_c = 14.0 + 0.45t_m \quad\quad\quad\quad (1.48)$$

Equation (1.48) applies to a range of $35 > t_m > 10$ with an accuracy of $\pm 2\,K$ in t_c.

Adaptive models are useful to provide data that may be used in design calculations, specifying building temperature set points and energy assessments for a building.

1.5   Thermal discomfort

Although the body may be thermally neutral in a particular environment, i.e. the thermal comfort equation is satisfied, there may not be thermal comfort if one part of the body is warm and another is cold. This local thermal discomfort can be caused by an asymmetric radiant field (e.g. cold windows or warm heaters), by contact with a warm or cold floor, by vertical air temperature gradient or by a local convective cooling of the body (draught). The first three sources of local discomfort are briefly discussed here but more emphasis will be directed towards the results from studies on the effect of draught, since this is a common occurrence in naturally or mechanically ventilated buildings in winter.

1.5.1   Asymmetrical thermal radiation

Non-uniform or asymmetrical thermal radiation in a space may be caused by cold/hot windows, walls, ceilings, solar patches in the room or heating panels. In office and residential buildings asymmetric radiation is mainly due to cold windows and hot ceilings whereas in factory and food storage buildings it can be due to infrared heaters, hot or cold equipment and so forth. The radiation asymmetry can be described by a parameter called radiant temperature asymmetry, Δt_{pr}, which is defined as the difference between the plane radiant temperatures of the two opposite sides of a small plane element. In this context, the plane radiant temperature, t_{pr}, is the uniform temperature of an enclosure in which the incident radiant flux on one side of a small plane element is the same as in the existing environment. The plane radiant temperature is a parameter which describes the effect of thermal radiation in one direction as compared with the mean radiant temperature, $\bar{t}_r$, which describes radiation from all surrounding surfaces.

For surfaces of high thermal emissivity ($\varepsilon \approx 1$), as is the case for most building materials, the plane radiant temperature, t_{pr}, can be calculated from a knowledge of the temperatures of all the surrounding surfaces and their respective view factors with respect to a small plane element at which t_{pr} is calculated, i.e.:

$$t_{pr}^4 = t_1^4 F_{p-1} + t_2^4 F_{p-2} + \cdots + t_n^4 F_{p-n} \quad\quad\quad\quad (1.49)$$

where t_{pr} = plane radiant temperature (K); t_n = temperature of surface n (K); F_{p-n} = view factor between a small plane element and surface, n and $\sum F_{p-n} = 1$.

Since the sum of all the view factors is unity, for small differences between the temperatures of the enclosure surfaces, the plane radiant temperature becomes the mean value of the surface temperatures weighted by the respective view factor for each surface. Hence, equation (1.49) may be linearized as follows:

$$t_{pr} = t_1 F_{p-1} + t_2 F_{p-2} + \cdots + t_n F_{p-n} \tag{1.50}$$

where all the temperatures are in °C. The view factors may be obtained from heat transfer textbooks or CIBSE or ASHRAE Handbooks.

The effect of thermal radiation asymmetry on human discomfort has been studied by several investigators, e.g. McIntyre [5] and Olesen [30]. Figures 1.11 and 1.12 show results of a study carried out at the Technical University of Denmark by Fanger *et al.* on seated and lightly clothed ($I_{cl} = 0.6$ clo) subjects exposed to a heated ceiling and a cold wall in a climatic chamber [30]. The subjects were otherwise in a state of thermal equilibrium with the environment in the chamber, i.e. the people were only dissatisfied with the radiant asymmetry. The effect of radiant temperature asymmetry, Δt_{pr}, on the percentage of dissatisfied due to a general warm or cold sensation is shown in these two figures for the heated ceiling and cold window respectively. It may be concluded from this study that people are more sensitive to asymmetric radiation caused by a warm ceiling than by a cold vertical surface. The influence of a cold ceiling and a warm wall was also studied but people were found to be less sensitive to these forms of radiant asymmetry.

The work of Fanger *et al.* on radiant asymmetry resulted in establishing limits for plane radiant temperatures by ISO Standard 7730 and ASHRAE Standard 55-1992. The limit in the vertical direction is 5 K and in the horizontal direction is 10 K. Both of these limits refer to either a small horizontal plane (vertical asymmetry) or a small vertical plane (horizontal asymmetry) 0.6 m above the floor. From Figures 1.11 and 1.12 these limits correspond to a percentage of dissatisfied of 7% and 5% for the vertical and horizontal asymmetry respectively.

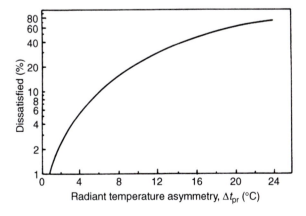

Figure 1.11 Effect of radiant temperature asymmetry due to a heated ceiling on the percentage of dissatisfied.

1.5.2   *Vertical air temperature gradient*

The air temperature in enclosures is not usually constant but increases vertically from floor to ceiling or varies horizontally from an external wall or window to an internal wall. The vertical temperature gradient in particular is a source of discomfort to a seated or a standing person. If this gradient is sufficiently large, local warm discomfort can occur at the head, and/or cold discomfort can occur at the feet, although the body as a whole may be thermally neutral. Figure 1.13 shows the effect of vertical air temperature gradient on the percentage dissatisfied from an investigation in a climate chamber [31]. Seated subjects were individually exposed to various air temperature differences between head (1.1 m above the floor) and ankle (0.1 m above the floor) whilst they were in thermal neutrality. ISO Standard 7730 recommends a maximum air temperature gradient of 3 K between 1.1 and 0.1 m above the floor, which corresponds to a percentage of dissatisfied of 5% according to Figure 1.13. However, the ASHRAE Standard 55-1992 recommends the same temperature gradient between 1.7

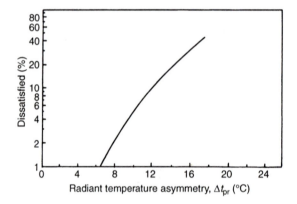

Figure 1.12 Effect of radiant temperature asymmetry due to a cold wall or window on the percentage of dissatisfied.

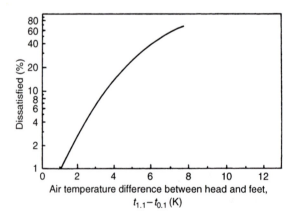

Figure 1.13 Effect of vertical air temperature difference between 1.1 and 0.1 m above floor on the percentage of dissatisfied.

and 0.1 m above the floor, which relates to a standing, not a seated, person as in the case of the ISO limit.

1.5.3   Warm or cold floor

In winter, the sensation of cold feet is a very common cause of thermal discomfort among sedentary people in residences and offices. This can be due to a low floor temperature and/or the general thermal state of the person. The effect of cold or warm floors on the comfort of people with bare feet and with footwear has been extensively investigated [31–33]. In studying the effect of cold/warm floors the subjects were kept in thermal neutrality, so dissatisfaction was only related to discomfort due to cold or warm feet. For bare feet, the flooring material was significant in determining an acceptable floor temperature. The recommended temperature range for a carpeted floor is 21–28 °C giving an expected percentage of dissatisfied of 15%. The flooring material was not found to be significant for people with footwear. An optimum floor temperature of 25 °C was obtained for a sedentary (seated) person and 23 °C for a standing or walking person. When the results for seated and standing persons were combined, Figure 1.14 was obtained which shows an optimum floor temperature of 24 °C corresponding to a percentage of dissatisfied of 6%. A value of 10% dissatisfied gives a floor temperature range of 19.5–28 °C.

ISO Standard 7730 recommends a floor temperature range of 19–26 °C for light, mainly sedentary activity in winter and floor heating system design temperature of 29 °C. The floor temperature range recommended in ASHRAE Standard 55-1992 is 18–29 °C. These ranges are applicable to people wearing appropriate indoor footwear.

1.5.4   Draught

Draught is defined as an undesired local cooling of the human body caused by air move-ment. It is one of the most frequent causes of complaint in heated or cooled buildings and in transport vehicles. As mentioned earlier in this chapter the thermoreceptors in the skin are very sensitive to changes in skin temperature. Draughts produce a cooling effect of the skin by convection, which is dependent on the temperature difference

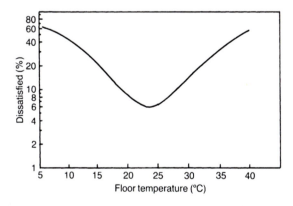

Figure 1.14 Effect of floor temperature on the percentage of dissatisfied with footwear.

between the air and skin, air speed and the magnitude of the fluctuations in the air speed, i.e. turbulence level.

Occupants who are subjected to draughts in winter tend to elevate the room temperature to counteract the cooling sensation thereby increasing the energy consumption. In extreme cases ventilation systems are shut off or air supply outlets are blocked off with a consequent deterioration of the indoor air quality. Draughts are caused by a high mean air speed and/or high turbulence intensity (defined as the ratio of the standard deviation of the velocity fluctuation to the mean velocity). Serious draught complaints can often occur at mean speeds lower than those recommended by standards (see Section 1.3) when the turbulence intensity is high.

Effect of mean air speed   The few early studies of the effect of draught were mainly concerned with the mean air speed and excluded the effect of turbulence. One of the earliest investigations was carried out by Houghton *et al.* [34] in 1938 in which ten male, medium-clothed subjects were each subjected to a jet of air from a duct positioned a short distance from the back of the neck or the ankles. The fall in skin temperature of the neck and ankles produced by a combination of air speed and air temperature was correlated with the percentage of subjects dissatisfied. A 1.8 K fall in skin temperature corresponded to 10% dissatisfied and a 2.4 K fall to 20% dissatisfied. Figure 1.15 shows the variation of air speed with air temperature for the neck region corresponding to 20% dissatisfied. It was also found that the fall in ankle temperature required to produce discomfort was the same as the neck, but larger speeds were needed to produce the same effect. The results of this investigation were later adopted to develop the ASHRAE effective draught temperature (see Chapter 6).

McIntyre [35] carried out a series of experiments on lightly clothed subjects in a chamber of temperature 21 or 23 °C with a jet of low-turbulence air from a 150-mm-diameter nozzle directed at the cheek from a distance of 300 mm. The subjects were given a 2 min exposure and the jet temperature was varied between 17 and 23 °C. Speeds of 0.35 m s^{-1} and higher were detected by all subjects whatever jet temperature was selected (i.e. 17–23 °C). However, at speeds below 0.35 m s^{-1} the

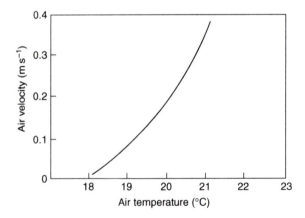

Figure 1.15 Variation of air speed with temperature at the back of the neck to produce 20% dissatisfaction.

percentage of people who were able to detect the draught increased as the jet temperature decreased. Jet speeds of $0.15\,\mathrm{m\,s^{-1}}$ or lower at $23\,^{\circ}\mathrm{C}$ were undetected by more than 20% of the subjects.

The effect of low-turbulence draught on the heads of 50 lightly clothed subjects (0.8 clo) was investigated by Mayer and Schwab [36] in a climatic chamber. The air was supplied to the head of each seated subject from four different directions: vertically upward, vertically downward, horizontally from the front and horizontally from the back. The air temperature was maintained at $23\,^{\circ}\mathrm{C}$ and was also equal to the mean radiant temperature. Five air speeds of 0.1, 0.2, 0.3, 0.4 and $0.45\,\mathrm{m\,s^{-1}}$ were used in the experiments whilst the turbulence intensity was unchanged at 5%. Each person was exposed to each speed for 10 min when he or she was thermally neutral. The results of percentage of dissatisfied because of draught against mean air velocity are shown in Figures 1.16 and 1.17. The percentage of dissatisfied due to draught at the face directed upward and downward is given in Figure 1.16 which shows a higher dissatisfaction with an upward draught. Heat transfer measurements from the heated head of a manikin subjected to upward and downward draughts at the face produced higher heat transfer coefficients for the upward draught than for the downward draught, which supports the subjects' sensation to this type of draught. This may be explained by the presence of an upward natural convective current due to buoyancy which enhances an upward-directed draught but diminishes a downward-directed draught at the face. The results for draught at the neck were found to lie between those shown in Figure 1.16 but no difference was found between an upward and a downward air motion. The dissatisfaction votes for draught towards the face and draught towards the neck are shown in Figure 1.17. In common with the findings of other investigations these results show the neck to be more sensitive to draught than the face.

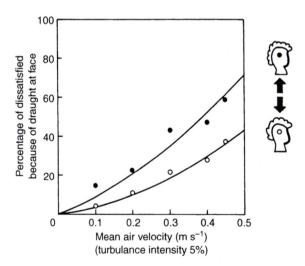

Figure 1.16 Effect of air speed on the percentage of dissatisfied by an upward and downward draught at the face.

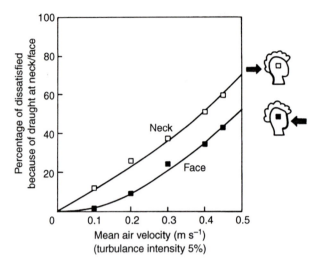

Figure 1.17 Effect of air speed on the percentage of dissatisfied by a horizontal draught towards the face and neck.

The experiments on the effect of draught reported so far were conducted with a low turbulence intensity air motion, i.e. approximately 5%. Measurements in numerous mechanically ventilated buildings have indicated turbulence intensities greater than 30% at a height of 1.1 m above the floor and in excess of 20% at a height of 0.1 m above the floor [37, 38]. These values refer to the highest mean speeds measured in the spaces investigated (≈ 0.3 m s^{-1}) and much higher intensities were measured at lower mean speeds. Measured air speeds and turbulence intensities in a number of naturally ventilated buildings with different heating systems were found to be slightly lower than those in the mechanically ventilated buildings, but the turbulence intensity was still more than 25% at a height of 1.1 m above the floor for the higher mean speed range.

The results of these studies prompted Fanger and his associates [39] to repeat the earlier experiments on the effect of draught using turbulence intensities of similar values as those obtained from the field measurements. They exposed 100 seated subjects (50 male and 50 female) in a climatic chamber to a turbulent air flow typical of real ventilated spaces. During the experiments the mean speed was varied from 0.05 to 0.4 m s^{-1} at air temperatures of 20, 23 and 26 °C. The subject's body as a whole was kept thermally neutral by modifying clothing ensembles to give a range of insulation from 0.58 to 0.91 clo. The turbulence intensity of the draught was in the range 30–65%. The subjects were exposed to draught at the back of the neck but limited tests were also carried out with the feet and elbow exposed to draught. Measurements of air velocity and temperature were taken at a distance of 0.15 m from the part of the body subjected to draught. Figure 1.18 shows the percentage of subjects who felt draught at the head region as a function of the mean speed at the neck. The head region comprises the head, neck, shoulders and upper part of the back. The scale used in the abscissa represents the square root of the speed since the convective heat transfer rate is approximately proportional to the square root of the mean velocity.

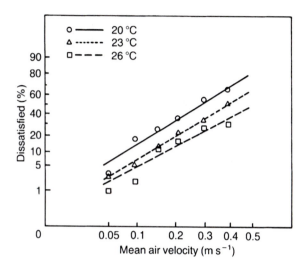

Figure 1.18 Effect of mean draught speed on the percentage of dissatisfied due to the feeling of draught at the head region.

The effect of the temperature of draught on the percentage dissatisfied is significant. The line for the percentage of dissatisfied for a temperature of 23 °C is similar to the curve in Figure 1.17 representing the results from Mayer and Schwab [36] at the same temperature but much lower turbulence intensity. This is rather surprising since higher turbulence intensity produces a greater cooling sensation as will become more obvious later in this section.

Based on the above measurements, a draught chart which identifies the percentage of subjects dissatisfied due to draughts in ventilated spaces was produced, as shown in Figure 1.19. The chart excludes discomfort due to draughts of speed below 0.04 m s^{-1} since it has been argued that at such low speeds the source of discomfort is something other than draught. The chart is represented by the expression:

$$PD = 13,800\{[(\bar{v} - 0.04)/(t_a - 13.7) + 0.0293]^2 - 8.57 \times 10^{-4}\} \qquad (1.51)$$

where PD = percentage of dissatisfied; $\bar{v}$ = mean air speed (m s^{-1}) and t_a = air temperature (°C).

The effect of draught on 50 subjects wearing clothing of 0.86 clo was studied by Berglund and Fobelets [40] in an environmental chamber. The draught was generated by a variable-speed fan producing an air speed in the range of 0.05–0.5 m s^{-1} at a distance 0.3 m upstream of the subjects. The turbulence intensity varied from about 44% at the lowest speed down to 5% at the highest speed. The air temperature was varied between 19 and 26 °C. A linear regression analysis of the test data produced the following expression for the percentage experiencing draught (PED):

$$PED = 113(\bar{v} - 0.05) - 2.15t_a + 46 \qquad (1.52)$$

Equations (1.51) and (1.52) are plotted in Figure 1.20 for air temperatures of 20 and 24 °C for comparison. Equation (1.51) due to Fanger and Christensen [39] produces

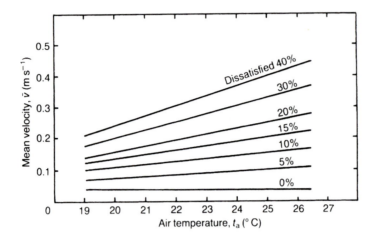

Figure 1.19 Draught chart for sedentary people wearing normal indoor clothing.

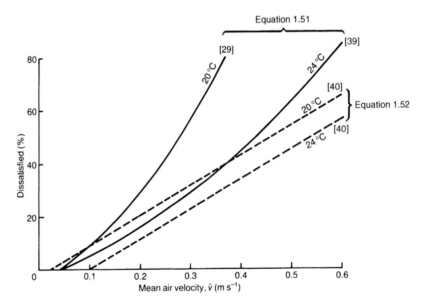

Figure 1.20 Comparison of percentage of dissatisfied due to draught using equations (1.51) and (1.52).

a higher PD than equation (1.52) of Berglund and Fobelets. The difference may be attributed to the larger turbulence intensity used in Fanger and Christensen's study (between 30% and 65%) and possibly also the difference in the questionnaires used in the two investigations.

ASHRAE Standard 55-1992 and the ISO Standard 7730 specify maximum mean air speeds of $0.15\,\mathrm{m\,s^{-1}}$ in winter and $0.25\,\mathrm{m\,s^{-1}}$ in summer. Taking the higher limit, it can be seen from the draught chart in Figure 1.19 that more than 25% of the occupants may be dissatisfied if this limit is used in design. This may explain why draught complaints are so common in ventilated spaces.

Effect of turbulence   The rate of change of skin temperature provides a stimulus for the thermoreceptors in the skin to initiate signals to the brain which senses this change [41]. Using an electrical analogue computer to simulate the human skin, Madsen [42] measured the heat flow through the receptors when the simulated skin was exposed to sinusoidal air velocities of the same amplitude but varying frequency. He obtained a maximum heat flow through the thermoreceptors at a frequency of 0.5 Hz. Fanger and Pederson [43] conducted experiments on 16 sedentary subjects exposed at the back of the neck to fluctuating air speeds with varying amplitudes and frequencies and different air temperatures. Each subject was exposed to mean speeds in the range $0.1–0.3 \, \mathrm{m\,s^{-1}}$, frequencies of $0–0.83 \, \mathrm{Hz}$ and turbulence intensities of 60–90%. Figure 1.21 shows the results for a mean air speed of $0.3 \, \mathrm{m\,s^{-1}}$ and in a remarkable agreement with Madsen's predictions maximum discomfort occurred at frequencies between 0.3 and 0.5 Hz. These studies have shown that a turbulent air velocity is less comfortable than a laminar air velocity of the same average value. Mayer [44] carried out laboratory measurements of the convective heat transfer coefficient of a heated artificial head under a range of air speeds and turbulence intensities. The heat transfer coefficient increased nearly fourfold when the standard deviation of the fluctuating velocity increased from 0 (laminar flow) to $0.22 \, \mathrm{m\,s^{-1}}$.

In occupied spaces the air movement is rather random and not well defined which is characteristic of turbulent flow. Typical air velocity signals in three rooms with different heating systems are shown in Figure 1.22 [40]. The small velocity fluctuations close to the floor in the case of the floor heating system are due to the downdraught over the floor from a window. The higher fluctuations in the radiator-heated room are caused by the interaction of convective currents produced by the radiator with downdraught from a window, since the signal was measured close to the window. Except for the difference in the turbulence intensity (*TI*) between the two signals the mean velocity is almost the same.

Measurements of air velocities in 12 mechanically ventilated buildings with air change rates of 4 and 8 per h are reported in [37]. Similar measurements were performed in 22 buildings of various types including offices, schools, lecture rooms, meeting rooms, auditoria, clean rooms, industrial halls and a swimming pool [38].

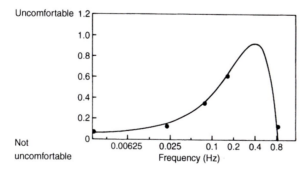

Figure 1.21 Degree of discomfort as affected by the frequency of air velocity fluctuation.

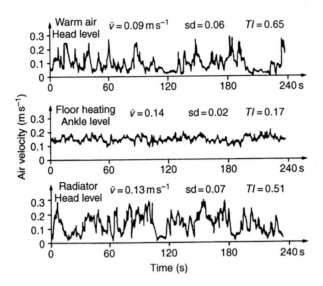

Figure 1.22 Air velocity signals in three rooms with different heating systems.

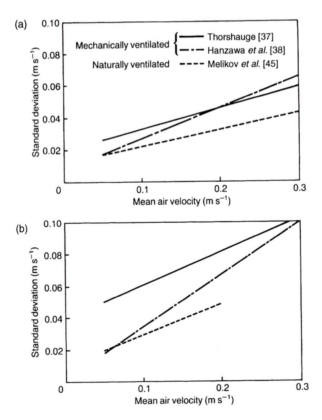

Figure 1.23 Variation of standard deviation with mean velocity in mechanically and naturally ventilated buildings: (a) 0.1 m above floor and (b) 1.1 m above floor.

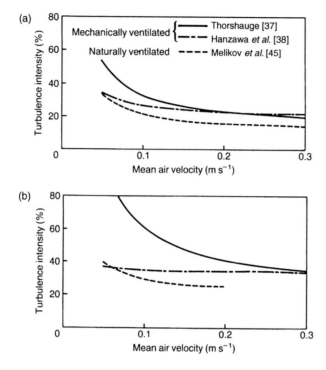

Figure 1.24 Variation of turbulence intensity with mean velocity in mechanically and naturally ventilated buildings: (a) 0.1 m above floor and (b) 1.1 m above floor.

The air change rates for these buildings were wide ranging: from 1.6 ach for a swimming pool to 25 ach for a clean room.

Velocity measurements were also carried out in eight naturally ventilated and heated buildings [45]. The results from these three investigations are shown in Figures 1.23 and 1.24 for heights of 0.1 and 1.1 m above the floor. Figure 1.23 shows that the standard deviation of the velocity fluctuations increases linearly with increasing mean speed, with higher fluctuations present in the mechanically ventilated buildings. The turbulence intensity (see Figure 1.24) is higher at low mean speeds and gradually decreases with mean speed until an asymptotic value is attained. These results represent a range of turbulence intensity of about 15–80%. The mean speeds in the naturally ventilated buildings were generally lower than in the mechanically ventilated buildings.

The effect of turbulence intensity on the sensation of draught was investigated in experiments carried out on 50 seated subjects (25 male and 25 female) in a draught chamber and each subject exposed to three levels of turbulence intensity ($TI < 12\%$, $20\% < TI < 35\%$ and $TI > 55\%$) from behind with a constant air temperature of 23 °C [46]. These TI levels have the same ranges as those obtained in field measurements. Apart from local draught discomfort the subjects were dressed to achieve thermal neutrality with the environment. In each experiment the sedentary subjects were exposed to six mean air speeds ranging from 0.05 to 0.4 m s^{-1}. The air velocity was measured at the neck level (1.1 m above the floor) and at a distance of 0.15 m from the neck as in previous experiments. The effect of turbulence intensity on

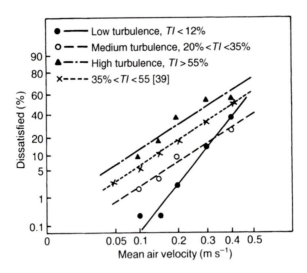

Figure 1.25 The percentage of dissatisfied due to draught at head region as a function of mean speed and turbulence intensity.

the percentage of dissatisfied is shown in Figure 1.25. The feeling of draught is significantly influenced by the turbulence intensity of the air stream that is directed towards the back of the neck. For a given percentage of dissatisfied a significantly higher mean air speed may be permitted if the *TI* is low. The influence of *TI* on women was slightly higher than on men but only at the lower range of mean air speed.

An empirical expression for the percentage of dissatisfied as a result of draught in the head region taking into consideration air temperature, mean air speed and turbulence intensity is given below [39, 46]:

$$PD = (34 - t_a)(\bar{v} - 0.05)^b (a + c\bar{v}TI) \qquad (1.53)$$

where PD = percentage of dissatisfied; t_a = air temperature (°C); $\bar{v}$ = mean air speed (m s^{-1}); TI = turbulence intensity (%); a = constant = 3.143; b = index = 0.6223 and c = constant = 0.3696. The ranges of the three parameters used in equation (1.53) are $20 < t_a < 26$ °C, $0.05 < \bar{v} < 0.4$ m s^{-1} and $0 < TI < 70$%. In this expression, when $\bar{v} < 0.05$ m s^{-1} a value of $v = 0.05$ m s^{-1} is used and if $PD > 100$% a value of 100% is assumed.

A three-dimensional representation of equation (1.53) is shown in Figure 1.26 for three values of the percentage of dissatisfied, i.e. $PD = 10$%, 15% and 20%. For 10% and 20% PD the relationship between air temperature and mean air speed is shown in Figure 1.27 for values of TI between 0% and 60%.

The draught model represented by equation (1.53) and Figures 1.26 and 1.27 may be used to calculate the percentage of dissatisfied as a result of draught at the head region if $t_a, \bar{v}$ and TI can be measured or calculated (see Chapters 8 and 9) at a height of 1.1 m above the floor. It may be extended to other heights such as ankles (0.1 m) or elbows (0.6 m) but may tend to overestimate the effect of draught on these normally clothed regions of the body.

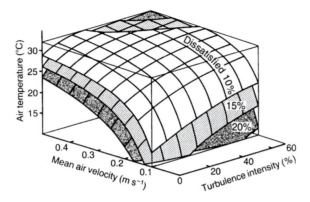

Figure 1.26 A three-dimensional representation of the three parameters influencing the percentage of dissatisfied due to draught at the head region.

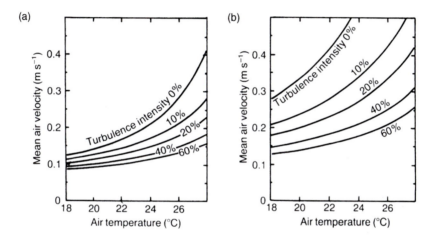

Figure 1.27 Relationship between mean air speed and temperature for different values of turbulence intensity to produce 10% and 20% dissatisfied as a result of draught: (a) 10% dissatisfied and (b) 20% dissatisfied.

Figure 1.28 shows values of non-fluctuating equivalent air velocities which will cause the same percentage of dissatisfied due to draught as fluctuating air flows with the same mean air speed and temperature but with $TI > 0$ obtained using the draught model given by equation (1.53). The effect of turbulence intensity on the percentage of dissatisfied for the comfort limits of operative temperature and mean air speeds specified by the thermal comfort standards for winter and summer is given in Figure 1.29. The winter conditions for comfort recommended by ISO 7730 and ASHRAE 55-1992 are: operative temperature between 20 and 24 °C and mean air speed less than 0.15 m s^{-1}. The summer conditions are: operative temperature between 23 and 26 °C and mean air speed less than 0.25 m s^{-1}. In Figure 1.29, the effect of TI on PD for the lower temperature limits for winter and summer is presented which also shows the equivalent limits specified by the German Standard DIN 1946 [47]. A horizontal line at 15% PD

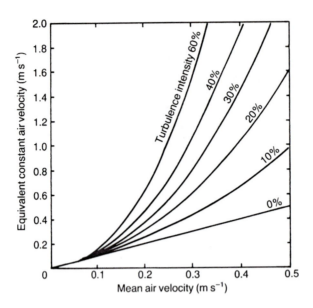

Figure 1.28 Equivalent air velocity of non-fluctuating flow which produces the same percentage of dissatisfied as the actual fluctuating flow.

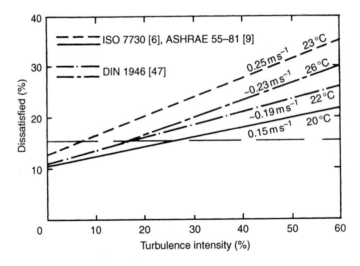

Figure 1.29 Effect of turbulence intensity on the percentage of dissatisfied for the temperature and air speed limits recommended in comfort standards.

is drawn on the figure to evaluate the values of *TI* that correspond to this value of *PD*. For the ISO and ASHRAE limits more than 15% dissatisfied among the occupants will result if the turbulence intensity in the room exceeds 7% in the summer and 25% in the winter. The DIN Standard 1946 value for summer and winter is approximately 17% which is within the ISO and ASHRAE Standards ranges. Although these standards do not specify limits for turbulence intensity for the purpose of providing a comfortable environment, the results given in Figure 1.29 clearly demonstrate that 15% *PD* can be easily exceeded in mechanically ventilated rooms where turbulence intensities can be as high as 50% at a mean speed of $0.15 \, \text{m s}^{-1}$ and about 35% at a mean speed of $0.25 \, \text{m s}^{-1}$ (see Figure 1.24). It is obvious that future comfort standards ought to incorporate the effect of the turbulence intensity of room air on human comfort.

1.6   Indoor air quality

The phrase 'sick building syndrome' is almost synonymous with poor indoor air quality and a wide range of professions (medics, psychologists, building designers, engineers, etc.) have in recent years attempted to identify the cause of this relatively new phrase in the dictionary of buildings.

Poor indoor air quality affects people in three areas:

- Comfort: e.g. stuffy, odorous environment.
- Acute health effects: e.g. burning eyes, chest symptoms, transmission of airborne disease.
- Chronic (delayed) health effects: may take decades to appear.

In addition, recent studies have shown that there is a direct link between the quality of room air and productivity [48].

Buildings that have direct influence on people's health and well-being are known as 'sick buildings'. Major factors associated with these buildings are the lack of fresh air and poor ventilation systems. In addition to externally generated pollution from vehicle emission, process discharge, building ventilation exhaust, boiler flues, etc. that is carried inside a building by ventilation air, there are many other types of pollutants present in indoor air. The sources of some of these are:

- Internal: the building itself, furnishing, occupants, equipment, etc.
- The ground: such as radon, methane, etc.

Problems related to the indoor air quality have increased in the last two decades primarily due to the following causes:

- Increase in building air tightness.
- Reduction in ventilation rates to reduce energy consumption.
- Increase in the use of computers, printers, photocopiers and other equipment in offices.
- The increasing use of textile floor covering and furnishing with high emission rate of pollutants.
- Increase in the use of air-conditioning.
- Lack of maintenance.

The main sources of internally produced pollution in buildings are the occupants, building materials, furnishing, equipment and processes carried out inside buildings. The pollutants that are most commonly found in indoor air are: odour, carbon dioxide, formaldehyde, volatile organic compounds (VOCs and TVOCs), tobacco smoke, ozone, radon, nitrogen oxides and aerosols. Although water vapour is not harmful to humans, when present with high concentrations could cause health problems indirectly. In most cases, the concentration of these pollutants is below the threshold limit values (TLVs) to show immediate effects although their long-term effects are often unknown. The effect of individual pollutants on humans is discussed in ASHRAE Standard 62-1999 [49] and HSE Guidelines [50].

The evidence for poor indoor air quality available so far appears to point towards poorly ventilated buildings and these are not necessarily buildings starved of outdoor air but rather the quality of the air supplied to them is poor. In fact, studies in naturally and mechanically ventilated buildings have shown that more complaints related to building sickness (such as nasal blockage and dry throat, headache, rhinitis and lethargy) were found among occupants of the mechanically ventilated buildings [1, 51]. It is believed that these complaints are caused more by pollutants emitted internally by building materials, furnishings, equipment and ventilation plants than by the occupants of the building themselves [52]. Indeed, the poorer indoor air quality in the case of mechanically ventilated buildings seems to support the argument that poorly maintained ventilation plants are major contributors to building sickness. The sources of indoor air pollution are discussed in Chapter 2 but in this section people's perception of indoor pollution concentrations will be presented.

1.6.1   *Olfactory stimuli*

The human olfactory system is usually superior to existing contaminant concentration measuring instruments and it can detect minute concentration levels that are sometimes undetectable by instruments. The olfactory receptors which are sensitive to odorants are located in the upper nasal passages, i.e. the olfactory cleft. The surrounding tissues in the nasal passages contain other receptors that respond to airborne vapours, the so-called common chemical sense. These receptors or nerve endings are stimulated by vapours that are pungent, e.g. ammonia. For all practical purposes, olfaction and the common chemical sense operate as a single perceptual system. The olfactory receptors are connected to the olfactory bulb in the brain which coordinates the passage of information it receives from the receptors to other regions of the body.

The perception of odorant by people is subjective and it is therefore difficult to define threshold limits for odours. The quality of air is usually determined by people's sensation to various odours present in the air. Most organic substances are olfactory stimuli but only a few inorganic substances are. Therefore, any subjective evaluation of the air quality using human olfaction will not include all sources of pollution. Some very harmful pollutants, e.g. carbon monoxide and radon, cannot be perceived by people even at high concentrations. The sensitivity of the human olfactory system to odour perception has been utilized by Fanger in developing a new approach for evaluating the intensity and concentration of pollutants as will be discussed below.

1.6.2   *Intensity and concentration of pollutants*

The relation between perceived odour intensity, s, and odour concentration, c, is in the form of the following power law [53]:

$$s = kc^n \tag{1.54}$$

where the units of concentration, c, are parts per million (p.p.m.) and the units of intensity, s, are arbitrary. The intensity scales which are used for odour detection are comparative with reference to a particular odour, such as butanol, or as a ratio of different concentrations. The value of the exponent, n, in equation (1.54) varies from one odorant to another with typical values between 0.2 and 0.7. Equation (1.54) is usually plotted on log–log graph paper to give a straight line of slope, n. Thus:

$$\log_{10} s = \log_{10} k + n \log_{10} c \tag{1.55}$$

Olfactometers are used to derive the index n and the constant k for each odorant. Further information on odour measurement can be found in the ASHRAE Handbook [53].

Existing ventilation standards for non-industrial buildings specify outdoor ventilation rates for different buildings per occupant or square metre of floor area assuming a certain occupancy density (see Chapter 2). Since the work of Yaglou *et al.* [54] in the 1930s on human bioeffluents, it has been assumed implicitly that humans are the main source of pollution in a building. This is now disputed, as will become more obvious later. In the late 1980s, Fanger [55] introduced two units, the *olf* (derived from the Latin word *olfactus*, i.e. olfaction) to quantify air pollution sources (cf. concentration) and the *decipol* (derived from the Latin word *pollutio*, i.e. pollution) to quantify the air pollution perceived by a person (cf. intensity). One olf is defined as the emission rate of air pollutants (bioeffluents) from a standard person. This is a relative unit based on subjective evaluation of the odorants by trained or untrained observers and includes both the olfactory and the chemical senses. The unit can also be used to express the strength of other pollution sources as equivalent to a number of standard persons (olfs) required to cause the same dissatisfaction as the actual source of pollution. The perceived intensity of air pollution caused by one standard person (olfs) ventilated by $1 \, \text{l s}^{-1}$ of unpolluted air is 1 pol. More conveniently, Fanger suggested the use of the unit decipol which is 0.1 pol. One decipol, therefore, is defined as the pollution caused by one standard person (1 olf) ventilated by $10 \, \text{l s}^{-1}$ of unpolluted air. The system of units is only applicable to odorant pollutants and cannot be used for evaluating odourless pollutants. Furthermore, it has been shown to be affected by the training level of the panel of observers, cultural differences among panel members, type of pollution perceived, etc. Nevertheless, the olf/decipol method has much to offer and these units have been adopted by a European pre Standard for ventilation [56] for specifying ventilation rates in design purposes.

1.6.3 *Acceptable concentration of pollutants*

Using the results from previous studies on bioeffluents involving more than 1000 sedentary men and women (1 met), Fanger obtained a correlation between the percentage of dissatisfied (*PD*) and the ventilation rate in $\text{l s}^{-1} \text{olf}^{-1}$ (Figure 1.30) [55]. The air quality was judged by 168 men and women just after entering the space. They were then asked to vote for either acceptable or unacceptable indoor air quality. The results shown in Figure 1.30 have been corrected for pollution from other sources in the buildings. The curve in the figure is described by the formula:

$$PD = 395 \exp\left(-1.83 \dot{v}^{0.25}\right) \tag{1.56}$$

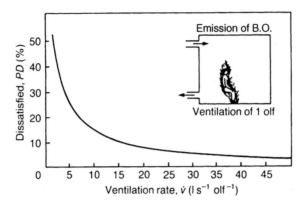

Figure 1.30 Percentage of dissatisfied as a function of ventilation rate for 1 olf.

Table 1.3 Olf values corresponding to various human activities [57]

Activity	No. of olfs
Sedentary person (1 met)	1
Active person (4 met)	5
Very active person (6 met)	11
Smoker (during smoking)	25
Smoker (average)	6

where PD is the percentage of dissatisfied, and $\dot{v}$ is the outdoor air flow rate for 1 olf in $1 \, \mathrm{s^{-1} \, olf^{-1}}$. Equation (1.56) applies to $\dot{v} > 0.32 \, \mathrm{ls^{-1} \, olf^{-1}}$, but for lower values of $\dot{v}$, PD is taken as 100%.

Figure 1.30 shows that for a ventilation rate of $10 \, \mathrm{ls^{-1} \, olf^{-1}}$ 15% dissatisfaction is expected. These results may also be used to predict the ventilation rates corresponding to other activities for a given PD by multiplying the ventilation rates obtained from Figure 1.30 or equation (1.56) by the number of olfs produced by each person at the new activity levels. Fanger gives the olf values produced by people at different activities, as presented in Table 1.3 [57]. The results in Figure 1.30 can be plotted in terms of perceived air pollution, C, to produce Figure 1.31 and the following formula:

$$PD = 395 \exp \left(-3.25 C^{-0.25}\right) \tag{1.57}$$

where C is in decipols. Equation (1.57) applies for $C \leq 31.3$ decipol, and for higher values of C PD is taken as 100%.

In believing that pollution levels in sick buildings are not exclusively caused by bio-effluent emission from the occupants, Fanger *et al.* [58] carried out an extensive survey of 20 randomly selected office buildings and assembly halls in Copenhagen. The spaces were visited three times by 54 trained observers (27 men and 27 women) who assessed the odour intensity using a six-point scale ranging from 0 (no odour) to 5 (overpowering odour). During the first visit the buildings were unoccupied and unventilated (except by air infiltration); during the second visit the buildings were unoccupied but

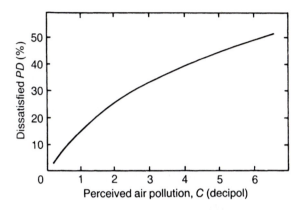

Figure 1.31 Percentage of dissatisfied as a function of the perceived air pollution in decipol.

mechanically ventilated; and during the final visit the buildings were occupied and ventilated. From the recorded percentage of dissatisfied in each building for the three test categories, it was possible to quantify the pollution concentration in olfs per square metre of floor area due to the occupants, the materials used in the buildings and the ventilation system. Although large differences in concentrations from the three pollution sources were experienced from one building to another, on average, for each occupant in 15 offices, 6–7 olf were from pollution sources other than the occupants. An average of 1–2 olf was from materials in the space, 3 olf from the ventilation system and 2 olf were caused by tobacco smoking. As a result, more than 30% of the subjects (observers) found the indoor air quality of the offices unacceptable even though the average ventilation rate was $25 \, \mathrm{l \, s^{-1}}$ per occupant which is much higher than the rates recommended by ventilation standards such as ASHRAE 62-1999 [49] and BS 5925 [59] (see also Chapter 2). Because of the high pollution concentrations produced by other sources the equivalent ventilation rate for the offices was only $4 \, \mathrm{l \, s^{-1} \, olf^{-1}}$.

1.6.4 Criteria for ventilation rates

In terms of the olf and decipol units a pollution balance for a ventilated enclosure can be written as:

$$C_i = C_o + 10 G / \dot{V} \tag{1.58}$$

where C_i = perceived air pollution within the enclosure (decipol); C_o = perceived pollution of outdoor air (decipol); G = generation of pollution in the enclosure and the ventilation system (olf); and $\dot{V}$ = outdoor air supply rate to the enclosure $(\mathrm{l \, s^{-1}})$. For design purposes, a ventilation rate, $\dot{V}$, is usually calculated to achieve an acceptable air pollution level within the enclosure and, in this case, equation (1.58) is rearranged to give:

$$\dot{V} = 10 G / (C_i - C_o) \tag{1.59}$$

If a given percentage of dissatisfied, PD, is required within the enclosure, C_i may be written in terms of PD by rearranging equation (1.57) to give:

$$C_i = 112/(5.98 - \ln PD)^4 \tag{1.60}$$

Equations (1.59) and (1.60) can be used to calculate a ventilation rate for an enclosure to achieve a desired percentage of satisfaction (e.g. 80% is recommended in ASHRAE Standard 62-1999) if the pollution intensity of the supply air, C_o, is known and the concentration of pollution generated internally and from the ventilation plant, G, is also known. The perceived outdoor air pollution is not usually included in the pollution balance except in large cities with known high pollution intensities. Outdoor concentrations, C_o, between 0.05 and 0.3 decipol have been reported for cities with moderate air pollution levels [58]. The other quantity which needs evaluation is the pollution generation rate within the enclosure, G. Currently, data on G are very scarce and a detailed analysis of the olf value of common building materials, furnishings and even personal clothing ensembles in addition to ventilation plant components is required to obtain an accurate assessment of acceptable ventilation rates. Based on field measurements of 15 office buildings, Fanger produced a pollution load table in olf per square metre of floor area as given in Table 1.4 [57]. The table gives an average pollution load for present building stock with 40% smoking of $0.7 \, \mathrm{olf \, m^{-2}}$ and for low-pollution buildings (without smoking) of $0.2 \, \mathrm{olf \, m^{-2}}$. Both these values are for an occupancy density of 0.1 persons $\mathrm{m^{-2}}$.

This data and the data from more recent sources have been used to develop a European pre Standard for ventilation [56]. Ventilation rates as well as room operative temperature, t_o, and sound pressure levels are specified for three categories of buildings, A, B and C, depending on the quality of the internal environment required in the building. This document is yet to be accepted by member states as a standard because of some concerns on implementing what amounts to be fairly strict criteria for the design and operation of buildings and the uncertainty regarding the use of the olf and decipol units for calculating ventilation rates.

It has been assumed so far that a perfect mixing of ventilation air with room air exists however, if the ventilation efficiency, E (see Chapter 2) is known then equation (1.59)

Table 1.4 Pollution loads for office buildings [57]

Pollution source	Load (olf m^{-2})
Occupants (0.1 persons m^{-2})	
Bioeffluent	0.1
Smoking	
20% smokers	0.1
40% smokers	0.2
60% smokers	0.3
Materials and ventilation systems	
Average of existing building stock	0.4
Low-pollution buildings	0.1
Total load	
Average in existing buildings with 40% smokers	0.7
Low-pollution building without smoking	0.2

Table 1.5 A comparison between ventilation rates for office buildings for occupancy density 0.1 persons m^{-2} with smoking

Source of ventilation rate	Ventilation rate (l s^{-1} m^{-2})
Equation (1.60)	5.0
ASHRAE Standard 62-1999 [49]	0.7*
BS 5925 [59]	1.3
DIN 1946 Standard [47], large office	1.9

Note
* Occupancy density is 0.07 persons m^{-2}.

is written as:

$$\dot{V} = 10G/[E(C_i - C_o)] \tag{1.61}$$

Similarly, transient conditions can be calculated in the case of transient occupancy if required.

Table 1.5 gives a comparison between the ventilation rates for office buildings with smoking, recommended by well-known ventilation standards, and that predicted using equations (1.59) and (1.60) for 80% satisfaction (1.4 decipol) and a pollution load of 0.7 olf m^{-2} from Table 1.4. The same predicted value will be obtained if Figure 1.30 is used instead of the equations. Because of the large pollution load from materials and ventilation plants, the predicted rate is some three to seven times higher than that recommended by the three standards. For the recommended rates to be acceptable by 80% of the occupants a reduction in the pollution load due to the materials and plant will be required. An increase in ventilation rates to 5 l s^{-1} m^{-2} as predicted by equation (1.59) will not be a recommended option because of the greater energy requirements and the increased risk of draught.

1.6.5 Effect of temperature and humidity on perceived air quality

Recent studies on quality of indoor air have shown that there is a correlation between people's acceptability or the perceived air quality in decipol and the enthalpy of the surrounding air. The enthalpy of air is determined by its dry-bulb temperature and moisture content or relative humidity, i.e. the sum of it's sensible and latent components. Initial studies have been carried out in a climatic chamber [60] by exposing people to clean air and air polluted by building materials under a range of air temperature and relative humidity (18–28 °C and 30–70% rh). In these studies the odour intensity was found to be almost independent of temperature and humidity; however, a significant impact of temperature and humidity on the perceived air quality was found. The air was perceived by people as less acceptable with increasing enthalpy (increasing temperature and humidity). Surprisingly, this correlation was found to be linear for a number of different pollution sources, as depicted in Figure 1.32 [60].

In another study involving a real office space, the impact of air temperature and humidity on the perceived air quality was studied in the ranges 20–26 °C and 40–60% rh with ventilation rates of 3.5 and 10 l s^{-1} per person occupying the office [61]. In this case the pollution sources in the office were building materials and

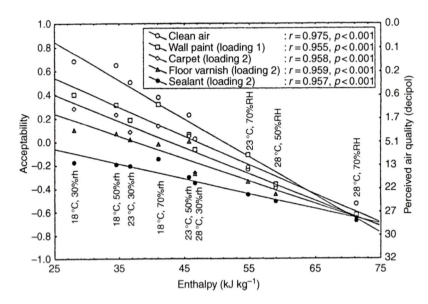

Figure 1.32 Acceptability score of air with enthalpy for different pollution sources – environmental chamber results.

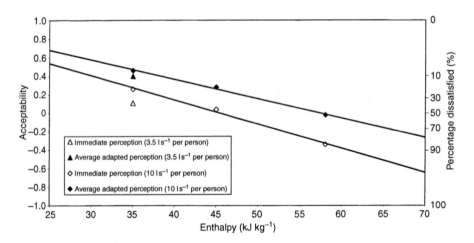

Figure 1.33 Acceptability score of air with enthalpy for different pollution sources – results from actual office.

bioeffluents. The subjects who assessed the air quality in the office were exposed to the pollutants for 4.6 h whilst performing simulated office work. This study confirmed the previous environmental chamber studies by observing a linear correlation between the acceptability of indoor air and air enthalpy, see Figure 1.33 [61]. It was concluded that the impact of a decrement of ventilation rate from 10 to $3.5\,\mathrm{l\,s^{-1}}$ per person on the air quality could be offset by a decrement of enthalpy from $45\,\mathrm{kJ\,kg^{-1}}$ (23 °C,

50% rh) to $35\,kJ\,kg^{-1}$ (20 °C, 40% rh). The correlation between ventilation rates and air enthalpy could have important energy implication.

References

1. Robertson, A. S. *et al.* (1985) Comparison of health problems related to work and environmental measurements in two office buildings with different ventilation systems. *Br. Med. J.*, **291**, 373–6.
2. Lind, A. R. (1963) A physiological criterion for setting thermal environmental limits for everybody's work. *J. Appl. Physiol.*, **18**, 51–6.
3. Olesen, B. W. (1982) Thermal comfort. *Brüel and Kjaer Tech. Rev.*, No. 2.
4. Fanger, P. O. (1972) *Thermal Comfort*, McGraw-Hill, New York.
5. McIntyre, D. A. (1980) *Indoor Climate*, Applied Science Publishers, Barking.
6. ISO 7730 (1994) Moderate thermal environments – determination of the PMV and PPD indices and specification of the conditions for thermal comfort, 2nd edn, International Standards Organisation, Geneva.
7. *CIBSE Guide C – Reference Data*(2001) Section C1: Properties of humid air, Chartered Institution of Building Services Engineers, London.
8. Nishi, Y., Gonzalez, R. R. and Gagge, A. P. (1975) Direct measurement of clothing heat transfer properties during sensible and insensible heat exchange with thermal environment. *ASHRAE Trans.*, **81** (2), 183–99.
9. ASHRAE Standard 55-1992 (1992) Thermal environmental conditions for human occupancy, American Society of Heating, Refrigeration and Air-Conditioning Engineers, Atlanta, GA.
10. Houghton, F. C. and Yaglou, C. P. (1923) ASHVE Research Report No. 673: determination of the comfort zone. *ASHVE Trans.*, **29**, 361.
11. Gagge, A. P. (1973) Rational indices of man's thermal environment and their use with a 2-node model of his temperature regulation. *Fed. Proc.*, **32**, 1572–82.
12. *ASHRAE Fundamentals Handbook* (2001) Ch. 8: Thermal comfort, American Society of Heating, Refrigeration and Air-Conditioning Engineers, Atlanta, GA.
13. *CIBSE Guide A* (1999) Environmental design. Chartered Institution of Building Services Engineers, London.
14. Awbi, H. B. and Savin, S. J. (1984) Air distribution methods for domestic warm air heating systems using low grade heat sources – tests with low level and high level air supply terminals. *Commission of the European Communities, Rep. No. EUR 9237 EN, Brussels.*
15. Thellier, F., Cordier, A., Galeou, M., Fudym, O. (1991) Comfort analysis as a criterion for energy management, *Proc. Building Simulation '91, Nice, France*, pp. 619–22.
16. Conceição, E. Z. E. (1998) Integral simulation of the human thermal system, *Proc. Rsoomvent '98, Stockholm*, Vol. 2, pp. 151–8.
17. Conceição, E. Z. E. (2000) Evaluation of thermal comfort and local discomfort conditions by numerical modelling of human and clothing thermal systems, in *Air Distribution in Rooms – Roomvent 2000*, H.B. Awbi (ed.), Vol. I, pp. 131–6, Elsevier, Oxford, UK.
18. Humphreys, M. A. (1976) Field studies of thermal comfort compared and applied. *Build. Serv. Eng.*, **44**, 5–27.
19. Humphreys, M. A. (1981) The dependence of comfortable temperatures upon indoor and outdoor climates, *Bioengineering, Thermal Physiology and Comfort*, K. Cena and J. A. Clark (eds), pp. 229–50, Elsevier, Amsterdam.
20. Auliciems, A. (1977) Thermal comfort criteria for indoor design temperatures in the Australian winter. *Arch. Sci. Rev.*, December, 86–90.
21. Fishman, D. S. and Pimbert, S. L. (1979) Responses to the thermal environment in offices. *Build. Ser. and Environ. Eng.*, January, 10 and 11.

22. Schiller, G. E. *et al.* (1988) A field study of thermal environments and comfort in office buildings. *ASHRAE Trans.*, **94** (2), 280–308.
23. ASHRAE (1990) Field studies of thermal comfort and the indoor environment. *ASHRAE Trans.*, **96**(1), 609–33 and 847–72.
24. Nicol, F., Humphreys, M., Sykes, O. and Roaf, S. (eds) (1995) *Standard for Thermal Comfort*, Chapman & Hall, London.
25. Cena, K. and de Dear, R. (1998) Field study of occupant comfort and office thermal environment in hot-arid climate. *ASHRAE Report RP-921*, American Society of Heating, Refrigeration and Air-Conditioning Engineers, Atlanta, GA.
26. Moving thermal comfort standards into the 21st Century. *Conference Proceedings*, Windsor, UK, 5–8 April 2001.
27. Williamson, T. J. *et al.* (1995) Comfort, preferences, or design data, in *Standards for Thermal Comfort*, F. Nicol *et al.* (eds), pp. 149–56, Chapman & Hall, London.
28. Nicol, F. (1995) Thermal comfort and temperature standards in Pakistan, in Standards for Thermal Comfort, F. Nicol *et al.* (eds), pp. 50–8, Chapman & Hall, London.
29. Brager, G. S. and de Dear, R. J. (2001) Climate, comfort and natural ventilation: a new adaptive comfort standard for ASHRAE Standard 55, *Proc. Conference on Moving Thermal Comfort Standards*, 5–8 April 2001, Windsor, UK.
30. Olesen, B. W. (1985) Local thermal discomfort. *Brüel and Kjaer Tech. Rev.*, No. 1.
31. Olesen, B. W., Scholer, M. and Fanger, P. O. (1979) Vertical air temperature differences and comfort, in *Indoor Climate* (P. O. Fanger and O. Valbjorn (eds), Danish Building Research Institute, Copenhagen, pp. 561–79.
32. Olesen, B. W. (1977) Thermal comfort requirements for floors occupied by people with bare feet. *ASHRAE Trans.*, **83** (2).
33. Olesen, B. W. (1977) Thermal comfort requirements for floors, *Proc. Meet. of Commission B1, B2 and El of the International Institute of Refrigeration, Belgrade*, pp. 337–43.
34. Houghton, F. C., Gutberlet, C. and Witkowski, E. (1938) Draft temperatures and velocities in relation to skin temperature and feeling of warmth. *ASHVE Trans.*, **44**, 289–308.
35. McIntyre, D. A. (1979) The effect of air movement on thermal comfort and sensation, in *Indoor Climate*, P. O. Fanger and O. Valbjorn (eds), Danish Building Research Institute, Copenhagen, pp. 541–60.
36. Mayer, E. and Schwab, R. (1988) Direction of low turbulent air flow and thermal comfort. *Proc. Healthy Buildings '88 (Vol. 2, Planning, Physics and Climatic Technology for Healthier Buildings)*, Stockholm, pp. 577–82.
37. Thorshauge, J. (1982) Air–velocity fluctuations in the occupied zone of ventilated spaces. *ASHRAE Trans.*, **88** (2), 753–64.
38. Hanzawa, H., Melikov, A. K. and Fanger, P. O. (1987) Airflow characteristics in the occupied zone of ventilated spaces. *ASHRAE Trans.*, **93** (1), 524–39.
39. Fanger, P. O. and Christensen, N. K. (1986) Perception of draught in ventilated spaces. *Ergonomics*, 29(2), 215–35.
40. Berglund, L. G. and Fobelets, A. P. R. (1987) Subjective human response to low-level air currents and asymmetric radiation. *ASHRAE Trans.*, **93** (1), 497–523.
41. Hensel, H. (1981) *Thermoreception and Temperature Regulation*, Academic Press, London.
42. Madsen, T. L. (1977) Limits for draught and asymmetric radiation in relation to human thermal well being, *Proc. Meet. of Commission B1, B2 and El of the International Institute of Refrigeration*, Belgrade.
43. Fanger, P. O. and Pederson, C. J. K. (1977) Discomfort due to air velocities in spaces, *Proc. Meet. of Commission B1, B2 and El of the International Institute of Refrigeration*, Belgrade, pp. 271–8.

44. Mayer, E. (1987) Physical causes of draft: some new findings. *ASHRAE Trans.*, 93 (1), 540–8.

45. Melikov, A. K., Hanzawa, H. and Fanger, P. O. (1988) Airflow characteristics in the occupied zone of heated spaces without mechanical ventilation. *ASHRAE Trans.*, 94 (1), 52–70.

46. Fanger, P. O. *et al.* (1988) Air turbulence and sensation of draught. *Energy Build.*, 12, 21–39.

47. DIN 1946 (1994) Teil 2, Raumlufttechnik Gesundheitstechnische Anforderungen (VDI-Lütungsregeln), Deutsches Institut für Normung, Berlin.

48. Fanger, P. O. (2000) Provide good air quality for people and improve their productivity, in *Air Distribution in Rooms (Proc. of ROOMVENT 2000)*, H. B. Awbi (ed.), Vol. 1, pp. 1–5.

49. ASHRAE Standard 62-1999 (1999) Ventilation for acceptable indoor air quality, American Society of Heating Refrigeration and Air-Conditioning Engineers, Atlanta, GA.

50. HSE Guidelines EH40/95 (1995) Occupational exposure limits, Health and Safety Executives, HSMO, London.

51. Ashley, S. (1986) Sick buildings. *Build. Serv.*, February, 25–30.

52. Bishop, V. L., Custer, D. E. and Vogel, R. H. (1985) The sick building syndrome: what it is, and how to prevent it. *Nat. Saf Health News*, 132 (6), 31–8.

53. *ASHRAE Fundamentals Handbook* (2001) Ch. 13: Odors, American Society of Heating, Refrigeration and Air-Conditioning Engineers, Atlanta, GA.

54. Yaglou, C. P., Riley, E. C. and Coggins, D. I. (1936) Ventilation requirements. *ASHVE Trans.*, 42, 133–62.

55. Fanger, P. O. (1988) Introduction of the olf and decipol units to quantify air pollution perceived by humans indoors and outdoors. *Energy Build.*, 12, 1–6.

56. CEN Standard pr ENV 1752 (1997) Design criteria for spaces in different types of buildings, European Committee for Standardization, Brussels.

57. Fanger, P. O. (1988) A comfort equation for indoor air quality and ventilation, *Proc. Healthy Buildings '88*, B. Berglund and T. Lindvall (eds), Stockholm, Vol. 1, pp. 39–51.

58. Fanger, P. O. *et al.* (1988) Air pollution sources in offices and assembly halls quantified by the olf unit. *Energy Build.*, 12, 7–19.

59. BS 5925 (1991) Code of Practice for ventilation principles and designing for natural ventilation, British Standard Institution, London.

60. Fang, L., Clausen, G. and Fanger, P. O. (1998) Impact of temperature and humidity on the perception of indoor air quality. *Indoor Air*, 8, 80–90.

61. Fang, L. *et al.* (1999) Field study on the impact of temperature, humidity and ventilation on perceived air quality, *Proc. 8th Int. Conference on Indoor Air Quality and Climate (Indoor Air '99)*, Edinburgh, August 1999, UK, Vol. 2, pp. 107–12.

2 Ventilation requirements

2.1 Introduction

Nowadays, ventilation occupies an important position in the building design process as building occupants expect good standards of indoor air quality and comfort. People have become more aware of the effect of the indoor environment on health as a result of media publicity surrounding building related sickness (BRS) and the sick building syndrome (SBS). SBS is fundamentally a complaint about the indoor air quality of a building which appears to be more common in air-conditioned buildings than in naturally ventilated buildings [1]. Building related sickness comprises the sensation of stuffy, stale and unacceptable indoor air, irritation of mucous membranes, headache, lethargy and so forth. These problems have been intensified in the last three decades as a result of a global reduction in air infiltration to buildings and fluctuation in the recommended supply rate of outdoor air to buildings in order to conserve energy.

In the past, internally produced pollution was wholly attributed to human bioeffluents [2] but more recent study led by Fanger in Denmark [3, 4] has identified other sources such as building materials, furnishings and even ventilation plants which, when combined, can produce a contamination concentration much greater than human bioeffluents. Most ventilation standards [5–7] still specify outdoor air flow rates that are only adequate to dilute human bioeffluents and contaminants produced by some activities (e.g. tobacco smoking) to acceptable concentrations, but they tend to ignore contaminations generated by other sources in the building as in many cases these contaminants are not easily quantifiable.

In today's technological society about 90% of the time is spent in an indoor environment (i.e. home, office, factory, transport vehicles, recreational buildings, etc.) and hence the provision of contaminant-free outdoor air to buildings is becoming a necessity. The outdoor air serves a number of purposes such as:

- human respiration which requires 0.1–$0.9\,l\,s^{-1}$ per person depending on metabolic rate;
- dilution of gaseous contaminants to achieve acceptable short-term exposure limits for carbon dioxide, odour and vapours of harmful chemical compounds;
- control of aerosols inside buildings using (filtered) outdoor air with lower aerosol concentration;
- control of internal humidity as outside air normally has lower moisture content;
- promoting air movement by proper air distribution design to provide comfort for the occupants.

There are many types of exposure limits in use for the workplace depending on the duration a person is exposed to the pollutant. Hence, a dose of a particular pollutant is conveniently expressed in terms of minutes, hours or years. The most widely used limits are: the threshold limit value (TLV) that should never be exceeded, the maximum exposure limit (MEL), the short-term exposure limit (STEL), the occupational exposure limit (OEL) and the long-term exposure limit (LTEL). Industrial exposures that are normally specified by the Health and Safety Executive in the UK or the Occupational Safety and Health Administration in the USA are based on 40-h working week with 8–10 h per working day.

In this chapter the main indoor pollutants are identified and the rates of outdoor air flow rates necessary to achieve acceptable pollutant concentrations will be postulated with reference to various ventilation guides and standards. The effectiveness of ventilation systems in removing indoor contaminants is reviewed with reference to the concept of ventilation effectiveness or efficiency.

2.2   Indoor contaminants

Before a ventilation rate can be specified to dilute or extract indoor contaminants it is necessary to identify those contaminants and their sources within the building as well as to establish acceptable concentrations in indoor air. Since more than 8000 chemical species have been identified in the indoor environment [8], this would appear a daunting task. Furthermore, very little scientific data exist on the potential health effect of most of these chemicals either on an individual basis or as aggregates. ASHRAE Standard 62-1999 [6] and the Health and Safety Executive in the UK [9] provide comprehensive lists of chemicals which are known to be present in domestic, commercial and industrial environments and give information on acceptable contaminant levels for both long- and short-term exposures. However, most of these pollutants are unlikely to be present in sufficient concentration in normal buildings and therefore only the most common pollutants are given in Table 2.1 which has been obtained from [6, 10, 11]. In this section, six pollutants of special concern to indoor air quality will be examined in some detail. These are tobacco smoke, formaldehyde, volatile organic compounds (VOCs), radon, ozone and aerosols. In addition, odours, carbon dioxide and moisture which have significant effects on indoor air quality, particularly in densely occupied spaces, are also investigated. Acceptable indoor concentrations of some other pollutants are also given in Table 2.1. However, more detailed treatment of indoor air pollutants can be found in the book by Meyer [12].

2.2.1   Odour

Odour is associated with occupancy, cooking, bathroom activities and waste, and odour pollution is associated with comfort rather than health effects. The human sense of smell permits perception of very low concentrations of odours but the sensitivity varies between individuals. People who are exposed to odours for a long period of time become less sensitive to them. Body odour is emitted by all people as a result of sweat and sebaceous secretion through the skin and also the human digestive system. Odour dilution to acceptable levels is usually achieved by supplying outdoor air to the occupied space. Yaglou *et al.* [2] established, through subjective tests on human bioeffluents, that the intensity of odour perceived by people entering an occupied room from

Table 2.1 Air quality guidelines [6, 10, 11]

Pollutant	Long-term exposure		Short-term exposure		Occupational exposure	
	Level (μg m^{-3})	Averaging time (years)	Level (μg m^{-3})	Averaging time (h)	Level (mg m^{-3})	Averaging time (h)
Sulphur dioxide (SO$_2$)	80	1	365	24	13	8
Particulate matter	75[a]	1	260[c]	24	5	8
Particulates (PM$_{10}$)	50[b]	1	150[a]	24	10	Instantaneous
Carbon monoxide (CO)	—	—	40,000[c] 10,000[c]	1 9	55	8
Ozone	—	—	235	1	0.2	8
Hydrocarbons	—	—	160	3	—	—
Formaldehyde (HCHO)	—	—	—	—	0.12	Continuous
Nitrogen dioxide (NO$_2$)	100	1	—	—	9	Instantaneous
Nitric oxide (NO)	—	—	—	—	30	8
Ammonia (NH$_3$)	500	1	7000	—	—	—
Acetone (CH$_3$COCH$_3$)	7000	24 h	24,000	0.5	—	—
Dichloroethane (CH$_3$CHCL$_2$)	2000	24 h	6000	0.5	—	—
Ethylacetate (CH$_2$COOC$_2$H$_5$)	14,000	24 h	42,000	0.5	—	—
Trichlorethylene (CH$_3$CCl$_3$)	2000	1	16,000	0.5	—	—
Mercury (Hg)	2	24 h	42,000	0.5	—	—
Lead (Pb)	1.5	0.25	—	—	—	—
Radon progeny	0.015 WL[d]	1	—	—	1 WL[d]	Instantaneous

Notes
a Geometric mean.
b Arithmetic mean.
c Not exceeded more than once per year.
d 1 WL = 1.3×10^5 Me V l^{-1} of air.

relatively odour-free air decreases with the logarithm of the outdoor air supply rate (i.e. concentration), as shown in Figure 2.1. This figure shows that for a space allowance of 5.7 m^3 (200 ft^3) per person, adult occupancy requires an outdoor air flow rate of 7.6 l s^{-1} per person and child occupancy (age group 7–14 years) requires 9.9 l s^{-1} per person to achieve a moderate odour intensity of two. Yaglou also found that the occupation density, m^2 per person, is an important parameter affecting odour intensity. As shown in Figure 2.2, to maintain an acceptable concentration of body odour (intensity index of two) the outdoor air flow rate is strongly dependent upon the occupation density of the space particularly for small density values. The results given in Figures 2.1 and 2.2 were obtained from tests in a room of approximately 3 m ceiling height [2]. However, no correlation between air supply rate and volume has been found by other researchers, which raises doubts regarding the applicability of Yaglou's results.

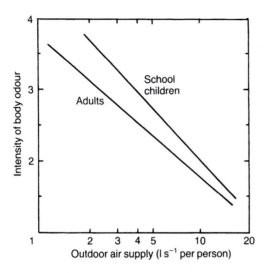

Figure 2.1 Effect of outdoor air supply rate on body odour intensity using the scale;
1 definite, 2 moderate, 3 strong, 4 very strong.

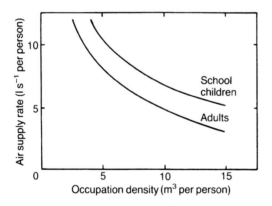

Figure 2.2 Influence of occupation density on body odour intensity.

2.2.2 *Carbon dioxide*

The rate of production of carbon dioxide (CO_2) by human respiration, G, is related
to the metabolic rate by the equation [13]:

$$G = 4 \times 10^{-5} \quad MA \tag{2.1}$$

where $G = CO_2$ production ($\mathrm{l\,s^{-1}}$); M = metabolic rate ($\mathrm{W\,m^{-2}}$) and A = body
surface area ($\mathrm{m^2}$).

An average sedentary adult ($M = 70\,\mathrm{W\,m^{-2}}$ and $A = 1.8\,\mathrm{m^2}$) produces about
$0.005\,\mathrm{l\,s^{-1}}$ ($18\,\mathrm{l\,h^{-1}}$) of CO_2 by respiration. Expired air contains about 4.4% by
volume of CO_2. It is easy to see from these figures why CO_2 concentration is very

Table 2.2 Outdoor air requirements for respiration [5]

Activity (adult male)	Metabolic rate, M (W)	Flow rate to maintain room CO_2 concentration at a given value assuming 0.04% CO_2 in fresh air $(l\,s^{-1})$	
		0.5% CO_2	0.25% CO_2
Seated quietly	100	0.8	1.8
Lightwork	160–320	1.3–2.6	2.8–5.6
Moderate work	320–480	2.6–3.9	5.6–8.4
Heavy work	480–650	3.9–5.3	8.4–11.4
Very heavy work	650–800	5.3–6.4	11.4–14.0

Note
These values are based on CO_2 production rate of $7.2 \times 10^{-5}\,M$ $(l\,s^{-1})$.

often used as an indicator for the control of outdoor air flow rates to a building, what is commonly known as *Demand Control Ventilation*. The presence of people in a room increases the concentration of CO_2 from an average outdoor air value of about 0.04% to a value which is dependent upon the population density, outdoor air flow rate and efficiency of the ventilation system. Unlike some other contaminants (e.g. tobacco smoke) CO_2 cannot be filtered, adsorbed or absorbed and it is therefore a good measure of the staleness of indoor air. The maximum concentration of CO_2 for 8 h occupation which is recommended by various ventilation standards is 0.5% although it has been reported [14] and now generally acknowledged that concentrations below 0.1% (1000 parts per million, p.p.m.) are required to avoid discomfort and headache. The outdoor air flow rates required for respiration and to maintain 0.5% CO_2 concentration for different metabolic rates are given in Table 2.2. This table assumes a perfect mixing of CO_2 with room air and if this is difficult to achieve in practice then higher ventilation rates than those given in the table will be needed.

2.2.3 Tobacco smoke

Tobacco smoking produces undesirable odour, particularly to non-smokers, and some of the smoke constituents can irritate the eyes and the nasal passages, acrolein in particular. Other products such as tar, nicotine and carbon monoxide can have serious health effects on smokers and also non-smokers exposed to a poorly ventilated environment (passive smoking). There have been studies [10] indicating that indirect or passive smoking can lead to lung cancer. Recommendations for outdoor air flow rates to dilute tobacco smoke vary considerably [15], but for an average smoker consuming 1.3 cigarettes per hour (17 cigarettes in a 13-h day) BS 5925 [5] recommends $7\,l\,s^{-1}$ per smoker in addition to the air flow rates required for other pollutants. Because of the large air flow requirement, the smoking population should be estimated at the design stage of a ventilation system. For large spaces such as theatres, public meeting rooms and large open-plan offices, statistical data of the smoking population can be used to estimate flow rates. However, in small rooms larger air flow rates will be needed to account for a denser smoking population or a larger smoking rate per person.

2.2.4   *Formaldehyde*

Formaldehyde (HCHO) is a chemical that has become ubiquitous in today's technological environments. It is extensively used as a preservative in cosmetics, toiletries and food packaging with up to 1% concentrations. About half of the formaldehyde currently produced is consumed in the production of urea- and phenol-formaldehyde resins that are used as bonding and laminating agents, as adhesives in compressed wood products, and as plastic foam insulation and packaging products. The energy conservation measures of the 1970s and 1980s have led to a sharp increase in the use of urea-formaldehyde foam insulation (UFFI) in buildings because of its good thermal insulation properties and its ease of moulding into any shape. The use of UFFI as a wall-cavity fill has been found to be particularly attractive. In the foaming process of UFFI a partially polymerized UF resin is brought into contact with an acid hardening agent in the presence of a foaming agent and the injection of compressed air. This results in a foam of small detergent bubbles that quickly gel and form a self-supporting polymer.

Because of the awareness of the adverse health effect from exposure to formaldehyde UFFI is not commonly used nowadays. However, formaldehyde resins are widely used in the manufacture of building materials and furnishings. The inexpensive UF resin is the most commonly used adhesive in the production of plywood, woodchip board, hardboard, plasterboard and as a binder in the production of fibreglass insulation. These products are extensively used as building materials and in the construction of furniture using plastic or wood veneer surfaces. Formaldehyde polymers are also used in the manufacture of wallpaper, carpets and textiles. Formaldehyde is also present in consumer products such as paper products, cosmetics, toiletries, glues and room deodorizers. Combustion appliances also generate appreciable amounts of formaldehyde and tobacco smoke contains as much as 40 p.p.m.

The emission of formaldehyde from UF foam (0.5% by weight of free formaldehyde is a necessary ingredient) is characterized by an initial peak release followed by a low-level continuous release. The initial high release rate is attributed to the presence of free formaldehyde in the foam and methylol. Upon exhaustion of these two sources a further but chronic release of formaldehyde is associated with the degradation of the UF polymer. The emission rate of formaldehyde from floor and wall materials, and wallpaper coverings to a flowing air stream is reported by Walsh *et al.* [10]. The data which was obtained from tests in a wind-tunnel-type apparatus are given in Table 2.3. These results show that woodchip board containing UF adhesive is the highest emitter of formaldehyde among the materials tested.

Table 2.3 Typical formaldehyde emission rate [10]

Material	Emission, E $(\mathrm{mg\,h^{-2}\,m^{-2}})$
Woodchip boards	0.46–1.69
Compressed cellulose boards (e.g. hardboard)	0.17–0.51
Plasterboards	0–0.13
Wallpapers	0–0.28
Carpets	0
Curtains	0

Formaldehyde can enter the body through inhalation, ingestion or skin adsorption. Most of that inhaled is absorbed in the upper respiratory tract. Once in the body, formaldehyde rapidly reacts with tissues containing hydrogen in the form of amino acids, proteins, DNA and others to form stable and unstable products and subsequently causes damage to the body tissues, i.e. it is genotoxic. Formaldehyde is a strong irritant that produces a variety of symptoms depending on the mode, duration and concentration of the exposure. Some studies have indicated that formaldehyde is a carcinogen in rats and mice but similar studies in humans have been inconclusive. However, the available data seems to suggest that formaldehyde is a direct-acting carcinogen and likely to pose a carcinogenic risk to humans also [10].

The concentration of formaldehyde in room air depends on the area of emitting surface, total air volume, air change rate and other parameters such as temperature, humidity and age of formaldehyde source. For a given formaldehyde source, the concentration of formaldehyde in air, c (p.p.m.), is related to these quantities through the equation:

$$c = AE/(\rho NV) \tag{2.2}$$

where A = area of formaldehyde-emitting surface (m^2); E = net formaldehyde emission rate from surface (mg m^{-2} h^{-1}); ρ = density of room air (kg m^{-3}); N = change rate of room air (h^{-1}) and V = air volume of room (m^3).

Equation (2.2) is based on a steady-state mass balance with no formaldehyde sinks present. It is only valid when the emission rate is not influenced by room air concentration of formaldehyde, i.e. unsuppressed emission. It does not apply to the cases of very low or zero change rates as, in the latter for example, the formaldehyde concentration will increase to its equilibrium value and the emission rate will then fall to zero (assuming no sinks); this becomes a fully suppressed emission. The other extreme condition occurs at a very high air change rate when the formaldehyde emission will reach a maximum value and the indoor air concentration will reach the outdoor air value, i.e. a completely unsuppressed emission. For a moderate air change rate some emission suppression can be expected, in which case an increase in air change rate will produce less than a proportionate decrease in formaldehyde concentration because of the resultant increase in the emission rate at source. This situation is illustrated in the data of Table 2.4 which was obtained from measurements in a modern energy-efficient house in Oak Ridge, Tennessee, USA [10]. A nearly threefold increase in air change rate (from 0.29 to 0.83 per hour) produced only about a 20% reduction in formaldehyde concentration.

Table 2.4 Effect of air change rate in a modern house on formaldehyde concentration

Air change rate (h^{-1})	Formaldehyde concentration (p.p.m.)	
	Mean	*Standard deviation*
0.29 (no heating, fan off)	0.332	±0.021
0.83 (no heating, fan on)	0.279	±0.017
0.83 (warm air heating on, fan on)	0.353	±0.022

Most current ventilation standards allow a maximum exposure limit to formaldehyde of about 0.1 p.p.m. or $\mu g\,m^{-3}$. This threshold limit value has been predicted by considering the irritating nature of formaldehyde, i.e. when acute health effects are experienced. However, this limit is not acknowledged to provide a protection of the health of individuals exposed, particularly those who are sensitive or sensitizable [10]. Furthermore, in field measurements this limit has been found to be exceeded for several types of dwellings particularly those insulated with UFFI. ASHRAE Standard 62-1999 [6] recommends a concentration of 0.4 p.p.m. indoors.

2.2.5 Volatile organic compounds

Volatile organic compounds are produced indoors from a variety of sources. There is, however, no clear definition of the classes of VOCs present in indoor air, though researchers define these as compounds having boiling points between 50 and 260 °C. Although formaldehyde is considered a VOC, it is usually dealt with separately because it requires different measuring techniques than those used for most other VOCs. In indoor air measurements VOCs are often reported as total volatile organic compounds (TVOCs). These are usually given as the sum of the concentrations of the individual VOCs. Research on the health effects of VOC is relatively new and there is little information available on the effect of long-term exposure to most known VOCs.

In most buildings the concentration of VOCs is not sufficiently large to be able to establish their health risk. Field studies in some European countries did not find a positive correlation between measured indoor air TVOC concentrations and SBS prevalence. There are, therefore, no established LTEL or STEL for TVOC in indoor air, although Molhave [16] conducted laboratory studies of the responses of human subjects exposed to controlled concentrations of 22 VOCs mixture. As a result of these studies we may be able to classify the exposure effect of VOCs as shown in Table 2.5. However, the concentrations in most buildings are usually much lower than those given in the table.

2.2.6 Ozone

Ozone (O_3) is naturally present in outdoor air, but its concentration is dependent on altitude and climate. It is also produced indoors by electrostatic appliances and office machines, such as photocopiers and laser printers. Ozone is considered as one of the most toxic pollutants regulated in indoor air. It has the potential for adverse, acute and chronic effects on humans if present in high concentrations. It appears to cause significant physiological and pathological changes in both animals and humans at exposure concentrations that are within the range found in polluted

Table 2.5 Classification of VOC exposure effects

Concentration range ($\mu g\,m^{-3}$)	Description
<200	Comfort range
200–3000	Multifactorial exposure range
3000–25,000	Discomfort range
>25,000	Toxic range

indoor air. Research conducted on animals and human volunteers under controlled laboratory conditions have shown significant lung function changes in response to exposures in the range 0.1–0.4 p.p.m. ($\approx$200–800 μg m^{-3}) for 1–2 h [17]. The World Health Organisation (WHO) recommends a maximum concentration of 100 μg m^{-3} or $\approx$50 p.p.b. (parts per billion) for 8-h exposure and ASHRAE Standard 62-1999 suggests the same limits for continuous exposure.

2.2.7   Radon

Radon (Rn) is a naturally occurring radioactive gas which arises from the decay of radium (Ra) present in small amounts in the earth's upper crust and in building materials. Radium originates from the decay chain of uranium (U). Radon itself produces a series of short-lived radioactive decay 'daughters', two of which, ^{222}Rn and ^{220}Rn, emit alpha particles. A fraction of the radon atoms emanate into the air and the alpha particles emitted from them normally present no health hazard because of their very short penetration depth into body tissues. However, if radon or its progeny are inhaled the alpha particles emitted may damage the lining of the lungs and pose the risk of lung cancer. Most of the health damage is caused by the radon daughters (radioactive) and not radon itself which is an inert gas. The daughters, which are electrically charged ions, attach themselves to dust particles in the air which may then be deposited in the lungs if inhaled.

The concentration of radon in the atmosphere is measured in picocuries per litre (pCi l^{-1}) or bequerels per cubic metre (Bq m^{-3}) where 1 pCi l^{-1} = 37 Bq m^{-3}. The concentration of radon daughters is evaluated in terms of the working level (WL), a unit that has been developed from investigations of radon exposure in uranium mines. The WL is defined as exposure to an atmosphere that contains any combination of radon daughters such that the total alpha particle emission, as a result of a complete decay, per litre of air is 1.3×10^5 MeV. This value corresponds to the potential alpha energy associated with the daughters in equilibrium with 100 pCi l^{-1} of radon, i.e. 1 WL = 100 pCi l^{-1}. The effect of exposure in health effects' assessment is evaluated using the working level month (WLM) which is defined as an exposure to 1 WL for 170 h (one working month). The continuous exposure, as in living environment, to 1 WL for a year (8760 h) would correspond to a total exposure of 51 WLM.

The concentration levels of Rn in buildings depend on the geological history of the site and the origin of the materials used in the construction. Hence, a wide variation in concentration levels can exist in any one country. A survey of 403 American houses reported by Ryan [18] produced an average concentration of Rn progeny of 0.0066 WL on the ground floors and 0.0127 WL in the basement. However, owing to the short monitoring periods used in the survey and because the soil considered was known to contain background ^{226}Ra concentrations, there may have been inherent biases in these data. In another study of 87 British dwellings, spread widely by location and type, Cliff [19] obtained a mean rate of emanation of Rn progeny into the air of living rooms of 0.0043 WL (a range of 0.0016–0.0471 WL depending on air change rate). Because of this disparity between Rn concentration, a value of 0.01 WL is usually used as a guide for the purpose of calculating ventilation rates. However, the choice of this concentration value is further complicated because full equilibrium of Rn daughters is rarely achieved in practice. If 50% equilibrium is assumed a maximum permissible level (MPL) of Rn progeny concentration of 0.02 WL or 74 Bq m^{-3} may be tolerated.

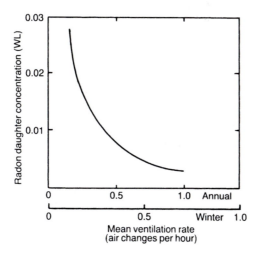

Figure 2.3 The effect of ventilation rate on radon daughter concentration in an average British house.

Nevertheless, the activity level can be influenced by a number of factors such as the radioactivity of the soil and building materials, the proportion of liberated Rn which diffuses into room air, the rate of deposition of Rn daughters on solid surfaces and the ventilation rate. For an Rn progeny emanation of 0.0054 WL the effect of air change rate on Rn daughter concentration is shown in Figure 2.3 which represents a survey of British dwellings [19]. A mean annual air change rate of 0.5 h^{-1} produces a concentration of 0.01 WL. For 8760 h in a year and assuming 80% of the time is spent indoors, this gives an annual exposure of 0.52 WLM per year or 36 WLM during a lifespan of 70 years. Relative risk estimates based on uranium mining surveys are in the range of 0.4–1% per WLM as a percentage increase in lung cancer risk [10]. A lifetime exposure (70 years) to 36 WLM will represent an increase in lung cancer risk to the occupants of these dwellings in the range 14–36%. Some countries specify MPL values for Rn exposure. In Sweden for example, the MPL Rn progeny is 70 Bq m^{-3} or 0.019 WL [14] which, according to Figure 2.3, corresponds to an air change rate of about 0.25 h^{-1}. A lower ventilation rate will clearly raise the MPL beyond the maximum permissible level. ASHRAE Standard 62-1999 recommends an MPL value of 0.04 WL.

2.2.8 *Aerosols*

Outdoor air pollution is a complex mixture of smokes, mists, fumes, granular particles biogenic particles and natural and synthetic fibres. When suspended in air, these particulates are called *aerosols*. An aerosol is a liquid or solid particle which is in a quasi-stable suspension in air. A sample of atmospheric dust usually consists of soot, silica, clay, lint and plant fibres, metallic fragments, living organisms such as pollen, mould spores, viruses, bacteria, etc. A large proportion of indoor aerosols originate from outdoor sources which penetrate the building envelope through cracks and openings by the action of wind, stack and ventilation systems. If a mechanical ventilation system is used the concentration of indoor airborne particles can be reduced by air filtration. The efficiency of most air filtration systems are highest for large or ultra-fine particles, however, the efficiency is lowest for fine particles, i.e. $<1\,\mu m$.

Aerosol diameters below $0.01\,\mu m$ are usually formed from combustion of fuels and radon progeny; those up to $0.1\,\mu m$ are produced by coking and cigarette smokes; $0.1–10\,\mu m$ is typically airborne dust, microorganisms and allergens; and particles of $100\,\mu m$ and larger are airborne soil, pollens and allergens. Figure 2.4 shows particle size ranges for aerosols. Concentrations of aerosols are generally at a maximum at sub-micrometer sizes and decrease rapidly as the aerosols size increases beyond $1\,\mu m$. In indoor air, particles of diameter greater than about $75\,\mu m$ settle out rapidly and are termed grit, but particles smaller than $50\,\mu m$ may remain suspended in air and constitute an aerosol. However, aerosols are never completely stable and particles are eventually deposited on internal surfaces at a rate which is inversely related to the size of particle. Very fine aerosols may remain suspended for many weeks, whereas larger aerosols may be deposited within minutes. The deposition of particles is influenced by static electricity, air currents and temperature gradients in the room, as these cause convective air currents that help the deposition of particles on cold surfaces (thermal deposition), particularly those particles below $2.5\,\mu m$ diameter (i.e. $PM_{2.5}$).

The concentration of particulate matter in the atmosphere is described in terms of either the number or mass of particles per unit volume. A common unit of particle count is m.p.p.c.f. (millions of particles per cubic foot) and of mass density is $mg\,m^{-3}$. The relation between mass density, and particle count depends on the density of the particles' material and their diameter, but for mineral dust an approximate conversion is $1\,mg\,m^{-3} = 6\,m.p.p.c.f.$ The TLV for particulate matter is $75\,\mu g\,m^{-3}$ for annual exposure and $260\,\mu g\,m^{-3}$ for 24 h exposure.

The effect on health of aerosols is mainly due to biogenic pollutants such as fungi, moulds, mites, bacteria, viruses and pollens. Indoor dust, which is a mixture of human and animal dander (skin flakes), fibrous material from textiles, organic particles and mites, is also a potential allergen and a cause of bronchial asthma and allergic rhinitis for about 1% of the population. Pollens are the most widespread allergies which cause hay fever seasonally. The main sources of pollen are outdoor vegetation, but their indoor concentrations can be reduced by air cleaners in the ventilation systems. Similarly, outdoor sources are the major contributors to indoor fungi spores although their indoor concentration can be higher than it is outdoors. Fine particulates (i.e. $PM_{2.5}$ and smaller) constitute the greatest long-term health risk, as these tend to be retained within the pulmonary region and lodge in tiny air spaces in the lungs.

Respiratory transmission of virus diseases occurs via person to person through airborne particle exchange. Diseases such as influenza, rubella (measles) and varicella (chickenpox) are widely transmitted by aerosols containing these viruses in indoor air. On the other hand, bacterial infectious diseases, such as Legionnaires' disease, are non-communicable infections emanating from sources in the building structure or services. In particular, outbreaks of Legionnaires' disease is mainly in buildings with air-conditioning installations that use water, such as water cooling towers or humidifiers. The bacterial agent *Legionella pneumophila* is multi-infectious involving the lungs (causing pneumonia) and intestine (causing vomiting and diarrhoea) and has an average incubation period of 5–6 days. Exposure to this bacterium will cause between 1% and 7% infections which have a case-fatality rate of about 15%. Recent epidemics have been associated with aerosols from cooling towers and evaporative condensers.

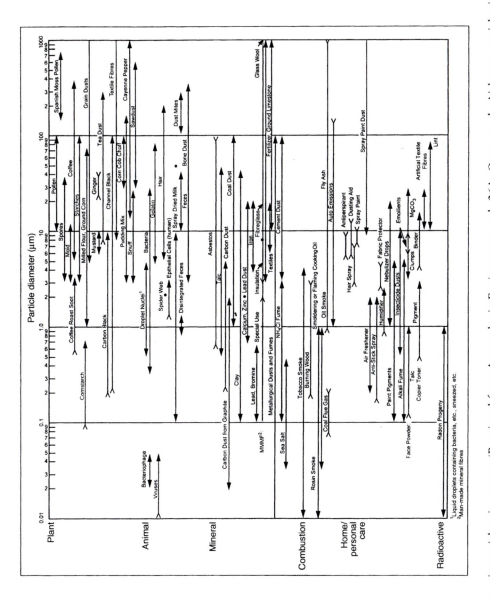

Figure 2.4 Indoor air particles size ranges (Reprinted from Atmospheric Environment, vol. 26A, Owen *et al.*, Airborne particle sizes and sources found in indoor air, 2149–62, copyright (1992), with permission of Elsevier Science).

The addition of oxidizing agents to cooling water and good plant maintenance practice can reduce the risk of exposure to *Legionella* [21]. However, as a result of publicized epidemics there has been a move away from water-cooled to air-cooled towers and condensers in the air-conditioning industry.

2.2.9 Water vapour

The concentration of water vapour in indoor air could influence the occupants' comfort and their well-being. There are three ways of defining the level of water vapour in air as follows.

Specific humidity (ω)   This is the ratio of mass of water vapour present in the air to the mass of dry air, i.e. it is the moisture content of air. It is expressed in $kg\,kg^{-1}$ or $g\,kg^{-1}$ of dry air. Thus:

$$\omega = m_v/m_a \tag{2.3}$$

where m_v and m_a are the mass of vapour and dry air respectively. Assuming both vapour and dry air behave as perfect gases (for a vapour this is almost the case if superheated) and applying the perfect gas law, it can be shown that:

$$\omega = 0.622p_v/p_a = 0.622p_v/(p - p_v) \tag{2.4}$$

where p_v and p_a are the partial pressures of vapour and dry air respectively and p is the total pressure of the mixture in pascals.

Relative humidity (ϕ)   This is the ratio of the actual vapour pressure to the saturation pressure at the same air temperature and is expressed as a percentage:

$$\phi = p_v/p_s \tag{2.5}$$

where p_s is the saturation pressure, Pa.

Percentage saturation (ψ)   This is the ratio of the actual specific humidity to the specific humidity of saturated air at the same temperature:

$$\psi = \omega/\omega_s \tag{2.6}$$

The percentage saturation is related to the relative humidity through the expression:

$$\psi = \phi[(p - p_s)/(p - p_v)] \tag{2.7}$$

Because the ratio $(p - p_s)/(p - p_v) \approx 1.0$ the percentage saturation is almost equal to the relative humidity.

Of the three definitions only the first (i.e. ω) represents a true measure of concentration of water vapour. The other two are relative measures of concentration.

The relation between air temperature, specific humidity, percentage saturation and specific enthalpy of the mixture is usually presented graphically on a psychrometric chart such as that shown in Figure 2.5 due to CIBSE.

The control of water vapour in the air is important for a number of reasons. It is widely believed among medical practitioners, for example, that low humidity

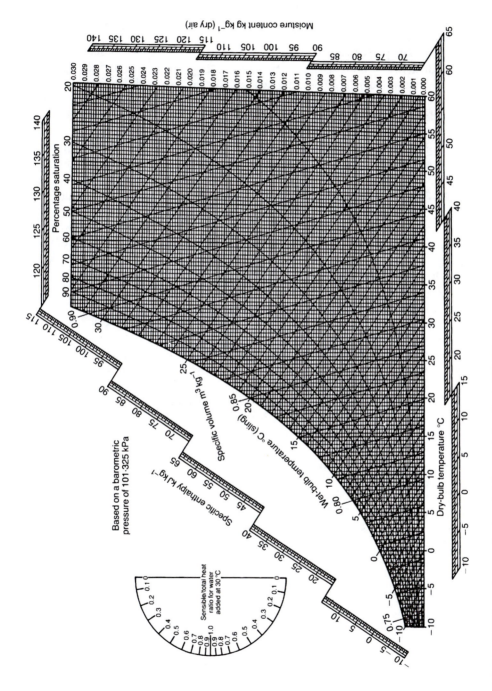

Figure 2.5 The CIBSE psychrometric chart (with permission from CIBSE).

could contribute towards increased risks of infection of the respiratory tract. This is attributed to the reduced flow of mucus in the nose or its complete drying out. The mucus layer in the nose serves as a trap to particles present in the inhaled air, including bacteria and viruses carried by dust particles, and prevents them entering the lungs. The foreign matter is cleared by a steady flow of the mucus bed outside the nose. Low humidity is also responsible for electrostatic shocks as a result of a reduction in the electric conductivity of floor coverings such as carpets. This can cause a build-up of static electricity up to about 2 kV and when contact with a metallic object is made a noticeable shock occurs but it is not usually harmful. However, even these low-energy charges could cause some problems in areas where electronic components are in use, such as computer rooms, or where inflammable gases are present.

High humidity levels can cause human discomfort by the inhibition of sweat evaporation through the skin. In inadequately ventilated spaces, increased humidity can cause stuffiness and smells and, if prolonged, a musty smell forms as a result of fungal growth. In Chapter 1 it was shown that high humidity can contribute to the perception of poor air quality. In winter, high indoor humidity can be responsible for water vapour condensing to liquid droplets on cold surfaces such as windows and cold external walls. The temperature at which condensation occurs, the dewpoint t_d, is dependent on the specific humidity of the air, i.e. the higher ω is the higher t_d is. The value of t_d for a given ω can be obtained from the psychrometric chart shown in Figure 2.5. Condensation can also occur on the surface or inside of a wall if the local temperature is equal to or below the dewpoint. The latter is called interstitial condensation and it can cause degradation of the wall fabric in severe cases, which is more commonly associated with unvented cavity walls where the cavity temperature on the outer leaf in winter could be considerably lower than the dewpoint.

Given suitable humidity conditions and nourishment, spores of mould and fungi, which are present in the air, can germinate and grow on surfaces over a wide range of temperatures and certainly between 0 and 20 °C. Some mould spores can germinate at 80–85% relative humidity and can spread if the humidity is over 70% for long periods [22]. This normally forms a musty smell and can be a health hazard causing allergies and illnesses including asthma, rhinitis and conjunctivitis.

Moisture is continuously emitted by building occupants and processes carried out indoors. Some emission rates from household activities are given in Table 2.6 and typical moisture production values per day within dwelling are given in Table 2.7 [5].

The build-up of moisture indoors can be reduced by ventilation because outdoor air normally has lower moisture content than indoor air particularly in winter when high moisture levels can lead to condensation on cold surfaces. If a homogeneous mixing of outdoor air with indoor air is assumed (i.e. 100% ventilation efficiency), the ventilation rate required to achieve a certain humidity level indoors can be calculated from a simple moisture balance equation for the building. The moisture content of outdoor air and moisture adsorption or desorption are also included in the equation to give [23]:

$$\dot{V}_o = (m_g - m_f)/[\rho_o(\omega_i - \omega_o)] \tag{2.8}$$

where $\dot{V}_o$ = volume flow rate of outdoor air (m^3 s^{-1}); ρ_o = density of outdoor air (kg m^{-3}); ω_o = specific humidity of outdoor air (kg kg^{-1}); ω_i = specific humidity indoors (kg kg^{-1}); m_g = rate of moisture generation within the building (kg s^{-1}) and m_f = rate of moisture diffusion through the building fabric (kg s^{-1}).

Table 2.6 Typical moisture emission rates [5]

Process	Emission rate
Adult occupation	
Sleeping	$0.04\,kg\,h^{-1}$ per person
Active	$0.05\,kg\,h^{-1}$ per person
Cooking	
Electricity	2.0 kg per day
Gas	3.0 kg per day
Dishwashing	0.4 kg per day
Bathing/washing	0.4 kg per person per day
Washing clothes	0.5 kg per day
Drying clothes indoor	1.5 kg per person per day
(e.g. unvented tumble drier)	

Table 2.7 Typical moisture production for households [5]

Number of persons in household	Daily moisture emission (kg)		
	Dry occupancy[a]	Moist occupancy[b]	Wet occupancy[c]
1	3.5	6	9
2	4	8	11
3	4	9	12
4	5	10	14
5	6	11	15
6	7	12	16

Notes
a Dry occupancy: where there is proper use of ventilation, it includes those buildings unoccupied during the day.
b Moist occupancy: where internal humidities are above normal; likely to have poor ventilation; possibly a family with children.
c Wet occupancy: ventilation hardly ever used; high moisture generation; probably a family with young children.

Using standard atmosphere pressure, the density ρ_o can be written in terms of the air temperature as follows:

$$\rho_o = p_o/RT_o = 353/RT_o$$

where p_o = standard atmosphere pressure = 101.325 kPa; R = gas constant for air = $0.287 + 0.461\omega \approx 0.287\,kJ\,kg^{-1}$ assuming ω for outdoor air to be small and T_o = temperature of outdoor air (K).

Substituting for ρ_o, equation (2.8) becomes:

$$\dot{V}_o = 0.00283T_o[(m_g - m_f)/(\omega_i - \omega_o)] \qquad (2.9)$$

The above equation ignores the existence of cross-ventilation with neighbouring zones or buildings. Values of moisture generation rates m_g can be obtained from Tables 2.6 and 2.7 and the indoor and outdoor specific humidities, ω_i and ω_o, may be derived from psychrometric charts (Figure 2.5) or tables if the temperature and percentage saturation are known. Alternatively, the approximate formula below may be used to calculate ω given the air temperature, t_a (°C), and the relative humidity, ϕ, as a fraction:

$$\omega = \phi \exp[11.56 - 4030/(t_a + 235)] \tag{2.10}$$

The diffusion of water vapour through the building fabric may be obtained using Fick's law:

$$m_f = \Delta p_v \sum_{i=1}^{n} A_i/R_i \tag{2.11}$$

where R_i = vapour resistance of fabric i (kN s kg^{-1}); A_i = surface area of fabric i (m^2) and Δp_v = difference in vapour pressure between inside and outside (kPa). Values of R_i for different building materials may be obtained from the CIBSE Guide [24] or the ASHRAE Handbook [25].

Under normal ventilation rates, the moisture diffusion through the fabric material is small in comparison to the moisture transfer by ventilation [26] and equation (2.9) reduces to:

$$\dot{V}_o = 0.002\,83\,T_o m_g/(\omega_i - \omega_o) \tag{2.12}$$

The risk of condensation on the internal surfaces of the building fabric may be evaluated using a psychrometric chart if the surface temperature is known. This can easily be calculated using the heat conduction equation:

$$t_s = t_i - (t_i - t_o)U/h_i \tag{2.13}$$

where t_s = internal surface temperature (°C); t_i = indoor air temperature (°C); t_o = outdoor air temperature (°C); U = heat transmittance coefficient for the fabric (W m^{-2} K^{-1}) and h_i = inside surface heat transfer coefficient (W m^{-2} K^{-1}).

The value of h_i depends on the position of the surface, the direction of heat flow and the temperature difference between the surface and air. The CIBSE Guide, section A3 [27] gives 3.0, 1.5 and 4.3 W m^{-2} K^{-1} for walls, floors and ceilings respectively. More accurate values can be found in [28, 29]. Condensation will occur on the surface when t_s calculated from equation (2.13) is equal to or lower than the dewpoint of indoor air. The dewpoint may be obtained from a psychrometric chart (i.e. saturation temperature) or may be calculated approximately using the expression:

$$t_d = 4030(t_a + 235)/[4030 - (t_a + 235)\ln\phi] - 235 \tag{2.14}$$

where t_d = dewpoint (°C) and ϕ = relative humidity as a fraction.

Procedures for the calculation of moist air properties can be found in [30]. These procedures involve the solution of a number of equations to generate the required property according to the Gibbs Phase Rule and are particularly suitable for use in a computer program.

2.3 Ventilation rates

In Section 2.2, the main indoor air pollutants were presented and some guidelines on tolerable concentration levels were specified. Data on the effect of ventilation rate on the concentration of each pollutant was also given. In this section the procedures used for determining acceptable or recommended ventilation rates for continuous and transient occupancies are described for a specified outdoor air quality. Minimum ventilation rates for various pollutants can also be found in an Annex IX report of the International Energy Agency [31].

2.3.1 Quality of ventilating air

Ventilation is the replacement of polluted or stale air indoors by 'fresh' or unpolluted air from outside. Although generally the building fabric could sometimes act as a 'filter' to infiltrated air, certain types of pollutants have been found to have higher concentrations inside than outdoors. This may be due either because such pollutants are produced internally by the occupants, their activities or building materials and other contents, or that air intakes are located in contaminated regions of the building envelope, such as close to traffic, air exhaust, exhausts from combustion or kitchen appliances, etc. Therefore, acceptable levels of contaminants in the outdoor air to be used for the purpose of building ventilation must be specified. The ASHRAE Handbook of Fundamentals [32] gives possible concentrations of several indoor pollutants, their sources and indoor/outdoor concentration ratios; these are reproduced in Table 2.8. The table shows that the concentrations of carbon monoxide, particulates, nitrogen dioxide, radon and carbon dioxide are much higher indoors than outdoors. However, sulphur dioxide, sulphate and ozone are at lower concentrations indoors than they are outdoors. The values in Table 2.8 are only typical values and they are not expected to be representative for every town or city. More specific site information of pollutants concentration can be obtained from national air pollution monitoring agencies and some of the information is found on the Internet, such as [33] in the UK. If the outdoor concentration of pollutants is higher than that given in Table 2.1 or in the ASHRAE Standard 62-1999 [6], treatment of outdoor air will be necessary before it is supplied to enclosed spaces. This treatment could be in the form of filtration, absorption, adsorption and so forth. Air-cleaning equipment commonly used in ventilation may be found in the ASHRAE Handbook of HVAC Systems and Equipment [34] or the CIBSE Guide B2 [35].

In addition to air filtration, it may sometimes be necessary to vary the moisture content of ventilation air in order to control the relative humidity indoor since this has some impact on IAQ. In winter, the moisture content of outdoor air is low and it may be necessary to humidify it by injecting steam or spraying water droplets before supplying it to the space. On the other hand, dehumidification of outdoor air may be required in summer by passing it over a cooling coil or a chilled-water spray dehumidifier.

2.3.2 Sources of external air pollution

External air pollution can be divided into: background levels which apply over a very large area and vary very slowly with time; urban concentrations which cover spatial distances of between 5 and 50 km and vary with time in the order of several hours; neighbourhood concentrations over distances of 2 km or less, which vary more quickly

Table 2.8 Sources, possible concentrations and indoor-to-outdoor (I/O) concentration ratios of some indoor pollutants [32]

Pollutant	Sources of indoor pollution	Possible indoor concentration*	I/O concentration ratio	Location
Carbon monoxide	Combustion equipment engines, faulty heating systems	100 p.p.m.	$\gg 1$	Skating rinks, offices, homes, cars, shops
Respirable particles	Stoves, fireplaces, cigarettes, condensation of volatiles, aerosol sprays, resuspension cooking	100–500 $\mu g\,m^{-3}$	$\gg 1$	Homes, offices, cars public facilities, bars, restaurants
Organic vapours	Combustion, solvents, resin products, pesticides, aerosol sprays	NA	>1	Homes, restaurants, public facilities, offices, hospitals
Nitrogen dioxide	Combustion, gas stoves, water heaters, dryers, cigarettes, engines	200–1000 $\mu g\,m^{-3}$	$\gg 1$	Homes, skating rinks
Sulphur dioxide	Heating system	20 $\mu g\,m^{-3}$	<1	Removal inside
Total suspended particles without smoking	Combustion, resuspension, heating system	100 $\mu g\,m^{-3}$	1	Homes, offices, transportations, restaurants
Sulphate	Matches, gas stoves	5 $\mu g\,m^{-3}$	<1	Removal inside
Formaldehyde	Insulation, product binders, particleboard	0.05–1.0 p.p.m.	>1	Homes, offices
Radon and progeny	Building materials, groundwater, soil	0.1–100 $mCi\,m^{-3}$	$\gg 1$	Homes, buildings
Asbestos	Fireproofing	$<10^6$ fibres per m^3	1	Homes, schools, offices
Mineral and synthetic fibres	Products, cloth, rugs, wallboard	NA	—	Homes, schools, offices
Carbon dioxide	Combustion, humans, pets	3000 p.p.m.	$\gg 1$	Homes, hospitals, schools, offices, public facilities
Viable organisms	Humans, pets, rodents, insects, plants, fungi, humidifiers, air-conditioners	NA	>1	Homes, hospitals, schools, offices, public facilities
Ozone	Electric arcing	0.02 p.p.m.	<1	Aeroplanes
	UV light sources	0.2 p.p.m.	>1	Offices

Notes

NA indicates it is not appropriate to list a concentration.

* Concentrations listed are only those reported indoors. Both higher and lower concentrations have been measured. No averaging times are given.

with time; and local pollution levels which show the impact of specific local pollutant sources and are affected by neighbouring buildings or streets as well as time. Figure 2.6 [36] shows how these various pollution components combine to produce the actual pollutant concentration at a particular location.

The main sources of external air pollution are described next.

Traffic pollution

Pollution from traffic is now the most dominant and widespread source of urban pollution and so all buildings are subject to it. However, a building will attenuate the variation in external pollution thus reducing the internal peak concentration relative to that outside [37]. Since traffic pollution is a ground-based source, its concentration tends to decrease with height, although ozone seems to be an exception where it is found to be greater at roof level than at street level [38]. This may be due to the fact that ozone from traffic is a by-product of NO_2 which involves a time delay in its formation. The variation of traffic pollution with height is not easily predictable but in general it is lower at roof level than at other heights. However, local wind flows and the landscape can cause significant variations in local pollution concentrations. As an example, Figure 2.7 shows two possible scenarios; one of pollution trapped between neighbouring buildings when the buildings' spacing (or street width) to their heights is small and the other the 'wind flushing' of the pollution when the spacing to height is sufficiently large.

Traffic pollution also varies with traffic density and congestion. It is lowest for a free-flowing traffic at a speed of 60–$80\,\mathrm{km\,h^{-1}}$ and highest for slowly moving traffic. Therefore, it is to be expected that traffic pollution is highest near road junctions, traffic

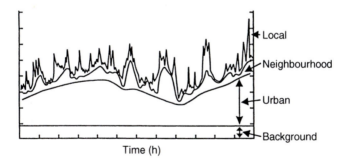

Figure 2.6 Variation of pollution components with time [36].

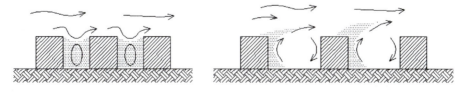

Figure 2.7 Effect of building spacing (street width) to building height on traffic pollution dispersion.

signals and streets with speed reduction systems. Hence, there is benefit in placing air intakes at roof height away from a busy street or known source of pollution [36].

Pollution from combustion appliances

The main pollutants from a gas-fired combustion appliances are CO, CO_2, NO and NO_2 and in addition oil and solid fuel boilers will emit SO_2 and particulates. Generally, flues from boilers and other combustion appliances are positioned above roof height to reduce the risk of flue gases entering the building through windows and ventilation openings. However, depending on the surrounding buildings and wind direction, plume downwash or cross-contamination can occur, as illustrated in Figure 2.8. It is therefore imperative that pollution sources on the building under consideration as well as adjoining buildings are identified and considered for their potential contamination of the building and particularly air intakes.

Pollution from industrial processes

Industrial processes can produce a variety of air, water and ground pollutants. In most countries these come under the control of the country's environmental protection agency. With respect to air pollution, discharges from stacks are required to be at a certain height and speed above the polluting and surrounding buildings. However, downwash and cross-contamination can occur as illustrated in Figure 2.8 and these problems must be identified and dealt with to ensure that the air supply to a building is not contaminated by effluents from other industrial buildings and plants. Guidance and calculation procedure for determining the height of stacks can be found in [38].

Pollution from ventilation exhausts and other equipment

Air extracts from buildings are often positioned on the same façade of the building as air intakes to balance the effect of wind pressure on the supply and extract fan systems and in some cases to facilitate the use of heat recovery from the two air streams. However, this poses the risk of cross-contamination whereby exhaust air enters the supply air intake. In most cases, the exhaust air may not contain toxic pollutants but even then the re-ingestion of these pollutants into the supply air will result in a build up of pollutants inside the building and the potential of poor IAQ. In some situations, however, such as extracts from kitchens, fume cupboards, etc. the short circuiting can have serious IAQ problems. Exhaust from some roof-top heat rejection HVAC

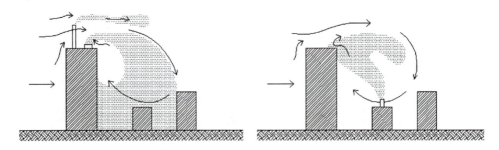

Figure 2.8 Plume downwash and cross-contamination from adjacent buildings.

equipment, such as water cooling towers, can have serious health problems if they enter into the air supply, notably legionella infections. Exhaust air from other heat rejection equipment if it enters the air supply can increase the air supply temperature and hence the ventilation cooling load. Therefore, attention to the location of such equipment relative to the air intakes and the prevailing wind direction need to be considered at the design stage. More information on the location of air intakes in a building can be found in [37].

2.3.3 *Recommended ventilation rates for continuous occupancy*

Until recently the concentration of carbon dioxide was considered by many as the criterion for admitting fresh air into a building. As it has been explained in Chapter 1, the suitability of odour and carbon dioxide concentrations as indices of air quality is disputed by results from more recent research data. In modern buildings, other pollutants can be just as or even more important than CO_2 in terms of quantities produced and their impact on human health.

Over the years, ventilation guides have continually revised the recommended fresh air supply rates. Over the past 160 years in the USA, recommended ventilation rates have swung from 2.5 l per person to nearly 15 l per person and back down to 2 l per person, see Figure 2.9. If anything, this shows our lack of knowledge of the optimum fresh air rate that a designer is required to provide a building with. This is partly due to design changes, technological development, changes in lifestyle and also to the relative cost of energy during any one period of time when these ventilation rates were specified.

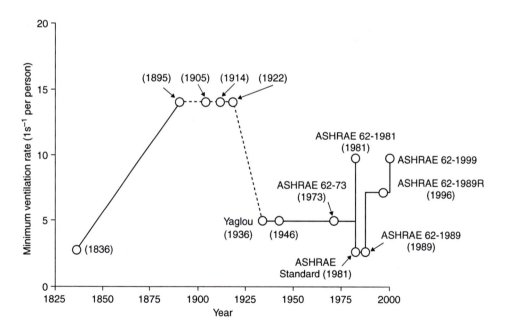

Figure 2.9 Changes in the minimum ventilation rates in the USA.

So how much ventilation does a room or building need? First we have to define what a ventilation rate is. The ventilation rate required for a given room or a building is determined to satisfy both health and comfort criteria. The health criterion should take into consideration the exposure of the occupants to indoor pollutants which will involve the identification of the pollutants, their sources, source strengths and a knowledge of the STEL or LTEL for the pollutants. These limits are used to estimate the ventilation rate required to obtain the pollutant concentration that can be tolerated for a specified time period. Where the location of the pollution sources can be identified, the preferred approach would be for the removal of such pollutants at source. However, in most cases the pollutions sources are varied and difficult to identify and dilution of their concentration is usually the only way of maintaining the concentration at the required level. The relationship between ventilation rate and pollution concentration is an inverse one, i.e. a lower concentration for larger ventilation rate and vice versa, as illustrated in Figure 2.10.

The comfort criterion, however, will produce ventilation rates that can minimize the effect of odour and sensory irritants from occupants' bioeffluents, occupants' activities and pollutants emitted from the building, the building systems and furnishings. This is usually used in domestic buildings, office buildings, public buildings, etc. while the health criterion is applied to industrial buildings. Despite their different chemical composition and sensory effects, studies have shown that pollutants can have additive impact called 'agonism' both in terms of smell and irritation. However, details of how agonism can be assessed are not available and as an approximation it is suggested that all source strengths (due to people and buildings) be added for the calculation of a design ventilation rate.

The ventilation rates needed to control pollution concentration in indoor air to the required level for each contaminant have been presented in Section 2.2. However, indoor air contains a large number of different contaminants and the ventilation rate needed to maintain acceptable levels of all these pollutants is difficult to predict because the perception and biological effect of the collective pollutants on the occupants may be different from those obtained from measurements made on each pollutant separately.

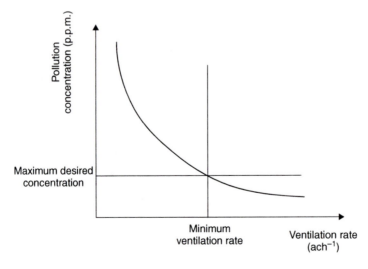

Figure 2.10 Variation of indoor pollution concentration with outdoor ventilation rates.

A well-known example is the lack of perception of body odour in a tobacco smoke environment. Because of this uncertainty, the largest ventilation rate calculated from the data presented in Section 2.2 is generally used to specify a ventilation rate for the enclosure which contains several contaminants.

In the past, odour and tobacco smoke concentrations were used as the main pollutants for determining ventilation rates. The minimum and recommended ventilation rates by CIBSE in the UK [39] and ASHRAE in the USA [6] are given in Tables 2.9 and 2.10. The CIBSE table gives ventilation rates for four categories: non-smoking, some smoking, heavy smoking and very heavy smoking. However, ASHRAE Standard 62-1999 (Table 2.10) gives a rate for each space type which is based on the outdoor concentrations for SO_2, PM_{10} particles, CO, O_3, NO_2, and Lead as those given in Table 2.1. The Standard gives two methods of calculation: the Ventilation Rate (VR) Procedure and the Indoor Air Quality (IAQ) Procedure. The VR Procedure prescribes the rate at which ventilation air must be delivered to the space and produces the minimum ventilation rate necessary for each space. The IAQ Procedure is a performance method for achieving the maximum permissible concentrations of certain notable contaminants in indoor air but does not prescribe ventilation rates. In practice

Table 2.9 Outdoor air supply rates recommended by CIBSE for sedentary occupants [39]

Level of smoking	Proportion of occupants that smoke (%)	Outdoor air supply rate ($l\,s^{-1}$ per person)
No smoking	0	8
Some smoking	25	16
Heavy smoking	45	24
Very heavy smoking	75	36

Table 2.10 Outdoor air supply rates recommended by ASHRAE [6]

Application	Estimated maximum occupancy (persons per $100\,m^2$)	Required air flow rate ($l\,s^{-1}$ per person)
Office spaces	7	10
Office conference rooms	70	10
Auditoria	150	8
Classrooms	50	8
Dining rooms	70	10
Bars and cocktail lounges	100	15
Ballrooms and discos	100	13
Smoking lounges	70	30*
Hospital patient rooms	10	13
Malls and arcades	20	1
Retail stores	5~30	0.5~1.5+
Swimming pools (pool and deck area)	—	2.5+
Residences (living areas)	—	0.35§

Notes
* Local exhaust recommended.
+ These values are given in $l\,s^{-1}$ per m^2 floor area.
§ This value is given as air change per hour (but not less than $7.5\,l\,s^{-1}$ per person).

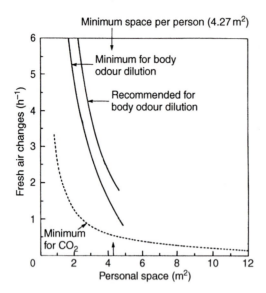

Figure 2.11 Outdoor air flow rates for different pollutants.

the ventilation rates produced by the VR Procedure are generally lower than the IAQ Procedure. Further details of the two procedures are given in [6].

In recent years, CO_2 TLVs have been widely used for the control of outdoor air supply rate to an enclosure using CO_2 sensors to control outdoor air dampers and fans. However, indoor air quality measurements by Fanger [3] on occupied buildings have shown CO_2 concentration to be a poor predictor of air quality perceived by people entering a space from outside. Similar findings were also obtained for concentrations of carbon monoxide, particulates and volatile organic compounds.

Air change rates per hour against occupancy density in metre square per person are shown in Figure 2.11 for an office space using body odour, smoking and CO_2 TLVs. It can be seen that CO_2 concentration requires the least air change rate.

Moisture production indoors is directly related to the type of activity as well as the number of occupants. Moisture generation in dwellings per person normally far exceeds that in commercial buildings for example. As Table 2.6 shows, most of the moisture in dwellings is produced by cooking, bathing and the washing and drying of clothing. Particular attention is therefore required to ventilate dwellings to dilute not only pollutants but also water vapour concentration.

Until recently outdoor air supply to most dwellings was provided by fortuitous ventilation through cracks and openings in the building fabric. However, as a result of replacement of open fires by electric or balanced flue gas central heating and widespread draught proofing of doors, windows, etc., dwellings have become less well ventilated. The result is an increase in dampness and mould-related problems in poorly heated dwellings in cold climates in particular. Therefore, in addition to improved fabric insulation and more heating, well-designed natural or mechanical ventilation systems are required to control high moisture levels in airtight dwellings. The loss of heat through ventilation is inevitable. The effect of air change rate per hour on the

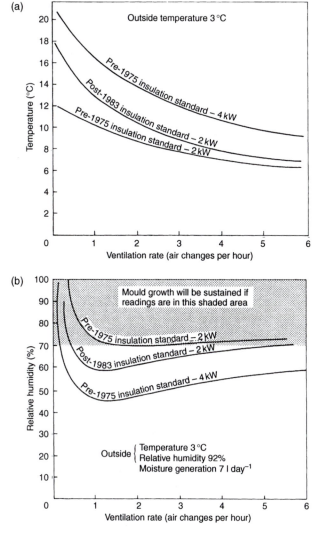

Figure 2.12 Effect of ventilation rate on house temperatures and relative humidities for different insulation standards and energy input. Reproduced from BRE Digest 297 [22].

indoor air temperature and relative humidity of a dwelling with two insulation standards and two heating loads is illustrated in Figure 2.12 [22]. The pre-1975 insulation corresponds to a U value for walls of about $1.6\,\mathrm{Wm^2\,K^{-1}}$ and for the roof of about $2.6\,\mathrm{Wm^{-2}\,K^{-1}}$, but the post-1983 insulation corresponds to 0.6 and $0.35\,\mathrm{Wm^{-2}\,K^{-1}}$ respectively. It is not surprising that in each case an increased ventilation rate produces lower internal air temperatures. However, the relative humidity initially decreases as a result of a moderate increase in ventilation and then increases at higher ventilation rates. A minimum relative humidity is achieved for this particular dwelling at an air change rate of about 1 ach. No improvement in relative humidity can be achieved with higher ventilation rates. In the lower insulation level and low heat input house the humidity remains above the 70% threshold limit for all ventilation rates. With the higher heat input or with improved insulation the relative humidity is below 70% for ventilation rates in excess of 0.5 ach.

Kitchens and bathrooms are critical condensation zones, particularly on glazing, as a large amount of moisture is generated within them over short periods. Additional heating of these zones does not normally alleviate the problem because of the saturation of the air with moisture. The problem can be solved by providing very high ventilation rates during the activity periods (i.e. cooking, washing and bathing) using mechanical ventilation. Extract fans, either manually operated or controlled by humidistats, are effective measures of controlling the spread of moisture from these zones to the rest of the dwelling [40]. Alternatively, a balanced supply and extract ventilation system with a heat recovery unit can be an effective method of controlling the humidity in dwellings without the penalty of high energy loss [23].

2.3.4   *Ventilation rates for transient occupancy*

The ventilation rates for enclosures which have transient or variable occupancy, such as classrooms, auditoria and other public buildings, may be varied to maintain acceptable contaminant concentrations at all times. Large amounts of energy can be saved by adjusting the ventilation rates using dampers or stopping and starting fans instead of maintaining the design flow rate all the time. In general, a pollutant can be present indoors before the start of occupancy, it can be produced by people, processes or materials within the building, or it can be supplied by the outdoor air. Any combination of these sources of pollution is also possible.

Transient ventilation rates

To evaluate the variation of ventilation rate or the concentration level of a pollutant with time, a solution of the mass balance equation of the pollutant over the whole enclosure is required. If a time step dt is assumed, the change in the concentration of indoor air is dc and this represents the quantity of pollutant generated inside plus the quantity carried by ventilation air into the enclosure minus the quantity leaving the enclosure through the extract air, i.e.:

$$V \, dc = (G + \dot{V}c_e - \dot{V}c) \, dt$$

or

$$V \, dc/dt = G + \dot{V}(c_e - c) \tag{2.15}$$

where V = effective volume of enclosure (m^3 or m^3 per person); $\dot{V}$ = outdoor air supply rate ($m^3 \, s^{-1}$ or $m^3 \, s^{-1}$ per person); c_e = external concentration of pollutant; c = internal concentration of pollutant at time, t (s) and G = volume of pollutant generated ($m^3 \, s^{-1}$ or $m^3 \, s^{-1}$ per person).

Assuming perfect mixing and no density changes through the enclosure, integration of equation (2.15) gives the indoor concentration at time t:

$$c = [(\dot{V}c_e + G)/(\dot{V} + G)] \times \{1 - \exp[-(\dot{V} + G)t/V]\} + c_0 \exp[-(\dot{V} + G)t/V] \tag{2.16}$$

where c_0 is the indoor concentration at time, $t = 0$.

A simplified form of the general equation (2.16) can be written for different conditions of practical interest as shown below.

1   Initial concentration in the enclosure is zero ($c_0 = 0$): substituting for $c_0 = 0$ in equation (2.16) gives:

$$c = [(\dot{V}c_e + G)/(\dot{V} + G)] \times \{1 - \exp[-(\dot{V} + G)t/V]\} \qquad (2.17)$$

2   The concentration in outdoor air is zero ($c_0 = 0$) and the initial concentration is also zero ($c_0 = 0$): the concentration equation for this case, after some rearrangement, simplifies to:

$$c = [1/(1 + \dot{V}/G)] \times \{1 - \exp[-(1 + \dot{V}/G)Gt/V]\} \qquad (2.18)$$

This equation is presented graphically in Figure 2.13 which shows a family of curves each representing the ratio $\dot{V}/G$. Given the quantities V, $\dot{V}$, G and time t the concentration indoors can be readily obtained from the figure.

3   The concentration in outdoor air is zero ($c_e = 0$) and there is no indoor contaminant generation ($G = 0$): in this case equation (2.16) simplifies to:

$$c = c_0 \exp(-Nt) \qquad (2.19)$$

where $N = \dot{V}/V$ is the air change rate for the space, i.e. number of air changes per second if t is in seconds or number of changes per hour if t is in hours. This is the simple decay equation which is commonly employed in measuring ventilation rates through a building using a tracer gas (see Chapter 3).

4   Steady-state condition ($t \to \infty$): equilibrium in indoor concentration is reached as $t \to \infty$ to give a final concentration:

$$c_\infty = (\dot{V}c_e + G)/(\dot{V} + G) \qquad (2.20)$$

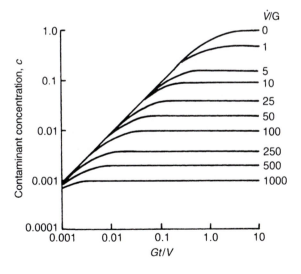

Figure 2.13 The increase of indoor contaminant concentration with time for $c_0 = c_e = 0$.

This equation shows that the final concentration c_∞ is independent of the enclosure volume, V. However, the value of V will affect the rate at which $c \to c_\infty$. The initial concentration c_0 also has no influence on the final concentration.

Lag and lead times

The control of ventilation systems may be necessary for spaces with transient or variable occupancy. The operation of such systems may lag or should lead occupancy depending on the source of contaminants and the variation in occupancy pattern. Control systems are now available which can be employed for regulating the outdoor air supply rate according to the concentrations of indoor contaminants, i.e. *Demand Control Ventilation*. Air quality sensors (see Chapter 9) can be installed in the ventilated space or in the extract ductwork and made to control outside air dampers or switch fans to maintain the set-point concentration. Although CO_2 controllers are also used for this purpose it is shown in Chapter 1 that CO_2 concentration in a space may not always be a good indicator of the air quality in the space. In addition to varying air flow rates, outdoor air control can also be used to provide lead and lag operating times for the ventilation system in the case of transient occupancy.

Lag time   When contaminants are associated only with human occupancy and are dissipated by natural means during unoccupied periods, the operation of the ventilation system can be delayed until the indoor concentration of the contaminants reaches the acceptable limit for minimum ventilation requirements under steady-state conditions. In this case, the supply of outdoor air may lag occupancy. The contaminant concentration, c, in the absence of outdoor air supply is given by:

$$c = (G/V)t \tag{2.21}$$

where the symbols have the same meaning as before. In the absence of outdoor air contamination the steady-state concentration can be obtained from equation (2.20) by letting $c_e = 0$ and assuming $G \ll \dot{V}$. Thus:

$$c_\infty = G/\dot{V} \tag{2.22}$$

The maximum ventilation delay time after occupancy occurs when $c = c_\infty$, i.e.:

$$t = V/\dot{V} \tag{2.23}$$

Equation (2.23) is plotted in Figure 2.14 as the ventilation rate $\dot{V}$ in $1\,s^{-1}$ per person against time t in hours, for different values of space volume V in m^3 per person.

   Lead time   When indoor-generated contaminants are not associated with the occupants or their activities but are produced by other indoor sources (e.g. materials and processes), the outdoor air supply may be shut off during unoccupied periods providing that the contaminants do not present a short-term health hazard. In such cases, the supply of outdoor air should lead occupancy to provide acceptable indoor air quality at the start of occupancy. The lead time can be calculated from equation (2.16) for no external contamination and the assumption that $G \ll \dot{V}$. In this case, equation (2.16) becomes:

$$c = (G/\dot{V})[1 - \exp(-\dot{V}t/V)] + c_0 \exp(-\dot{V}t/V)$$

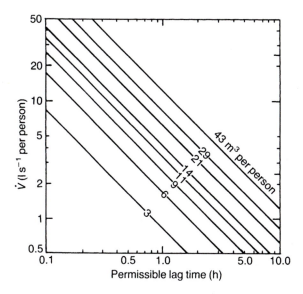

Figure 2.14 Maximum permissible ventilation lag time.

Hence,

$$\exp(-\dot{V}t/V) = (c\dot{V} - G)/(c_0\dot{V} - G)$$

or

$$t = (V/\dot{V})\ln[(c_0\dot{V} - G)/(c\dot{V} - G)] = (V/\dot{V})\ln[(c_0\dot{V}/G - 1)/(c\dot{V}/G - 1)]$$

But from equation (2.22) the steady-state concentration is:

$$c_\infty = G/\dot{V}$$

Substituting this in the equation for t, it becomes:

$$t = (V/\dot{V})\ln[(c_0/c_\infty - 1)/(c/c_\infty - 1)]$$

or

$$t = (V/\dot{V})\ln[(c_0/c_\infty - 1)/X] \tag{2.24}$$

where $X = c/c_\infty, - 1$, which is a fraction of the increase in concentration above the steady-state value. Equation (2.24) shows that to achieve a finite lead time X must be greater than zero. However, if large values of X are used to shorten the lead time, the occupants will be exposed to concentrations higher than the threshold values by X at the start of occupancy. In practice a compromise must be reached between the lead time and the initial concentration that is acceptable. The ASHRAE Standard [6] recommends a value of $X = 0.25$. The lead time calculated by using equation (2.24) for $X = 0.25$ and $c_0 = 10c_\infty$, is shown in Figure 2.15 with the ventilation rate $\dot{V}$ in $1\,\mathrm{s}^{-1}$ per person.

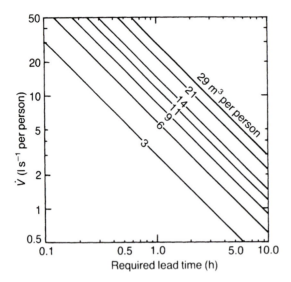

Figure 2.15 Minimum ventilation time required before occupancy of space.

2.4   Air change effectiveness and age of air

In deriving the concentration equation (2.16) and its simplified versions, it has been assumed that a perfect mixing of the supply air with room air occurs, i.e. perfect dilution of the indoor contaminants. In reality this is very rarely the case and, invariably, the supply air does not mix perfectly with the air in the occupied zone, which can lead to outdoor air being exhausted before picking up its full share of indoor contaminants, i.e. short circuiting. As a result, different concentration rates will exist in the occupied zone and to achieve the threshold limit value a larger air supply rate will be required.

The effectiveness of an air distribution system in supplying outdoor air to a room is expressed by the term 'air change effectiveness' or sometimes called 'air exchange efficiency'. Another useful concept in determining the length of time the outdoor air supplied by a system remains in a ventilated space is the 'age of air'. On the other hand, the ability of the air distribution system of removing internally produced pollutants is referred to as the 'ventilation effectiveness' or sometimes called the *'ventilation efficiency'*. These and related concepts which have become standard terms in ventilation research and applications will be briefly considered here but the interested reader may refer to the book by Etheridge and Sandberg for further details [41].

2.4.1   *Age of air*

Etheridge and Sandberg [41] considered the movement of particles in the air (e.g. air, contaminants, tracer particles, etc.) of a room by defining for a particle:

- the time that has elapsed since the particle has entered the room, i.e. the *internal age*, τ_i
- the remaining time for the particle in the room, i.e. the *residual life time*, τ_{rl}
- the total time the particle remains in the room from the time it enters until it leaves, i.e. the *residence time*, τ_r.

Hence,

$$\tau_r = \tau_i + \tau_{rl} \tag{2.25}$$

The residence time is an indicator of how long a particle remains in the room. In particular, for a particle entering the room at the air supply point, its age is zero and for one leaving the room through the exhaust opening its age is the same as the residence time, as the residual life time is zero at the exhaust. Similarly, a mean age and a mean residence time may be defined for a particular group of particles, such as a particular pollutant or a tracer gas. However, these parameters do not provide information on how effective the air distribution system is in removing the internally produced pollutants.

In practice the age of air and residence time can be evaluated experimentally using tracer gas techniques (step-up or step-down methods) or by solving the age equation in a computational fluid dynamic code (see Chapters 3 and 8 respectively). The local mean age of air at a particular point, p, in a room, $\bar{\tau}_p$, can be calculated by integrating the local tracer concentration, $C_p(t)$, with time and dividing this by the initial concentration at $t = 0, C(0)$, namely:

$$\bar{\tau}_p = \frac{\int_0^\infty C_p(t)\,dt}{C(0)} \tag{2.26}$$

Equation (2.26) can be applied to calculate the local mean age of air at certain points of interest in the room, such as the breathing zone. Sometimes it is useful to know the mean age of air in the room as whole, $\langle\bar{\tau}\rangle$. This can be quantified by measuring the tracer concentration at the exhaust air point and integrating this with time, namely:

$$\langle\bar{\tau}\rangle = \frac{\int_0^\infty t C_e(t)\,dt}{\int_0^\infty C_e(t)\,dt} \tag{2.27}$$

where $C_e(t)$ = concentration at the exhaust at time t.

If the mean age is required for a certain zone in the room, e.g. the occupied zone, the measurement of tracer concentration is required at a number of points within the zone and the local mean age at each point is calculated using equation (2.26). The mean value for the whole zone can then be calculated by taking an arithmetic or geometric mean of all the values. The interested reader may refer to reference [42].

2.4.2 Air change effectiveness

The 'local air change effectiveness' (LACE), sometimes called 'local air exchange index', is the effectiveness of the air distribution system of delivering the supply air to a particular point in the room. On the other hand, the 'mean air change effectiveness' (MACE), or the mean air exchange index, is a measure of effectiveness of ventilation air delivery to the room. LACE is defined by:

$$E_p = \frac{\tau_n}{2\bar{\tau}_p} \tag{2.28}$$

where τ_n is the nominal time constant for the room in seconds which is found from a knowledge of the air supply rate and the room volume using:

$$\tau_n = V/\dot{V} \tag{2.29}$$

the air change rate, N, is the reciprocal of the nominal time constant, i.e.:

$$N = 1/\tau_n = \dot{V}/V \quad s^{-1}$$

or

$$N = 1/\tau_n = 3600\dot{V}/V \quad h^{-1} \tag{2.30}$$

The room mean air change effectiveness is represented by:

$$\bar{E} = \frac{\tau_n}{2\langle\bar{\tau}\rangle} \tag{2.31}$$

2.4.3   Ventilation effectiveness

This is the effectiveness of an air distribution system in removing internally generated pollutants or heat from the ventilated space. There are two values: one refers to the distribution of pollutants which is defined in terms of the concentration of the pollutant in the room [43] and the other refers to the temperature distribution which is defined in terms of the temperature in the room [44]. The local ventilation effectiveness expresses how the system's ventilation ability varies between different parts of a room. It can be expressed either as a local relative effectiveness or as an average or overall relative effectiveness for the whole occupied zone.

The local ventilation effectiveness for the removal of pollutants, ε_c, is expressed as:

$$\varepsilon_c = (c_e - c_\infty)/(c_p - c_\infty) \tag{2.32}$$

and the overall ventilation effectiveness for the removal of pollutants, $\bar{\varepsilon}_c$, is expressed as:

$$\bar{\varepsilon}_c = (c_e - c_\infty)/(\bar{c} - c_\infty) \tag{2.33}$$

where c = contaminant concentration at a point (p.p.m.); $\bar{c}$ = mean concentration in the occupied zone (p.p.m.); c_∞ = contaminant concentration in the outdoor supply air (p.p.m.) and c_e = contaminant concentration in the exhaust air (p.p.m.). For steady-state situations, the concentration in the exhaust, c_e, is obtained using:

$$c_e = \dot{m}_c/\dot{m}_a$$

where $\dot{m}_c$ is the rate of discharge from the pollution source (kg s^{-1} or m^3 s^{-1}) and $\dot{m}_a$ is the rate of supply of outside air (kg s^{-1} or m^3 s^{-1}).

The definition for pollutant removal effectiveness given by equation (2.32) is a relative value, i.e. represents a ratio of the value at one point relative to another point. If one, however, is interested in the ability of the ventilation system in reducing the pollution concentration at a point with time, then another definition referred to as the absolute ventilation efficiency, ε_a, is used which is expressed by [43]:

$$\varepsilon_a = (c_o - c_t)/(c_o - c_\infty) \tag{2.34}$$

where c_o is the initial concentration at a point and c_t is concentration at the same point after time, t, seconds.

The relative ventilation effectiveness for pollutant removal is a measure of pollutant dispersion and does not take into account either the absolute concentration levels or changes in concentration from initial values. The value of ε_c is always positive and can be less than, equal to or greater than one depending on the position in the room and the method of air distribution used. However, the absolute ventilation efficiency, ε_a, represents the change in concentration as a result of change in the ventilation rate and it is always less than one.

To overcome the effect of imperfect dilution of indoor pollution by the outdoor air, an air supply rate greater than that given in the concentration equations (2.16)–(2.24) will be required. This is expressed quantitatively by replacing $\dot{V}$ in these equations by $\bar{\varepsilon}_c \dot{V}$. The value of $\bar{\varepsilon}_c$ is clearly dependent on the type of air distribution system which is used to supply and extract the air to the room (see Section 2.5). The ventilation effectiveness for heat distribution or removal is dealt with in Section 2.6.

2.4.4 Purging flow

Although the concept of mean age of air is useful and the parameters involved are relatively easy to measure and calculate, it does not express the contaminant removal capability of a ventilation system. Therefore, the concept of the 'local purging flow rate' is used to describe the nature of the contaminant purging process at a point or a region within a ventilated enclosure. This concept originates from the study of chemical reactors involving concentration of chemical or biological species but was first introduced to ventilation applications by Sandberg and Sjöberg [45]. When applied to ventilated enclosures it can characterize the local pollutant-purging capability of ventilation air. A small purging flow rate at a point or a region suggests that that point or region is weakly connected with the ventilation process whereas a large value would suggest the reverse. Although it was originally introduced to quantify the net flow rate at which a pollutant present at a point is removed by the local air flow, it can also be used to characterize the distribution of ventilation air in an enclosure. The 'local purging flow rate' is sometimes defined as the net flow rate at which air is supplied from the inlet to a point or region within the enclosure. In other words, this concept can be used to quantify the rate at which a passive pollutant (one that moves with the air particles) is 'flushed' out from a point or a homogeneously mixed region, or the rate at which ventilation air is supplied to the point or region.

It can be seen from the above definition that the purging flow rate refers to local properties and is therefore usually referred to as the 'local' purging flow rate, U_p. To define U_p, a small control volume that conforms to mass continuity is considered such that, within this control volume, a source of pollutant emitting a mass flow rate of m_p will result in a pollutant concentration of c_p leaving the control volume, i.e.

$$U_p = m_p / c_p \tag{2.35}$$

In principle the local purging flow rate may be determined by releasing a short burst of a known mass, M, of a tracer gas at the point of interest and measuring the local concentration of the tracer with time as given by the following equation:

$$U_p = \frac{M}{\int_0^\infty c(t)\,\mathrm{d}t} \tag{2.36}$$

However, in practice it may be possible to measure U_p at one or two points or small regions but it would be a difficult task to do in a large enclosure where multiple zones are involved. In a single zone, it would be possible to inject a tracer gas at a known rate, m, measure the spatial average concentration $\langle c \rangle$ and apply the expression below to obtain the mean purging flow rate, $\langle U_p \rangle$, corresponding to that zone:

$$\langle U_p \rangle = m / \langle c \rangle \tag{2.37}$$

The accuracy in the value of $\langle U_p \rangle$ depends on how uniform the pollution is dispersed in the zone and on the accuracy of measuring the mean concentration, $\langle c \rangle$, in the zone. For the purpose of finding the mean purging flow rate for the zone, it would be possible to achieve a uniform tracer concentration within a particular zone, e.g. a room, by artificial mixing of the tracer using a fan. However, care must be exercised in this case because such a mixing process could interfere with the natural air movement in the zone.

If a complete mixing of the tracer gas with air is achievable in each zone of a multi-zone building, then the principle of local purging flow may be extended to determine interzonal flow rates. The mass conservation equations (air or tracer flow) will, in this case, result into a flow matrix that contains terms for the interzonal flows and the flows between the zones and outdoor. For a building of n zones, e.g. n separate rooms, there will be $(n^2 - n)$ internal zonal flow terms and $2n$ flow terms between the zones and outside, i.e. $n(n+1)$ flow terms and the same number of equations. Hence, for a small house consisting of five rooms there will a total of 30 equations to solve to give the same number of flow rates for air or tracer gas. This will require repeating the measurement n, or in this case 30, times or using n different tracer gases simultaneously. Using a large number of tracer gases requires a large capital expenditure on tracer measuring equipment whereas using a single tracer is a very time consuming task and therefore, in practice, interzonal flows is limited to a few number of zones. The methods used in the measurement are explained in Chapter 3 and the interested reader may also refer to other publications on the subject, such as the AIVC TN 34 [46].

Purging flows and interzonal flow rates can also be calculated using computational fluid dynamics and applying statistical methods, such as a Markov chain model. However, the calculation tasks involved for three-dimensional geometry with a number of zones are huge and these are currently restricted to simple cases of two-dimensional enclosures, see e.g. [47–49].

2.5   Types of ventilation systems

2.5.1   *Local exhaust ventilation (LEV)*

This method of ventilation is based on the principle of capturing the contaminant at source before it spreads into the room air. An LEV capture device or extract hood is used and its shape and flow characteristics are important elements of this method of ventilation. Some guidelines for the design of extract hoods are given in Chapter 5 but for more detailed analysis of the principles of hood design interested readers should refer to specialist texts on the subject such as Goodfellow and Tähti [50], Hayashi *et al.* [51] and a BSRIA Technical Note [52].

Local exhaust ventilation is the most effective method of contaminant extraction and it is widely used in industrial ventilation particularly where localized sources of

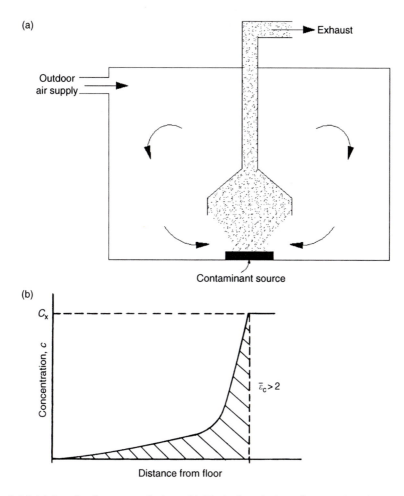

Figure 2.16 (a) Local exhaust ventilation. (b) Typical variation of contaminant concentration between the air supply and exhaust points in LEV system.

hazardous contaminants are present. It is rarely used in commercial or public buildings except perhaps in some rooms where certain processes are present, such as kitchens. Figure 2.16(a) illustrates the air movement in a room caused by an extract hood and Figure 2.16(b) shows the expected rise in the contaminant concentration with distance from the supply air point (assumed at zero concentration) to the extract duct. As shown, the rapid increase in concentration occurs close to the extract hood and this contributes to the large value of the overall relative ventilation efficiency of this system. Values of $\bar{\varepsilon}_c$ as high as 10 may be achieved with good hood design practice.

2.5.2 Piston ventilation (PV)

Piston flow represents a unidirectional flow of air in which outdoor air propels the contaminated room air ahead of it like a front. The room air is continuously 'swept' by outdoor air and little spread of contamination generated within the room takes place before this is carried by the outdoor air to the extract duct. This method of ventilation is employed in 'clean rooms', but to be effective the air turbulence must be reduced to a minimum so that contaminant dispersion is minimized. This is normally performed in practice by supplying the air through 'laminar flow' panels containing high-efficiency particulate air (HEPA) filters or ultra-low penetration air (ULPA) filters placed on the ceiling (down-flow clean room) or on a wall (cross-flow clean room) at a velocity between 0.4 and 0.5 m s^{-1}. HEPA filters have a filtration efficiency in excess of 99.97% on 0.3 μm particles whereas ULPA filters have a minimum efficiency of 99.999% on 0.12 μm particles. Figure 2.17(a) is a sketch of a down-flow clean room using piston ventilation and Figure 2.17(b) shows the increase in contamination concentration with distance in the direction of flow if uniform generation of contaminants is assumed throughout the room and mixing is ignored. The linear increase in concentration with distance produces an overall relative ventilation efficiency, $\bar{\varepsilon}_c = 2$ or a room mean air change effectiveness, $\bar{E} = 2$.

Clean rooms are used in the semiconductor industry, pharmaceutical industry, hospitals and so forth, and there are many classes of clean rooms depending on the number of airborne particulates that can be permitted in a unit volume of air. ASHRAE [53] uses the ISO Standard 14644 definition of the class of a clean spaces which is based

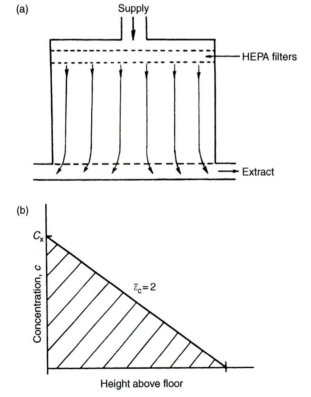

Figure 2.17 (a) Piston ventilation in a down-flow clean room. (b) Increase of concentration with distance in piston ventilation.

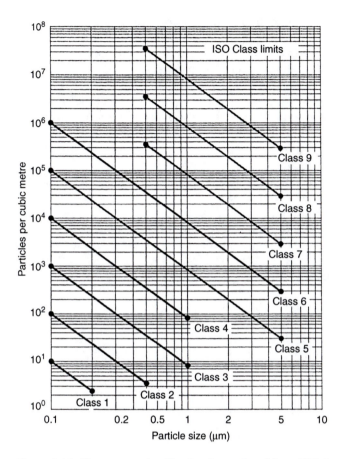

Figure 2.18 Clean-room classification (reproduced from ISO Standard 14644).
Copyright 1999, American Society of Heating, Refrigerating + Air-conditioning Engineers, inc. www.ashrae.org. Reprinted by permission from 1999 ASHRAE Handbook – HVAC Applications. 'The tems and definitions taken from ISO 14644-1:1999 Cleanrooms and associated controlled environments, Fig. A.1, are reproduced with the permission of the International Organization for Standardization, ISO. This standard can be obtained from any ISO member and from the Web site of ISO Central Secretariat at the following address: www.iso.org. Copyright remains with ISO'.

on the maximum number of particles of a given size range that exist in a cubic metre of air. As shown in Figure 2.18, nine classes are given ranging from Class 1 corresponding to a total of 10 particles per m^3 of size 0.15 μm or less and Class 9 which corresponds to a total of 3.5×10^7 particles of size 0.4 μm or less. Additional design data for clean spaces can be found in the ASHRAE Handbook [53] and BS 5295 [54].

2.5.3 Displacement ventilation (DV)

As in piston ventilation this method also relies on the displacement of room air by a fresh supply of outdoor air but in a less discreet way than in piston ventilation.

Displacement ventilation is used in buildings with large occupancy and internal heat gains where mainly cooling is required. Unlike piston ventilation where the driving force is mainly the momentum of supply air, here momentum is usually small and buoyancy is the dominant force in the creation of room air movement. There are upward and downward displacement ventilation systems and the design procedures used in both systems are described in Chapter 6. However, the upward system is more widely used than the downward system because it produces better indoor air quality for the occupants. In this system, outdoor air or a mixture of outdoor and recirculated air is delivered close to the floor of the room at a low velocity (e.g. 0.25–$0.35\,\mathrm{m\,s^{-1}}$) which rises through the room by a combination of momentum and buoyancy forces, the latter caused by the presence of people or warm surfaces such as IT equipment in offices. Temperature and concentration stratification then develop with cool, clean air at low level and warm, contaminated air at a higher level above the occupants. This is also discussed in Section 4.6.5. As described in Chapter 6, depending on the application, different methods of supplying the air at low level can be used.

The flow rate, $\dot{V}_c$, of a convective current created by a hot or cold surface of height H can be calculated using the formula below [55]:

For a laminar boundary layer:

$$\dot{V}_c = 2.87 \times 10^{-3} l (t_w - t_a)^{1/4} H^{3/4} \quad \mathrm{m^3\,s^{-1}} \tag{2.38a}$$

For a turbulent boundary layer:

$$\dot{V}_c = 2.75 \times 10^{-3} l (t_w - t_a)^{2/5} H^{6/5} \quad \mathrm{m^3\,s^{-1}} \tag{2.38b}$$

where t_w = surface temperature (°C); t_a = temperature of surrounding room air (°C); H = height of surface (m) and l = length of surface (m). Equations (2.38a) and (2.38b) apply to walls, windows and other vertical surfaces covered by a boundary layer.

The flow rate of a plume rising from a heat source, $\dot{V}_p$, can be predicted using the formula below [56] or the plume equations given in Chapter 4:

$$\dot{V}_p = 0.0061 P^{1/3} (y + d)^{5/3} \quad \mathrm{m^3\,s^{-1}} \tag{2.39}$$

where P = convective heat output from source (W); y = distance above the heat source (m) and d = diameter of heat source (m). To avoid recirculation of contaminated air into the occupied zone the total air supply rate near the floor must equal the sum of the flow rates due to convective currents from vertical surfaces, plumes and outdoor air requirement for the occupants, i.e.:

$$\dot{V} = \dot{V}_c + \dot{V}_p + \dot{V}_o \tag{2.40}$$

where $\dot{V}_o$ = outdoor air flow rate required by occupants ($\mathrm{m^3\,s^{-1}}$).

If $\dot{V}_c$ is a downdraught from a cold surface, such as a window, then it has a negative value.

In the 1980s, upward displacement systems were developed in Scandinavian countries [57] and have subsequently spread to many countries as a result of their superior ventilation effectiveness and energy efficiency. In one commonly used system nowadays, air from a low-level wall or above floor units is supplied over the floor, 'flooding' the occupied zone and then rising towards the ceiling, where it is extracted, by the

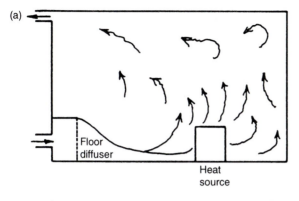

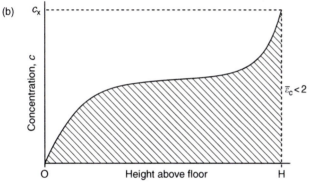

Figure 2.19 (a) Flow produced by floor displacement ventilation. (b) Typical increase in concentration with distance for floor displacement ventilation.

convective currents created by people or heat sources in the room. This is achieved by supplying low-velocity ($<0.5\,\mathrm{m\,s^{-1}}$) and low-turbulence air current with a temperature only a few degrees lower than the room temperature, i.e. in cooling mode. The flow produced by this system and the increase in concentration with distance are depicted in Figures 2.19(a) and (b). The overall relative ventilation efficiency of an upward displacement system is less than the piston ventilation and is usually within the range 1.4–1.7 [58].

A displacement system on its own is not suitable for heating as the outdoor supply air will rise towards the ceiling before it fills the occupied zone. For this purpose a separate heating system is needed with the displacement system used as a fresh air supply at or slightly below room temperature. In addition, the amount of cooling that a floor current displacement system could handle is limited to about $30–35\,\mathrm{W\,m^{-2}}$ of floor area [42]. Larger cooling loads are therefore dealt with by supplementing the cooling using an additional cooling system, such as a chilled beam or chilled ceiling panels. In these supplementary cooling systems chilled water is circulated in a long finned heat exchanger (chilled beam) or in an unfinned ceiling panel. Careful design considerations are required to avoid the downflow currents from the beam or panel counteracting the displacement flow from the floor thus breaking the upward displacement flow sought with displacement ventilation [59].

2.5.4   *Mixing ventilation (MV)*

This method requires that the supply air and room air are mixed well by the actions of the supply jet momentum and buoyancy. The mixing can be achieved by various methods as described in Chapter 6. This is still the most widely used method in ventilation and air-conditioning systems because, with a proper design, it can be used for heating and cooling as well as ventilation and it can cope with much larger room loads than in displacement ventilation. The supply air is used to dilute the concentration of contaminants in the room (hence the name dilution ventilation, which is also used) but not to displace them as in the previous methods. However, unless the sources of pollution are evenly distributed throughout the space, high concentration levels are always found near these sources. Because the outdoor air is usually supplied at high level (i.e. above the occupied zone) the concentration levels above the occupied zone are generally lower than those existing within the zone, which results in an overall relative ventilation efficiency and mean air change effectiveness of less than one. In addition, stagnant zones of high concentration levels and/or short-circuiting of supply air to the extract points can cause a sharp drop in the ventilation efficiency. To avoid these occurrences, the design procedures described in Chapter 6 should be followed.

Figures 2.20(a) and (b) show the expected air movement and concentration spread for a ceiling supply mixing system.

(a)

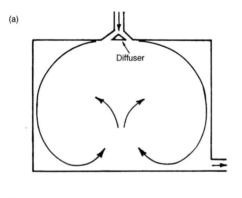

(b)

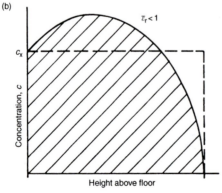

Figure 2.20 (a) Air movement in a mixing ventilation system. (b) Typical increase in concentration with distance for a mixing ventilation system.

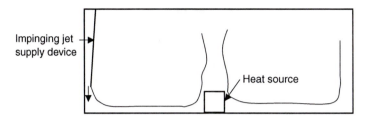

Figure 2.21 Impinging jet ventilation.

2.5.5 *Impinging jet ventilation (IJV)*

It was shown that in displacement ventilation, cool air supplied at low level is entrained by plumes rising from heat sources in the room. To create an effective air movement in the occupied zone in such systems, there should be a balance between the momentum in the supply air and thermal (buoyancy) forces due to heat sources. In the low momentum displacement flow, the buoyancy forces created by heat sources have a tendency to take over and thus often causing poor ventilation efficiency in some zones of the room. Another disadvantage of a displacement system is that it can only be used for cooling and is not suitable for winter heating. Recently, a new method of air distribution has been developed that is based on the impinging jet principle [60] in which case a jet of air with high momentum is supplied downwards onto the floor (see Figure 2.21). As the jet impinges onto the floor it spreads over a large area causing the jet momentum to recede but still has a sufficient force to reach long distances. Unlike the displacement system which 'floods' the floor with supply air, the resulting flow from an IJV is a very thin layer of air over the floor. This method has the advantages of both the mixing and displacement ventilation systems and has lower momentum than mixing and higher momentum than wall displacement ventilation, i.e. when the air terminal unit is wall mounted at low level. Although higher momentum than wall DV, IJV produces a similar flow field and has, therefore, promising applications [60, 61].

2.6 Energy requirement for ventilation

2.6.1 *Latent and sensible energy*

The temperature and moisture content of outdoor air are seldom appropriate for room supply. It is therefore necessary to heat, cool, humidify or dehumidify the outdoor air before it is supplied directly to the room or mixed with recirculated air. The total energy (sensible + latent) required to heat or cool air is given by:

$$\dot{Q}_t = \dot{m}_a(h_1 - h_2) \tag{2.41}$$

where $\dot{Q}_t$ = total energy (kW); m_a = mass flow rate of dry air (kg s^{-1}); and h_1 and h_2 are the enthalpies before and after the heater or cooler (kJ kg^{-1} dry air). The mass of dry air flow can be determined from the volume flow rate and the density or specific volume of dry air. Thus:

$$\dot{m}_a = \rho \dot{V} = \dot{V}/\upsilon \tag{2.42}$$

where v can be obtained from a psychrometric chart. Similarly h_1 and h_2 are obtained from a psychrometric chart given two properties of the air (see Figure 2.5).

The sensible energy is given by:

$$\dot{Q}_s = \dot{m}_a C_p (t_1 - t_2) \tag{2.43}$$

where C_p = specific heat of dry air $(kJ\,kg^{-1}\,K^{-1})$; t_1 and t_2 are the dry-bulb temperatures of the air (°C).

The latent enthalpy is calculated either from the difference between $\dot{Q}_t$ and $\dot{Q}_s$ or directly using:

$$\dot{Q}_l = \dot{m}_a (\omega_1 - \omega_2) h_{fg} \tag{2.44}$$

where ω_1 and ω_2 are the specific humidities (moisture contents) of the air across the humidifier or dehumidifier and h_{fg} is the specific enthalpy of water evaporation $(kJ\,kg^{-1})$.

The thermal energy contribution of fan power is sensible only and can be obtained by calculating the rise in the air temperature across the fan and then using equation (2.43). The temperature rise across a fan is:

$$\Delta t = \Delta p / (\eta \rho C_p) \tag{2.45}$$

where Δt = temperature increase due to fan power (K); Δp = fan total pressure (kPa); and η = total efficiency of fan. The total efficiency of a particular fan is not constant but varies with the flow rate delivered by the fan. Maximum efficiency should ideally correspond to the design flow rate for the system. There are many varieties of fans used in ventilation systems and this is not a topic that will be dealt with in this book. Interested readers should consult the ASHRAE Handbook [62] or the book by Osborne [63] on fans.

2.6.2   *Air distribution effectiveness*

The local ventilation effectiveness for the temperature distribution in a ventilated space may be defined for a local point or for the whole space as in the case of the ventilation effectiveness for the distribution of pollutants (see Section 2.6.3). The local ventilation effectiveness for temperature distribution, ε_t, is expressed as:

$$\varepsilon_t = (t_e - t_\infty)/(t_p - t_\infty) \tag{2.46}$$

and the overall ventilation effectiveness for temperature distribution, $\bar{\varepsilon}_t$, is expressed as:

$$\bar{\varepsilon}_t = (t_e - t_\infty)/(\bar{t} - t_\infty) \tag{2.47}$$

where t_p = air temperature at a point (°C); $\bar{t}$ = mean air temperature in the occupied zone (°C); t_∞ = temperature of outdoor supply air (°C); t_e = temperature of exhaust air (°C).

An efficient ventilation system is one that can achieve good thermal comfort and air quality in the occupied zone. Although high values of $\bar{\varepsilon}_c$ and $\bar{\varepsilon}_t$ represent a high performance in terms of heat distribution and contaminant removal, these quantities alone do not give a good indication of the thermal comfort and air quality in the occupied zone. One approach is to combine $\bar{\varepsilon}_c$ with a subjective parameter representing the perception of indoor air quality and $\bar{\varepsilon}_t$ with a parameter representing thermal perception. Fanger's Percentage of Dissatisfied (PD) for air quality and (PPD) the Predicted Percentage of Dissatisfied (PPD) with the thermal environment would be suitable parameters to use. These parameters are explained in Chapter 1 with expressions given for each. The overall ventilation effectiveness for contaminants and temperature have been combined with PD and PPD to produce new numbers for air quality and thermal comfort, see Awbi and Gan [44]:

$$N_c = \frac{\bar{\varepsilon}_c}{PD} \tag{2.48}$$

$$N_t = \frac{\bar{\varepsilon}_t}{PPD} \tag{2.49}$$

where N_c and N_t are the air distribution numbers for air quality and thermal comfort respectively. Expressions for PPD and PD are given by equations (1.39) and (1.57) respectively.

An efficient ventilation system is one that produces the greatest values of N_c and N_t that may be achieved. The two numbers can be combined into a single parameter that determines the 'global' effectiveness of an air distribution system in providing indoor air quality and thermal comfort. N_c and N_t may be combined in at least two possible ways to produce what Awbi and Gan [44] called a Ventilation Parameter (VP) or the Air Distribution Index (ADI):

$$VP = N_c/N_t \tag{2.50}$$

$$ADI = \sqrt{N_c N_t} \tag{2.51}$$

A value of $VP = 1.0$ indicates the existence of a balance between the way contaminants and heat are distributed in the ventilated space. A value of $VP \ll 1$ suggests that the effectiveness for contaminants is much worse than that for temperature and a $VP \gg 1$ suggests the opposite. However, a $VP \approx 1$ is not sufficient to ensure that both contaminants and heat are distributed effectively by the ventilation system. An additional requirement is that both N_c and N_t are large.

The ADI does not suffer from the same anomaly as VP as it represents the square root of the products of N_c and N_t and a high value of ADI also signifies large values of both N_c and N_t. If a ventilation system is to be designed for acceptable values for PD and PPD of 10% then a good air distribution system would achieve a value of $ADI \geq 10$. However, it has been shown [44, 64] that a large value of VP (also applies to ADI) would not guarantee the existence of a good air distribution system as unfavourable local conditions may still be present. This is because these parameters represent mean values and do not explain large local variations in room air movement such as the presence of local draughts or high or low temperature regions within the occupied zone. These parameters should therefore be applied together with room air movement and temperature distribution measurements or predictions to assess the global as well as the local conditions in the room.

2.6.3   *Ventilation energy savings*

In modern and retrofit buildings, ventilation is probably the greatest component of the total energy consumption. This is usually in the range of 30–60% of the building energy consumption. The large proportion of ventilation energy is due to three main reasons. Firstly, modern buildings are generally well insulated and, therefore, the heat gain or loss through the fabric is low. Secondly, modern building materials and furnishings emit large amounts of VOCs and TVOCs and so it becomes necessary to dilute their concentration by supplying greater ventilation rates to these buildings. The third cause is as a result of the recent concern regarding the sick building syndrome and other building related illnesses which have influenced HVAC designers to improve the indoor air quality by specifying greater quantities of fresh air supply. As a result of these factors, the contribution of energy required for heating or cooling ventilation air to the total energy consumption for the building has increased. There are, however, practical means of reducing the ventilating air energy requirement, some of which are briefly described here.

Room temperature control

Both ventilating air and fabric energy consumption can be reduced if the set point temperature in the building is reduced during the heating season and increased during the cooling season. It has been estimated that a reduction of 1 K in internal temperature from that recommended by ISO 7730 will reduce the energy consumption in UK buildings by 6% [65]. Similar reductions have been estimated for Finland [66]. Field studies of thermal comfort have shown that up to 2.4 K reduction in indoor temperature from that specified in comfort standards can be tolerated by the building occupants without adverse affects on comfort. This is due to the fact that current standards, such as ISO 7730 and ASHRAE Standard 55, recommend indoor temperatures which are based on laboratory studies but as has been shown in Chapter 1, in real buildings, clothing habits and activity levels are known to be different from those under ideal test conditions.

Ventilation system balancing

Improper balancing of mechanical ventilation systems can result in increased fresh air rates to some zones and a reduction to others. This could not only cause discomfort due to draught in over-ventilated zones and poor air quality in under-ventilated zones, but could also increase the ventilation energy consumption. It is, therefore, important to check the actual delivery of fresh air to different zones of the building during commissioning and routine maintenance to ensure optimum operation of the ventilation system. Another source of energy wastage is leakage from ventilation ducts. This can also be reduced by specifying better ducts and good quality control measures during installation.

Heat recovery

In mechanically ventilated buildings, heat recovery from ventilation air is the single most important means of reducing ventilation energy consumption. Many different types of heat recovery systems are available for transferring energy from the exhaust air to the supply air or vice versa, e.g. see [67]. The most commonly used systems are the regenerative type (thermal wheel), plate heat exchanger and the run-around coil. Heat recovery of up to 70% can be achieved depending on the system used and the enthalpy or temperature difference between supply and exhaust air. Heat recovery is an essential part of a fully outdoor air supply ventilation system, i.e. one that uses no recirculation.

Demand controlled ventilation

Demand controlled ventilation (DCV) is a method of controlling fresh air supply to a room according to the pollution load present in the room. Although there are many pollutants that could be produced in the room due to building materials, furnishing, equipment and people's activities, it is impractical to use all these different pollutants to control the amount of fresh air supply to the room. Usually the concentration in the room of a few types of pollutants are controlled by the DCV system. These are carbon dioxide (CO_2), TVOCs, tobacco smoke and moisture but, though may not always be an ideal pollutant to base outdoor air supply rates on, for most buildings the carbon dioxide concentration in the room is used to control the quantity of outdoor air supply by using carbon dioxide sensors which control the fresh air dampers. By controlling the fresh air supply to achieve a maximum allowable carbon dioxide concentration, say 1000 p.p.m., it would be possible to reduce the ventilation rate during low or no occupancy, thus saving energy.

Although considerable amounts of research results on DCV have been accumulated, there is still a lack of experience in the installation and operation of DCV systems. Another drawback of DCV is the fact that usually only one source of pollutant (e.g. CO_2) is used for controlling fresh air rate but in normal buildings there is usually a combination of pollutants produced at different rates depending on the activities within the buildings.

User control ventilation

When designing a conventional heating, ventilation or air-conditioning system, the energy and the air charge rate requirements are normally based respectively on a heat and pollution concentration balance over the whole space for a 'typical' day. However, there is evidence to suggest (see Chapter 1) that the neutral or comfort temperature for occupants can vary substantially within the same building depending on clothing and activity of the occupants. Therefore, maintaining a whole building at the same temperature and the same outdoor air supply rate can be energy wasteful. A substantial saving in energy may be achieved by individual and automatic control of the local environment by providing personalized or task-conditioning systems which can be controlled by a single user according to his or her needs. Although the capital cost of such systems is currently higher than conventional systems, this may be outweighed by the improved comfort and increased productivity of the occupants, in addition to energy saving for heating, cooling and ventilation.

References

1. Robertson, A. S. *et al.* (1985) Comparison of health problems related to work and environment measurements in two office buildings with different ventilation systems. *Br. Med. J.*, **291**, 373–6.

2. Yaglou, C. P., Riley, B. C. and Coggins, D. J. (1936) Ventilation requirements. *Trans. ASHVE*, **42**, 133–62.

3. Fanger, P. O. *et al.* (1988) Air pollution sources in offices and assembly halls, quantified by the olf unit. *Energy Build.*, **12**, 7–19.

4. Fanger, P. O. (1988) A comfort equation for indoor air quality and ventilation, *Proc. Healthy Buildings '88* (Vol.1, *State of the Art Reviews*), Stockholm, pp. 39–51.

5. BS 5925 (1991) Code of practice for: ventilation principles and designing for natural ventilation, British Standards Institution, London.

6. ASHRAE Standard 62-1999 (1999) Ventilation for acceptable indoor air quality, American Society of Heating, Refrigeration and Air-Conditioning Engineers, Atlanta, GA.

7. DIN 1946: Teil 2 (1983) Raumlufttechnik Gesundheitstechnische Anforderungen (VDI-Lüftungsregeln), Deutsches Institut für Normung, Berlin.

8. Goodfellow, H. D. (1988) Keynote address: ventilation – past present and future, *Proc. 2nd Int. Symp. on Ventilation for Contaminant Control (Ventilation '88)*, London, 20–23 September 1988, pp. 5–11.

9. Occupational exposure limits – guide note EH 40/90 (1990) Health and Safety Executive, HMSO, London.

10. Walsh, P. J., Dudney, C. S. and Copenhaver, E. D. (1984) *Indoor Air Quality*, CRC Press, Boca Raton, FL.

11. Hind, W. C. (1999) *Aerosol Technology: Properties, Behavior, and Measurement of Airborne Particles*, John Wiley, New York.

12. Meyer, B. (1983) *Indoor Air Quality*, Addison-Wesley, Reading, Massachusetts.

13. McIntyre, D. A. (1980) *Indoor Climate*, Applied Science Publishers, Barking.

14. Sundell, J. (1982) Guidelines for NORDIC building regulations regarding indoor air quality. *Environ. Int.*, **8**, 17–20.

15. Brundrert, G. W. (1975) Ventilation requirements in rooms occupied by smokers: A review. Electricity Council Research Centre, UK, Rep. ECRC/M870.

16. Molhave, L. (1990) Volatile organic compounds, indoor air, quality and health, *Proc. of 5th Int. Conf. on Indoor Air Quality and Climate, Indoor Air '90*, **5**, 15–34.

17. Godish, T. (1995) *Sick Buildings: Definition, Diagnosis and Mitigation*, CRC Press, Boca Raton, FL.

18. Ryan, M. T. (1981) Radiological impacts of uranium recovery in the phosphate industry (eds R. O. Chester and C. T. Gorten, Jr.). *Nucl. Saf.*, **22**, 70–7.

19. Cliff, K. D. (1978) Population exposure to the short lived daughters of radon-222 in Great Britain. *Radiol. Prot. Bull*, no. 22, 18–23, HMSO, London.

20. Owen *et al.* (1992) Airborne particle sizes and sources found in indoor air, *Atmospheric Environment*, **26A**(12), 2149–62.

21. CIBSE (1987) Minimising the risk of legionnaires' disease. *CIBSE Tech. Memo.* TM13, London.

22. BRE Digest 297 (1985) Surface condensation and mould growth in traditionally built dwellings, Building Research Establishment, Watford, England.

23. Awbi, H. B. and Allwinkle, S. J. (1986) Domestic ventilation with heat recovery to improve indoor air quality. *Energy Build.*, **9**, 305–12.

24. *CIBSE Guide Environmental Design* (1999) Section A10: Moisture transfer and condensation, Chartered Institution of Building Services Engineers, London.

25. *ASHRAE Handbook of Fundamentals* (2001) Ch. 25: Thermal and water vapor transmission data, American Society of Heating Refrigeration and Air-Conditioning Engineers, Atlanta, GA.

26. BS 5250 (1989) Control of condensation in buildings, British Standards Institution, London.

27. *CIBSE Guide* (1999) Section A3: Thermal properties of building structures, Chartered Institution of Building Services Engineers, London.

28. Awbi, H. B. and Hatton, A. (1999) Natural convection from heated room surfaces, *Energy Build.*, 30, 233–44.

29. Awbi, H. B. and Hatton, A. (2000) Mixed convection from heated room surfaces, *Energy Build.*, 32, 153–66.

30. Devres, Y. O. (1994) Psychrometric properties of humid air: calculation procedure, *Applied Energy*, 48, 1–18.

31. IEA Annex IX (1987) *Minimum Ventilation Rates*, Final report, International Energy Agency.

32. *ASHRAE Handbook of Fundamentals* (2001) Ch. 12: Air contaminants, American Society of Heating, Refrigeration and Air-Conditioning Engineers, Atlanta, GA.

33. Air Quality Archive: http://www.aeat.co.uk/netcen/aqarchive/archome.html.

34. *ASHRAE Handbook of HVAC Systems and Equipment* (2000) Ch. 24: Air cleaners for particulate contaminants, American Society of Heating, Refrigeration and Air-Conditioning Engineers, Atlanta, GA.

35. *CIBSE Guide B2* (2001) Ventilation and air conditioning, Chartered Institution of Building Services Engineers, London.

36. Hall, D. J. *et al.* (1996) Exposure of buildings to pollutants in urban areas – a review of the contributions from different sources, Report No. CR 209/96, Building Research Establishment, UK.

37. Kukadia, V., Pike, J. and White, M. (1997) Air pollution and natural ventilation in urban environment, *Proc. CIBSE National Conference*, 1997, Chartered Institution of Building Services Engineers, London.

38. CIBSE Tech. Mem. 21 (1999) *Minimising Pollution at Air Inlets*, Chartered Institution of Building Services Engineers, London.

39. *CIBSE Guide* (1999) Section A1: Environmental criteria for design, Chartered Institution of Building Services Engineers, London.

40. Cornish, J. P., Sanders, C. H. and Garratt, J. (1985) The effectiveness of remedies to surface condensation and mould, *Workshop on Condensation and Energy Problems: A Search for an International Strategy*, 23–25 September, Leuven.

41. Etheridge, D. and Sandberg, M. (1996) *Building Ventilation: Theory and Measurements*, John Wiley, England.

42. Xing, H., Hatton, A. and Awbi, H. B. (2001) A study of the air quality in the breathing zone in a room with displacement ventilation, *Building and Environment*, 36, 809–20.

43. Sandberg, M. (1981) What is ventilation efficiency? *Building and Environment*, 16, 123–35.

44. Awbi, H. B. and Gan, G. (1993) Evaluation of the overall performance of room air distribution, *Proc. Indoor Air '93*, 5, 283–8, Helsinki.

45. Sandberg, M. and Sjöberg, M. (1983) The use of moment for assessing air quality in ventilated rooms, *Building and Environment*, 18, 181–97.

46. Roulet, C-A. and Vandaele, L. (1991) Flow patterns within buildings – Measurements techniques, TN 34, Air Infiltration and Ventilation centre, Belgium.

47. Davidson, L. and Olsson, E. (1987) Calculation of age and local purging flow rate in rooms, *Building and Environment*, 22, 111–27.

48. Peng, S-H. and Davidson, L. (1997) Towards the determination of regional purging flow rate, *Building and Environment*, 32, 513–25.

49. Cho, Y. and Awbi, H. B. (2000) The effect of object positions on ventilation performance, *Proc. Roomvent 2000, Air Distribution in Rooms: Ventilation for Health and Sustainable Environment*, H. B. Awbi (ed.), 1, 433–8, Elsevier.

50. Goodfellow, H. D. and Tähti, E. (eds) (2000) *Industrial Ventilation: Design Guidebook*, Academic Press, San Diego, California, USA.

51. Hayashi, T., Howell, R. H., Shibata, M. and Tsuji, K. (1987) *Industrial Ventilation and Air Conditioning*, CRC Press, Boca Raton, FL.

52. BSRIA (1985) Design guidelines for exhaust hoods. *BSRIA Tech. Note* TN 3/85, Building Services Research and Information Association, Bracknell, UK.

53. *ASHRAE Handbook of Applications* (1999) Ch. 15: Clean spaces, American Society of Heating, Refrigeration and Air-Conditioning Engineers, Atlanta, GA.

54. BS 5295: Part 1 (1989) Environmental cleanliness in enclosed spaces, British Standard Institution, London.

55. Mundt, E. (1996) The performance of displacement ventilation systems, *Report for BFR Project # 920937-0*, Royal Institute of Technology, Stockholm.

56. Baturin, V. V. (1972) *Fundamentals of Industrial Ventilation*, Pergamon, Oxford.

57. Danielsson, P. O. (1988) Displacement ventilation – a spreading disease or a blessing?, *Proc. 2nd Int. Symp. on Ventilation for Contaminant Control (Ventilation '88)*, London, 20–23 September 1988, pp. 471–80.

58. Appleby, P. (1989) Displacement ventilation: a design guide. *Build. Serv.*, April, 63–6.

59. Alamdari, F. (1998) Displacement ventilation and cooled ceilings, *Proc. Roomvent '98*, 1, 197–204, Stockholm, 1998.

60. Karimipanah, T. and Awbi, H. B. (2002) Theoretical and experimental investigation of impinging jet ventilation and comparison with wall displacement flow, *Building and Environment*, 37, 1329–42.

61. Karimipanah, T., Sandberg, M. and Awbi, H. B. (2000) A comparative study of air distribution systems in classrooms, *Proc. Roomvent 2000, Air Distribution in Rooms: Ventilation for Health and Sustainable Environment*, H. B. Awbi (ed.), 2, 1013–18, Elsevier.

62. *ASHRAE Handbook of HVAC Systems an Equipment* (2000) Ch. 18: Fans, American Society of Heating, Refrigeration and Air-Conditioning Engineers, Atlanta, GA.

63. Osborne, W. C. (1977) *Fans*, Pergamon, Oxford.

64. Awbi, H. B. (1998) Energy efficient room air distribution, *Renewable Energy*, 15, 293–9.

65. Croome, D. J., Gan, G., Swaid, H. and Awbi, H. B. (1993) Energy implications of thermal comfort standards, *Proc. Building Design Technology and Wellbeing in Temperate Climates*, Brussels.

66. Seppanen, O. (1994) Good energy economy and indoor climate – Conflicting requirements?, *Proc. European Conference on Energy Performance and Indoor Climate in Buildings*, Lyon, France.

67. Liddament, M. (1996) A Guide to Energy Efficient Ventilation, *Air Infiltration and Ventilation Centre*, International Network for Information on Ventilation, Brussels, Belgium (www.aivc.org).

3 Air infiltration calculation and measurement

3.1 Introduction

Since the mid-1970s there has been an urgent need to reduce the use of energy in the heating and cooling of buildings worldwide. In the developing countries, the energy consumption for maintaining an acceptable environment in buildings constitutes by far the largest part of the total energy demand for these countries. In the developed countries, the proportion of energy usage in buildings is larger than that used in industrial processes or transport. It is not surprising, therefore, that numerous measures have been taken by governments, groups and individuals to reduce the usage of energy for heating and cooling buildings. This has recently been manifested by international agreements between countries, e.g. the Kyoto Protocol, 1997. A major initiative undertaken by industrial and developing countries alike was the reduction of heat transmission through the building fabric by increasing the insulation standard – in some countries by legislation and provision of government grants. As the insulation qualities of buildings have improved the proportion of energy lost by air exfiltration through the building fabric has steadily increased, reaching in some cases half the total energy requirement for the building. Although some measures of reducing air infiltration/exfiltration rates have been taken, there has generally been a lack of concerted effort on this front by government agencies. Although there is now much legislation on fabric insulation, currently only some Nordic countries have produced legislation concerning the airtightness of the building envelope. This lack of coordination stems partly from the disagreement on the minimum amount of outside air required to enter a building to achieve acceptable indoor air quality for the occupants. The major ventilation standards have frequently been revised during the last three decades and studies by Professor P. O. Fanger in Denmark, reported in Chapter 2, have shed new light on what is an acceptable ventilation rate. Whatever the outcome of this debate, a ventilation engineer is required to assess the air leakage characteristics of a building and be able to design and specify ventilation openings as required.

This chapter covers three areas of air infiltration. The first concerns the air flow characteristics of the building fabric and building components and how the air flow through them is calculated. The second area deals with models that have been developed for the calculation of air infiltration through the envelope of a building. And the third area is on the measurement of air infiltration through the whole building envelope. The combined knowledge developed in this chapter will be applied in the design of passive and natural ventilation systems in Chapter 7.

3.2   Air leakage characteristics of buildings

3.2.1   *Flow through openings*

Typical air leakage paths in a house are illustrated in Figure 3.1. In general, the air leakage rate through a building envelope is dependent on: (i) the size and distribution of leakage paths; (ii) the flow characteristics of the leakage paths and (iii) the pressure difference across the leakage paths.

   To evaluate the air flow through a building component or envelope it is essential that all these three influencing factors be known or can be determined by one of the methods described later in this section. The main concept of air leakage estimation is based on the mass balance of air across the whole building envelope. The air flow through a building component or envelope is very complex, but an estimate of the flow rate may be made using one of two basic equations depending on the type of opening present. These are the equations for fully turbulent flow and the equation for laminar or transitional flow.

Large openings

For an opening of relatively large free area, such as a vent or a large crack, the flow tends to be approximately turbulent under normal pressures. In this case the flow rate, Q, is proportional to the square root of the pressure difference and can be evaluated using the standard orifice flow equation:

$$Q = C_d A \sqrt{2\Delta p/\rho_o} \tag{3.1}$$

where C_d = discharge coefficient of opening; A = flow area (m^2); Δp = pressure difference across opening (Pa); and ρ_o = air density at a reference temperature and pressure, T_o and $p_o (\text{kg m}^{-3})$.

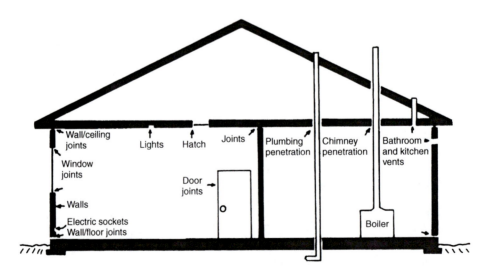

Figure 3.1 Typical air leakage paths in a house.

In sharp-edge orifice flow the discharge coefficient is almost independent of the Reynolds number and has a value of 0.61. However, for most building openings this constancy of C_d is not observed [1] because of the geometry of the openings and the variation in pressure difference (i.e. Reynolds number) with the environmental conditions inside and outside the building. BS 5925 [2] recommends using equation (3.1) for openings of typical dimensions larger than approximately 10 mm. The effective leakage area, $C_d A$, can be determined by means of a building pressurization or depressurization test (see Section 3.4) or may be taken from tables given in the ASHRAE Handbook [3] for different building components.

Small openings

For extremely narrow openings (cracks) with deep flow paths (such as mortar joints and tight-fitting components) the flow within the openings is essentially laminar or viscous. In such cases the flow rate is given by the Couette flow equation:

$$Q = [bh^3/(12\mu L)]\Delta p \quad \text{m}^3\,\text{s}^{-1} \tag{3.2}$$

where b = length of crack (m); h = height of crack (m); L = depth of crack in flow direction (m) and μ = absolute viscosity of air (Pa s). For wider cracks the flow is usually neither laminar nor fully turbulent but in the transition region. It is therefore appropriate to lump equations (3.1) and (3.2) into a single power-law equation of the form:

$$Q = kL(\Delta p)^n \quad \text{m}^3\,\text{s}^{-1} \tag{3.3}$$

where k = flow coefficient ($\text{m}^3\,\text{s}^{-1}\,\text{m}^{-1}\,\text{Pa}^{-1}$); L = length of crack (m) and n = flow exponent. This is normally called the 'crack flow equation' in which k is a function of the crack geometry and n is dependent on the flow regime and acquires a value of 0.5 for fully turbulent flow and 1.0 for laminar flow. However, in practice the value of n for cracks or adventitious openings tends to be between 0.6 and 0.7. A range of values of k for cracks formed around closed windows are given in Table 3.1 [2]. These should be used with a value of $n = 0.67$.

Table 3.1 Values of k in equation (3.3) for windows [2]

Window type	k for unweatherstripped mean (range)	k for weatherstripped mean (range)
Timber		
Side-hung casement	0.23 (1.19–0.04)	0.03 (0.10–0.01)
Top-hung casement	1.08 (1.38–0.88)	0.42 (1.22–0.11)
Centre-pivoted	0.8 (1.25–0.04)	0.02
Metal		
Side-hung casement	0.31 (0.45–0.21)	0.27 (0.29–0.14)
Top-hung casement	0.32 (0.55–0.18)	—
Vertically sliding	0.45 (1.20–0.20)	0.18 (0.34–0.04)
Horizontally sliding	0.22 (0.43–0.12)	—

Note
The values of k in this table are in $l\,\text{s}^{-1}\,\text{m}^{-1}$ of crack length for an applied pressure difference of 1 Pa.

Because equation (3.3) is not dimensionally homogeneous it lacks generality of application.

It has been suggested [4] that a quadratic form of equation (3.3) which is dimensionally homogenous provides a more accurate assessment of the flow through a crack. Such an equation has the form:

$$\Delta p = \alpha Q + \beta Q^2 \tag{3.4}$$

where α and β are flow coefficients. The first term on the right-hand side of equation (3.4) represents laminar flow and the second represents turbulent flow. The former becomes more significant at low flow rates and the latter is more significant at large flow rates. The values of α and β can be obtained from experimental tests on the openings and values for some openings can be found in [5].

Both the power-law and quadratic forms of the crack flow equations have been used extensively in the calculation of air infiltration through a building envelope. These equations are normally written in a form suitable for a network of nodes, linking the external pressure points to the internal pressure points, suitable for solution by a computer. This is achieved by an iteration process in which an arbitrarily guessed internal pressure value is successively improved until a flow balance is achieved across the building fabric. This procedure is explained in more detail in Section 3.3.2. Liddament [6] has compared the air infiltration rates predicted by the power-law and quadratic equations with measurements in a family dwelling and showed that the two equations produced identical results for large pressure differences ($\Delta p > 20\,\mathrm{Pa}$) but the power-law was superior at low pressure differences, i.e. of the order existing under normal environmental conditions.

3.2.2 Wind pressure

The pressure difference across an opening or crack is the driving force for air leakage. This is produced by the action of wind and temperature difference (stack pressure) and by the operation of any mechanical ventilation systems that are present.

With reference to the static pressure of the wind upstream of an opening, the time-mean pressure due to the wind flow, p_w, on to or away from a surface is given by:

$$p_w = 0.5\,C_p \rho_0 v^2 \quad \mathrm{Pa} \tag{3.5}$$

where C_p = static pressure coefficient and v = wind speed at datum level, usually height of building or opening ($\mathrm{m\,s^{-1}}$). The pressure coefficient is normally derived from pressure measurements in wind tunnels using reduced-scale models of buildings or building components or pressure measurements in the actual buildings. The value of C_p at a point on the building surface is determined by:

- the building geometry;
- the wind velocity (i.e. speed and direction) relative to the building;
- the exposure of the building, i.e. its location relative to other buildings and the topography and roughness of the terrain in the wind direction.

For a building with sharp corners C_p is almost independent of the wind speed (i.e. of the Reynolds number) because the flow separation points normally occur at the

Table 3.2 Terrain factors for equation (3.6) [2]

Terrain	c	a
Open flat country	0.68	0.17
Country with scattered wind breaks	0.52	0.20
Urban	0.35	0.25
City	0.21	0.33

sharp edges. This may not be the case for round buildings where the position of the separation point can be affected by the wind speed. Values of C_p for simple building geometry may be obtained from [2, 7 or 8].

3.2.3 Wind speed

The other quantity which requires evaluation in order to calculate the wind pressure using equation (3.5) is the wind speed. Natural wind has a highly turbulent and gusting character and even if a time-mean speed is used this is found to vary with height from the ground as well as the roughness of the terrain over which the wind passes. The time-mean wind-speed profile can be determined using the following expression [2]:

$$v/v_r = cH^a \qquad (3.6)$$

where v = mean wind speed at height H above the ground (m s^{-1}); v_r = mean wind speed measured at a weather station normally at a height of 10 m above the ground (m s^{-1}) and c and a are factors which depend on the terrain which are given in Table 3.2.

To evaluate the wind speed at height H it is necessary to know the value of v_r for the required location. This may either be obtained from a local weather station or from wind contour maps of the country. Normally, v_r represents the hourly mean wind speed which is exceeded 50% of the time at a particular site.

However, wind speeds exceeded by different ratios may sometimes be used in which case wind loading standards (e.g. [7]) should be consulted.

Another formula of the power-law type (equation (3.6)) has the form [9]:

$$v/v' = \alpha(H/10)^\gamma /[\alpha'(H'/10)^{\gamma'}] \qquad (3.7)$$

where v' = measured wind speed (m s^{-1}); H' = height of wind-speed measurement (m); H = height of building (m) and α and γ are the terrain parameters with the unprimed values referring to the location of the building and the primed values referring to the location of the weather station. Values of α and γ are given in Table 3.3.

A solution to the turbulent boundary layer equation which includes pressure gradient, Coriolis force (which arises from a reference to axes fixed on the rotating earth) and the Reynolds stresses (see Chapter 8) gives the following logarithmic expression for the wind profile [10]:

$$v = Kv_r \qquad (3.8)$$

where:

$$K = \left[\ln(\tilde{H}/b) + CfH\right] / \left[\ln(10/b) + 10Cf\right] \qquad (3.9)$$

Table 3.3 Terrain parameters for equation (3.7) [9]

Terrain description	α	γ
Ocean or other body of water with at least 5 km of unrestricted expanse	0.10	1.30
Flat terrain with some isolated obstacles, e.g. buildings or trees well separated from each other	0.15	1.00
Rural area with low buildings, trees, etc.	0.20	0.85
Urban, industrial or forest areas	0.25	0.67
Centre of large city	0.35	0.47

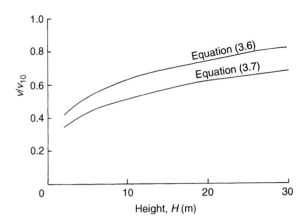

Figure 3.2 A comparison between the wind speed calculated for urban terrain using two different expressions.

In equation (3.9), b is the surface roughness height which is dependent on the terrain, C is a terrain factor and f a Coriolis parameter (approximately $1 \times 10^{-4} \, \text{rad s}^{-1}$), and $\tilde{H}$ is the effective height of the building equal to $H - d$, where d is a displacement height which is the effective height of the obstacles surrounding the building. For heights $\tilde{H} < 30\,\text{m}$ in open-country terrain or greater in rough terrains, equation (3.9) simplifies to:

$$K = \ln(\tilde{H}/b)/\ln(10/b) \tag{3.10}$$

Figure 3.2 shows a comparison between the values of v/v_{10} calculated using equations (3.6) and (3.7) for buildings of height up to 30 m in an urban location. Equation (3.8) requires knowledge of the local terrain such as average height of surrounding buildings and the terrain surface roughness parameter, and therefore it is not included here in the comparison. Equation (3.6) gives larger values of wind speeds than equation (3.7).

Equation (3.5) does not make any allowance for the fluctuating pressure produced by the fluctuations in wind speed. In some situations the transient pressure components can produce a net air leakage through an opening significantly greater than that calculated from equation (3.5) using time-mean speeds and pressure coefficients. Although some attempts have been made to account for the transient pressure components (e.g. [8]) the techniques thus far developed have yet to be verified.

3.2.4 Stack pressure

The pressure due to buoyancy (stack pressure) is an additional component which controls the air leakage through the envelope of a building. This arises from the difference in temperature, hence density, between the air inside and that outside of a building. The variation of air density with temperature produces pressure gradients both within the internal and external zones and across the building fabric. When the inside air temperature is greater than that outside, cooler outside air leaks into the building through openings at the lower parts of the building and warm inside air escapes through openings at a higher level. A reversal of this flow direction occurs when the inside air temperature is lower than that outside. The height at which transition between inflow and outflow occurs is the neutral plane where the pressures inside and outside are equal (Figure 3.3). In practice, the position of the neutral plane is a function of the overall distribution and flow characteristics of the openings, which is seldom known, and the stack pressure is usually expressed relative to the position of the lowest opening or a convenient datum in the building.

The pressure due to buoyancy (stack pressure) is given by:

$$p_s = p_0 - \rho g y \tag{3.11}$$

where p_0 = pressure at a datum height h_0 (Pa); ρ = air density (kg m^{-3}); g = acceleration due to gravity (m s^{-2}) and y = height measured from datum (m).

The vertical pressure is given by:

$$dp/dy = -\rho g = -\rho_0 g T_0 / T \tag{3.12}$$

where ρ_0 = air density at reference temperature T_0 (kg m^{-3}) and T = air temperature (K).

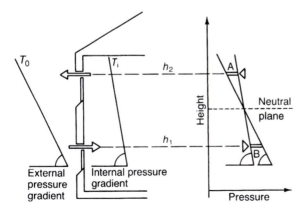

Figure 3.3 Buoyancy-induced air leakage through two vertical openings.

From equation (3.12) it is clear that the stack pressure decreases with height. Using this equation the stack pressure difference between two vertical openings separated by a vertical distance h is given by (Figure 3.3):

$$p_s = -\rho_0 g h(1 - T_0/T_i) \tag{3.13}$$

where T_0 = outdoor air temperature (K) and T_i = internal air temperature (K). Equation (3.13) produces a pressure difference between two openings 5 m apart with an internal to external temperature difference of 20 K of 4.3 Pa which is comparable with the pressure exerted by a wind of speed $5\,\text{m\,s}^{-1}$ on to a building of similar height.

3.2.5   Mechanical air infiltration

The air infiltration (or exfiltration) rate created by a mechanical extract (or supply) fan on a small building can be determined by matching the building's air leakage curve with the fan's characteristic curve in a similar way as in fan and duct system matching. Figure 3.4 illustrates this for both a supply and an extract fan. Using such plots the air infiltration rate due to mechanical ventilation can be determined.

3.2.6   Total air infiltration

The total pressure acting on an opening or the building envelope as a whole is the sum of the pressure due to wind, stack and mechanical ventilation (fan pressure) taking into consideration the sign of each pressure component. The pressures due to the effect of wind, stack and fan and resulting air flow rates are illustrated qualitatively in Figure 3.5. A chimney has a similar effect to an extract fan, i.e. it causes a decrease in the internal pressure. However, since the flow through an opening is proportional to $(\Delta p)^n$ (see equation (3.3)) the total flow through an opening may be approximated

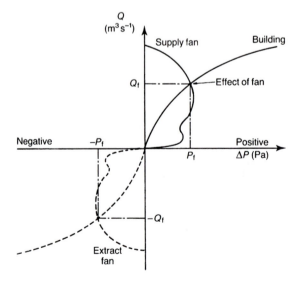

Figure 3.4 Matching fan curve and building air leakage curve.

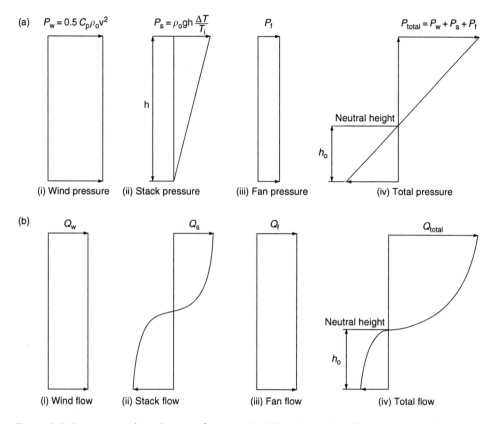

Figure 3.5 Pressures and resulting air flows in a building due to the effect of wind, stack and fan.

by [11]:

$$Q_t = \left[Q_w^{1/n} + Q_s^{1/n} + Q_{fu}^{1/n} \right]^n \tag{3.14}$$

where Q_{fu} is the flow rate due to the unbalanced mechanical ventilator, i.e. a supply or an extract fan only, which can be obtained using the procedure given in Figure 3.4. Equation (3.14) is equivalent to the simple summation of pressure difference across an opening due to wind, stack effect and fan. The value of n varies from 0.5 to 1.0 but a value of 0.67 is a good approximation for most air leakage paths. The equation can be used to calculate the total air infiltration due to wind, stack and mechanical extract (or forced draught) where the forced extract creates a pressure imbalance across the building envelope. When a balanced mechanical ventilator is used, such as an air-to-air heat recovery unit with a supply fan and an equal flow extract fan, the additional air flow does not affect the total pressure across the building envelope. For a balanced mechanical ventilator the total flow rate is calculated using:

$$Q_t = Q_{fb} + \left[Q_w^{1/n} + Q_s^{1/n} \right]^n \tag{3.15}$$

where Q_{fb} is the flow rate provided by the balanced fan ventilation system.

The above method of calculating the total air infiltration assumes that the internal pressure in the building is influenced by wind and stack effect in the same way which is not always the case in practice. The internal pressure results from the addition of individual wind and stack pressures, which is a non-linear combination (see equations (3.5) and (3.13)), and the true internal pressure can only be obtained by iterative solutions. To avoid the use of complex iterative models, several simple methods of superposing the stack and wind effects have been proposed [12]. Walker and Wilson [12] considered four popular superposition methods and compared the measured air infiltration with that predicted by the four methods. The constant concentration tracer gas technique (see Section 3.4.2) was used to measure the air infiltration in two houses where the infiltration due to wind and stack were isolated, i.e. at low wind speeds ($<1.5\,\mathrm{m\,s^{-1}}$) the dominant pressure was assumed to be the stack pressure and at higher wind speeds the dominant pressure was assumed to be due to wind alone. The four models of combining Q_s and Q_w tested were:

(i) The simple linearity method:

$$Q_t = Q_w + Q_s \tag{3.16}$$

(ii) The quadrature method:

$$Q_t = [Q_w^2 + Q_s^2]^{1/2} \tag{3.17}$$

(iii) The simple pressure addition method:

$$Q_t = [Q_w^{1/n} + Q_s^{1/n}]^n \tag{3.18}$$

(iv) The method of interaction between stack and wind effects, the so-called interaction method or the AIM-2 (Alberta Infiltration Model) superposition method:

$$Q_t = [Q_w^{1/n} + Q_s^{1/n} + B(Q_sQ_w)^{(1/2n)}]^n \tag{3.19}$$

where $B(Q_sQ_w)^{1/2n}$ is the interaction term with $B \approx -0.33$ according to the measurements in [12]. This term, which retains dimensional consistency in equation (3.19), disappears when either the stack or wind effects dominate and has the greatest value when they are equal, which is $B = (2^{(1/2n)} - 2)$.

Walker and Wilson [12] found that although equation (3.17) does not account for the interaction of wind and stack effect, it nevertheless produced similar results to equation (3.19) which does account for interaction. The physically unrealistic quadrature equation (3.17) also produced similar results to equations (3.18) and (3.19), however the simple linearity equation (3.16) was found to over-predict the air infiltration by up to about 50%. On the basis of this study, Walker and Wilson recommended the use of the pressure addition method (equation (3.17)) because of its simplicity and reliability in predicting the combined effect of wind and stack.

3.3   Air infiltration calculation and modelling

The information given in Section 3.2 will, in principle, enable the calculation of air infiltration rate to zones within a building, or the whole building if considered a single zone, providing that the following quantities are known or can be evaluated:

1 the wind speed and direction at the desired location;
2 the internal and external air temperatures;
3 the position and flow characteristics of all openings;
4 the pressure distribution over the building for the wind direction under consideration.

In practice, it is difficult, if not impossible, to determine all these quantities accurately and a simplification of the problem will be necessary to reach a solution based principally on the equations presented in the previous section. There are two types of solutions that could be applied to calculate the air infiltration rate to a building: (i) empirical models and (ii) network models. These types of solution are considered in the following sections with reference to the more well-known models.

3.3.1 *Empirical air infiltration models*

These are simplified procedures based on empirical data which can produce estimates of air infiltration rates in essentially single-zone buildings such as residential premises and small industrial or commercial buildings. The methods are considered adequate for sizing heating or cooling plants.

ASHRAE method

An empirical equation which combines the air flow due to buoyancy and wind in quadrature (see equation (3.17)) is used to calculate the total air infiltration across the whole building envelope. It is based on knowledge of the total effective leakage area of the building and the values of some empirical constants derived from a database of air infiltration measurements. The effective leakage area can be either determined from pressurization or depressurization tests on the building or evaluated from tables given in [3]. The bulk air flow rate in a single-zone building is given by the equation:

$$Q = A\sqrt{(a\Delta T + bv_r^2)} \quad \mathrm{m^3\,h^{-1}} \tag{3.20}$$

where A = the total effective leakage area of the building ($\mathrm{cm^2}$); a = stack coefficient ($\mathrm{m^6\,h^{-2}\,cm^{-4}\,K^{-1}}$); b = wind coefficient ($\mathrm{m^4\,s^2\,h^{-2}\,cm^{-4}}$); ΔT = average inside-outside temperature difference (K) and v_r = mean wind speed measured at a local weather station ($\mathrm{m\,s^{-1}}$).

The value of a depends on the building height. This is given as: 0.00188 for a one-storey building; 0.00376 for a two-storey building and 0.00564 for a three-storey building.

The values for the wind coefficient, b, are dependent upon the building height and wind shielding category. Table 3.4 gives values of b for one-, two- and three-storey buildings and five classes of wind shielding.

Example 3.1 Estimate the average air change rate during the heating season in a two-storey house with an air volume of $300\,\mathrm{m^3}$ and a total leakage area of $800\,\mathrm{cm^2}$, assuming an average wind speed during the heating season of $5\,\mathrm{m\,s^{-1}}$ and an internal–external temperature difference of 22 K. The house is located in a suburban area of a large city.

Table 3.4 Wind coefficient, *b*, in equation (3.20) [3]

Shielding class	Building height (storeys)		
	One	Two	Three
I	0.00413	0.00544	0.00640
II	0.00319	0.00421	0.00495
III	0.00226	0.00299	0.00351
IV	0.00135	0.00178	0.00209
V	0.00041	0.00054	0.00063

Local shielding classes

Class	Description
I	No obstructions or local shielding
II	Light local shielding by few obstructions
III	Moderate local shielding due to other buildings of similar height
IV	Heavy shielding with obstructions taller than the building, e.g. suburban areas
V	Very heavy shielding by adjacent large obstructions, e.g. urban areas

Equation (3.20) is used to evaluate the air infiltration rate:

$$Q = v(a\Delta T + bv_r^2)$$

For a two-storey house $a = 0.00376$ and for a class IV shielding (suburban), Table 3.4 gives a value of 0.00178 for b. Hence:

$$Q = 800\,v\,(0.00376 \times 22 + 0.00178 \times 5^2) = 285.3\,\mathrm{m}^3\,\mathrm{h}^{-1}$$

The air change rate is $285.3/300 = 0.95\,\mathrm{h}^{-1}$.

British Standards method

Assuming two-dimensional flow through a building and ignoring internal partitions, BS 5925 [2] gives tables of formulae for calculating the air infiltration rate due to wind, buoyancy and combined wind and buoyancy for openings on opposite walls and openings on the same wall. Tables 3.5 and 3.6 show schematically the expected air flow patterns for different conditions and give the formulae used for calculating the air infiltration rate for each case. Figure 3.6 shows the effect of the opening angle of a window on the value of $J(\theta)$ in the equation of Table 3.6(c).

The formulae given in Tables 3.5 and 3.6 illustrate a number of general characteristics of large openings used for the calculation of natural ventilation in a single zone of a building. These can be summarized as follows:

1   The total area of a number of openings in parallel across which the same pressure difference is applied can be obtained by a simple summation of all the areas.
2   The total area of a number of openings in series across which the same pressure difference is applied can be obtained by adding the inverse squares and taking the inverse of the square root of the total.
3   When wind and stack effects are of the same order of magnitude their interaction is complicated, but to a first approximation the larger of the two rates may be taken.

Table 3.5 Formulae for cross-ventilation [2]

Conditions	Schematic representation	Formula
(a) Wind only		$Q_w = C_d A_w V (\Delta C_p)^{1/2}$ $\dfrac{1}{A_w^2} = \dfrac{1}{(A_1 + A_2)^2} + \dfrac{1}{(A_3 + A_4)^2}$
(b) Temperature difference only		$Q_b = C_d A_b \left(\dfrac{2\Delta\theta g H_1}{\overline{T}}\right)^{1/2}$ $\dfrac{1}{A_b^2} - \dfrac{1}{(A_1 + A_3)^2} + \dfrac{1}{(A_2 + A_4)^2}$ $\overline{T} = \dfrac{1}{2}(T_e + T_i)$
(c) Wind and temperature difference together		$Q = Q_b$ for $\dfrac{V}{\sqrt{(\Delta T)}} < 0.26 \left(\dfrac{A_b}{A_w}\right)^{1/2} \left(\dfrac{H_1}{\Delta C_p}\right)^{1/2}$ $Q = Q_w$ for $\dfrac{V}{\sqrt{(\Delta T)}} > 0.26 \left(\dfrac{A_b}{A_w}\right)^{1/2} \left(\dfrac{H_1}{\Delta C_p}\right)^{1/2}$ $\Delta T = T_i - T_e$

Table 3.6 Formulae for single sided wall opening [2]

Conditions	Schematic representation	Formula
(a) Due to wind	 Plan	$Q = 0.025\, AV$
(b) Due to temperature difference with two openings		$Q = C_d A \left[\dfrac{\epsilon\sqrt{2}}{(1+\epsilon)(1+\epsilon^2)^{1/2}}\right] \left(\dfrac{\Delta T g H_1}{\overline{T}}\right)^{1/2}$ $\epsilon = \dfrac{A_1}{A_2}; \quad A = A_1 + A_2$
(c) Due to temperature difference with one opening		$Q = C_d \dfrac{A}{3} \left(\dfrac{\Delta T g H_2}{\overline{T}}\right)^{1/2}$ If an opening light is present $Q = C_d \dfrac{A}{3} J(\theta) \left(\dfrac{\Delta T g H_2}{\overline{T}}\right)^{1/2}$ where $J(\theta)$ is given in Figure 3.6

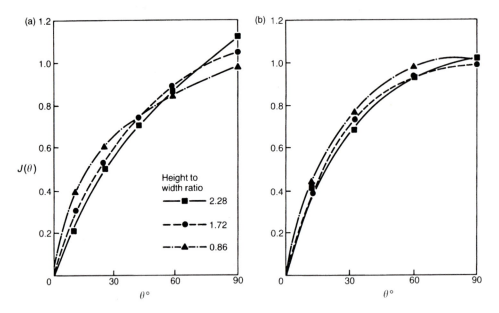

Figure 3.6 Effect of window opening on $J(\theta)$. (a) Side-mounted casement windows and (b) centre pivoted windows [2].

3.3.2   *Network air infiltration models*

These are theoretically based models which are more rigorous than the empirical models described in the previous section. They provide estimates of air infiltration rates and are used for air infiltration calculations in a single-zone or a multi-zone building. A single-zone building is one which has no internal partitioning such that the internal conditions may be assumed homogeneous throughout. Although such buildings rarely exist this approximation can in practice be made in situations where the effect of internal partitions have small influences on the air movement inside the building, such as houses or large enclosures. A multi-zone modelling is necessary where a building consists of a number of well defined zones, such as large or high-rise buildings. In this case each internal zone is assumed to be homogeneous and is influenced by the conditions in adjacent zones as well as outside. These two types of calculation models are described below.

Single-zone models

In a single-zone model the interior of the building is described by a single, well-mixed zone at uniform pressure and temperature, i.e. one pressure node. This internal pressure node is connected to either a single external node or a number of external nodes at different pressures, see Figure 3.7. Single-zone models demand less effort than the multi-zone models but they do not provide any information on the pattern of distribution of air infiltration around the building envelope. Two such models that have been validated by the Air Infiltration and Ventilation Centre (AIVC) for a range of dwellings and climatic conditions [13] are described here.

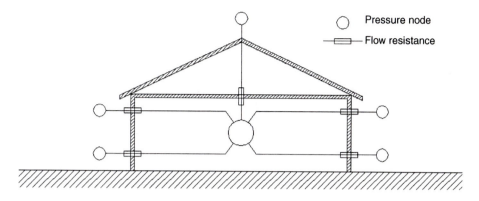

Figure 3.7 A single-zone flow model.

Building Research Establishment (BRE) model   The BRE model [14] provides a method for relating the air infiltration rate for any given internal and external conditions to the leakage characteristics of the building as determined by a pressurization test. The total air infiltration rate through a building due to ambient conditions is described by the following power-law expression:

$$Q_t = Q_p[\rho_e v^2/\Delta p_p]^n F_t(Ar, \theta) \tag{3.21}$$

where Q_t = air infiltration rate due to wind and buoyancy effects (m^3 s^{-1}); Q_p = air flow rate obtained during a pressurization test at an arbitrarily chosen reference pressure, normally at $\Delta p_p = 50$ Pa (m^3 s^{-1}); ρ_e = density of external air (kg m^{-3}); v = wind speed at roof ridge height (m s^{-1}); Δp_p = internal–external pressure difference for the pressurization test, normally = 50 Pa; n = leakage exponent (0.5–0.7) obtained from pressurization tests; $F_t(Ar, \theta)$ = total infiltration function which depends on the Archimedes number, Ar, and wind direction, θ; θ = wind direction which determines the pressure distribution over the building surfaces (°); $Ar = \Delta Tgh/(T_i v^2)$ is the building Archimedes number; ΔT = internal–external temperature difference (K); T_i = internal temperature (K) and h = height of ventilated space (m). The wind speed at h is calculated using equation (3.6) and values of c and a are obtained from Table 3.2.

The air infiltration due to the wind action alone, Q_w, is obtained using:

$$Q_w = Q_p[\rho_e v^2/\Delta p_p]^n F_w(\theta) \tag{3.22}$$

where $F_w(\theta)$ is the wind infiltration function which is a function of wind direction. The air infiltration due to the stack effect only, Q_s, is determined using:

$$Q_s = Q_p[\Delta T \rho_e gh/(T_i \Delta p_p)]^n F_s \tag{3.23}$$

where F_s is the stack infiltration function which depends on the physical characteristics of the building.

Warren and Webb [14] have calculated values of F_t, F_w and F_s for three typical British houses: detached, semi-detached and centre terrace. They used the pressure

Table 3.7 Values of F_w and F_s for the BRE model [14]

House type	n	F_s		$F_w(0°\ wind)$		$F_w(90°\ wind)$		$F_w(270°\ wind)$	
		(i)	(ii)	(i)	(ii)	(i)	(ii)	(i)	(ii)
Detached	0.5	0.26	—	0.17	—	0.20	—	—	—
	0.6	0.23	0.26	0.15	0.13	0.18	0.16	—	—
	0.7	0.20	—	0.13	—	0.16	—	—	—
Semi-detached	0.5	0.26	—	0.16	—	0.18	—	0.12	—
	0.6	0.23	0.27	0.15	0.15	0.16	0.18	0.10	0.10
	0.7	0.20	—	0.14	—	0.15	—	0.08	—
Centre terrace	0.5	0.26	—	0.20	—	0.13	—	—	—
	0.6	0.23	0.27	0.18	0.18	0.10	0.08	—	—
	0.7	0.20	—	0.16	—	0.08	—	—	—

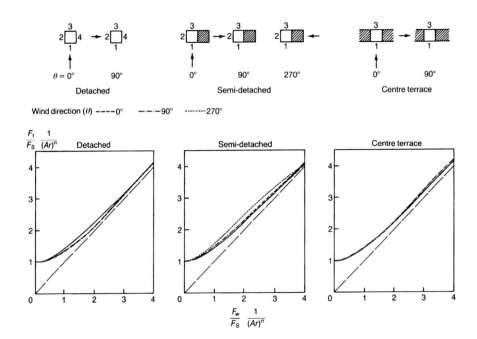

Figure 3.8 Variation of $(F_t/F_s)/Ar^n$ with $(F_w/F_s)/Ar^n$ and wind direction, θ, for three house types ($n = 0.6$), reference [14].

data given in [7] for the purpose and assumed that Q_p is uniformly distributed over the exposed surfaces of the house in each case. The values of F_w and F_s are given in Table 3.7 and those for F_t are shown in Figure 3.8. Comparisons between calculated and measured air infiltration rates for four house types have been carried out by the AIVC [13]. A typical comparison for a concrete basement, two-floor timber-frame house (Maugwil) of volume 342 m³ is shown in Figure 3.9.

The following assumptions have been made in formulating the BRE model:

1 The volume of the building for infiltration purposes is represented by a rectangular parallelopiped of height h.

2 The pressure generated by the wind is assumed to be uniform across each face of the building and inferred directly from wind data for the particular building and wind direction using:

$$p = p_o + 0.5C_p\rho v^2$$

where p = wind pressure (Pa); p_o = free stream static pressure (Pa) and C_p = pressure coefficient for the face.

3 The leakage, Q_p, at reference pressure difference, Δp_p, and the exponent, n, are derived from pressurization tests between a pressure range of 10–60 Pa.

4 The air leakage through the building envelope is assumed to be uniformly distributed across each external surface, but the total leakage in a pressurization test, Q_p, may be distributed in any chosen proportions among the surfaces.

5 The same exponent, n, is assumed to apply to all leakage paths.

6 Party walls and solid floors are assumed to be impermeable.

7 If the under-floor space is ventilated, its surface pressure is obtained by determining the area-weighted mean of the pressures of the exposed walls.

Lawrence Berkeley Laboratory (LBL) model   The LBL model [9, 11] was developed for a single-zone building and its main feature is simplicity; hence precise detail of the building envelope is excluded. The building is approximated by a single rectangular structure through which the air infiltration rate is described by the equation for large openings (equation (3.1)).

Thus:

$$Q = A_o\sqrt{\frac{2\Delta p}{\rho}} \tag{3.1}$$

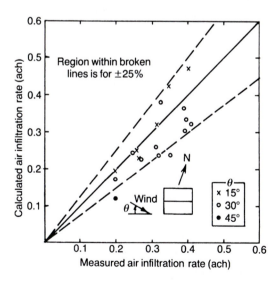

Figure 3.9 Comparison between calculated and measured air infiltration rates for a two-storey house using the BRE model [13].

where $A_o = C_d A$ is the total effective leakage area (m^2) which is determined from a building pressurization test or calculated from data given in [3] or [8]. In this model, the air infiltration due to wind and stack are calculated separately and then added in quadrature (equation (3.17)) to obtain the total infiltration rate.

The total effective leakage area, A_o, is calculated from equation (3.1) in which case Q is calculated from the crack flow equation (3.3) for the building at a reference pressure difference, Δp_{ref}, i.e.:

$$A_o = K_t(\Delta p_{ref})^n / \sqrt{(2\Delta p_{ref}/\rho)} \tag{3.24a}$$

where K_t is the total flow coefficient for the building and n is the flow exponent, both of which are obtained from pressurization test data. Δp_{ref} has been chosen arbitrarily to be 4 Pa in which case equation (3.24a) reduces to:

$$A_o = K_t(4)^n / \sqrt{(8/\rho)} \tag{3.24b}$$

The wind-induced infiltration is determined using the expression:

$$Q_w = f_w^* A_o v' \tag{3.25}$$

where $f_w^* =$ the 'reduced' wind parameter and $v' =$ the wind speed measured at a weather station (m s^{-1}).

The reduced wind parameter, f_w^*, is given by the following expression:

$$f_w^* = C' \sqrt[3]{(1-R)} \left[\alpha(H/10)^\gamma / \{\alpha'(H'/10)^{\gamma'}\} \right] \tag{3.26}$$

where the quantity in the square brackets represents the ratio of wind speed at the building height H to the reference wind speed at height H' (i.e. the right-hand side of equation (3.7)), C' is the generalized shielding coefficient and R is the fraction of the total effective leakage area that is horizontal. Values of C' for various shielding classes are given in Table 3.8 and values of α and γ (or α' and γ') are given in Table 3.3.

The horizontal area ratio is given by:

$$R = (A_c + A_f)/A_o \tag{3.27}$$

where $A_c =$ ceiling leakage area (m^2) and $A_f =$ floor leakage area (m^2). The infiltration due to the stack effect is:

$$Q_s = f_s^* A_o \sqrt{(\Delta T)} \tag{3.28}$$

Table 3.8 Generalized shielding coefficient for the LBL model [11]

Shielding class	C'	Description
I	0.34	No obstructions or local shielding whatsoever
II	0.30	Light local shielding with few obstructions
III	0.25	Moderate local shielding, some obstructions within two house heights
IV	0.19	Heavy shielding, obstructions around most of perimeter
V	0.11	Very heavy shielding, large obstruction surrounding perimeter within two house heights

where ΔT is the inside–outside temperature difference (K) and f_s^* is the 'reduced' stack parameter expressed as:

$$f_s^* = [(1 + 0.5R)/3] \times \{1 - [X^2/(2 - R)^2]\}\sqrt{(gH/T_i)} \tag{3.29}$$

where $X = (A_c - A_f)/A_o$ which is the difference between ceiling and floor leakage areas; T_i = internal temperature (K) and g = acceleration due to gravity (m s^{-2}). The total infiltration is obtained by summing the flow due to wind, stack and mechanical ventilation in quadrature by putting $n = 0.5$ in equation (3.14) for an unbalanced mechanical ventilation system or by putting $n = 0.5$ in equation (3.15) for a balanced system.

The advantage of the LBL model is the relative simplicity in the calculation of wind and stack infiltration rates. However, the method requires knowledge of the distribution of leakage area on the horizontal and vertical surfaces of the building which cannot normally be evaluated from pressurization tests. Furthermore, allowance for large openings such as vents, doors, etc. cannot be made and therefore their effect cannot be modelled. The accuracy of this method is estimated to be about ±25% as depicted in Figure 3.10 which compares measurements with predictions using the LBL model for the same house used to assess the BRE model in Figure 3.9.

Example 3.2 A two-storey house situated in an urban area has a plan area of 10 m × 7.5 m, an eaves height of 5 m and a ridge height of 7 m. The effective air volume inside the house is 350 m^3. A pressurization test carried out on the house produced a flow coefficient and an exponent of 195 cm^2 Pa^{-1} and 0.67 respectively. Calculate the number of air changes for a reference wind speed of 15 m s^{-1} and internal and external air temperatures of 22 and 0 °C respectively.

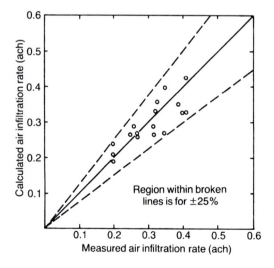

Figure 3.10 Comparison between calculated and measured air infiltration rates for a two-storey house using the LBL model [13].

The effective leakage area for the house is obtained from equation (3.24b).

$$A_o = K_t(4)^n/\sqrt{(8/\rho)}$$

The wind-induced air infiltration rate is calculated using equation (3.25).

$$Q_w = f_w^* A_o v'$$

where the reduced wind parameter f_w^* is obtained from equation (3.26).

$$f_w^* = C'\sqrt[3]{(1-R)}\left[\alpha(H/10)^\gamma/\{\alpha'(H'/10)^{\gamma'}\}\right]$$

In this equation, R is the ratio of leakage area which is horizontal and which may be calculated if the total leakage area is assumed evenly distributed across the house envelope:

Surface area of walls $= 35 \times 5 = 175\,\text{m}^2$
Surface area of floor and ceiling $= 2 \times 10 \times 7.5 = 150\,\text{m}^2$

Hence:

$$R = 175/(175 + 150) = 0.538$$

From Table 3.3 for an urban area, $\gamma = 0.25$ and $\alpha = 0.67$. From Table 3.8 for a shielding class IV, $C' = 0.19$. Hence:

$$f_w^* = 0.19\sqrt[3]{(1 - 0.538)}[0.67(7/10)^{0.25}/(0.67(10/10)^{0.25})]$$

$$= 0.134$$

$$Q_w = 0.134 \times 191.1 \times 10^{-4} \times 15$$

$$= 0.0386\,\text{m}^3\,\text{s}^{-1}$$

The stack infiltration is given by equation (3.28):

$$Q_s = f_s^* A_o \sqrt{(\Delta T)}$$

where the reduced stack parameter, f_s^*, is given by equation (3.29):

$$f_s^* = [(1 + 0.5R)/3]\{1 - [X^2/(2 - R)^2]\}\sqrt{(gH/T_i)}$$

Since the floor leakage area is assumed equal to the ceiling leakage area, $X = 0$. Hence:

$$f_s^* = [(1 + 0.5 \times 0.538)/3]\sqrt{(9.806 \times 7/295)}$$

$$= 0.204$$

$$Q_s = 0.204 \times 191.1 \times 10^{-4} \times 22$$

$$= 0.0183\,\text{m}^3\,\text{s}^{-1}$$

Total air infiltration rate:

$$Q_t = \sqrt{(Q_w^2 + Q_s^2)}$$
$$= \sqrt{[(0.0386)^2 + (0.0183)^2]}$$
$$= 0.0427\, \text{m}^3\, \text{s}^{-1}$$

The air change rate, N, is equal to $0.0427 \times 3600/350 = 0.44\ \text{h}^{-1}$.

Multi-zone models

The single-zone models of predicting air infiltration rates described above are based on simplified assumptions to what is essentially a complex flow phenomenon. Consequently, these models can only be expected to predict infiltration rates in a building that can be considered as a single zone, such as a house or a small building. There are many instances when the single-zone approach will be of little value and therefore consideration must be given to the flow within internal zones as well as the flow through the external envelope of the building. In a multi-zone building, the flow through the external envelope is also affected by the flow resistance of the internal zones and the prediction of infiltration rates requires a multi-zone network analysis. This is a grid based system in which the nodes are the rooms or zones of the building and the connection between two nodes represents a flow path of a given resistance. The flow resistances are doors and windows (open or shut), leakage paths of building surfaces, and ducts or ventilation components. Each node in the network represents a pressure value which is usually known for the nodes outside the building but must be determined for the internal nodes. Because of the non-linear dependency of the volume flow rate on the pressure difference (see equations (3.1) and (3.3)) the internal pressures can only be determined by an iterative solution of the flow equations. This is normally done using a computer.

The advantage of network models, besides the prediction of air infiltration to large buildings, is the ability to calculate interconnecting mass flows between different zones. This is significant for the following reasons:

1   The exchange of external air with individual zones inside the building which may be required for combustion or as fresh air to occupants.
2   The transport of contaminants and airborne particles between zones which is important in hospitals and factories.
3   The spread of smoke in the event of fire inside the building.

In a multi-zone building, the pressure nodes may be connected in series or in parallel (Figure 3.11). The flow between any two nodes can be estimated by applying the crack flow equation (3.3) provided that k and n are known for each leakage path. Equation (3.3) may be written as follows:

$$Q_i = k_i(\Delta p_i)^{n_i} \tag{3.30}$$

where subscript i refers to the ith flow path and k is the flow coefficient in $\text{m}^3\, \text{s}^{-1}\, \text{Pa}^{-1}$.

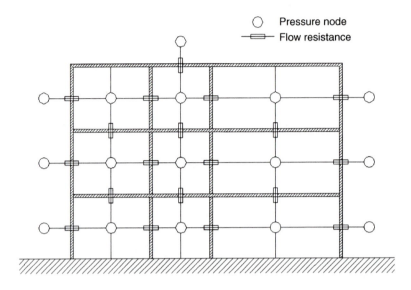

Figure 3.11 Network flow analysis for a multi-zone building.

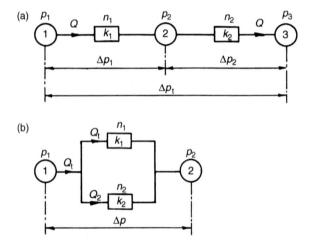

Figure 3.12 Types of air leakage paths. (a) Two leakage paths in series and (b) two leakage
paths in parallel.

Considering two flow paths in series, as in Figure 3.12(a), the flow equations for
this case are:

$$\Delta p_t = \Delta p_1 + \Delta p_2 \tag{3.31}$$

$$Q = k_1(\Delta p_1)^{n_1} = k_2(\Delta p_2)^{n_2} \tag{3.32}$$

By substituting equation (3.32) into (3.31) we obtain:

$$\Delta p_t = (Q/k_1)^{1/n_1} + (Q/k_2)^{1/n_2} \tag{3.33}$$

Given the values of k_1, k_2, n_1 and n_2, the flow rate through two leakage paths in series can be calculated by solving equations (3.32) and (3.33) iteratively because of their non-linearity.

For two leakage paths in parallel, as in Figure 3.12(b), the flow equations are:

$$Q_t = Q_1 + Q_2 \tag{3.34}$$

$$Q_1 = k_1(\Delta p)^{n_1} \tag{3.35}$$

$$Q_2 = k_2(\Delta p)^{n_2} \tag{3.36}$$

Hence:

$$Q_t = k_1(\Delta p)^{n_1} + k_2(\Delta p)^{n_2} \tag{3.37}$$

The total flow rate, Q_t, can be calculated by an iterative solution of the above equations.

Treating each zone of a building as a separate flow system, which may be connected to other zones if required, a zone may be represented by a one-junction flow network as shown in Figure 3.13. Applying the mass balance equation for j flow paths gives:

$$\sum_{i=1}^{j} \rho_i Q_i = 0 \tag{3.38}$$

where Q_i = volumetric flow rate through ith path ($\mathrm{m^3\,s^{-1}}$) and ρ_i = density of air flowing through ith path ($\mathrm{kg\,m^{-3}}$).

By substituting equation (3.30) into (3.38) we obtain:

$$\sum_{i=1}^{j} \rho_i k_i (p_i - p)^{n_i} = 0 \tag{3.39}$$

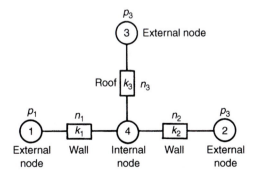

Figure 3.13 A single-zone flow network.

where p is the internal pressure in the zone. To avoid exponentiating a negative number when $p_i < p$, equation (3.39) is usually written in the following form:

$$\sum_{i=1}^{i} \rho_i k_i \underbrace{|p_i - p|^{n_i}}_{\text{I}} \left[\underbrace{\frac{p_i - p}{|p_i - p|}}_{\text{II}} \right] = 0 \tag{3.40}$$

In equation (3.40) term I expresses the absolute value of the external–internal pressure difference across each flow path and term II restores the sign of the flow direction which was lost in term I, i.e.:

for $p_i > p$,   term II $= 1$

for $p_i < p$,   term II $= -1$

The sign convention for the flow direction is positive for infiltration into the zone and negative for exfiltration out of the zone. Given the values of k_i and n_i for each flow path and pressure p_i, equation (3.40) can be solved iteratively to calculate the internal pressure, p, for the zone from which Q_i can be determined for each leakage path using equation (3.30). The total air infiltration to the zone is then given by:

$$Q_t = \sum_{i=1}^{j} Q_i \tag{3.41}$$

The hourly air change rate for the zone is then given as:

$$N = 3600 \; Q_t / V \tag{3.42}$$

where $N =$ air change rate per hour and $V =$ internal volume of zone (m^3).

The flow produced by an unbalanced ventilation system Q_{fu} (e.g. extract or supply fan) can be easily incorporated into equation (3.38) to give:

$$\sum_{i=1}^{j} \rho_i (Q_i + Q_{fu}) = 0 \tag{3.43}$$

in which case Q_{fu} is negative for extract ventilation and positive for supply and the air density, ρ_i, in the former represents that of the internal air and in the latter that of the external air. If the flow rate, Q_{fu}, is influenced by the pressure difference across the zone, then Q_{fu} can be specified as a function of this pressure. A balanced ventilation system has no influence on the internal pressure of the zone and in this case the total infiltration rate, Q_t, is given by:

$$Q_t = Q_{fb} + \sum_{i=1}^{j} Q_i \tag{3.44}$$

The above analysis applies to a single zone in a multi-zone building. It can be used to calculate the air infiltration into the zone through any number of flow paths terminating within the zone. The same procedure is also applicable for calculating the air infiltration through a number of zones in a building. For the mth zone with a total of j_m flow paths, the mass balance equation becomes:

$$\sum_{i_m=1}^{j_m} \rho_{im} Q_{im} = 0 \tag{3.45}$$

In this case equation (3.40) can be written as:

$$\sum_{i_m=1}^{j_m} \rho_{im} k_{im} |p_{im} - p_m|^{n_{im}} \left[\frac{p_{im} - p_m}{|p_{im} - p_m|} \right] = 0 \tag{3.46}$$

where k_{im} = flow coefficient of the ith flow path of the mth zone; n_{im} = flow exponent of the ith flow path of the mth zone; p_{im} = pressure of zone adjacent to the mth node across which the ith flow path connects and p_m = internal pressure of the mth zone.

Equation (3.46) must apply to each zone, but for a total of q zones the mass balance equation becomes:

$$\sum_{m=1}^{q} \sum_{i_m=1}^{j_m} \rho_{im} k_{im} |p_{im} - p_m|^{n_{im}} \left[\frac{p_{im} - p_m}{|p_{im} - p_m|} \right] = 0 \tag{3.47}$$

In a multi-zone solution involving q zones there are an equal number of internal pressures p_m. This will generate a number of simultaneous equations which must be solved before an air infiltration rate for the building or for a particular zone can be determined. Mechanical ventilation for each zone is analysed in the same way as presented earlier for a single-zone but care should be exercised as supply and extract terminals are sometimes positioned in different zones.

A number of computer algorithms have been devised for solving the air flow through single- and multi-zone buildings, some of these are described in AIVC publications [8, 13]. In Appendix A five well-known models are briefly described. The models are:

AIDA single-zone model [15, 16]
LBL single-zone model [9, 11]
AIOLIS multi-zone model [17]
COMIS multi-zone model [18]
CONTAM multi-zone model [19]

3.4   Measurement of air infiltration

Measurement techniques are the fundamental means of acquiring a greater understanding of air infiltration and ventilation in existing buildings as they provide a quantitative assessment of the air flow characteristics of these buildings. In the last few decades many advances have been made in the measurement techniques for air infiltration as a result of great effort by ventilation researchers. There are now techniques available for measuring the air infiltration rates into a building or components of a building under both simulated and actual environmental conditions. Advances have also been made in quantifying the air exchange rate between the internal zones of a building. The location and distribution of the air leakage paths and the air leakage characteristics of specific building components can be evaluated. Such information has been implemented in a number of the air infiltration models described in Section 3.3 and in Appendix A.

In many countries the evaluation of the airtightness of the building envelope has become a routine test and in some a mandatory exercise. This section describes the main techniques which are currently used for measuring the airtightness and the air infiltration rates of buildings. A fuller treatment of these measuring techniques can be found in the AIVC Guide [20].

3.4.1   Measurement of airtightness

The main reason for conducting measurements on building fabric and building component airtightness is to evaluate their air flow characteristics under controlled conditions and without the influence of climatic parameters. These characteristics are needed for calculating air infiltration rates and interzonal flows under a variety of environmental and building-usage conditions using the models described in Section 3.3.

Building components

Building component airtightness measurements can be performed under controlled laboratory conditions or on site [21, 22]. In both cases this involves setting up a collection chamber over the interior face of the building component as illustrated in Figure 3.14. In both cases air is supplied from the chamber at a rate required to maintain a specific static pressure difference across the specimen and the resultant flow rate is measured. The results from such tests are usually presented in terms of air leakage per unit area $(m^3 h^{-1} m^{-2})$ or leakage per unit crack length $(m^3 h^{-1} m^{-1})$, both specified for 1 Pa pressure difference (see Table 3.1) or at another specified pressure difference. Because of the low flow rates associated with building components a pressure difference of 200 Pa is sometimes used [22].

The air flow characteristics of building components can be accurately determined under laboratory test conditions. The advantages of laboratory tests are that large numbers of specimens can be examined under similar conditions which are unaffected by changes in climatic conditions. The disadvantages of laboratory tests, however, are

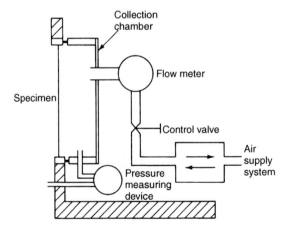

Figure 3.14 Airtightness test arrangement of a building component.

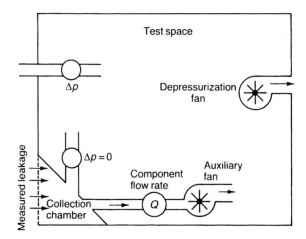

Figure 3.15 Site measurement of building component airtightness.

that the results obtained may be significantly different from those obtained on site on seemingly identical components. This is inevitable because of the difference in quality control between the laboratory and the building site. In the case of site tests there is usually a difference between the pressure inside the building and that outside which could cause air leakage across the collection chamber itself and errors in the measured air flow rates will then be introduced if the test chamber in Figure 3.14 is used. These errors can be greatly reduced by balancing the pressure in the collection chamber to that in the room containing the building component. This can be achieved by employing an auxiliary fan in the room connected to the collection chamber (Figure 3.15) to maintain zero pressure difference between the chamber and the room. In Figure 3.15 the pressure difference across the building component is created by the depressurization fan only.

Building envelope

To determine the air leakage characteristics of a building envelope two methods are in use, namely the steady-state pressurization/depressurization technique (commonly called the 'blower-door' test) and a more elaborate dynamic pressurization technique. Although the former has been in use for many years and is the basis of several national standards, e.g. reference [23] and [24], the latter is more recently developed and as yet not extensively used probably because of its more complex nature and unsuitability for testing the leakage of large enclosures, which are more commonly tested for air leakage.

Steady-state pressurization This method consists of mechanical pressurization or depressurization of a building and measurements of the resulting air flow rates at a given internal–external static pressure difference. An apparatus with variable speed pressurization or depressurization fan, pressure gauges and a flow meter is normally used for this purpose. Such a unit is commonly known as a 'blower door' and is available commercially. Figure 3.16(a) shows a blower door and Figure 3.16(b) shows

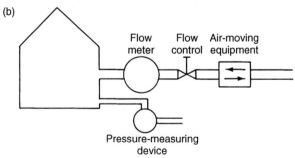

Figure 3.16 (a) Blower door and (b) steady-state airtightness test arrangement for a building.

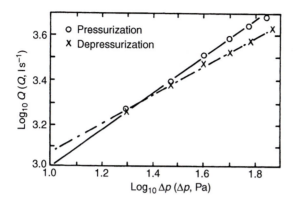

Figure 3.17 Typical air leakage log plot.

a schematic of a steady-state pressurization/depressurization testing arrangement. To reduce the influence of climatic conditions (i.e. internal–external temperature difference and fluctuating wind conditions), these tests are normally performed at a minimum pressure difference of 10 Pa and up to a maximum of 75 Pa. These pressures are far in excess of naturally produced pressures due to stack and wind effects. The results produced by these tests are often used for the purpose of comparing airtightness of different buildings and therefore air leakage rates are normally quoted at $\Delta p = 50$ Pa, although for large buildings such a large pressure may not easily be achieved and a value of $\Delta p = 25$ Pa is sometimes used. The technique usually involves replacing an external door (without necessarily removing the door from its hinges) with the blower-door to which a variable-speed fan and a flow measuring device are attached. A static pressure tapping on the inside surface of the panel is usually used to measure the internal–external pressure difference by connecting it to a manometer or a pressure transducer, the other end being exposed to the external pressure.

Because of the nature of air leakage paths in a building envelope the flow area in a pressurization test may be different from that in a depressurization test. This can be caused by windows and doors being forced against their frames under pressurization and released under depressurization. Furthermore the flow paths in the two directions may not be the same. For this reason the two tests are usually performed and the average leakage characteristics from the two tests are used. Typical pressurization and depressurization test results are shown in Figure 3.17. Because of its simplicity, this technique is also employed to evaluate the influence of refurbishment of the building envelope (such as window or door replacement, draughtproofing, etc.) and it is often used in conjunction with thermography or smoke tests to provide a qualitative assessment of the distribution of air leakage paths through the building envelope. In carrying out airtightness tests, attention should be given to seal large openings such as vents, flue stacks, etc. In buildings which are equipped with a mechanical ventilation system it is possible to employ the supply or extract fan to perform the pressurization or depressurization tests. However, care ought to be exercised to ensure that the ducting system which is not being used is completely sealed by shutting all dampers and sealing air terminal devices in that system.

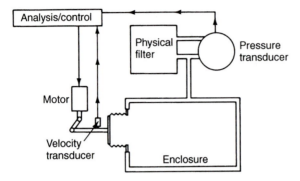

Figure 3.18 A dynamic (AC) airtightness test arrangement for a building.

Dynamic pressurization   To be performed accurately, steady-state pressurization/ depressurization tests require stable climatic conditions. They do not produce repeatable results under gusty weather conditions because of the superimposition of the fluctuating wind pressure on the fan pressure. As a result, steady-state pressurization/ depressurization tests are conducted at pressure differences $\Delta p > 10\,\text{Pa}$ which are considerably greater than those normally produced by climatic conditions (0–5 Pa). The calculation of the air leakage area of a building at low pressures (i.e. $\Delta p < 10\,\text{Pa}$) requires extrapolation of the measured data beyond the measurement range (typically $10 < \Delta p < 60\,\text{Pa}$), which causes uncertainties in the results. To overcome these difficulties a dynamic (AC) pressurization technique has been developed [25] which can be used to measure building airtightness at small pressure differences ($\approx 4\,\text{Pa}$). A schematic of a dynamic pressurization apparatus is shown in Figure 3.18. This apparatus uses a piston driven by a variable-speed electric motor to generate a sinusoidal change in the building's air volume. As a result a periodic pressure difference is created across the building envelope which can be distinguished from the naturally occurring random pressure fluctuations. The airtightness of a building envelope affects both the amplitude and phase of the pressure change in the building which is produced by the periodic change in the volume. If a building is completely airtight and rigid the pressure signal will be determined by the piston displacement and the volume of the building. Hence, any deviation from this predicted pressure can be attributed to air leakage through the building envelope. The measured changes in volume and pressure response are then used to calculate the air leakage through the envelope. A typical frequency range for this device is 0.01–10 Hz and a volume displacement of 1–200 l will be suitable for residential buildings [25].

This technique has been used in several test houses and the results were compared with steady-state pressurization tests. The leakage areas obtained by the dynamic tests were always lower than those from the steady-state measurements. No confirmation of the accuracy of the two methods was given by Modera and Sherman [25] but it should be pointed out that the steady-state results were extrapolated to a pressure difference of 4 Pa to correspond with that for the dynamic tests. This extrapolation beyond the measuring range could have contributed to the difference in values.

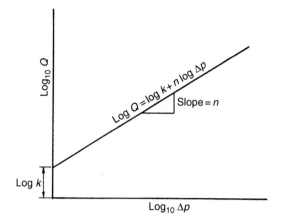

Figure 3.19 Presentation of pressurization/depressurization test results.

Presentation of airtightness measurements

In general, the crack flow equation (3.3) applies to both individual building components and building envelope air leakages, i.e.:

$$Q = k(\Delta p)^n$$

In a pressurization or depressurization test, values of the flow rate, Q, are measured at corresponding pressure differences, Δp. Hence a logarithmic plot of Q against Δp produces a straight line which has a slope of n and an intercept on the Q axis of log k, as shown in Figure 3.19. Such a plot will provide values of the air leakage characteristics of the building component or the envelope over the range of pressures examined.

The air leakage corresponding to a pressure difference of 50 Pa, Q_{50}, is significant for a building envelope as this value has been adopted as a standard for comparing the airtightness of building envelopes. For a building component a higher pressure is sometimes used such as 200 Pa.

Because a building envelope has numerous leakage paths of different shapes, sizes and flow characteristics, an equivalent leakage area of a building is sometimes quoted. This is the equivalent area of a sharp-edge orifice which produces the same flow rate as the building envelope under an equal pressure difference. It can be calculated using equation (3.24a) for a given reference pressure difference, Δp_{ref}. This area is sometimes used in air infiltration models and building airtightness standards for the calculation of air infiltration rates.

The equivalent leakage area, A_o, is also represented as a specific leakage area, A_s, or a normalized leakage area A_n, which are defined as follows:

$$A_s = A_o/\text{floor area}$$

$$A_n = A_o/\text{envelope area}$$

A description of the airtightness measuring standard is given in the AIVC Guide [20]. The accuracy of measuring the air flow corresponding to a pressure difference of ± 5 Pa is estimated to be about 5% by most of these standards.

An air leakage test using the methods described earlier does not provide a value of the expected air infiltration rate into the building because it is normally conducted for a pressure differential which is much greater than that present under normal temperature and wind conditions. The air leakage under existing prevailing conditions can be obtained using a tracer gas test as described in the following section. Most air tightness tests are carried out over a range of pressures, including the pressure of 50 Pa. Using data from a large number of tests on dwellings it has been found that the average infiltration rate is approximately equal to 1/20 of the air leakage corresponding to 50 Pa pressure difference, i.e. the infiltration rate expressed as air changes per hour (*ACH*) may be obtained using:

$$ACH = Q_{50}/(20\ V)\quad \text{ach}^{-1} \tag{3.48}$$

where Q_{50} is the measured air leakage at a pressure of 50 Pa ($m^3\,s^{-1}$) and V is the internal volume of the building (m^3). There is at present no formula similar to equation (3.48) which can be used for large buildings. However, work at the Building Research Establishment, UK [24] involving tracer gas measurements and airtighness tests on the same buildings, or a combination of measurements and predictions, has led to following relationship for large non-domestic buildings:

$$ACH = Q_{50}/(60\,S)\quad \text{ach}^{-1} \tag{3.49}$$

where S is the surface area of the external walls and roof (m^2). Equation (3.49) is suitable for large buildings with average standard of construction under typical weather conditions.

3.4.2 Measurement of air leakage rate

The direct measurement of the air leakage rate from a building requires the release and subsequent monitoring of the concentration levels of a tracer gas. Normally an inert or non-reactive gas which ideally has a density close to that of air is uniformly mixed in the space to be examined and the concentration of the gas is sampled over the monitoring period. A good tracer gas should be non-toxic, chemically stable, not absorbable or adsorbable by building materials and furnishings, and one that exists at low concentration in the atmosphere. Common tracer gases are nitrous oxide (N_2O), sulphurhexafluoride (SF_6), carbon dioxide (CO_2), methane (CH_4), ethane (C_2H_6) and helium (He). Methane and ethane are combustible gases and should be used in concentrations not exceeding 2000 p.p.m. and 1500 p.p.m. (parts per million by volume) respectively. However, a concentration of 10–1000 p.p.m. is considered satisfactory for most gases.

The rate of change in the concentration of a tracer gas is given by the amount of gas leaving the space *minus* the amount of gas entering the space *plus* that generated within the space. The general tracer mass balance equation (refer to Chapter 2) is:

$$V dc/dt = G + Q(c_e - c) \tag{3.50}$$

where V = effective volume of enclosure (m^3); Q = air volume flow rate through the enclosure ($m^3\,s^{-1}$); c_e = external concentration of tracer gas; c = internal concentration of tracer gas at time t and G = generation rate of tracer gas within the enclosure ($m^3 s^{-1}$). Equation (3.50) is the basic mass balance formula for tracer gas measurements. There are three methods of applying this equation to determine the air leakage rate in a building and these are considered here.

Concentration decay method

This is the most common of the three methods and requires the least sophisticated equipment to perform a test. Tracer gas is initially injected into the test space (e.g. from a gas cylinder) and the gas is allowed to mix with the air in the space, sometimes assisted by a fan or the air handling system in mechanically ventilated buildings. After proper mixing, the tracer concentration is monitored using a suitable detector (gas analyzer) over the required time period. Two techniques of monitoring the tracer decay with time are normally used: a site analysis approach in which the tracer gas analyzer is placed in the building being tested, and a 'grab sampling' approach which involves taking samples of air in the building and analysing them at a later time in a laboratory (see also Chapter 9).

In the grab sampling technique air samples are taken from the space being tested at regular intervals (a minimum of two) during a test using syringes, flexible bottles, air bags or detector tubes. These samples are analysed in a laboratory and the concentrations at different time intervals are determined. The main advantage of this method is that non-technical personnel can perform the site tasks, thus allowing trained laboratory staff to perform the gas analysis using more accurate equipment than that usually available on site.

The site analysis technique involves the sampling and analysis of tracer gas concentrations on site using a measurement set-up similar to that shown in Figure 3.20. Sample tubes which can be located at different positions in the building can be connected to the gas analyzer either individually through a selector box or altogether to a chamber connected to the analyzer. The former connection enables air leakage rates to be determined for different zones of a building whilst the latter is more suitable for measuring air leakage in a large space in which the tracer concentration may not be uniformly mixed. There are two main categories of gas analysis equipment that are frequently used: infrared absorption spectroscopy and gas chromatography, see Section 9.10 for further details. These usually produce a digital or analogue readout and an analogue signal in the form of a voltage which is proportional to the tracer gas concentration. The voltage output can be connected to a chart recorder or a data logger or a microcomputer with a printer/plotter to produce a logarithmic plot of concentration with time. A typical plot is shown in Figure 3.21 from which the air change rate for the space can be determined directly as shown here.

Following the cessation of gas injection, assuming that no incidental sources of the tracer gas are present within the building and neglecting the tracer concentration in the external air, equation (3.50) reduces to:

$$V dc/dt = -Qc \tag{3.51}$$

Rearranging this equation and integrating, assuming a constant air flow rate, Q, we find:

$$\int_{c_0}^{c} \frac{dc}{c} = -\frac{Q}{V} \int_{0}^{t} dt$$

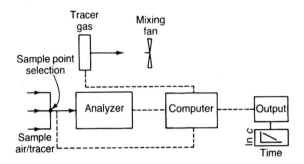

Figure 3.20 Tracer decay site analysis scheme.

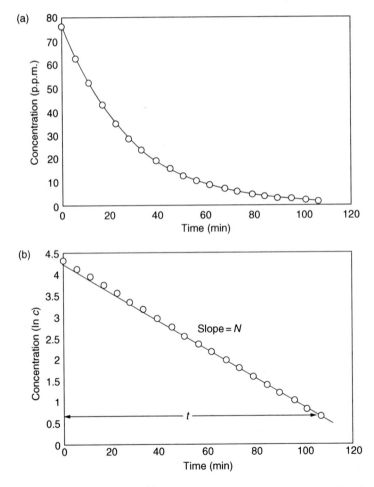

Figure 3.21 Typical tracer gas decay rate measurement in a ventilated room. (a) Measured tracer decay and (b) logarithmic plot of (a).

hence:

$$\ln c - \ln c_0 = -(Q/V)t \tag{3.52}$$

where c_0 = the concentration of tracer gas at time $t = 0$ and $Q/V = N$ = air change rate per unit time. Alternatively equation (3.52) can be written as:

$$c = c_0 \exp[-(Q/V)t] \tag{3.53}$$

From equation (3.53) it is clear that a plot of $\ln c$ with t produces a straight line of (negative) slope Q/V which is the air change rate during the measurement period.

Constant tracer injection method

A constant tracer gas injection rate into the test space also provides a method of air infiltration measurement in the space. The building is initially charged with tracer gas and then the injection rate is set to a constant value that produces an easily measurable concentration. A constant injection rate can be achieved by using a gas cylinder with a needle valve that can be set to give the desired flow rate, by using a passive permeation source which emits tracer at essentially a constant rate, or by utilizing a leak-proof gas bag equipped with a small pump. In this method it is necessary to monitor both injection rate and concentration in the space. The measurement of tracer concentration is similar to the decay method in which case either a site analysis or a grab sample approach is used. For site analysis an apparatus such as that illustrated in Figure 3.22 is used which is equipped with both injection and concentration monitoring devices. In this case a microcomputer is used to control the injection rate and perform the concentration analysis. The theory of this method which is also based on the tracer mass balance equation (3.50) is given below.

Neglecting external tracer concentration, equation (3.50) reduces to:

$$V dc/dt = G - Qc \tag{3.54}$$

A solution of this equation for tracer concentration in the test space gives:

$$c = G/Q + (c_0 - G/Q) \exp[-(Q/V)t] \tag{3.55}$$

Assuming a constant air infiltration rate, Q, and if no tracer is present before the test, equation (3.55) becomes:

$$c = (G/Q)[1 - \exp(-Nt)] \tag{3.56}$$

where $N = Q/V$ = air change rate per unit time.

For a constant value of N a finite time is required to achieve a constant concentration in the test space determined by the value of the exponential function in the square brackets in equation (3.56). Plots of this function with time are shown in Figure 3.23 for different air change rates N. Once the concentration reaches equilibrium the air flow rate will be given by:

$$Q = \overline{G}/c \tag{3.57}$$

where $\overline{G}$ = the steady rate of tracer injection ($m^3\ s^{-1}$).

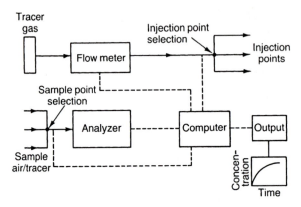

Figure 3.22 Constant injection site analysis scheme.

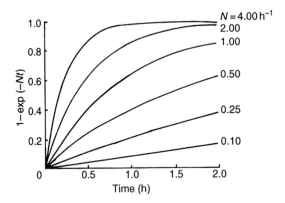

Figure 3.23 Variation of the exponential function $[1 - \exp(-Nt)]$ in equation (3.56) with time for different air change rates.

This equation shows that by achieving a measurable constant concentration rate, c, in the space the air infiltration rate into a building can be evaluated. In practice, a constant concentration rate can only be achieved if the flow rate, Q, is constant which may not always be the case. For this reason concentration samples are usually taken over a sampling period (typically 15 min) and averaged to obtain the mean concentration.

Constant concentration method

Conceptually, this is the simplest of the three methods used for air infiltration measurement but to apply it in practice requires expertise and advanced equipment. The method relies on maintaining a constant level of concentration in the test space throughout the monitoring period. This requirement simplifies the tracer mass balance equation (3.54) to:

$$G - Q\bar{c} = 0$$

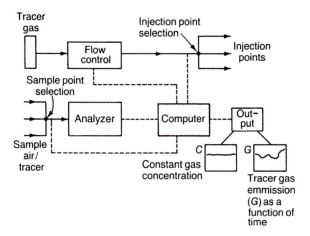

Figure 3.24 Constant tracer concentration scheme.

i.e.:

$$Q = G/\bar{c} \tag{3.58}$$

where $\bar{c}$ = time-mean concentration, p.p.m.

At a first glance, it would appear that equation (3.58) is the same as equation (3.57), however the difference between them is that in equation (3.57) a constant concentration is maintained by varying the gas injection rate whereas in equation (3.57) a constant injection rate is used. In equation (3.58) the air flow rate is directly proportional to the tracer gas injection rate required to maintain a given concentration, $\bar{c}$.

To implement the constant concentration concept in practice, it is necessary to use an apparatus that can control the rate of gas injection, G, to maintain a given level of concentration throughout the whole enclosure. This requires a periodic sampling of the air on site and control of the gas injection rate to achieve a constant concentration. A scheme such as that illustrated in Figure 3.24 can be used for this purpose. A major advantage of this scheme, if fully automated, is that it can be left unattended in the test building to gather data which can then be stored on a computer disk for analysis at a later stage or sent directly to a remote computer via a standard telephone modem. This method can also be used to measure the air change rate of individual zones in a building and, if short time intervals are used between the air sampling and gas injection, it is possible to obtain a continuous measurement of the air change rate in the zones.

Further details of air infiltration measurement procedures, test equipment and their characteristics can be found in [20] and [26]. Most air infiltration rate standards specify a range of 5–20% as the accuracy for measuring the air flow rates using the methods described in this section.

3.4.3 *Measurement of interzonal air leakage*

The air leakage measurement techniques described in the previous section relate to a single, well-mixed enclosure and are adequate for determining the rate of outside air ingress required by the occupants and for sizing heating or cooling plants. In recent years there has been concerted efforts nationally and internationally to reduce indoor air pollution because of its associated health effects. To study the indoor air quality in a building requires the evaluation of outside air leakage into each zone as well as the leakage between internal zones of the building. At domestic level for example, a measure of the rate of water vapour transport from the kitchen and bathroom into other rooms is needed to assess the risk of surface condensation in colder zones such as bedrooms and unventilated areas of the dwelling. Other situations where knowledge of interzonal flows is needed may be in industrial buildings where pollution generated by processes is significant, in hospital buildings where germs may be present in certain zones, etc. In such cases interzonal air flow evaluation is required to assess the risks and provide means of isolating the polluting zones.

The measurement of interzonal flows and individual zone air change rates can also be made using the tracer gas methods described earlier. Depending on the design of the building and how it is used, it is divided into a number of zones or cells and the flow rate through each cell is measured separately using a well-mixed tracer gas inside the cell. The exchange of tracer in the cell occurs with the external environment as well as with adjacent cells. Therefore, the concentration mass balance equation must include all the cells in the building. Neglecting the tracer concentration in the atmosphere and assuming n cells, the concentration equation for the ith cell ($i = 1$ to n) is given by:

$$V_i \frac{dc_i}{dt} = G_i + \left[\sum_{j=1}^{n} Q_{ij} c_j (1 - \delta_{ij}) \right] - \left[Q_{ie} c_i + \sum_{j=1}^{n} Q_{ij} c_i (1 - \delta_{ij}) \right] \tag{3.59}$$

where V_i = effective volume of cell i (m^3); c_i = tracer gas concentration in cell i at time t (p.p.m.); c_j = tracer gas concentration in cell j ($j = 1$ to n) at time t (p.p.m.); G_i = generation rate of tracer gas within cell i (cm^3 s^{-1}); Q_{ij}, Q_{ji} = volume of air flow rate between cells i and j and j and i respectively where in general $Q_{ij} \neq Q_{ji}$ (m^3 s^{-1}); Q_{ie} = volume flow rate from cell i to the external environment (m^3 s^{-1}) and δ_{ij} = Kronecker delta which is 0 when $i \neq j$ and 1 when $i = j$. The expression (3.59) represents a set of n equations, but the number of unknown flow rates involved in these equations is $n(n + 1)$ as each of $(n + 1)$ zones (including the external environment) exchanges air with the other n zones. Thus to solve for all the flow rates a total of $n(n + 1)$ equations is needed. Since continuity of air flow exists for each cell (i.e. the net inflow and outflow must be zero) a set of n continuity equations can be obtained for n cells. These are given by the expression:

$$Q_{ei} + \sum_{j=1}^{n} Q_{ji}(1 - \delta_{ij}) = Q_{ie} + \sum_{j=1}^{n} Q_{ij}(1 - \delta_{ij}) \tag{3.60}$$

where Q_{ei} = volume flow rate from the external environment to cell i (m^3 s^{-1}).

If equations (3.59) and (3.60) are used (i.e. total of $2n$ equations), a further $(n^2 - n)$ equations will be needed to solve for all the flow rates, Q_{ij}. These additional equations can be generated by writing $(n - 1)$ independent sets of equations similar to equation (3.59). Sinden [27] suggested applying equation (3.59) to n different intervals of the concentration decay curve for each cell. This will produce n^2 equations which, together with n air flow equations (equation (3.60)), would yield the necessary $n(n+1)$ equations. In practice, the method may be applied by performing one experiment in which the concentration decays are measured in *n* cells simultaneously. Because of the arbitrary nature of choosing the *n* different intervals on the decay curve of each cell, it has been shown [28] that this approach can produce unrealistic values of the flow rate in the cells.

However, there are other techniques which may be used to determine interzonal flows, such as the following:

1 The use of constant tracer injection in a cell and measuring the tracer concentration in each cell, then repeating the measurements for *n* different injection rates. This will produce n^2 concentration equations which can be solved together with the air flow equation to give the flow rate in each cell.
2 Injecting a tracer into one cell and measuring the concentration decay in all the cells. The procedure is repeated by introducing the tracer in each of the other $(n-1)$ cells. This produces n^2 concentration equations which can be solved with the additional *n* flow equations.
3 Introducing *n* different non-reacting tracer gases simultaneously, one in each cell. The ensuing concentration evolutions of all the tracer gases are then recorded in each cell giving n^2 concentration equations. Any of the three methods described in Section 3.4.2 may be used to carry out the measurements in each cell.

There are advantages and disadvantages associated with each of the three methods. Methods 1 and 2 require the use of one tracer gas only, i.e. one gas analyzer can be used, and this makes them less expensive than the multi-tracer method 3 which requires *n* gas analyzers. However, the time required to perform the tests in methods 1 and 2 is considerably greater than that needed for method 3. This could introduce considerable errors in the measurements particularly if changes in the external and/or internal conditions occur during the monitoring periods. More detail on the measurement techniques using the three methods can be found in [20] and [28] but the concentration measuring techniques in each cell are similar to those described in Section 3.4.2.

References

1. Etheridge, D. W. (1977) Crack flow equations and scale effect. *Build. Environ.*, **12**, 181–9.
2. BS 5925 (1991) Code of practice for ventilation principles and designing for natural ventilation, British Standards Institution, London.
3. *ASHRAE Handbook of Fundamental* (2001) Ch. 26: Ventilation and infiltration, American Society of Heating, Refrigeration and Air-Conditioning Engineers, Atlanta, GA.
4. Thomas, D. A. and Dick, J. B. (1953) Air infiltration through gaps around windows. *JIHVE*, **21**, 85–97.
5. Baker, P. H., Sharples, S. and Ward, I. C. (1987) Air flow through cracks. *Build. Environ.*, **22**, 293–304.
6. Liddament, M. W. (1987) Power law rules – OK? *Air Infiltration Rev.* 8 (2), 4–6.
7. BS 6399-2 (1997) Ch. V – Loading for buildings: Code of practice for wind loads, British Standards Institution, London.
8. Liddament, M. W. (1986) Air infiltration calculation techniques – an application guide, *Air Infiltration and Ventilation Centre*, International Network for Information on Ventilation, Brussels, Belgium.
9. Sherman, M. H. and Grimsrud, D. T. (1980) Infiltration-pressurization correlation: simplified physical modeling. *ASHRAE Trans.*, **86** (2), 778–807.

10. ESDU Data Item No. 82026 (1990) Strong winds in the atmospheric boundary layer – Part 1: Mean hourly windspeeds, Engineering Sciences Data Unit International, London.
11. Sherman, M. H. and Grimsrud, D. T. (1980) Measurement of infiltration using fan pressurization and weather data, *Proc. 1st AIC Conf on Air Infiltration Instrumentation and Measuring Techniques*, 6–8 October 1980, Windsor, England.
12. Walker, I. S. and Wilson, D. J. (1993) Evaluating models for superposition of wind and stack effect in air infiltration, *Build. Environ.*, **28**, 201–10.
13. Liddament, M. and Allen, C. (1983) The validation and comparison of mathematical models of air infiltration. *Tech*. Note AIC 11, *Air Infiltration and Ventilation Centre*, International Network for Information on Ventilation, Brussels, Belgium (www.aivc.org).
14. Warren, P. R. and Webb, B. C. (1980) The relationship between tracer gas and pressurisation techniques in dwellings, *Proc. 1st AIC Conf. on Air Infiltration Instrumentation and Measuring Techniques*, 6–8 October 1980, Windsor, England.
15. Liddament, M. (1989) AIDA – An air infiltration development algorithm, *Air Infiltration Rev.*, **11** (1), 10–12.
16. Liddament, M. (1996) A guide to energy efficient ventilation, *Air Infiltration and Ventilation Centre*, International Network for Information on Ventilation, Brussels, Belgium (www.aivc.org).
17. Allard, F. (ed.) (1997) *Natural Ventilation in Buildings: A Design Handbook*, James and James (Science Publishers), London.
18. Feustel, H. E. and Smith, B. (eds.) (1997) *COMIS 3.0 User Guide*, Lawrence Berkeley Laboratory, Berkeley, California, USA.
19. Dols, W. S., Walton, G. N. and Denton, K. R. (2000) *CONTAMW 1.0 User Manual*, National Institute of Standards and Technology, Gaithersburg, MD, USA.
20. Charlesworth, P. S. (1988) Air exchange rate and airtightness measurement techniques – an applications guide, *Air Infiltration and Ventilation Centre*, International Network for Information on Ventilation, Brussels, Belgium (www.aivc.org).
21. ASTM Standard E783–93 (1993) *Test method for field measurement of air leakage through installed exterior windows and doors*, American Society for Testing and Materials, Philadelphia, PA, USA.
22. Thorogood, R. P. (1979) Resistance to air flow through external walls. *BRE IP 14/79*, Building Research Establishment, UK.
23. ASTM Standard E779–99 (1999) *Measuring air leakage by the fan pressurization method*, American Society for Testing and Materials, Philadelphia, PA, USA.
24. CIBSE TM 23 (2000) *Testing buildings for air leakage*, TM 23, Chartered Institution of Building Services Engineers, London.
25. Modera, M. P. and Sherman, M. H. (1985) AC pressurization: A technique for measuring the leakage area in residential buildings. *ASHRAE Trans.* **91** (2B), 120–32.
26. ASTM Standard E741–95 (1995) *Standard test method for determining air change in a single zone by means of gas dilution*, American Society for Testing and Materials, Philadelphia, PA, USA.
27. Sinden, F. W. (1978) Multi-chamber theory of air infiltration. *Build. Environ.*, **13**, 21–8.
28. Afonso, C. F. A., Maldonado, E. A. B. and Skåret, E. (1986) A single tracer-gas method to characterize multi-room air exchanges. *Energy Build.* **9**, 273–80.

4 Principles of air jets and plumes

4.1 Introduction

The jet of a fluid has been extensively studied for its numerous occurrences in engineering systems involving flow through an opening. The flow of a jet differs from other kinds of fluid flow because a jet is surrounded on one or more sides by a free boundary of the same fluid. The interaction between the flow within the jet and the boundary, referred to as entrainment, has a major influence on the development of the flow in the jet as will be discussed later in this chapter.

In most mechanically ventilated buildings, air jets are used to mix the processed air from the plant with the room air. This type of ventilation is usually referred to as 'mixing ventilation' as opposed to 'displacement ventilation' where the processed air displaces the room air. In mixing ventilation, the air jet is the main distribution medium of thermal energy, moisture and fresh air into a room. It is therefore vitally important that the jet is effectively mixed with the room air before the air extraction point in the room is reached. Failure to do so may render some regions of the room improperly ventilated with unacceptable temperature gradients and draughts. In addition, 'short-circuiting' of the supply air towards the extraction point may occur, causing not only discomfort to the occupants but also an increased energy consumption of the air processing plant.

Another important flow that has major influence on room air movement is the flow of 'plumes'. A plume is produced by hot or cold source in a room as opposed to a boundary layer, which is the flow close to hot/cold surfaces. The effect of plumes on room air movement is particularly significant in displacement ventilation in which case the flow momentum is small and plumes are the dominant forces that influence the room flow.

In this chapter the theory of air jets and the equations governing the flow in different types of air jet are presented. In ventilation studies, it is very seldom that the jet is isothermal and therefore the effect of buoyancy forces on the flow of the jet is also discussed and, where available, data are presented for use in air distribution design. Plume theory is treated after the jet theory is introduced.

4.2 Free air jet

A free air jet is a term used to describe a flow of air issuing from an opening or a nozzle into an air space where there are no solid boundaries to influence the flow pattern and where the static pressure within the jet is the same as the static pressure of the surrounding space. As the jet leaves the opening, a shear layer develops around

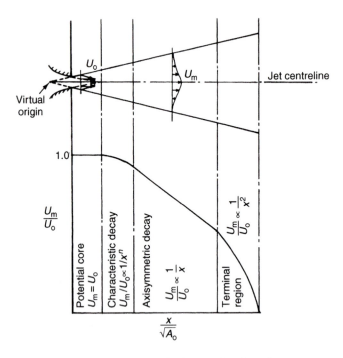

Figure 4.1 The four regions of a free jet (log–log plot).

its boundary as a result of the velocity discontinuity at this boundary. This is usually referred to as the 'free shear layer'. The thickness of the free shear layer increases with axial distance until the central region, usually called the 'potential core region', is completely consumed. Downstream of the core the flow becomes more turbulent and the centreline velocity decreases with distance. Depending on how the centreline velocity of the jet varies with distance from the opening, four zones may be identified for a free jet [1] (Figure 4.1).

Potential core region   This is the region immediately downstream of the supply opening where mixing of the jet fluid with the surrounding fluid is not complete. The length of the core depends on the type of opening and the turbulence of the air supply but usually it extends to 5–10 equivalent opening diameters. In this region the centreline velocity, U_m, is constant and equal to the supply velocity, U_o.

Characteristic decay region   After the consumption of the potential core by the free shear layer the centreline velocity begins to gradually decrease so that:

$$U_m/U_o \propto 1/x^n \tag{4.1}$$

where x is the distance from the supply and n is an index which has a value between 0.33 and 1.0. The extent of this region and the value of n depend on the shape of the supply opening and it is usually associated with large aspect ratio (ratio of length to height) openings, i.e. it is negligible for circular or square openings.

Axisymmetric decay region   This is a region dominated by a highly turbulent flow generated by viscous shear at the edge of the shear layer. For three-dimensional jets this is usually referred to as the 'fully developed flow region' where the spread angle of the jet is a constant whose value depends on the geometry of the opening. It is the predominant region for a jet discharging from a low aspect ratio opening where it extends to about 100 equivalent diameters. It is not so significant for high aspect ratio openings. The centreline velocity here decreases inversely with the distance from the opening, i.e.:

$$U_m/U_o \propto 1/x \tag{4.2}$$

Terminal region   This is a region of rapid diffusion and the jet becomes indistinguishable from the surrounding air. The centreline velocity decays with the square of the distance, i.e.:

$$U_m/U_o \propto 1/x^2 \tag{4.3}$$

A free jet may be produced by a circular, cylindrical, rectangular or infinitely long opening discharging air into a large enclosure containing a stagnant mass of air. The extent of each of these four regions will primarily depend on the geometry of the opening and, to a lesser extent, on the turbulence characteristics at the opening.

In practice, the potential core and the characteristic decay regions are dominant regions in a plane jet (two-dimensional free jet) and the potential core and the axisymmetric decay regions are the main regions in an axisymmetric jet (three-dimensional free jet). As the aspect ratio of the opening approaches ∞, the axisymmetric decay region diminishes and as it approaches 1, the characteristic decay region almost disappears.

To investigate the flow in the intermediate regions of a free jet experimental observations have shown [2] that the following principles can be applied to these regions.

1   The total momentum across any section of the jet, M_x, remains constant and it is equal to the initial momentum at the supply opening, M_o, thus:

$$M_x = M_o = \int_0^A \rho u^2 \, dA \tag{4.4}$$

$$M_o = \rho U_o^2 A_o \tag{4.5}$$

where u is the velocity component in the x direction for an element of area dA, ρ is the density of fluid, U_o is the supply velocity and A_o the effective area of the supply opening.

2   The profile of the u component of velocity across the jet has the same shape at different axial distances from the supply opening.

3   The static pressure across the jet is constant and equal to the surrounding pressure.

Applying these assumptions and a suitable turbulence model it is possible to obtain solutions to the flow in the intermediate regions of a free jet [2]. These solutions include expressions for the velocity profiles, the decay of centreline velocity, flow entrainment, etc.

4.2.1   Circular jet

A jet discharging from a circular opening is axisymmetric and without a characteristic decay region. The major part of the jet is the axisymmetric region and therefore the solution of the flow equations in this region only will be discussed. The Navier–Stokes equation for the flow in the x direction, in cylindrical coordinates, is [2]:

$$u\frac{\partial u}{\partial x} + v\frac{\partial u}{\partial r} = -\frac{1}{\rho}\frac{\partial p}{\partial x} + v\left(\frac{\partial^2 u}{\partial x^2} + \frac{1}{r}\frac{\partial u}{\partial r} + \frac{\partial^2 u}{\partial r^2}\right) - \left[\frac{\partial}{\partial x}(\overline{u'^2}) + \frac{\partial}{\partial r}(\overline{u'v'}) + \frac{\overline{u'v'}}{r}\right]$$

(4.6)

where u and v are the time-mean velocity components in the x and r directions and the superscript $(')$ indicates the fluctuating components (Figure 4.2). Using the same notation, the continuity equation is given by:

$$\frac{\partial}{\partial x}(ru) + \frac{\partial}{\partial r}(rv) = 0$$

(4.7)

For a high Reynolds number jet (i.e. $Re = 2r_0U_0/v > 3 \times 10^4$) the viscous stresses in the developed region are much smaller than the corresponding turbulent stresses and the second term on the right-hand side of equation (4.6) is neglected. Furthermore, if an isotropic turbulence is assumed so that:

$$\frac{\partial u'}{\partial x} \approx \frac{\partial v'}{\partial r}$$

and the pressure gradient term is neglected, equation (4.6) may be simplified to:

$$u\frac{\partial u}{\partial x} + v\frac{\partial u}{\partial r} = -\frac{1}{r}\frac{\partial}{\partial r}(r\overline{u'v'})$$

(4.8)

Substituting for the Reynolds shear stress term by:

$$\tau_t = -\rho\overline{u'v'}$$

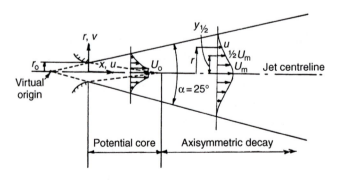

Figure 4.2 Schematic of a free circular jet.

the equation becomes:

$$u\frac{\partial u}{\partial x} + v\frac{\partial u}{\partial r} = \frac{1}{\rho r}\frac{\partial}{\partial r}(r\tau_t) \tag{4.9}$$

By integrating equation (4.9) from $r = 0$ to ∞ and noting that the turbulent shear stress at the two limits of integration is zero, it can be shown that:

$$\frac{d}{dx}\int_0^\infty 2\pi r d(\rho r u^2) = 0$$

This equation shows that the rate of change of x-momentum is zero or the momentum in the x-direction remains constant (cf. equation (4.4)).

Applying equations (4.7) and (4.9) and a suitable expression for the turbulent shear stress, τ_t, it is possible to develop equations for the velocity profile across the jet and for the decay of centreline velocity with distance from the supply opening. The expression for the turbulent shear stress is:

$$\tau_t = \mu_t \frac{\partial u}{\partial r}$$

where μ_t is the coefficient of turbulent viscosity. The first known solution of the above equations is that due to Tollmien [3] who used Prandtl's mixing length hypothesis to obtain an expression for μ_t, thus:

$$\mu_t = -\rho L_m^2 \left|\frac{\partial u}{\partial r}\right|$$

where L_m is called the mixing length. Further details of this model and other turbulence models are given in Chapter 8. Schlichting [4] used a Goertler-type solution, which assumes a value for μ_t directly proportional to the centreline velocity of the jet. The velocity profiles predicted by both solutions are compared with experimental results in Figure 4.3, where η is the ratio of the radius of the jet at a point, r, to the radius, $r_{1/2}$, where $u = 0.5\,U_m$. Whereas the Goertler-type solution shows a better agreement with measured velocities near the jet centreline, the Tollmien solution gives a superior prediction near the outer boundary. For a free circular jet, the value of $r_{1/2}$ is given by:

$$r_{1/2} = 0.1\,x \tag{4.10}$$

The two solutions produce centreline velocity decay equations of the form:

$$U_m/U_o = K_v/(x/d_o) \tag{4.11}$$

where K_v is a constant usually referred to as the throw constant and d_o is the effective diameter of the supply opening equal to $2r_o$. The value of K_v for the Tollmien solution is 7.32 and for the Goertler solution is 5.75, but Rajaratnam [2] recommends a value of 6.3. Using extensive experimental data for different free axisymmetric jets Baturin [5] obtained the velocity decay equation given below:

$$\frac{U_m}{U_o} = \frac{0.48}{(ax/d_o + 0.145)} \tag{4.12}$$

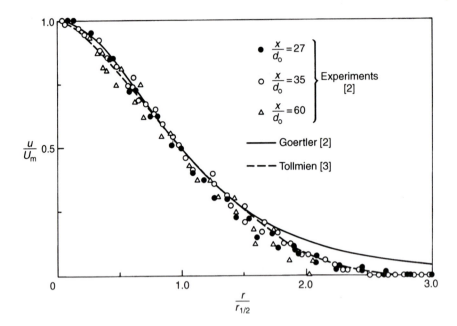

Figure 4.3 Non-dimensional velocity profile across a circular jet.

Table 4.1 Circular jet constant and spread angle [5]

Type of opening	a	α (degree)
Convergent nozzle	0.066–0.071	25–27
Cylindrical tube	0.076–0.08	29
Axial fan with guide vanes	0.12	44
90° angle with guide vanes	0.2	68
Cased axial fan with open grille	0.24	79
Swirl diffuser with eight vanes at 45° to jet centreline	0.27	85

where *a* is a constant whose value depends on the type of supply opening. The larger the value of *a* the faster is the decay of centreline velocity. Values of *a* for different openings are given in Table 4.1. Equations (4.11) and (4.12) are plotted on Figure 4.4 and Baturin's formula appears to produce results lying between those of Tollmien's and Goertler's lines.

Table 4.1 also gives the included angle of spread, α, for each jet, which represents the angle of the outer envelope of the jet where the velocity is zero. This angle is larger for jets having higher turbulence and swirl at the opening.

Experimental measurements by Ricou and Spalding [6] produced the expression given below for the flow entrainment of a circular jet:

$$Q/Q_o = 0.32(x/d_o) \qquad (4.13)$$

where Q is the volume flow rate at distance x and Q_o is the supply flow rate. The entrainment velocity (i.e. the vertical component at the jet boundary) can be

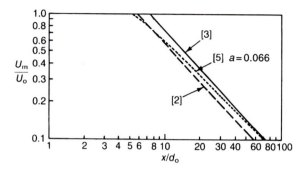

Figure 4.4 Centreline velocity decay for a circular jet.

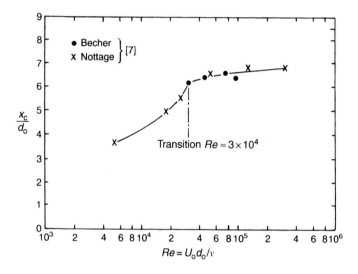

Figure 4.5 Effect of Reynolds number on core length for a free circular jet.

expressed by:

$$V_e = C_e U_m \tag{4.14}$$

where C_e is the entrainment coefficient which has a value of 0.026 for a circular jet [2].

The length of the core region is influenced by the Reynolds number and the turbulence intensity at the outlet. Figure 4.5 shows the effect of Re on the core length of a free circular jet based on the experimental results of Becher and Nottage given in [7]. The results suggest a critical Reynolds number of about 3×10^4. It is evident that for $Re > 3 \times 10^4$ the core length is almost independent of the value of Re. The same conclusions can also be made regarding the throw constant K_v. Available data, though insufficient, seems to suggest that the core length decreases with turbulence intensity for Reynolds numbers below the critical value but is unaffected for higher Reynolds numbers.

4.2.2 *Plane jet*

A plane jet is a free jet produced by an infinitely long rectangular slot so that the only lateral change in the properties of the jet occurs in a plane normal to the slot length. In practice, a jet issuing from a slot of aspect ratio, b/h, greater than 40 can be approximated by a plane jet, as variations in the direction of the slot length are negligible. The flow of a plane jet, which can be treated as two-dimensional, consists of a core region and a fully developed region reminiscent of the characteristic decay region for a three-dimensional jet discussed in Section 4.2.4.

Assuming x and y to be the axes along the centreline, and normal to the centreline respectively, the velocity profiles of a plane jet at different positions from the opening are shown in Figure 4.6(a) and (b). The measured velocity profiles of Förthmann [8] for an opening of height 30 mm and length 650 mm (b/h = 21.7) are depicted in Figure 4.6(a) and the data for the fully developed region are presented in dimensionless form in Figure 4.6(b) where $y_{1/2}$ is the distance from the jet axis where $u = 0.5\,U_m$.

The flow equations, in Cartesian coordinates, for a plane jet with zero pressure gradient are:

$$u\frac{\partial u}{\partial x} + v\frac{\partial u}{\partial y} = \frac{1}{\rho}\frac{\partial \tau_t}{\partial y} \tag{4.15}$$

$$\frac{\partial u}{\partial x} + \frac{\partial v}{\partial y} = 0 \tag{4.16}$$

In a similar manner to the circular jet equations (4.9) and (4.7), the solutions due to Tollmien and Goertler of equations (4.15) and (4.16) produce velocity profiles in close agreement with experimental data. The dimensionless velocity profile exhibits a Gaussian error curve, which may be represented by the expression:

$$u/U_m = \exp(-0.693\eta^2) \tag{4.17}$$

where $\eta = y/(y_{1/2})$ and $y_{1/2} = 0.1\,x$, which is identical to equation (4.10) for a circular jet, i.e. the growth of the two jets is similar. The decay of the centreline velocity of a plane jet in the developed region is represented by:

$$U_m/U_o = K_v/\sqrt{x/h} \tag{4.18}$$

where K_v for the Tollmien solution is 2.67 and for the Goertler solution is 2.40. Analysing experimental data from various sources, Rajaratnam [2] recommends a value of 2.47 for K_v. He also obtained the following expressions for entrainment flow and entrainment velocity:

$$Q/Q_o = 0.62\sqrt{(x/h)} \tag{4.19}$$

$$V_e = 0.053\,U_m \tag{4.20}$$

4.2.3 *Radial jet*

A radial jet is produced by the radial flow of a fluid between two closely spaced discs into a stagnant fluid (Figure 4.7). The development of the jet flow in the r and y directions is similar to other jets and the differential equations are similar to those for

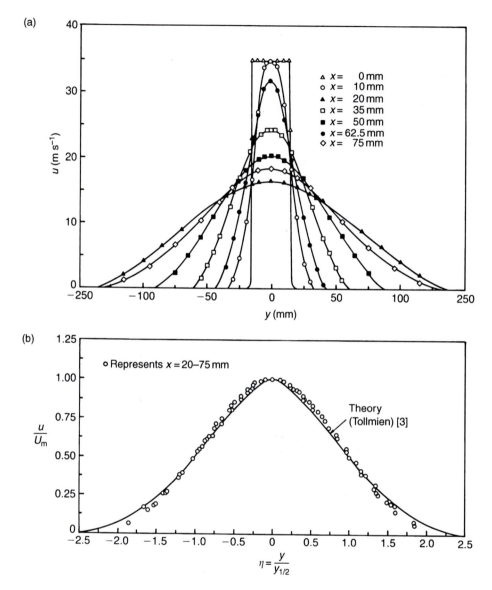

Figure 4.6 Velocity profiles for a plane jet [8]: (a) measured profiles and (b) non-dimensional profile.

a two-dimensional turbulent jet. The velocity profile in the fully developed region is the same as that for a plane jet. Applying Tollmien's solution and the experimental results of Heskestad [9], Rajaratnam [2] obtained the following velocity decay equation for a radial jet:

$$U_\mathrm{m}/U_\mathrm{o} = [2.47/(r/b)]\sqrt{(r_\mathrm{o}/b)} \tag{4.21}$$

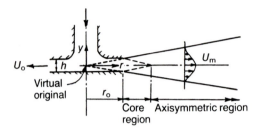

Figure 4.7 Schematic of a radial jet.

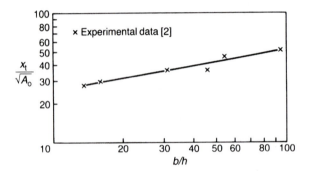

Figure 4.8 Transition point for rectangular jets ($13 < b/h < 100$).

where r_o is the radius of the disc, h the distance separating the two discs and r the radius from the centre of the disc. Becher [10] gives the equation below, which has been found to give similar results to equation (4.21):

$$U_m/U_o = 2.2\sqrt{\{r_o h/[r(r - r_o)]\}} \tag{4.22}$$

4.2.4   *Three-dimensional jet*

Jets that are neither plane nor circular exhibit three-dimensional flow characteristics, and are therefore referred to as three-dimensional jets. In practice, jets issuing from openings of aspect ratios $1 < b/h < 40$ usually fall in this category. Initially, the major axis decreases and the minor axis increases until they approach each other in width far downstream where the flow in the jet then tends toward axisymmetry. The velocity profiles in the core region in the two planes of symmetry are similar, but in the characteristic decay region only the profiles in the plane containing the minor axis are similar. For this reason algebraic solutions of three-dimensional jets do not produce sufficient information and empirical formulae are called for. In these jets the four regions discussed earlier (Figure 4.1) are present, but for most practical purposes the interest lies in the two middle regions. The length of the core and characteristic decay regions is dependent upon the aspect ratio of the supply opening. Using the data of Yevdjevich, given in Rajaratnam [2], it is shown in Figure 4.8 that the transition distance from the

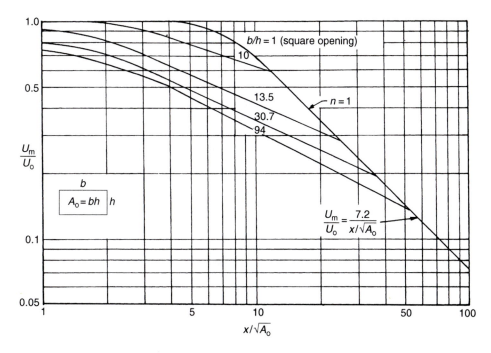

Figure 4.9 Decay of maximum velocity for free jets of different aspect ratios.

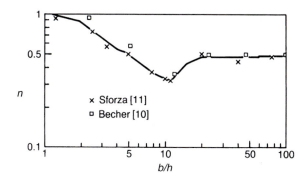

Figure 4.10 Effect of aspect ratio on characteristic region index.

characteristic decay region to the axisymmetric region, x_t, measured from the opening, increases with aspect ratio. These results can be used for $b/h > 13.5$, because for lower aspect ratios the size of the characteristic decay region becomes less significant.

The decay of the centreline velocity, U_m, with $x/\sqrt{A_o}$, where $A_o = bh$, is shown in Figure 4.9 for rectangular jets of various aspect ratios. The figure shows that the slope of the curves in the characteristic decay region depends on b/h whereas the slope in the axisymmetric region is the same for all aspect ratios and is -1. Figure 4.10 shows how the index n in equation (4.1) varies with b/h. For a plane jet $n = 0.5$ and for a square jet $n = 1.0$ (absence of characteristic decay) but between these limits n reaches a minimum value of 0.33 at $b/h = 10$, implying that the velocity decay for this aspect ratio is slower than that for a plane jet [11].

The velocity decay in the axisymmetric region can be expressed by:

$$U_m/U_o = K_v/(x/\sqrt{A_o}) \tag{4.23}$$

where K_v has a value 7.2 [2, 12].

4.2.5 Circular jet with swirl

By adding some swirl to a circular jet before it leaves the opening it is possible to achieve a rapid spread of the jet and faster velocity decay. These types of jet are now used extensively in floor air supplies where the room cooling load is large such as in data processing rooms. Beyond a certain critical value of swirl it is possible that the jet will attach to a surface, which is in the same plane as the opening. A solution of the flow equations shows that the sum of pressure and axial momentum, M, of the jet remains constant and also that the angular momentum, T, is preserved. Consequently, a swirl number is defined by:

$$S = T/(r_o M) \tag{4.24}$$

where r_o is the radius of the circular opening.

The experimental data of Chigier and Chervinsky [12] shows that the velocity profile in the axial direction can be approximated by:

$$u/U_m = \exp(-k\Lambda^2) \tag{4.25}$$

where Λ is the ratio of the radius to the axial distance from the virtual origin, r/x, and k is a constant whose value is dependent on the swirl number (Figure 4.11). The virtual origin for the jet is located at a distance of $2.3d_o$ behind the opening.

The decay of the axial velocity is expressed by:

$$U_m/U_o = C_u/(x/d_o) \tag{4.26}$$

This equation is the same as that for a circular jet except that C_u is dependent on the swirl number as shown in Figure 4.12.

The decay of the maximum angular velocity component, W_m, is expressed by:

$$W_m/W_o = C_w/(x/d_o)^2 \tag{4.27}$$

where W_o is the initial angular velocity and C_w is a constant that varies with S. Owing to the lack of experimental data there are no reliable values of C_w, except that values ranging from about 4 to 10.4 are quoted in Rajaratnam [2].

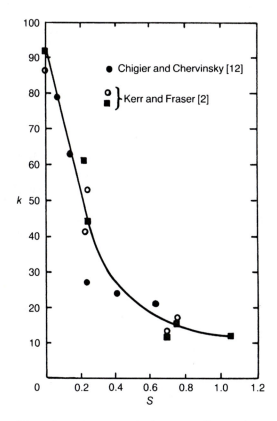

Figure 4.11 Variation of velocity profile constant with swirl number.

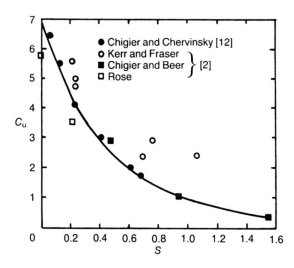

Figure 4.12 Variation of velocity decay constant with swirl number.

The flow entrainment rate is strongly influenced by the swirl and is expressed by:

$$Q/Q_o = (0.32 + 0.8S)x/d_o \tag{4.28}$$

where x is the distance from the opening and d_o is the opening diameter.

4.3 Wall jet

A wall jet is produced when the flow is bound by a flat surface on one side and the velocity at exit from the opening is parallel to the surface. After exit from the opening, the potential core is consumed at the point where the boundary layer growth on the surface meets the shear layer expansion on the free boundary. The flow downstream of the core then becomes fully developed. It is a common practice in mixing ventilation to use a wall jet discharging on the ceiling so that the high-velocity region is restricted to the ceiling thus freeing the occupied zone from draught. Under normal design conditions and due to the Coanda effect, the jet remains attached to the ceiling until the opposite wall is reached where it is then deflected downward into the occupied zone. If an infinite aspect ratio opening is used a two-dimensional or a plane wall jet is produced, otherwise the jet is three-dimensional. For a wall jet there are three main regions of maximum velocity decay, which are similar to those for a free jet, namely the potential core, the characteristic decay and the radial decay regions. The following sections describe some common types of wall jet and their characteristics.

4.3.1 Plane wall jet

A plane wall jet is one which issues from a slot of very large aspect ratio (i.e. $b/h > 40$) where any lateral change in the flow properties occurs in a plane normal to the slot length only. The only velocity decay regions present are the potential core and the characteristic decay regions (Figure 4.13).

The region from the wall to δ is known as the boundary layer and that above it is the free mixing region. The maximum velocity in the jet occurs at the point where the two regions meet. Experimental measurements [8] of the velocity profiles in the characteristic decay region, i.e. $x/h > 7$, show these profiles to be similar at all sections normal to the wall (Figure 4.14). The slot dimensions in this case were 650 mm × 30 mm ($b/h = 21.7$) but the slot was confined by walls on either side, in addition to the lower

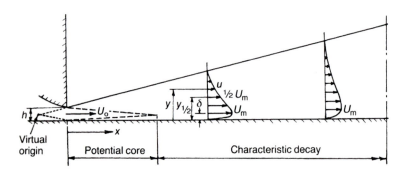

Figure 4.13 Schematic of a plane wall jet.

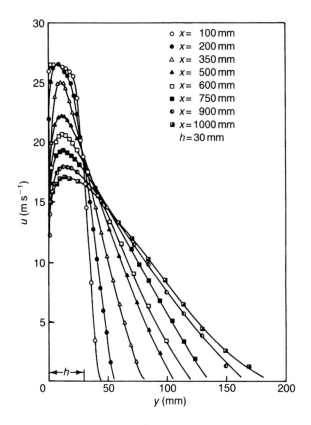

Figure 4.14 Velocity profiles across a plane wall jet [8].

surface, to produce a two-dimensional flow equivalent to $b/h \rightarrow \infty$. Rajaratnam [2] expressed the non-dimensional velocity profile by the empirical expression:

$$u/U_m = 1.48\eta^{1/7}[1 - \text{erf}(0.68\eta)] \tag{4.29}$$

where $\eta = y/y_{1/2}$ and $y_{1/2} = 0.068\,(x + 10h)$.

Schwarz and Cosart [13] obtained the expression:

$$u/U_m = \exp[-0.937(\eta - 0.14)^2] \tag{4.30}$$

These two expressions and the experimental results of Rajaratnam [2] are plotted in Figure 4.15 for comparison. Equation (4.29) gives better agreement with experimental data than equation (4.30), which predicts the velocity profile for $\eta \geq 0.14$.

The decay of the maximum jet velocity can be expressed by the equation below which has been derived from the experimental data of different sources [2]:

$$U_m/U_o = 3.50/\sqrt{(x/h)} \tag{4.31}$$

where x is the distance from the slot.

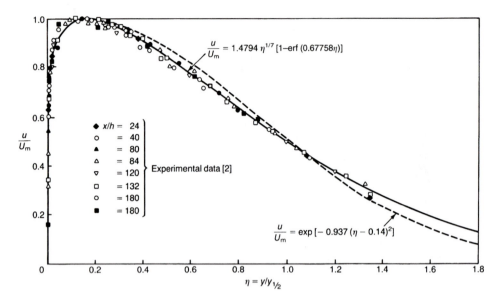

Figure 4.15 Non-dimensional velocity profile of a plane wall jet.

The entrainment flow for a plane wall jet is given by:

$$Q/Q_o = 0.248(x/h) \tag{4.32}$$

and the entrainment velocity is:

$$V_e = 0.035\, U_m \tag{4.33}$$

in which case the entrainment coefficient is 0.035 as compared with 0.053 for a plane jet (equation (4.20)). The growth of the boundary layer may be approximately expressed by [13]:

$$\delta/h = 0.068(x/h + 11.2) \tag{4.34}$$

and this was found always to be independent of the Reynolds number in the range 1.3×10^4 to 4.2×10^4 where Re is defined by $U_o h/\nu$.

4.3.2   Radial wall jet

Radial wall jets are produced by discharging a fluid between a circular disc and a flat surface, as shown in Figure 4.16(a), or by the impingement of a circular jet normal to the surface, as in Figure 4.16(b).

In the fully developed region of the wall jet in both cases, the shape of the velocity profile in the y direction is similar, and, as in other two-dimensional jets discussed earlier, a single non-dimensional profile can represent the velocity distribution. The decay of the maximum velocity in the radial direction can be represented by an expression which is similar to that for a free radial jet given by equation (4.21). For the jet

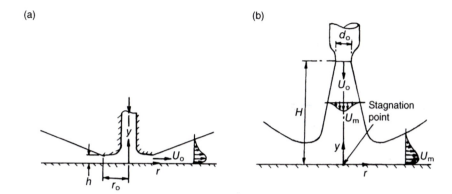

Figure 4.16 Schematic of a radial wall jet.

produced by a disc, experimental data [2] suggest the expression:

$$U_m/U_o = [3.5/(r/h)]\sqrt{(r_o/h)} \tag{4.35}$$

Becher [10], however, recommends the expression:

$$U_m/U_o = 3.1\sqrt{\{r_o h/[r(r-r_o)]\}} \tag{4.36}$$

These two equations produce very similar results.

4.3.3 *Impinging jet*

An impinging jet is the flow produced from an air supply opening situated at a distance from the surface with the flow directed towards the surface. The centerline of the opening may be either perpendicular or at an angle to the surface. In the case of a perpendicular outlet, the jet leaving the opening will impinge on the surface creating a radial wall jet. The flow field is divided into three regions: (I) a free jet region, (II) the impingement region and (III) the wall jet region as illustrated in Figure 4.17. There are also transitional zones between these regions.

The impingement of a turbulent jet on a flat plate (wall) has been widely studied with different configurations [14]. Beltaos and Rajaratnam [15] provide results for the impingement, free jet and wall jet regions. Their results suggest that the impingement of the jet is not affected by the presence of the plate over 75% of the distance between the nozzle and the plate and also the turbulent properties of the jet change from their equilibrium level close to the impingement region.

As the jet approaches the plate it will begin to 'feel' the presence of the plate at some distance from it. For example the mean velocity on the centreline is similar to a free jet up to some distance and then decreases (faster) to the zero value at the impingement point, see Figure 4.18 [14]. The distance from the plate where the centreline velocity starts to deviate from the free jet value is taken as the location of the end of the free jet region and the beginning of the impingement region. From the figure one can see that the effect of the plate is felt when the distance from the plate is less than 0.14h.

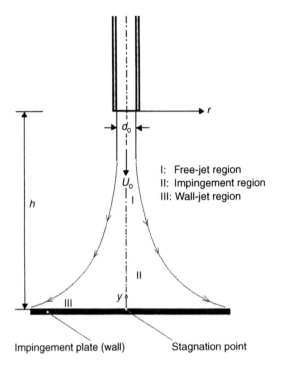

Figure 4.17 Impinging jet configuration.

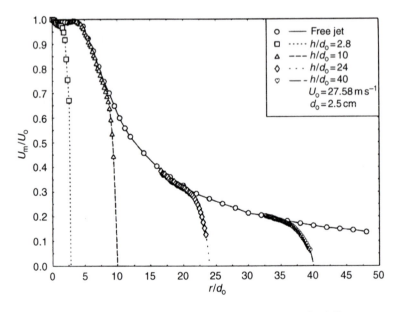

Figure 4.18 Decay of centreline velocity of an impinging jet with different impingement heights ($d_o = 25$ mm).

Becher [10] gives the expression below for the decay of the maximum velocity for an impinging jet:

$$U_m/U_o = 1.03/(r/d_o) \qquad (4.37)$$

where d_o is the diameter of the supply opening and r the radius measured from the jet centreline. It is interesting to note that the height of the nozzle above the surface, H, does not appear in equation (4.37). This equation is based on experimental data for values of h/d_o ranging from 8 to 24.

Based on measurements in a room, Karimipanah and Awbi [16] give the equation below for the decay of maximum velocity from the jet centreline:

$$U_m/U_o = 2.45/\left(r/\sqrt{A_o}\right)^{1.1} \qquad (4.38)$$

where A_o is the area of the supply opening. Equation (4.38) has an exponent of -1.1 whereas that of equation (4.37) is -1.0, which is the theoretical value for an ideal impinging jet. This difference in the value of the two exponents is due to friction losses and entrainment of room air by the jet and also the fact that the jet momentum can not be conserved in a confined enclosure, such as room.

4.3.4   Three-dimensional wall jet

The flow produced by a rectangular opening of aspect ratio $b/h < 40$ with one large side on a flat surface is three-dimensional and hence called a three-dimensional wall jet. This jet has three main regions, which are similar to the three-dimensional free jet discussed in Section 4.2.4. The potential core region extends from the opening to the point where the boundary layer meets the shear layer from the upper free boundary. The region starting from the end of the core to the point where the two shear layers from the two free sides meet is the characteristic decay region. The third region is usually

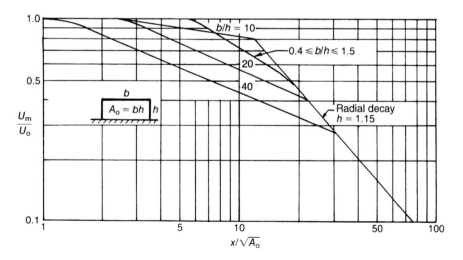

Figure 4.19 Maximum velocity decay for three-dimensional wall jets.

referred to as the radial-type decay region because the maximum velocity decays in a similar way to a radial wall jet.

In a symmetrical plane normal to the flat surface the velocity profiles are similar to a plane wall jet in both the characteristic decay and the radial-type regions [2, 11]. Similarity is also present in a plane parallel to the flat surface but only in the radial-type region. The decay of the maximum velocity for wall jets of different aspect ratios is shown in Figure 4.19. As in the case of three-dimensional free jets the velocity decay within and the length of the characteristic decay region are dependent on the aspect ratio of the opening. However, jets of all aspect ratios decay within the radial-type region in the same way with a decay index $n = 1.15$.

4.4   Effect of buoyancy

So far in this chapter the air jet has always been assumed to have the same temperature as the surrounding air, i.e. an isothermal jet. This condition is seldom realized in building ventilation and usually the temperature of the jet is either higher or lower than the surrounding air temperature depending on whether the jet is being used for heating or cooling the space, i.e. a non-isothermal jet. In this case, the diffusion of the jet will be influenced by the buoyancy forces as well as the inertia forces due to jet momentum.

The buoyancy force acting vertically on a unit volume of an air jet of temperature T and density ρ which are different from those of the surrounding fluid T_0 and ρ_0, is:

$$F = g(\rho_0 - \rho)$$

But for a gas $\rho \propto 1/T$. Therefore:

$$F = g\rho(\rho_0/\rho - 1) = g\rho(T/T_0 - 1) = g\rho\Delta T/T_0 \tag{4.39}$$

where $\Delta T = T - T_0$.

This force will produce a vertical acceleration $\dot{v}$ given by:

$$\dot{v} = F/\rho = g\Delta T/T_0 \tag{4.40}$$

If the vertical distance between this unit volume and a reference point is H, then the unit volume will move vertically by a velocity v relative to the reference point such that:

$$v^2 = 2H\dot{v}$$

Substituting for $\dot{v}$ from equation (4.40):

$$v^2 = 2gH\Delta T/T_0 \tag{4.41}$$

Equation (4.41) combines the inertia force and the buoyancy force acting vertically on a unit volume of a different temperature from the surroundings. The terms in this equation are usually combined in the form given below which represents a non-dimensional number of the buoyancy force to the inertia force usually referred to as the Archimedes number, Ar:

$$Ar = g\sqrt{A_0}\Delta T/(T_0 U_0^2) \tag{4.42}$$

where A_o is a reference area usually taken to be the effective area of the supply opening and U_o is the supply velocity. For a gas the cubic expansion coefficient $\beta \approx 1/T_o$, hence:

$$Ar = g\beta\sqrt{(A_o)}\Delta T/U_o^2 = Gr/(Re)^2 \tag{4.43}$$

where Gr is the Grashof number, $g\beta(A_o)^{1.5}\Delta T/\nu^2$.

The Archimedes number is analogous to the Froude number in hydraulics, which is normally defined as the square root of the ratio of the momentum force to the gravitational force acting on the fluid. The Archimedes number plays a significant part in determining the diffusion of a non-isothermal jet as will be shown below.

4.4.1 Non-isothermal horizontal jets

Defining a non-dimensional jet temperature decay ratio, θ_m, by:

$$\theta_m = (T_m - T_r)/(T_o - T_r) \tag{4.44}$$

where T_m is the maximum (or minimum) temperature across the jet, T_o is the temperature of the supply and T_r is a reference temperature (e.g. room temperature), four temperature decay regions can be identified, which are similar to the velocity decay regions discussed earlier in the chapter. However, the extents of these regions are different from those for the velocity decay and they usually occur before the corresponding velocity regions. Outside the core region the temperature profiles have the form of an error function but flatter than those for velocity because energy diffusion is more intensive than momentum diffusion. Figure 4.20 shows a normalized temperature profile across a plane jet obtained from the experimental results of [17]. In this case $y_{\theta/2}$ is the distance from the jet centreline where the temperature difference $\theta = 0.5\theta_m$ such that θ_m is given by equation (4.44) and θ is:

$$\theta = (T - T_r)/(T_o - T_r) \tag{4.45}$$

where T is the temperature at any point within the jet. The velocity profile is also depicted in Figure 4.20 and it is clear that the temperature profile is flatter than the velocity profile. In general the two profiles are related by the following equation:

$$u/U_m = (\theta/\theta_m)^2 \tag{4.46}$$

where η is based on the temperature profile i.e. $\eta = y/y_{\theta/2}$.

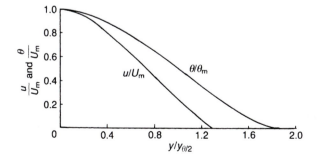

Figure 4.20 Non-dimensional temperature and velocity profiles for a plane jet [17].

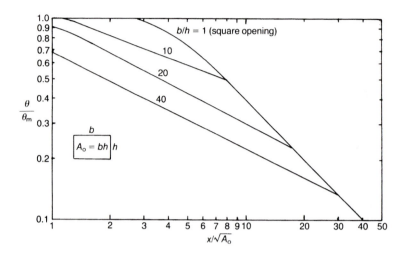

Figure 4.21 Decay of maximum temperature for free jets.

The decay of the maximum temperature θ_m with axial distance $x/\sqrt{A_o}$ is shown in Figure 4.21 for different aspect ratio free jets. These show some similarity with the velocity decays shown in Figure 4.9 except that they are steeper. For an axisymmetric jet the following expression due to Frean and Billington [7] may be used to calculate the temperature decay:

$$\theta_m = 8.1/\left(x/\sqrt{A_o}\right)^{1.25} \tag{4.47}$$

The velocity distribution in a non-isothermal jet is very similar to that in an isothermal jet, and to allow for the change in the diffusion characteristics of the jet due to buoyancy the axial distance may be corrected using the expression suggested by Laufer [18] which is:

$$\left(x/\sqrt{A_o}\right)_\theta = \sqrt{(\rho_m/\rho_r)}\left(x/\sqrt{A_o}\right) \tag{4.48}$$

where subscripts θ, m and r refer to temperature, maximum and reference, respectively. The equation can be used together with the isothermal velocity decay results presented earlier.

By describing the inertia and buoyancy forces acting on a non-isothermal jet, Koestel [19] derived an analytical representation of the path or trajectory of a horizontally projected circular jet. The drop of a cold jet or rise of a hot jet y (Figure 4.22) is given by:

$$y/d_o = [(a/b + 1)/(6K_v)](x/d_o)^3 Ar \tag{4.49}$$

where K_v is the throw constant (see equation (4.11)) and a/b is a function of the turbulent Prandtl number, $Pr_t = \mu C_p/\lambda$ (where C_p is the specific heat and λ is the thermal conductivity), given by:

$$b/a = 4/(1 + 1/Pr_t) - 1$$

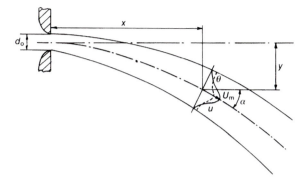

Figure 4.22 Path of a cold horizontal jet.

Using a commonly accepted value of $Pr_t = 0.7$ gives $b/a = 0.647$ or $a/b = 1.546$. Substituting the value of 6.5 for K_v, Koestel obtained:

$$y/d_o = 0.0652(x/d_o)^3 Ar \qquad (4.50)$$

where $Ar = g\beta d_o \Delta T/U_o^2$.

However, for a three-dimensional jet the value of K_v is 7.2 (see equation (4.23)) and representing d_o by $\sqrt{((4/\pi) A_o)}$, equation (4.49) for a three-dimensional jet becomes:

$$y/\sqrt{A_o} = 0.0522(x/\sqrt{A_o})^3 Ar \qquad (4.51)$$

Frean and Billington [7] carried out a regression analysis on their jet trajectory data for circular and rectangular orifices and on the data of Koestel [19] for circular nozzles and obtained the following expression for hot or cold jets:

$$y/\sqrt{A_o} = 0.226(x/\sqrt{A_o})^{2.61} Ar \qquad (4.52)$$

Equations (4.51) and (4.52) are plotted in Figure 4.23 and as shown the difference between them is quite significant. Koestel's equation gives a better estimate of jets produced by orifices whereas Frean and Billington's equation applies to different types of supply openings.

4.4.2 Non-isothermal free inclined jets

Figure 4.24 represents a hot jet issuing from a nozzle inclined at an angle, α, to the horizontal plane. For a circular jet the trajectory can be represented by the following equation due to Baturin [5]:

$$y/\sqrt{A_o} = x/\sqrt{(A_o)} \tan\alpha + 0.0354 Ar\sqrt{(T_o/T_r)}\{x/[\sqrt{(A_o)}\cos\alpha]\}^3 \qquad (4.53)$$

Equation (4.53) will also apply to a cold jet projecting downward. However, if a hot jet is projected downward or a cold jet projected upward by an angle α then the trajectory will have the shape given in Figure 4.25. The trajectory equation for this case can be obtained by substituting for α, in equation (4.53), by $-\alpha$. Thus:

$$y/\sqrt{A_o} = -x/\sqrt{(A_o)} \tan\alpha + 0.0354 Ar\sqrt{(T_o/T_r)}\{x/[\sqrt{(A_o)}\cos\alpha]\}^3 \qquad (4.54)$$

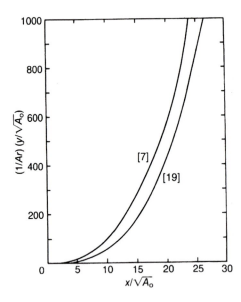

Figure 4.23 General trajectory of non-isothermal jets.

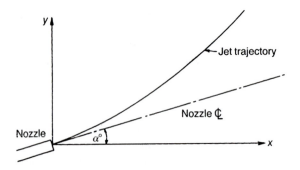

Figure 4.24 Trajectory of an inclined hot jet.

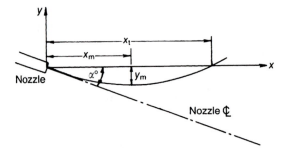

Figure 4.25 Trajectory of a hot jet inclined downward.

The distance from the supply, x_m, where the trajectory is at its lowest point, y_m, can be obtained by differentiating equation (4.54) and equating it to zero, giving:

$$x_m/\sqrt{A_o} = 5.434\left[(\sin\alpha\cos^2\alpha/Ar)\sqrt{(T_r/T_o)}\right]^{1/2} \tag{4.55}$$

Substituting this value of x_m in equation (4.54) gives the value of y_m. Thus:

$$y_m/\sqrt{A_o} = -3.623\left[(\sin^3\alpha/Ar)\sqrt{(T_r/T_o)}\right]^{1/2} \tag{4.56}$$

The jet trajectory intersects the horizontal axis from the supply when $y = 0$ in equation (4.54), i.e. at:

$$x_t/\sqrt{A_o} = 5.315\left[(\sin\alpha\cos^2\alpha/Ar)\sqrt{(T_r/T_o)}\right]^{1/2} \tag{4.57}$$

The maximum drop of a hot vertical jet can be determined by putting $\alpha = 90°$ in equation (4.56), giving:

$$\hat{y}_m/\sqrt{A_o} = -3.623\left[(1/Ar)\sqrt{(T_r/T_o)}\right]^{1/2} \tag{4.58}$$

4.4.3   Non-isothermal free vertical jets

The decay of maximum velocity of non-isothermal vertical jets was investigated by Regenscheit [20]. He studied two cases: one where the buoyancy force acts in the same direction as the inertia force and the other where buoyancy opposes inertia. The first situation occurs when a cold jet is projecting downwards from a ceiling or a hot jet is projecting upwards from a floor, whereas the second situation is found with a hot jet projecting downwards or a cold jet projecting upwards. For a circular jet Regenscheit obtained the following empirical equation for the decay of maximum velocity with distance from the outlet:

$$U_m/U_o = x_o/x \pm \sqrt{\{(Ar/m)[1 + \ln(2x/x_o)]\}} \tag{4.59}$$

where x_o = length of jet core; x = distance from outlet; $m = d_o/x_o$ which is a mixing number having a value between 0.2 and 0.3 for circular jets. In equation (4.59) the plus sign is taken when the buoyancy force assists the inertia force to increase the jet velocity and the minus sign is taken when buoyancy opposes inertia.

For a plane jet issuing from a slot the velocity decay equation is given as:

$$U_m/U_o = x_o/x \pm \sqrt{\{(Ar/m)[2.83\sqrt{x/x_o} - 1]\}} \tag{4.60}$$

where $m = h/x_o$ and h is the slot height.

4.4.4   Non-isothermal wall jets

The flow in a non-isothermal wall jet is dependent upon the relationship between the inertia, buoyancy and viscous forces acting on the jet. Albright and Scott [21] studied experimentally a cold, plane wall jet discharging vertically along a surface and derived the following expression for the non-dimensional temperature profile in the fully developed region:

$$\theta/\theta_m = \exp[-0.8(\eta - 0.1)^{1.4}] \tag{4.61}$$

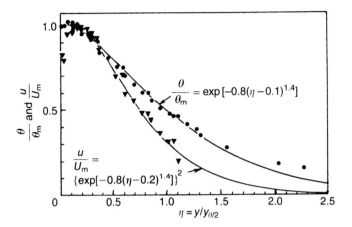

Figure 4.26 Temperature and velocity profiles for a non-isothermal plane wall jet [21].

Using the same characteristic dimension η as for the temperature profile, they also found that:

$$u/U_m = (\theta/\theta_m)^2$$

as for the case of a free jet discussed earlier. Figure 4.26 shows the temperature and velocity profiles for a wall jet.

The decay of the non-dimensional temperature at the edge of the thermal boundary layer can be described by the equation below [21] for $1050 < Re < 8055$:

$$\theta_m = 0.587(Re)^{0.224}(x/h)^{-0.6} \qquad (4.62)$$

Figure 4.27 represents plots of θ_m versus x/h for different Re using equation (4.62), and as can be observed the temperature decay is faster at lower Re. In this case the buoyancy force tends to accelerate the jet diffusion particularly for a jet of low inertia, i.e. low Re.

Nielson [22] carried out measurements on two types of rectangular wall-mounted diffusers placed close to the ceiling and supplying cool air to a test room. He measured the position at which the jet separated from the ceiling, x_s, due to the downward buoyancy force and obtained a linear correlation between x_s and $1/\sqrt{Ar}$. His results can be represented by the following relationship:

$$x_s/\sqrt{A_0} = 1.1/\sqrt{(Ar)} \qquad (4.63)$$

The experimental results in Figure 4.28 represent jet trajectories for a circular aperture discharging a hot air jet over a flat surface. For Ar (based on the discharge diameter d_o) up to 0.0097 the jet remains attached to the surface by the Coanda force but for higher values the jet separates from the surface. When $Ar = 0.054$ the jet breaks away from the surface as soon as it leaves the aperture.

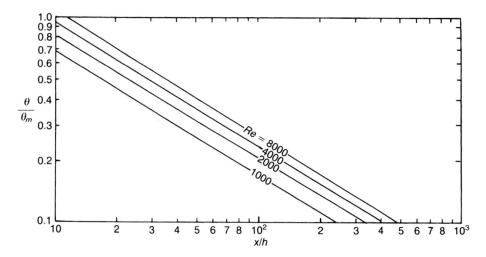

Figure 4.27 Temperature decay for a non-isothermal wall jet.

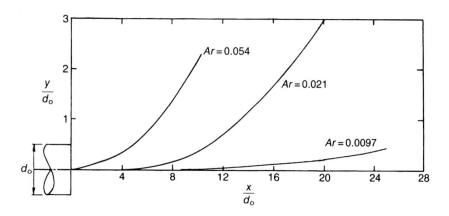

Figure 4.28 Trajectories for a hot circular wall jet [5].

Similar observations are also found for a cold jet issuing from a circular aperture over a ceiling.

4.5 Jet interference

4.5.1 Surface proximity

The proximity of a surface to one side of the supply outlet restricts the fluid entrainment to that side. This creates a pressure difference across the jet, which causes it to curve towards the surface, further reducing entrainment to that side of the jet. The curvature of the jet increases until it attaches to the surface. This phenomenon is usually referred to as the '*Coanda effect*'. The attachment of the jet to the surface only occurs when

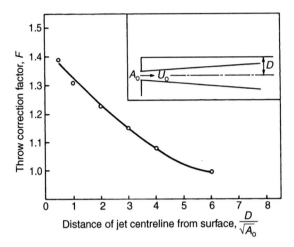

Figure 4.29 Throw correction factor for surface proximity [23].

the distance between the outlet and the surface is below a certain value, i.e. a critical distance D_c, otherwise the jet will propagate as a free jet. If the jet attaches to the surface the flow downstream of the attachment point (the point at which the static pressure is maximum) is similar to that of a wall jet because from that point onward entrainment takes place at the outer side of the jet only and a boundary layer forms on the surface. For an axisymmetric isothermal jet, Farquharson [23] found that the critical distance for jet attachment to the surface is $D_c = 6\sqrt{A_0}$. For $D < 6\sqrt{A_0}$ the throw constant K_v, in equation (4.23) becomes greater than the corresponding value for a free jet because entrainment occurs on one side only. Figure 4.29 shows the factor F by which K_v, should be multiplied to compensate for surface proximity.

The length of the recirculation (attachment) region, x_a, for a plane isothermal jet was investigated experimentally and theoretically by Sawyer [24]. Figure 4.30 shows the effect of the distance of the outlet from the surface, D, on the attachment distance measured from the outlet position. Sawyer's results and those of Bourque and Newmann [25] and Miller and Comings [26] can be represented by the straight line equation below over the range of D/h between 3 and 37:

$$x_a/h = 1.37(D/h) + 3.0 \tag{4.64}$$

Sandberg *et al.* [27] investigated the effect of opposing (negative) buoyancy force on the length of reattachment region of a jet from a 20 mm slot at a distance from the surface, $5 < D/h < 13$. The results show that the location of maximum pressure point on the surface (attachment point) was not affected by buoyancy for the range of Ar up to 1.71×10^{-3} that was investigated. However, the results produced an attachment length, x_a, larger than that given by equation (4.64), which is represented by:

$$x_a/h = 1.175(D/h) + 6.25 \tag{4.65}$$

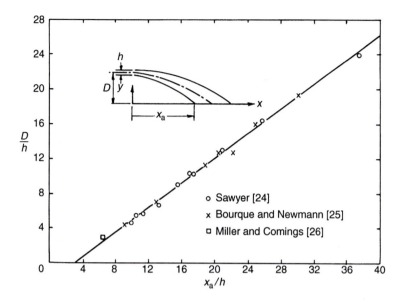

Figure 4.30 Effect of distance from surface on the attachment distance for a plane jet.

The flow of a plane jet issuing at an angle to a flat surface was also investigated experimentally by Bourque and Newmann [25] and theoretically by Sawyer [24]. The effect of the angle between the axis of the jet at the outlet and the surface, α, on the length of the circulation bubble, L_b, is depicted in Figure 4.31. The theoretical analysis due to Sawyer predicts no reattachment when $\alpha > 64°$. The various symbols in Figure 4.31 denote results for different Reynolds number and as can be seen some effect of Re on L_b is detectable.

4.5.2   Effect of opposite wall

A ceiling jet is often deflected downwards by an opposite wall to form a recirculation bubble in the corner as depicted in Figure 4.32. The effect of this deflection is detectable in the jet well ahead of the wall. In the testing of air diffusers, measurements close to an opposite wall are not recommended and diffuser testing standards, e.g. ISO 5219 [28] and BS 4773 [29], specify a minimum distance from the wall, x within which velocity measurements are not to be taken. This is either 0.5 or 1.0 m depending on the type of diffuser being tested. In practice, the determination of the interference distance x_i by measurement is not plausible because of the complexity of the flow close to a corner. However, using a numerical solution of the jet flow equations and the k–ε turbulence model (see Chapter 8), Awbi and Setrak [30] obtained a theoretical prediction of the distance from the supply of a plane wall jet for which the effect of the opposite wall is not detectable, x_f. They found that x_f is a function of the supply opening height, h, and the distance of the wall from the supply, L, i.e.:

$$x_f/h = 0.52(L/h)^{1.09}$$

(4.66)

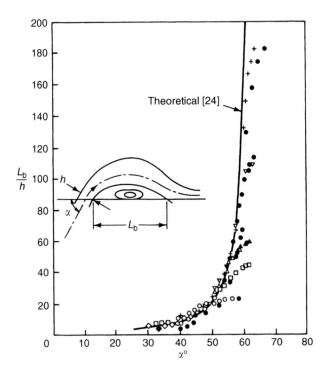

Figure 4.31 Effect of plane jet angle on recirculation bubble length [24].

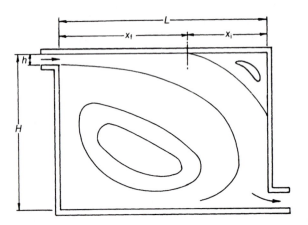

Figure 4.32 Effect of opposite wall.

The value of x_i can be obtained from:

$$x_i = L - x_f$$

The two standards [28] and [29] specify a value of $x_i = 0.5\,\text{m}$ for a two-dimensional slot diffuser in a room with two side walls. However, equation (4.66) indicates that x_i is not constant but will depend on the slot height and room length.

4.5.3 *Effect of obstacles*

When a wall jet encounters an obstacle attached to the surface it can take one of three courses. It can:

(i) almost be unaffected by the obstacle;
(ii) separates from the surface and reattaches downstream of the obstacle forming a recirculation bubble;
(iii) completely separates from the surface.

There are a number of factors that can determine the course a jet takes, such as the height of the obstacle relative to the height of the supply outlet, the distance of the obstacle from the outlet, the turbulence level of the air supply, the difference in temperature between the supply air and the surrounding air, etc. For a particular obstacle in the path of a particular wall jet, the distance of the obstacle from the outlet determines the course that the jet takes. Therefore, a distance is defined such that if the obstacle is placed closer to the outlet than this distance the jet will separate from the surface completely. This is called the critical distance, x_c. Figure 4.33 shows experimentally determined data of the effect of the height of the obstacle, d, on the critical distance for a plane wall jet. Holmes and Sachariewicz [31] used square-section obstacles and obtained different values of x_c depending on whether the obstacle is progressively moved away from the outlet or towards it, i.e. a hysteresis effect was found. The experiments of Söllner and Klinkenberg [32] were carried out by moving the obstacle towards the outlet until the critical distance is reached. Their results are in agreement with those of Holmes and Sachariewicz for the same case, as shown in Figure 4.33. The difference in the critical distance between the two cases is very significant at lower values of d/h and becomes less significant as the height ratio increases.

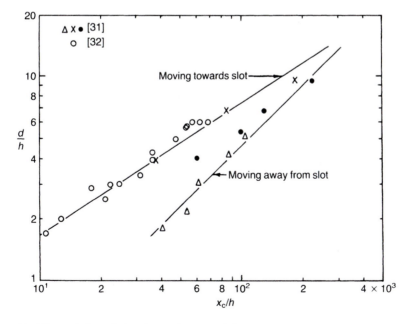

Figure 4.33 Effect of obstacle height on the critical distance.

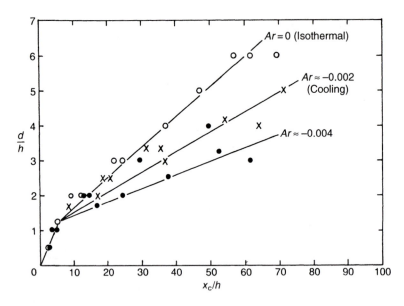

Figure 4.34 Effect of the Archimedes number on the critical distance for a plane wall jet [32].

Although Holmes and Sachariewicz's tests were carried out with non-isothermal wall jets, they could not detect any significant influence of the temperature difference on the critical distance. However, Söllner and Klinkenberg have found that for a cold jet over a ceiling with an obstacle the critical distance is not affected by the temperature difference for obstacles of small heights, but for larger obstacles they found that the larger the temperature difference between the supply air and the surrounding air, the greater is the critical distance. Their results are presented in non-dimensional form in Figure 4.34. In spite of the scatter, the effect of temperature difference is quite noticeable.

When the distance of an obstacle from the outlet, x_d, is greater than the critical distance, x_c, the jet reattaches to the surface downstream of the obstacle forming a recirculation bubble as depicted in Figure 4.35. In this case the displacement of the jet from the surface reaches a peak downstream of the obstacle before it reattaches again. The maximum vertical displacement of the jet trajectory, y_m, is shown in Figure 4.36. After reattachment downstream of the obstacle, the decay of the maximum velocity of the jet can be represented by the following expression [31]:

$$U_m/U_o = 1.02/\sqrt{[(x - x_d)/h]} \tag{4.67}$$

For an outlet of finite width, w, and a longer obstacle of length, l, some of the jet will spread over the length of the obstacle, deflecting at right angles at each end and causing a thinning of the jet and thus increasing the critical distance. For a three-dimensional outlet, Holmes and Sachariewicz [31] suggest the following empirical equation for the velocity decay:

$$U_m/U_o = 2.2/\sqrt{[(x - x_d)/h]}[1 - 0.785(l/w)]^{1/2} \tag{4.68}$$

which applies for $0.5 \leq l/w \leq 1.0$.

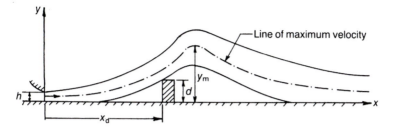

Figure 4.35 Trajectory of a reattached wall jet over an obstacle.

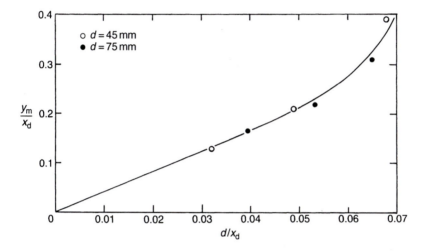

Figure 4.36 Maximum displacement of a plane wall jet due to an obstacle – jet reattaches to surface [31].

Awbi and Setrak [33] studied the effect of the aspect ratio of a two-dimensional obstacle, d/b, on the critical distance. Their experimental data is presented in Figure 4.37. As the depth of the obstacle, b, increases, the critical distance also increases. This is because, as the depth increases, reattachment to the surface is hindered.

Rajaratnam [2] studied the effect of surface roughness on the diffusion of an isothermal plane wall jet. He found that the velocity decay is enhanced by increasing the surface roughness. It may be expressed by:

$$U_m/U_o = C - 0.54 \log_{10}(x/d) \tag{4.69}$$

where d is the effective height of wall roughness and C is a non-dimensional coefficient which depends on the ratio d/b as shown in Figure 4.38. This expression was derived from measurements in the range $0.0046 \leq d/b \leq 0.126$ and $1.9 \times 10^4 < Re < 1 \times 10^5$ where Re is based on the outlet conditions.

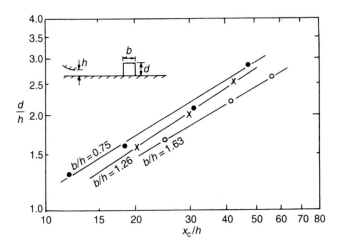

Figure 4.37 Effect of aspect ratio of obstacle on critical distance [33].

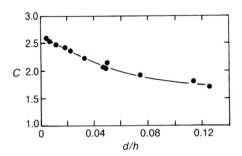

Figure 4.38 Effect of roughness ratio on the velocity decay coefficient [2].

4.5.4   Confluent jets

When jets issuing from different apertures in the same plane flow in parallel directions then at a certain distance downstream they coalesce and move as a single jet. The distance from the outlets at which the jets merge depends on the distance between the area of the outlets. Figure 4.39(a) and (b) shows velocity profiles measured by the author across three confluent isothermal jets issuing from three, sharp circular outlets of diameter $d = 100$ mm for two distances separating the outlets. Figure 4.39(a) is for a spacing $S = 2d$ (i.e. a spacing ratio $S/\sqrt{A_o} = 2.8$) and Figure 4.39(b) is for $S = 4d$ (i.e. $S/\sqrt{A_o} = 5.6$). In Figure 4.39(b) the three jets diffuse initially as separate jets until they interact at a distance $x/\sqrt{A_o} \approx 20$ downstream of the outlets. Similar observations were made by Baturin [5] for the same spacing ratio. The three jets in Figure 4.39(a) coalesce shortly after leaving the outlets and form a single jet, initially with three 'humps', before these merge together towards the centre to form a normal single-jet profile. Figure 4.40 shows velocity profiles for five circular outlets with $S = 2d$ which are similar to those in Figure 4.39(a) but cover a much wider flow field.

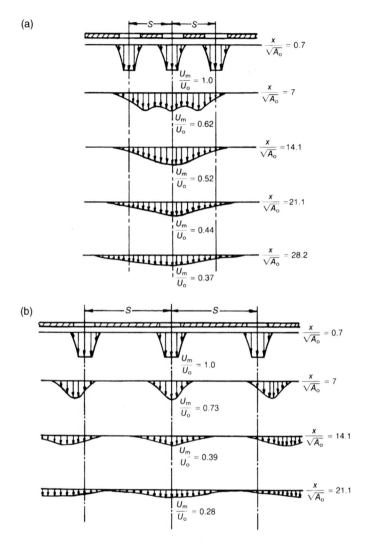

Figure 4.39 Velocity profiles for three confluent circular jets. (a) Closely spaced, $S/\sqrt{A_o} =$ 2.8, $Re = 8 \times 10^4$ and (b) widely spaced $S/\sqrt{A_o} = 5.6$, $Re = 5.8 \times 10^4$.

The decay of the maximum velocity in the flow for the above cases as well as for rectangular outlets of height 150 mm and width 50 mm are given in Figures 4.41 and 4.42. Figure 4.41 represents the results for three outlets with two spacing ratios $S/\sqrt{A_o} = 2.8$ and 5.6 and Figure 4.42 is for five outlets of $S/\sqrt{A_o} = 2.8$.

In Figure 4.41 the decay of maximum velocity in the case of the closely spaced jets is initially faster than in the case of the widely spaced jets, but further downstream the reverse happens. This rapid decay of velocity for the closely spaced jets is associated with the presence of a low pressure between the jets as they emerge from the outlet causing a swift coalescence of the air mass from the three outlets and a subsequent reduction in the maximum velocity of the jet. In this figure it can also be seen that

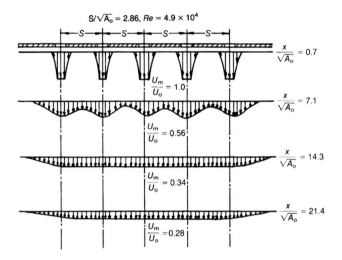

Figure 4.40 Velocity profiles for five confluent circular jets.

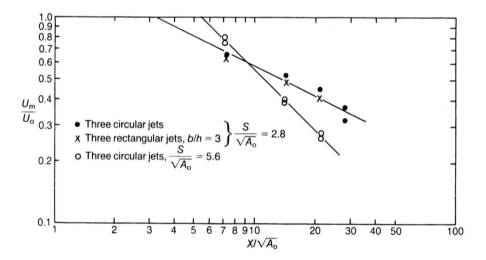

Figure 4.41 Velocity decay for three confluent jets.

the rectangular jets behave in a similar manner to the circular jets because these jets become axisymmetric a short distance from the outlet as was observed earlier in this chapter. Comparing Figures 4.41 and 4.42 for three and five jets but with equal spacing in both cases, it can be seen that the decay of the maximum velocity in the case of five jets is faster than that for three jets. This is probably due to the larger entrainment area associated with the five jets and the flatter velocity profile further downstream.

For a jet from an annular slot formed by two concentric tubes, the annular flow merges towards the axis of the tubes at a distance of about 5 diameters forming a velocity profile further downstream similar to that for a circular jet. Based on experimental data, Baturin [5] suggests using the following expression to obtain an equivalent

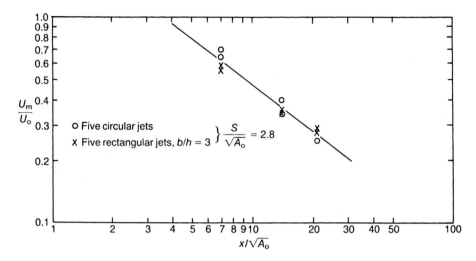

Figure 4.42 Velocity decay for five confluent jets.

velocity, U_e, and then using this with equation (4.11) for a circular jet to obtain the maximum velocity decay for an annular jet:

$$U_e = U_o\sqrt{[1 - (d_1/d_2)^2]} \qquad (4.70)$$

where U_o is the annular outlet velocity and d_1 and d_2 are the inner and outer diameters of the annulus. In equation (4.11) U_o is replaced by U_e calculated from equation (4.70) and d_o is replaced by d_2.

4.6 Plumes

Room air flow and thermal stratification is strongly influenced by thermal plumes produced by all types of heat sources present in rooms, e.g. heaters, computers, human bodies, hot or cold surfaces, etc. Their degree of influence on the flow depends on the type of air distribution system used. Most of the published work on thermal plumes in ventilated rooms has been focused on displacement ventilation since the effect of plumes in these cases is most prominent. Interaction of plumes is more pronounced if the room ventilation is almost purely natural or only weakly forced as it is in the case of the displacement ventilation. However, in recent research involving mixing ventilation [34], it has been found that thermal plumes can also have a significant influence on a ceiling jet and consequently the air movement in the room.

Plumes are produced by convection as a result of difference in temperature (buoyancy) between the fluid in contact with the heat source and the surrounding fluid, thereby the former fluid entraining fluid from the surroundings. Knowledge of the developing vertical temperature, velocity profiles and the flow rate in plumes is required for the evaluation of room air movement, thermal comfort and air quality. However, these parameters are not only dependent on the geometry and power of the plume source but also on the air movement and vertical temperature gradient in the room.

In addition, the room size, particularly the ceiling height, can have a major influence on the development of plumes.

4.6.1   *Plumes in neutral surroundings*

Early investigations considered the development of plumes in unbounded neutral surroundings, i.e. unlimited boundaries and uniform ambient temperature. Analytical expressions for velocity, temperature and airflow rate distribution along a plume were derived by applying momentum and energy conservation and assuming a Gaussian velocity and temperature distribution across the plume.

The mass flow in a plume increases with height as a result of entrainment of ambient air. The heat sources that produce the plume can be of many different shapes and configurations. Figure 4.43 shows typical configurations of the types of heat sources to be considered here. For a plume in a room *without* temperature stratification the following expressions may be used for the volume flow rate, q_y, in the plume [35]:

(i)   Vertical surface (laminar boundary layer):

$$q_H = 2.87 \times 10^{-3} l (T_w - T_a)^{1/4} H^{3/4} \tag{4.71a}$$

Vertical surface (turbulent boundary layer):

$$q_H = 2.75 \times 10^{-3} l (T_w - T_a)^{2/5} H^{6/5} \tag{4.71b}$$

(ii)   Horizontal surface: $q_y = 0.5 \times 10^{-3} l^{2/3} P_c^{1/3} (y + w)^{5/3}$ (4.72)

(iii)   Point source: $q_y = 5.5 \times 10^{-3} P_c^{1/3} (y + y_o)^{5/3}$ (4.73)

(iv)   Line source: $q_y = 14.0 \times 10^{-3} l^{2/3} P_c^{1/3} (y + y_o)^{5/3}$ (4.74)

where q_y and q_H = volume flow rate at a height y or H (m^3 s^{-1}); P_c = convective heat load component of the source (W); T_w = the surface temperature (°C); T_a = the air temperature, (°C); w = width of horizontal surface (m); l = length of surface or line source (m); H = height of vertical surface (m); y = height above source (m); y_o = distance between the top of the actual heat source and the virtual source (m) which is dependent upon the source geometry as follows:

$$y_o = C(d/2 + \delta) \tag{4.75}$$

where C is a constant = 4.18 for a point source and 3.8 for a line source; d is the heat source diameter (m); δ is the boundary layer thickness (m) at the *top* of the heat source which is 0 for a horizontal source and is given by $\delta = 0.048 \cdot (h/\Delta T)^{1/4}$ for a laminar boundary layer on a vertical source, or $\delta = 0.11 \, h^{0.7} \Delta T^{-0.1}$ for a turbulent boundary layer where h is source height (m) and ΔT is the temperature difference between the source surface and surroundings (K). For a small source, the virtual source distance, y_o, may be assumed to be 0.

Popiolek *et al.* [36] gives the following expression for the volume flow rate at a height y above plumes of different shapes (sphere, cylinder of $h/d = 2.7$ to 8.0, plates, radiator):

$$q_y = 6.0 \times 10^{-3} P_c^{1/3} (y + y_o)^{5/3} \tag{4.76}$$

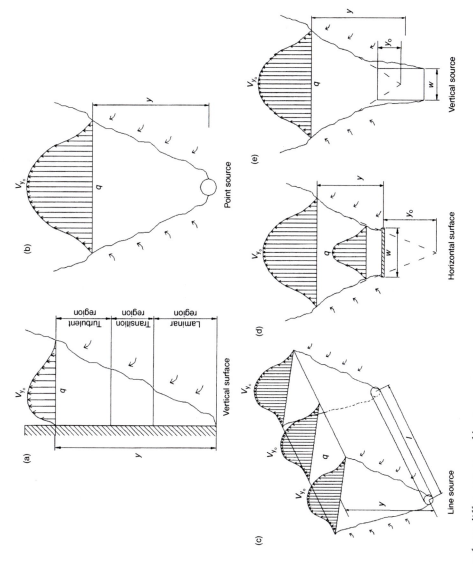

Figure 4.43 Plumes from different types of heat sources.

in which case y_0 is given by:

$$y_0 = 7.75\, w_y - y \tag{4.77}$$

where w_y is the width of plume's velocity profile at a distance y from the source. In the absence of firm data on the source, Morton *et al.* [37] suggest that the position of the virtual source be located at:

$$y_0 = 1.7 - 2.1d \tag{4.78}$$

below the real source where d is the source diameter or width. Alternatively, y_0 may be obtained using:

$$y_0 = d/[2\tan(\alpha/2)] \tag{4.79}$$

where α is the included angle of the plume which is normally taken to be 25°.

Hautalampi *et al.* [38] give the expression below for plume flow rate that was derived from experiments:

$$q_y = AP_c^{1/3}(y + y_0)^{5/3} \tag{4.80}$$

where the values of A and y_0 for different heat sources are given in Table 4.2.

The maximum velocity in the plume, v_{y_0}, and the temperature difference between the centre of plume and ambient, ΔT_{y_0}, are given by the expressions below [39, 40]:

(i) Point source: $$v_{y_0} = 0.128\, P_c^{1/3} y^{-1/3} \tag{4.81}$$

$$\Delta T_{y_0} = 0.329\, P_c^{2/3} y^{-5/3} \tag{4.82}$$

(ii) Line source: $$v_{y_0} = 0.067\, P_c^{1/3} \tag{4.83}$$

$$\Delta T_{y_0} = 0.094\, P_c^{2/3} y^{-1} \tag{4.84}$$

In the above expressions the source power P_c is the *convective* component only which may be expressed by:

$$P_c = K\, P_t \tag{4.85}$$

where K is the convective heat factor and P_t is the total power of the heat source. Table 4.3 gives typical values of K for different types of heat sources [35, 41].

Table 4.2 Values for A and y_0 in equation (4.80)

Heat source	$A \times 10^{-3}$	y_0 (m)
Cylinder	5.2	1.2
Radiator	4.4–5.6	1.0
Convector	1.5	4.5

Table 4.3 Values for the factor K in equation (4.85)

Heat source	K
People	0.5
Point source	0.8–1.0
Pipes and ducts	0.7–0.9
Extended surfaces	0.5
Small machines or components	0.4–0.6
Large machines or components	0.3–0.5

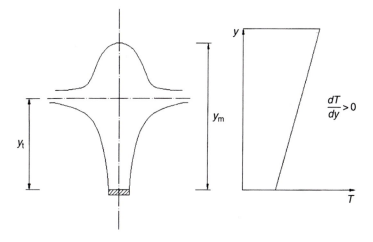

Figure 4.44 Plume in stratified surroundings.

4.6.2 *Plumes in surroundings with temperature stratification*

Because the driving force behind a plume is buoyancy due to temperature difference, a temperature gradient in the room, i.e. temperature stratification, will influence the plume development and its rise height. In this case, the velocity in the plume will be lower than for a similar plume in homogeneous surroundings because the increase in temperature with height will reduce the buoyancy force. As a result, the buoyancy force will disappear at a height, y_t, where the plume temperature and the surrounding temperature are equal. However, due to the momentum of the plume, it will continue to rise to a height, y_m, before it spreads laterally, see Figure 4.44.

Morton *et al.* [37] developed an analytical solution for an unbounded plume in stratified surroundings. Mundt [35] carried out theoretical and experimental investigations of thermal plumes in a room with displacement ventilation, basing her analysis on the theoretical work of Morton *et al.* In this case the low temperature ventilation air is supplied at floor level, while the air warmed by heat sources in the room is extracted at ceiling level. This is a typical stratified plume flow.

Assuming linear temperature stratification, i.e.:

$$dT/dy = By^n \qquad (4.86)$$

where B is a constant and n the stratification index, the heights y_t and y_m may be obtained from the equations below [35].

(i) Point source: $$y_t = 0.74 P_c^{1/4} \left(\frac{dT}{dy} \right)^{-3/8} \tag{4.87}$$

$$y_m = 0.98 P_c^{1/4} \left(\frac{dT}{dy} \right)^{-3/8} \tag{4.88}$$

(ii) Line source: $$y_t = 0.35 \left(\frac{P_c}{l} \right)^{1/3} \left(\frac{dT}{dy} \right)^{-1/2} \tag{4.89}$$

$$y_m = 0.51 \left(\frac{P_c}{l} \right)^{1/3} \left(\frac{dT}{dy} \right)^{-1/2} \tag{4.90}$$

where l is the source length (m) and P the convective power (W).
   The volume flow rate, q_y, is given by:

(i) Point source: $$q_y = 2.38 \, m \, P_c^{3/4} (dT/dy)^{-5/8} \tag{4.91}$$

where

$$m = 0.004 + 0.039 \, y' + 0.380 \, y'^2 - 0.062 \, y'^3 \tag{4.92}$$

$$y' = 2.86(y + y_o)(dT/dy)^{3/8} P_c^{-1/4} \tag{4.93}$$

Equation (4.91) applies for $y' < 2.125$ and if y' is within $2.8 > y' > 2.125$ then the difference in density is negligible and the solution based on temperature stratification becomes uncertain. In such a case, equation (4.73) for non-stratified surroundings should be used instead. When $y' = 2.8$ the plume would have reached its maximum height, y_m.

(i) Line source: $$q_y = 4.82 \, m \, (P_c/l)^{2/3} (dT/dy)^{-1/2} \tag{4.94}$$

where $m = 0.004 + 0.477 \, y' + 0.029 \, y'^2 - 0.018 \, y'^3$ $\tag{4.95}$

$$y' = 5.78(y + y_o)(dT/dy)^{1/2} P_c^{-1/3} \tag{4.96}$$

Equation (4.94) applies for $y' < 2.0$ and if y' is within the range $2.95 > y' > 2.0$ then the difference in density is negligible and the solution based on temperature stratification becomes uncertain. In such case, equation (4.74) for non-stratified surroundings should be used instead. When $y' = 2.95$ the plume would have reached its maximum height, y_m.

4.6.3   *Plume close to surfaces*

When a heat source is located close to a wall the plume from it may attach to the wall due to the Coanda effect, Figure 4.45(a). As a result, the entrained of room air will be less than the rate for a free plume. In fact, Kofoed [42] has shown that the plume flow, q_y, in this case is half that produced by a free plume of power $2P$. A similar situation

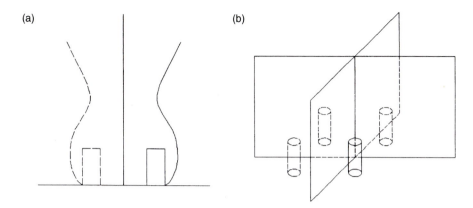

Figure 4.45 Heat source close to (a) a wall or (b) a corner.

will arise when a heat source is located at a corner, Figure 4.45(b), in which case the plume flow is quarter that produced by a free plume of power 4P. Hence, for a point source without temperature stratification, the flow rates are:

Plume close to a wall: $q_y = 3.5 \times 10^{-3} P_c^{1/3} (y + y_o)^{5/3}$ (4.97)

Plume in a corner: $q_y = 2.2 \times 10^{-3} P_c^{1/3} (y + y_o)^{5/3}$ (4.98)

4.6.4   *Plume interaction*

When a number of identical heat sources, N, are placed close to each other, the plumes will merge into a single plume and the flow of the combined plume, Q_{yN}, may be estimated using:

$$Q_{yN} = N^{1/3} q_y \qquad (4.99)$$

When a number of heat sources are placed close to each other then the individual plumes will merge to form one large plume as shown in Figure 4.46. In this case the equation (4.73) is used to calculate the flow of the combined plume but using the sum of *all* the plume powers.

4.6.5   *Plumes in confined spaces*

The plumes from heat sources entrain air from the surroundings in an upward movement, and downdraft from a cold surface may transport air down to the occupied zone. In studying these flows a neutral height, y_n, is defined for the room which corresponds to the height where the net upward flow rate of the plumes, i.e. $q_1 - q_2$, is equal to the air supply rate, q_o, in Figure 4.47. Thus, two zones in the room will be formed: a lower zone with uni-directional upward displacement flow and an upper recirculation flow zone. If the plume is contaminated then the contaminant will be taken up to the recirculation zone. The neutral height therefore, is the height in the room containing mainly the fresh air supply and the upper recirculation zone is the region where room contaminants are contained. It is important, therefore, that the neutral height

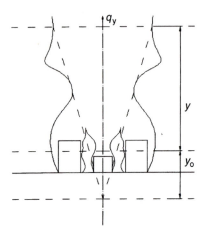

Figure 4.46 Interaction of several plumes.

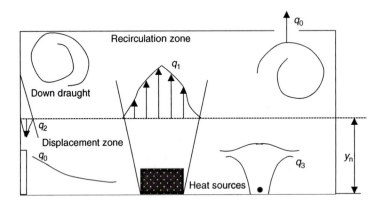

Figure 4.47 Displacement ventilation and thermal plumes.

is always maintained above the occupied zone (i.e. 1.8 m above the floor for standing occupancy and 1.1 m for seated occupancy) to avoid contaminant recirculation in this zone. Any plumes that have a maximum height, y_m, below the neutral height, y_n, will not influence the value of y_n in the room, such as the plume q_3 in Figure 4.47.

To determine the neutral height, y_n, in a room it is necessary to know the air supply rate and the flow rates in the plumes. The latter can be calculated using the method outlined in the previous sections. It is also possible to find y_n from measurements of temperature or contaminant concentration in the room [43]. As shown in Figure 4.48, y_n corresponds to the height where the air temperature in the room is equal to the mean temperature of all the walls or alternatively to the point in the room where the contaminant concentration starts to rise sharply from an initially uniform distribution, indicating the departure from the displacement zone and entrance into the upper recirculation zone. This region is indicated by δ in Figure 4.48.

Figure 4.49 shows a comparison between the values of neutral height calculated using the plume equations presented in the previous sections and measurement of air

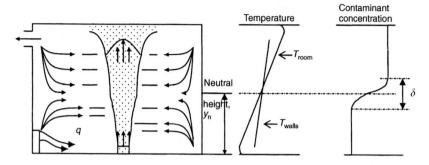

Figure 4.48 Distribution profiles of temperature and concentration in a room with displacement ventilation and a plume.

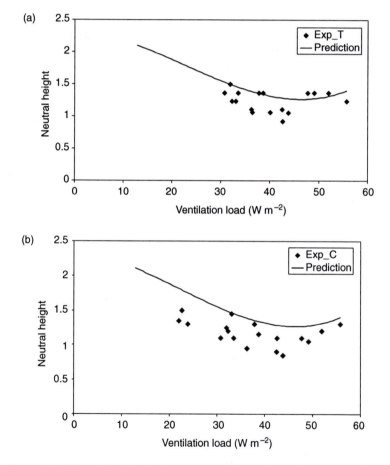

Figure 4.49 Neutral height in the room versus ventilation load. (a) Temperature measurement and (b) tracer gas concentration measurement.

an wall temperature (Figure 4.49(a)) and tracer gas concentration (Figure 4.49(b)) for different room loads. The room in this case is ventilated using different types of displacement systems (one wall-mounted flat air terminal at low-level, one wall-mounted semi-cylindrical air terminal at low-level or two floor-mounted swirl diffusers) and contained plumes due to four types of heat sources (computer simulator box, strip light, heated plate on a wall and a heated mannequin). The results are for two values of supply air flow rate of 25 and $35\,l\,s^{-1}$ (5 and $7\,ach^{-1}$). It is shown that the results based on the temperature measurement produces slightly better correlations with calculations using the plume flow equations than those based on concentration measurements. As can be seen the value of y_n decreases with increasing the ventilation load up to a value of about $45\,W\,m^{-2}$ and then gradually increases. This load represents the limit of the displacement flow in the room and for higher loads the circulation flow reaches the lower zone and the displacement flow starts to breakdown. The results in Figure 4.49 include data from the three ventilation systems mentioned earlier.

The neutral height is also influenced by the ventilation rate, q_0, and it will be expected that y_n will increase as q_0 increases as there will be more primary air for entrainment by the plume and, furthermore, the temperature gradient in the room will decrease thus increasing the buoyancy force.

Example 4.1 A computer monitor that may be approximated by 0.5 m cube is placed on a desk 0.7 above the floor produces a convective heat flux 150 W in a room ventilated by a low-level displacement system. It is required to estimate the plume flow rate at a height of 1.8 m above the floor and the maximum velocity and temperature of the plume at the same height (see Figure 4.50).

Applying the equations for a point source:

Plume flow rate:
$$q_y = 5.5 \times 10^{-3} P_c^{1/3}(y + y_0)^{5/3} \qquad (4.73)$$

Maximum velocity in plume:
$$v_{y_0} = 0.128\, P_c^{1/3} y^{-1/3} \qquad (4.81)$$

Temperature at the plume centre:
$$\Delta T_{y_0} = 0.329\, P_c^{2/3} y^{-5/3} \qquad (4.82)$$

The virtual source distance from the bottom of the box is:

$$y_0 = 1.7 - 2.1d \qquad (4.78)$$

From equation (4.78) the virtual source position $= 1.7 - 2.1 \times 0.5 = 0.65$ m. Distance between the top of box and virtual source position, $y_0 = 0.5 + 0.65 = 1.15$ m. The height from the box at which the values are calculated, $y = 1.8 - 0.5 - 0.7 = 0.6$ m. Using equation (4.73), $q_y = 5.5 \times 10^{-3} \times (150)^{1/3}(0.6 + 1.15)^{5/3} = 0.074\,m^3\,s^{-1}$ the flow rate in the plume at a height of 1.8 m above the floor.

Using equation (4.81), $v_{y_0} = 0.128 \times (150)^{1/3}(0.6)^{-1/3} = 0.805\,m\,s^{-1}$ which is the maximum velocity in the centre of plume at a height of 1.8 m above the floor.

Using equation (4.82), $\Delta T_{y_0} = 0.329 \times (150)^{2/3}(0.6)^{-5/3} = 21.7\,K$ which is the temperature difference between the centre of plume and the room at a height of 1.8 m above the floor.

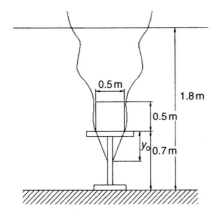

Figure 4.50 Configurations for Example 4.1.

Example **4.2** If the room in Example 4.2 is ventilated by supplying $0.05 \, \text{m}^3 \, \text{s}^{-1}$ of air at $18 \, °C$ at low level calculate the position of the neutral height in the room. Assume that the air extract is from the ceiling of height 3 m and neglect all heat sources except the box in Example 4.1.

The neutral height corresponds to the height where the flow in the plume is equal to the air supply to the room, i.e. $q_y = 0.05 \, \text{m}^3 \, \text{s}^{-1}$.

The temperature gradient in the room is calculated from the heat input and the flow rate as follows:

$$P_c = \rho \, q_y C_p \Delta T$$

i.e. $\Delta T = 150/(1.2 \times 0.05 \times 1000) = 0.4 \, \text{K} \, \text{m}^{-1}$ the temperature gradient in the room.

Because the temperature gradient is small, equation (4.73) may be used to find the neutral height, i.e.:

$$q_y = 5.5 \times 10^{-3} P_c^{1/3} (y + y_o)^{5/3}$$

Hence, $(y + y_o) = [0.05/(5.5 \times 10^{-3} \times 150^{1/3})]^{3/5} = 1.382 \, \text{m}$, $y = 1.382 - y_o = 1.382 - 0.65 = 0.735 \, \text{m}$ the position of the neutral height above the box. The neural height from the floor $= 0.735 + 0.5 + 0.7 = 1.935 \, \text{m}$.

References

1. Nevins, R. G. (1976) *Air Diffusion Dynamics – Theory, Design and Application*, Business News Publishing, Birmingham, MI, USA.
2. Rajaratnam, N. (1976) *Turbulent Jets*, Elsevier, Amsterdam.
3. Tollmien, W. (1926) Berechnung turbulenter Ausbreitungsvorgange. *ZAMM*, 6, 468–78 [English translation (1945) *NACA TM 1085*].
4. Schlichting, H. (1968) *Boundary-Layer Theory*, McGraw-Hill, New York.
5. Baturin, V. V. (1972) *Fundmentals of Industrial Ventilation*, Pergamon, Oxford.
6. Ricou, F. P. and Spalding, D. B. (1961) Measurements of entrainment by axisymmetrical turbulent jets. *J. Fluid Mech.*, 11, 21–32.

7. Frean, D. H. and Billington, N. S. (1955) The ventilation jet. *JIHVE*, **23**, 313–33.
8. Förthmann, E. (1934) Uber turbulente Strahlausbreitung. *mg. Arch.*, **5**, 42 [English translation (1936) *NACA TM789*].
9. Heskestad, G. (1966) Hot-wire measurements in a radial turbulent jet. *Trans. ASME, J. Appl. Mech.*, **33**, 417–24.
10. Becher, P. (1966) Air distribution in ventilated rooms. *JIHVE*, **34**, 219–27.
11. Sforza, P. M. (1977) Three-dimensional free jets and wall jets: applications to heating and ventilation, *Proc. UNESCO mt. Seminar on Heat Transfer in Buildings*, Dubrovnik, Yugoslavia, 29 August–2 September 1977, Vol. 4.
12. Chigier, N. A. and Chervinsky, A. (1967) Experimental investigation of swirling vortex motion in jets. *Trans. ASME, J. Appl. Mech.*, **34**, 443–51.
13. Schwarz, W. H. and Cosart, W. P. (1961) The two-dimensional turbulent wall-jet. *J. Fluid Mech.*, **10**, 481–95.
14. Karimipanah, T. (1996) *Turbulent jets in confined spaces*, PhD thesis, Royal Institute of Technology, Sweden.
15. Beltaos, S., and Rajaratnam, N. (1973) Plane turbulent impinging jets. *J. Hydraulic Research*, **11**, 29–59.
16. Karimipanah, T. and Awbi, H. B. (2002) Theoretical and experimental investigation of impinging jet ventilation and comparison with wall displacement ventilation. *Build. Envir.*, **37**, 1329–42.
17. O'Callaghan, P. W. *et al.* (1975) Velocity and temperature distributions for cold air jets issuing from linear slot vents into relatively warm air. *J. Mech. Eng. Sci.*, **17**, 139–49.
18. Laufer, J. (1969) Turbulent shear flows of variable density. *JAIAA*, **7**, 706–12.
19. Koestel, A. (1955) Paths of horizontally projected heated and chilled air jets. *Trans. ASHAF*, **61**, 213–32.
20. Regenscheit, B. (1970) Die Archimedes-Zahl-Kenzah zur beurteilung von Raumstromungen. *Gesund. mg.*, **91**(6), 1717.
21. Albright, L. D. and Scott, N. R. (1974) The low-speed non-isothermal wall jet. *J. Agric. Eng. Res.*, **19**, 25–34.
22. Nielsen, P. V. (1985) Measurement of the three-dimensional wall jet from different types of air diffusers, *Proc. CLIMA 2000*, Copenhagen, pp. 383–7.
23. Farquharson, I. M. C. (1952) The ventilating air jet. *JIHVE*, **19**, 449–69.
24. Sawyer, R. A. (1963) Two-dimensional reattaching jet flows including the effects of curvature on entrainment. *J. Fluid Mech.*, **17**, 481–98.
25. Bourque, C. and Newmann, B. G. (1960) Reattachment of a two-dimensional incompressible jet to an adjacent flat plate. *Aero. Quart.*, **11**, 201.
26. Miller, D. R. and Comings, E. W. (1960) Force–momentum fields in a dual-jet flow. *J. Fluid Mech.*, **7**, 237.
27. Sandberg, M., Wirén, B. and Claesson, L. (1992) Attachment of a cold jet to the ceiling – length of recirculation region and separation distance, *Proc. Air Distribution in Rooms (Roomvent '92)*, Aalborg, Denmark, Vol. 1, pp. 489–99.
28. ISO 5219 (1984) *Air Distribution and Air Diffusion – Laboratory Aerodynamic Testing and Rating for Air Terminal Devices*, International Standards Organization, Geneva.
29. BS 4773 (1991) *Methods for Testing and Rating Air Terminal Devices for Air Distribution Systems – Part 1: Aerodynamic Testing*, British Standards Institution, London.
30. Awbi, H. B. and Setrak, A. A. (1986) Numerical solution of ventilation air jet, *Proc. 5th Int. Symp. on the Use of Computers for Environmental Engineering Related to Buildings*, Bath, UK, pp. 236–46.
31. Holmes, M. J. and Sachariewicz, E. (1973) The effect of ceiling beams and light fittings on ventilating jets. *BSRIA Lab. Rep. No.79*, Building Services Research and Information Association, Bracknell, UK.

32. Söllner, G. and Klinkenberg, K. (1976) Leuchten als Storkorper im Luftstrom. *Heiz. Luftung Haustech.*, **27**, 442–8.
33. Awbi, H. B. and Setrak, A. A. (1987) Air jet interference due to ceiling-mounted obstacles, *Air Distribution in Ventilated Spaces Conf. (Roomvent '87)*, Stockholm.
34. Cho, Y. and Awbi, H. B. (2002) Effect of heat source location in a room on the ventilation performance, *Proc. Air Distribution in Rooms (Roomvent 2002)*, Copenhagen, pp. 445–8.
35. Mundt, E. (1996) The performance of displacement ventilation systems, *Report for BFR Project # 920937-0*, Royal Institute of Technology, Stockholm.
36. Popiolek, Z., Trzeciakiewwicz, S. and Mierzwinski, S. (1998) Improvement of a plume volume flux calculation method, *Proc. Air Distribution in Ventilated Rooms (Roomvent '98)*, Stockholm, Vol. 1, pp. 423–30.
37. Morton, B. R., Taylor, G. and Turner, J. S. (1956) Turbulent gravitational convection from maintained and instantaneous sources, *Proc. Royal Soc.*, Vol. 234A, p. 1.
38. Hautalampi, T., Sandberg, E. and Koskela, H. (1998) Behaviour of convective plumes with active displacement air flow patterns, *Proc. Air Distribution in Ventilated Rooms (Roomvent '98)*, Stockholm, Vol. 1, pp. 415–21.
39. Mierzwinski, S. (1981) Air motion and temperature distribution above a human body as a result of natural convection, A4-serien no. 45, Inst. för Uppv.- o Vent. Teknik, KTH, Stockholm.
40. Popiolek, Z. (1981) Problems of testing and mathematical modeling of plumes above human body and other extensive heat sources, A4-serien no. 54, Inst. för Uppv.- o Vent. Teknik, KTH, Stockholm.
41. Nielsen, P. V. (1993) *Displacement Ventilation – Theory and Design*, Aalborg University, Denmark, August 1993, ISSN 0902-8002 U9306.
42. Kofoed, P. (1991) Thermal plumes in ventilated rooms, PhD thesis, Aalborg University, Denmark.
43. Xing, H. and Awbi, H. B. (2002) Measurement and calculation of the neutral height in a room with displacement ventilation. *Build Envir.*, **37**, 961–7.

5 Air diffusion devices

5.1 Introduction

The air jets used in mechanically ventilated buildings are seldom supplied by simple nozzles or sharp openings and invariably air terminal devices (ATDs) are employed. The characteristics of air jets described in Chapter 4 will, to a certain extent, be also valid for jets created by ATDs. However, the air jets from ATDs are usually evaluated by the ATD manufacturer to enable reliable and easy selection of the device and to know its performance in practice.

ATDs are used in preference to simple openings for one or more of the following reasons:

- to direct the jet to the required position by means of vanes or baffles
- to be able to control the volume of air through the outlet
- to vary the throw of the jet by controlling the diffusion area of the jet.

However, ATDs have the following disadvantages:

- produce higher-pressure losses than a free outlet
- increase the aerodynamic noise
- involve additional cost.

In this chapter the aerodynamic characteristics of different types of ATDs are discussed. The criteria normally applied for the selection of supply and extract devices are also outlined.

5.2 Air diffusion glossary

In BS 4773 [1] ATD is defined as 'a component of the installation that is designed to ensure the predetermined movement of air into or from a treated space, e.g. grille, diffuser, etc.' From this statement, it is clear that on selecting an ATD for a particular application the aerodynamic as well as the geometric characteristics of the device are required. For an exhaust or extract ATD the aerodynamic characteristics usually imply knowing the total pressure loss caused by the insertion of the device in the duct opening and for a supply ATD they also imply knowing the envelope of the air jet.

Given below are definitions of the basic terms used in this chapter to describe the characteristics of ATDs.

Core area (A_c) The gross area containing all the openings of an ATD through which the air can pass.

Free area (A) This is the sum of the smallest areas of all the openings of an ATD through which air can pass.

Effective area (A_o) This is the net area of an ATD through which air passes and which may not equal its free area. It is given by:

$$A_o = C_d A \tag{5.1}$$

where A is the free area and C_d is the discharge coefficient, which usually varies between 0.65 and 0.9.

Free area ratio (R_a) This is the ratio of free area to core area.

Envelope   This is a geometrical surface of a jet that has the same prescribed velocity, Figure 5.1.

Throw (x_m) The distance between the plane of a supply ATD and a plane that is tangential to the jet envelope (defined for 0.25, 0.5, 0.75 or 1.0 m s^{-1}) and perpendicular to the initial jet direction, Figure 5.1. For a radial air supply the throw is known as the radius of diffusion, Figure 5.2.

Drop or rise (y_m) The vertical distance between the centre of a supply ATD core and a horizontal plane tangential to a specified envelope, Figure 5.1(a) and (b).

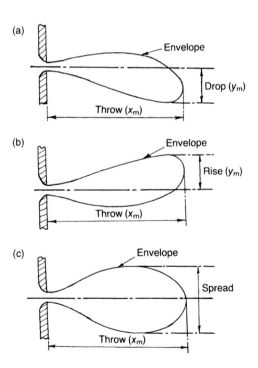

Figure 5.1 Jet envelope of a supply ATD. (a) Cold jet; (b) warm jet and (c) isothermal, warm and cold jets.

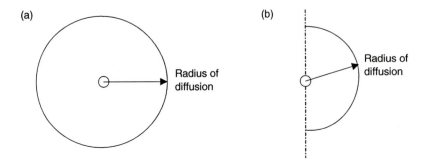

Figure 5.2 Radial air diffusion: (a) Circular ATD and (b) semi-circular ATD.

Spread (z_m)   The maximum horizontal distance between two vertical planes tangential to a specified envelope of a supply ATD and perpendicular to a plane through the core axis, Figure 5.1(c).

Core velocity (U_c)   The volume of air flow rate divided by the core area of the ATD.

Discharge velocity (U_o)   The average velocity at the discharge from an ATD opening, which is equal to $U_c/(C_d R_a)$.

Terminal velocity (U_t)   The maximum jet velocity at the end of the throw, usually specified as 0.25, 0.5, 0.75 or 1.0 m s^{-1}.

5.3   Performance of air terminal devices

5.3.1   *Pressure losses*

The total pressure loss coefficient, ξ, is based on the measurement of pressure loss when the ATD is tested in a special pressure test facility under standard conditions (e.g. [1] and [2]), and is defined as:

$$\xi = \Delta p_t/p_v = 1 + p_s/p_v \qquad (5.2)$$

where Δp_t is the total pressure drop measured across the ATD, p_s is the static pressure in the duct, which is used for the test, and p_v is the velocity pressure in the duct, which is given by:

$$p_v = 0.5\rho(\dot{V}/A_d)^2 \qquad (5.3)$$

where ρ is the air density (kg m^{-3}), $\dot{V}$ the air volume flow rate through the duct (m^3 s^{-1}), A_d is the nominal duct area (m^2) and the pressure is in pascals (Pa). Given the values of ξ and p_v the pressure loss Δp_t for the device is determined from equation (5.2).

For an extract ATD, the static pressure in the duct is usually measured at least 6 duct diameters downstream of the device [1, 2] and the static pressure in equation (5.2) should be corrected for duct friction losses using the equation recommended by ISO 5219:

$$p_s = p_{sm} - (0.02 L/D_h)p_v \qquad (5.4)$$

where p_{sm} is the static pressure measured in the duct at a distance L from the device and D_h is the duct hydraulic diameter given by: $D_h = 4 \times$ *Cross-sectional Area/Perimeter*.

5.3.2 *Throw constants for supply devices*

According to the geometry of their free area, ATDs are usually divided into four classes [2].

1 Devices from which the jet is essentially three-dimensional (3D), e.g. nozzles, grilles and registers.
2 Devices from which the jet flows radially along a surface, e.g. ceiling diffusers.
3 Devices from which the jet is essentially two-dimensional (2D), e.g. linear grilles, slots and linear diffusers.
4 Devices for generating buoyant flows, e.g. low-velocity air terminals.

The flow within modern ATDs can be very complex and the jets that they produce may not behave in the same way as those jets produced by simple openings, which were described in Chapter 4. Characterizing the jet flow by four regions (see Chapter 4), devices of class 1 have the core and the axisymmetric decay as the main regions whereas devices of classes 2 and 3 have the core and characteristic decay as the predominant regions. The selection of ATDs should be based on the data available for each particular ATD (nomogram) normally supplied by the manufacturer. ATDs are usually tested in accordance with the procedures laid out in ISO 5219 [2].

Large aspect ratio supply devices

The maximum velocity in the core region is constant and equal to the average initial discharge velocity from the device. In the characteristic decay region, which is the main region of practical interest for devices of classes 2 and 3, the maximum velocity, U_m, decays as follows:

$$U_m/U_o = K_v/\sqrt{(x/h_o)} \tag{5.5}$$

where U_o = average initial velocity at discharge from an opening (m s^{-1}) = $U_c/(C_d R_a)$ where C_d is the discharge coefficient for the opening, R_a is the free area ratio and U_c is the core velocity for the device; h_o = effective height of slot (m) = $C_d h$ where h is the free height (m); K_v = throw constant and x = distance from the device (m).

Values of K_v should be obtained from the supplier of the ATD, but values for typical devices given in the ASHRAE Handbook [3] are given in Table 5.1 for isothermal flow.

In this table, the ATD height is based on either the core or the duct area and therefore h_o in equation (5.5) should be based on the area given in Table 5.1 to achieve the correct throw.

To predict the air movement in a room the throw and drop of the jet are normally required. Using equation (5.5), the throw for a large aspect ratio ATD is given by:

$$x_m/h_o = K_v^2 U_o^2/U_m^2 \tag{5.6}$$

where U_m is the throw velocity usually taken as 0.25 or 0.5 m s^{-1}. The drop of an isothermal jet can be calculated if the spread angle, α, is known. The drop y_m is then given as:

$$y_m = x \tan(\alpha/2) \tag{5.7}$$

Table 5.1 Typical throw constants K_v for classes 2 and 3 devices [3]

ATD type	Discharge pattern	Area	K_v
High-level wall grilles	No deflection	Core	2.38
	Wide deflection	Core	2.04
High-level linear wall diffusers	Core width <100 mm	Core	2.10
	Core width >100 mm	Core	2.23
Low-level wall grilles	Upward, no deflection	Core	2.23
	Upward, wide deflection	Core	1.71
Skirting (base) board grilles	Upward, no deflection	Duct	2.10
	Upward, wide deflection	Duct	1.40
Floor grilles	No deflection	Core	2.15
	Wide deflection	Core	1.26
Linear ceiling diffusers	Horizontal along ceiling	Core	2.33
Circular ceiling diffusers	360° horizontal	Duct	1.06
Square ceiling diffusers	Four-way	Duct	1.93

Table 5.2 Typical throw constants K_v for class 1 devices

ATD type	K_v	
	U_o (m s^{-1}): 2.5–5	10–50
Free openings		
Circular or square	5.7	7.0
Rectangular, aspect ratio <40	4.9	6.0
Grilles and grids, free area >40%	4.6	5.7
Perforated panels		
Free area 3–5%	3.0	3.7
Free area 10–20%	4.0	4.9

where x is the distance from the opening. For a large aspect ratio slot, $a \approx 33°$, i.e. $y_m = 0.3\,x$.

Low aspect ratio supply devices

For ATDs of class 1 the main region of the jet is the axisymmetric decay within which the maximum velocity decays according to:

$$U_m/U_o = K_v/(x/\sqrt{A_o}) \tag{5.8}$$

where the effective area A_o is given by equation (5.1).

Typical values of K_v for low aspect ratio ATDs to be used with equation (5.8) are given in Table 5.2.

From equation (5.8) the throw for a low aspect ratio device is given by:

$$x_m/\sqrt{A_o} = K_v U_o/U_m \tag{5.9}$$

where U_m is the throw velocity, usually either 0.25 or 0.5 m s^{-1}.

The drop y_m can be calculated from equation (5.7) if the jet spread angle is known. Typical angles for a circular ATD are given in Table 4.1.

Perforated panel outlets

A jet of constant velocity core is formed from the coalescence of individual jets issuing from a panel perforation. This extends a distance of about 5 equivalent diameters of the outlet panel core. In this region, the core velocity, U_c, is given by [3]:

$$U_c = 1.2U_0\sqrt{(C_d R_a)} \tag{5.10}$$

where R_a is the ratio of free area to gross or core area of the panel and U_0 is the velocity of the jet at the *vena contracta* of the perforation, i.e.:

$$U_0 = U_{oc}/(C_d R_a)$$

where U_{oc} is the discharge velocity based on core area.

For a panel of sharp-edge holes, $C_d \approx 0.65$ and equation (5.10) reduces to:

$$U_c = U_0 \sqrt{R_a} \tag{5.11}$$

To determine the decay of a jet issuing from a perforated panel normal to its face, the core velocity obtained from equations (5.10) or (5.11) is used as the outlet velocity in equation (5.9).

5.3.3   Effect of buoyancy

The effect of buoyancy force on the trajectory of a hot or cold jet was considered in Chapter 4. Additional information such as throw and drop are also required to predict the air movement in a room. These are both affected by the buoyancy force, which is represented by the Archimedes number, Ar. Data for the throw and drop of hot or cold jets issuing from ATDs are usually given by the suppliers. However, in the absence of this data empirical equations can be used. Data for wall-mounted grilles discharging cold air into a free space produced the equations that follow for the throw and drop to $0.25 \, \mathrm{m \, s^{-1}}$ [4]:

$$x_m/\sqrt{A_c} = 3.9(Ar)^{-0.45} \tag{5.12}$$

$$y_m = 1.05x_m(Ar)^{0.27} \tag{5.13}$$

where A_c = gross (core) area of grille (m^2); Ar = Archimedes number = $\beta g \, \theta_0 \sqrt{A_c}/U_f^2$; θ_0 = room temperature − supply temperature = $t_r - t_0$ (K); U_f = face velocity at grille, $\dot{V}/A_c$ (m s^{-1}) and $\dot{V}$ = discharge volume flow rate (m^3 s^{-1}).

Equations (5.12) and (5.13) are not to be used for extremely low values of Ar where the air jet is almost isothermal. In this case, the equations for isothermal flow, i.e. equations (5.6) and (5.7), should be used instead.

5.3.4   Extract devices

The air movement of a ventilated room depends on the type and position of the supply ATD. However, the effect of extract inlet type and position in the room is often much less significant unless 'short-circuiting' of the supply air occurs or the extract point is below a plume, pollution source, etc. This is due to the fact that air approaches an

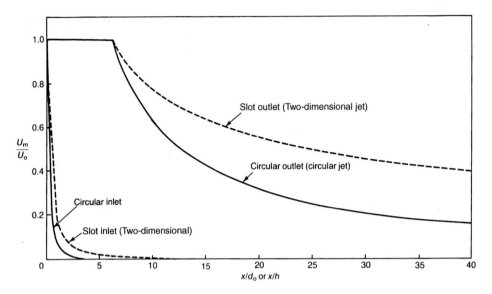

Figure 5.3 Comparison of the maximum velocity decay for circular and slot inlets with similar outlets.

extract opening from all directions, i.e. a large flow area upstream of the opening, and hence there is a rapid decrease of velocity with distance from the opening.

Figure 5.3 compares the variation of maximum velocity for a circular and a slot inlet with distance from the inlet with that corresponding to two similar outlets, i.e. jet flows. It is clear from this figure that at only a few diameters or slot heights from the inlet, very little air movement exists, whereas for the outlet the air movement is felt many diameters or slot heights downstream.

The effectiveness of local exhaust

In industrial applications where local exhausts are used to remove contaminants, it is important to know the value of the capture efficiency of the exhaust system. The pollutant capturing efficiency depends on the exhaust unit or hood position with respect to the contaminant source, the source type and the exhaust air flow rate in addition to the hood design. The *capture efficiency* is the fraction of contaminant generated outside the exhaust terminal that is captured by the exhaust. In general, there are three types of contaminant sources: (i) non-buoyant (diffusion) source; (ii) buoyant source and (iii) dynamic source. The non-buoyant source is charaterized by uniform diffusion due to concentration gradient in the room, which is affected by the concentration of contaminant in room air, room velocity and turbulence. The movement of the buoyant source is mainly due to temperature difference between the source and surroundings, forming a plume above the source. In the dynamic source, the contaminant moves in the space in the form of a jet or particle flow. In certain cases a source can be a combination of all these types.

Table 5.3 Typical ranges of capture velocity (V_c)

Contaminant release conditions	Typical application	V_c (m s^{-1})*
Zero velocity into still air	Evaporation from tanks, degreasing, planting	0.25–0.5
Low velocity into moderately still air	Container filling, low-speed conveyer transfers, welding	0.5–1.0
Active generation into a zone of significant air movement	Barrel filling, chute loading, cool shakeout	1.0–2.5
High-velocity generation into a zone of very high air movement	Grinding, abrasive blasting, tumbling, hot shakeout	2.5–10.0

Note
* The lower range of V_c applies to: favourable room air movement for capture, low toxicity contaminants, low or intermittent production and when a large hood and/or air flow is used. The upper range applies to: disturbing room air currents, high toxicity contaminants, high production and small hood or local control only.

The ratio between the local exhaust velocity and the velocity of the contaminant is a decisive factor in determining whether the exhaust device will capture the contaminant. The velocity of the exhaust air, which is necessary to capture a contaminant outside the exhaust opening and transport it into the opening, is called the '*capture velocity*', V_c. The concept of capture velocity is used to determine the required volumetric flow rate for an exhaust terminal or hood. For a certain flow rate the capture efficiency for a certain exhaust design will be higher than for another design if the capture velocity is higher. Typical values of capture velocity for some industrial processes are given in Table 5.3. These are based on experience under ideal conditions [5].

The target air exhaust flow rate, $\dot{V}^*$, may be determined using:

$$\dot{V}^* = k_h \, k_r \, \dot{V}_o \qquad (5.14)$$

where $\dot{V}_o = V_o \, A$ (m^3 s^{-1}); V_o = average air velocity at hood opening that ensures capture velocity at the point of contaminant release (m s^{-1}); A = hood opening area (m^2); k_h = dimensionless coefficient that depends on hood design and k_r = correction coefficient that depends on room air movement, usually >1.0.

Air movement near a point sink

The flow of a point sink can be used to approximate the air flow in the vicinity of an un-flanged exhaust opening. A point sink will draw air equally from all directions through an area equal to an imaginary sphere of radius, r. The radial velocity, v_r, of the sink is therefore:

$$v_r = \dot{V}/(4\pi r^2) \qquad (5.15)$$

If the sink is close to a surface then the flow area will be less than that for a free sink and consequently the radial velocity will be higher. In such cases equation (5.15) may be written in terms of the angle representing the flow area, α (radiance), i.e.:

$$v_r = \dot{V}/(\alpha r^2) \qquad (5.16)$$

Table 5.4 Flow angle (α) and capture radius (r_c) for a point sink at different locations

Sink arrangement	Flow angle, α (radiance)	Capture radius, r_c
Sink restricted by an infinite surface	4π	$r_c = \sqrt{\dfrac{\dot{V}}{4\pi V_c}}$
Sink restricted by two infinite perpendicular surfaces	2π	$r_c = \sqrt{\dfrac{\dot{V}}{2\pi V_c}}$
Sink in a corner	π	$r_c = \sqrt{\dfrac{\dot{V}}{\pi V_c}}$
	$\pi/2$	$r_c = \sqrt{\dfrac{2\dot{V}}{\pi V_c}}$

If the capture velocity, V_c, for a particular application is known (see Table 5.3) then the capture radius, r_c, for the exhaust is given by:

$$r_c = \sqrt{\frac{\dot{V}}{\alpha V_c}} \tag{5.17}$$

where α is the flow angle (radiance). Values of α and r_c for different point sink locations are given in Table 5.4.

Air movement near a line sink

A line sink creates a two-dimensional air flow in a radial direction. The radial velocity in this case is calculated using equation (5.18) in which the volume flow rate, $\dot{V}$, is taken as m^3 s^{-1} per metre length of sink. A free two-dimensional sink has a flow angle of 2π but if the sink is close to surfaces the flow angle, α, is reduced. The capture distance, r_c, is calculated using equation (5.19). Table 5.5 gives values of α and the

Table 5.5 Flow angle (α) and capture radius (r_c) for a line sink at different locations

Sink arrangement	Flow angle, α (radiance)	Capture radius, r_c
Free sink	2π	$r_c = \dfrac{\dot{V}}{2\pi V_c l}$
Sink outside the vertex of a 90° corner	$3\pi/2$	$r_c = \dfrac{2\dot{V}}{3\pi V_c l}$
Restricted by an infinite surface	π	$r_c = \dfrac{\dot{V}}{\pi V_c l}$
Sink in a right-angle corner	$\pi/2$	$r_c = \dfrac{2\dot{V}}{\pi V_c l}$

capture radius, r_c, for different arrangements of a two-dimensional sink.

$$v_r = \frac{\dot{V}}{\alpha r l} \tag{5.18}$$

$$r_c = \frac{\dot{V}}{V_c l} \tag{5.19}$$

where l is the length of sink (m).

Air movement near local exhausts

The air movement in the vicinity of exhaust hoods and canopies is quite complex and is dependent upon many factors. Generally, the air velocity distribution across a hood face is not uniform and is influenced by wake formation near the sides of the hood or flow contraction, both of which will reduce the effective face area of the hood opening. Hood and canopy design is a specialized area and the interested reader is referred to other references that specifically deal with local exhaust systems, e.g. references [6–9]. Here we shall consider the air movement for some basic shapes of exhaust hoods.

The variation of velocity with distance for various types of exhaust devices can be found in [6, 7]. For two-dimensional free slots of aspect ratio $R \geq 10$, the velocity at a distance x from the inlet, V_x, is given by:

$$\frac{V_x}{V_o} = \frac{1}{1 + 4\sqrt{(x/h)^3}} \tag{5.20}$$

where V_o is the mean velocity at the slot inlet and h is the slot height. For a flanged slot or a slot in a flat surface, e.g. wall or ceiling slot, the velocity upstream of the slot

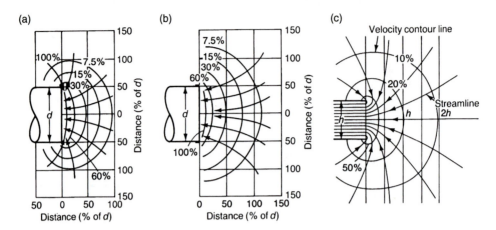

Figure 5.4 Velocity contours for circular and slot inlets. (a) Sharp-edged opening; (b) flanged opening; and (c) two-dimensional opening.

is approximately 33% higher than for a plain slot. The velocity in this case is given by:

$$\frac{V_x}{V_0} = \frac{1.33}{1 + 4\sqrt{(x/h)^3}} \tag{5.21}$$

In the case of a free rectangular inlet of aspect ratio $R < 10$, the velocity at a distance x from the slot is given by:

$$\frac{V_x}{V_0} = \frac{1}{1 + 10x^2(\sqrt{R}/A)} \tag{5.22}$$

where A is the inlet area. Similarly for a flanged or surface inlet the velocity increases as given by the equation:

$$\frac{V_x}{V_0} = \frac{1.33}{1 + 10x^2(\sqrt{R}/A)} \tag{5.23}$$

For a corner inlet with one large side along the line joining the two surfaces, e.g. a low-level wall extract, the velocity increases further and is given by:

$$\frac{V_x}{V_0} = \frac{2}{1 + 10x^2(\sqrt{R}/A)} \tag{5.24}$$

Typical constant velocity contours (*isovels*) for a free circular inlet, a flanged circular inlet and a free two-dimensional slot are shown in Figure 5.4. These contours have been found to be independent of Reynolds number for the range 3.5×10^{-4} to 10^5 [10].

5.4   Types of air terminal devices

The extensive progress in air distribution systems over the last two decades has resulted in the development of new types of supply ATDs. There are now in use many different types of supply ATDs as listed below:

• Grilles and registers
• Nozzles
• Diffusers
• Low-velocity terminals
• Impinging jet terminals
• Textile terminals
• High-impulse terminals
• Laminar flow panels.

The traditional, yet still commonly used, devices are the grilles and registers, which produce flow essentially perpendicular to the face of the device, and the different types of diffusers that can deflect or spread the flow in different directions. Nozzles have also been in use for a long time and are mainly suited for supplying air in large enclosures. Low-velocity displacement flow terminals are being increasingly used in many types of buildings. These are devices that supply low-velocity air at or slightly below room temperature at low level to produce an upward displacement flow in the room. The impinging jet terminals are relatively new and can be of various shapes and geometry but essentially they all supply air down towards the floor. Textile terminals are relatively new too and are used in situations where large cooling is needed at low air velocity. High-impulse terminals use high momentum air through small perforations in a circular air duct and can be used for heating and cooling applications. Laminar flow panels are specialized ATDs that are mainly used in certain industrial applications.

The choice of an ATD should be based upon the capability of the device to provide the desired air velocity and temperature distribution in the space with acceptable pressure losses, turbulence and noise levels. The types of air supply ATDs listed here will be described in the following sections.

5.4.1   *Grilles and registers*

There are many different types of grilles and registers usually made of aluminium, steel or a synthetic material with either fixed or adjustable vanes forming the core. The vanes, which are straight, curved or an aerofoil shape, are set horizontally or vertically in a single deflection grille or both in a double deflection grille. Adjustable vanes, if available, are used normally for varying the spread of the jet. The characteristics of grille and damper accessories (e.g. flap, multi-shutter, opposed blade) are normally provided by the device manufacturer. Popular types of grilles and registers are briefly described in the following.

Supply grilles   Supply air grilles having one or two sets of individually adjustable vanes are suitable for commercial and industrial application requiring adjustment of air flow and throw. They are usually used as low-level or high-level wall supply outlets. The vertical and horizontal vanes can normally be set to spread the flow by up to 45° in one or two directions. The throw of a supply grille almost halves when the vanes on either side are deflected by 45° (90° included angle) but at the same time the aerodynamic noise increases. Figure 5.5(a) shows a section through a supply grille with horizontal front vanes and vertical rear vanes. These grilles can be supplied

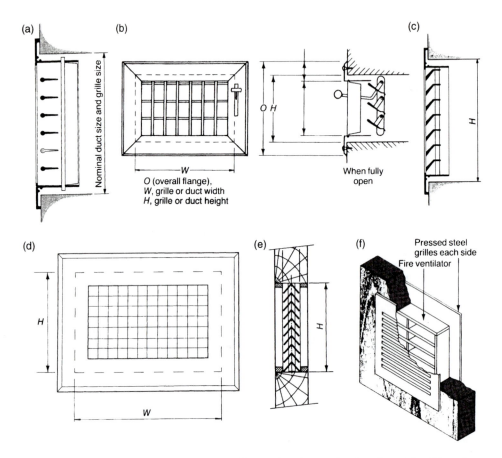

Figure 5.5 Common supply and extract grilles and registers. (a) Supply guide; (b) register; (c) extract grille; (d) egg-crate grille; (e) transfer grille and (f) fire transfer grille.

with opposed blade dampers, flap dampers, multi-shutter dampers or rhomboidal dampers for flow control. Figure 5.5(b) shows a typical domestic register with one set of individually adjustable vertical front blades and a set of rear horizontal linked damper blades.

Extract (exhaust) grilles   The grille core consists of fixed vanes or meshes in a wide range of shapes. Figure 5.5(c) shows a section though an extract grille with 45° fixed blades and Figure 5.5(d) shows a front view of an egg-crate core extract grille. Perforated plates are sometimes used for aesthetics and economy. Extract grilles can also be supplied with dampers. These grilles are used for wall extract inlets.

Sill and floor grilles   These grilles comprise an outer frame and a removable inner core with fixed blades retained by reinforcing bars across. They are usually designed for sill- or floor-level perimeter heating systems that can withstand light traffic. The flexible core varieties are convenient for room perimeter application. All these grilles may be used for air supply or extraction.

Linear grilles   These are large-aspect ratio grilles used for air supply and extraction as individual grilles or as a continuous array fitted to a wall, sill, bulkhead or ceiling.

They have fixed blades with or without an enclosing border and are available in a wide range of core styles. Because of the large-aspect ratio these grilles produce a large throw when used as air supply terminals.

Transfer grilles   These are usually constructed from inverted V-shaped blades for non-vision purposes and are suitable for door, partition or wall mounting where transfer or exhaust air passes. They normally produce larger pressure losses and noise levels than exhaust grilles. Figure 5.5(e) shows a section through a transfer grille. Acoustic transfer grilles are used in room partitions and doors where a reduction in noise transfer from one space to another is required. These are lined on the inside with acoustic damping material. Fire resistance transfer grilles, which are also used in doors and room partitions, see Figure 5.5(f), consist of an element of the same thickness as the door or partition with an intumescent medium that expands and carbonizes when exposed to temperatures in excess of about 150 °C and seals off the ventilation openings on either side of the element.

5.4.2   Nozzles

In industrial applications where large air flow rates and long throws are required, freely suspended nozzles are often used, as in the heating and cooling of auditoria, large halls and in factory heating systems. Most commercially available nozzles offer some flexibility in adjusting the discharge angle of the nozzle, which is useful in commissioning and when changing from heating to cooling mode or vice versa. A basic nozzle consists of a circular or rectangular duct opening with or without a core. In most applications, however, various types of devices are attached to the nozzle outlets to provide a better control of the air diffusion from the nozzle. The characteristics of the jet discharged from nozzles are treated in Chapter 4 and those for some circular nozzle types in use are given here.

Flared nozzle   A flared diffuser is used as an air supply nozzle in preference to a straight duct outlet when a lower throw is required, see Figure 5.6 (a). However, the included flare angle is usually limited to about 10° as a greater divergence of the flow creates instability in the air discharge and a tendency for the jet to separate from one side of the core and attach to an opposite side. This problem is usually overcome by inserting a baffle at the point where the diffuser meets the duct, Figure 5.6(b), or inserting one or more conical guide vanes in the diffuser, Figure 5.6(c). The maximum velocity decay for a three-core nozzle with equal flow areas at the duct end of each core is compared with that for a circular duct outlet in Figure 5.7. The effect of the cones on the diffusion of the jet is significant in which case the velocity ratio, U_m/U_o, at a distance $10d$ from the nozzle is 0.13 compared with 0.57 for the circular duct outlet. This is due to the extensive widening of the jet and hence a greater entrainment of the surrounding air in the case of the conical nozzle.

To avoid short duct bends in some applications, turning vanes are inserted at the duct outlet, Figure 5.6(d), so that a horizontal jet throw can be achieved. If adjustable turning vanes are used the jet can then be deflected in any desired direction.

Hemispherical nozzles   A wide spread of the air jet can be achieved with a nozzle in the form of a perforated hemisphere. The main jet flow is formed by the merger of smaller jets produced by the perforations, and hence the spread and velocity decay of the main jet are strongly influenced by the shape and distribution of the perforations

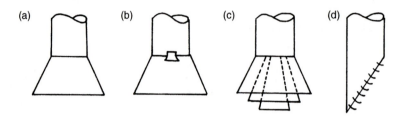

Figure 5.6 Types of flared nozzles. (a) Flared nozzle; (b) nozzle with a baffle; (c) nozzle with guide vanes and (d) nozzle with turning vanes.

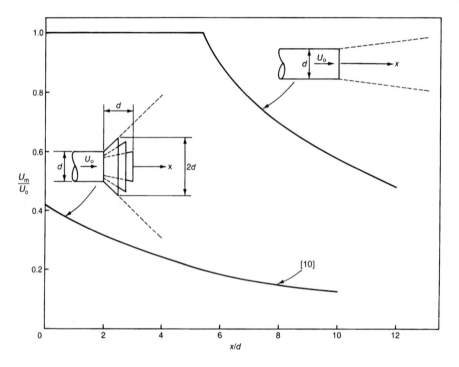

Figure 5.7 A comparison between the velocity decay for a three-core nozzle with a circular duct nozzle.

over the hemisphere. A nozzle with uniformly spaced circular holes over the hemisphere does not normally produce a radial flow, but instead the small jets merge towards the nozzle axis to produce a main jet flow similar to that emerging from a cylindrical duct but at a much larger pressure loss penalty. For such a nozzle, a pressure loss coefficient of $\xi \approx 11$ is possible compared with 1 for a cylindrical outlet. A wider spread with a lower pressure loss ($\xi = 1$–2) can be achieved by either increasing the diameter of the holes towards the base of the hemisphere or using petal-shaped slots, which are wider near the base and narrower towards the cap of the hemisphere. The decay of the maximum jet velocities for the two types of hemispherical nozzles mentioned here is compared in Figure 5.8 with the decay of a jet from a circular duct.

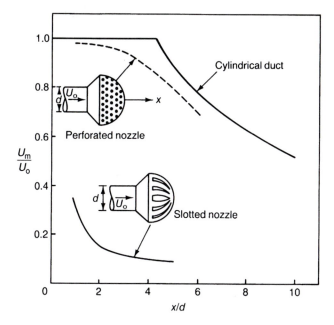

Figure 5.8 Velocity decays for hemispherical nozzles [10].

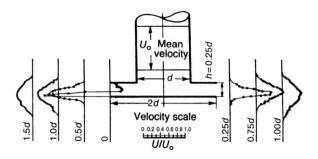

Figure 5.9 Single baffle nozzle.

There is little difference between the decay of the jet from a uniformly perforated hemispherical nozzle and that from a cylindrical duct outlet, but a slotted hemisphere gives a much faster decay due to the wide spread produced by a nozzle of this type.

Nozzles with baffles A baffle plate at the exit of a circular duct causes the flow to deflect by 90° and form a radial jet. This type of nozzle arrangement is used where a large dispersion of the air is desired radially. Test data for the type of baffle plate nozzle shown in Figure 5.9 are given in [10]. The diameter of the plate in this case is twice the duct diameter, which is sufficient to turn the jet completely by 90°.

Although a larger plate diameter will have little effect on the jet produced by this nozzle, a plate diameter smaller than $1.5d$ may not produce a complete turning of the flow. The spacing between the duct exit and the baffle affects the maximum exit

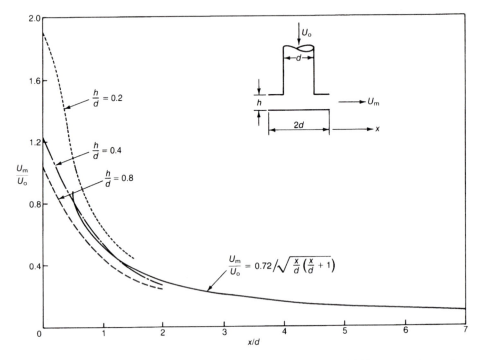

Figure 5.10 Decay of maximum velocity for a single baffle nozzle of different spacing ratio.

velocity, U_{mo}, but has little influence on the maximum velocity U_m for $x > 0.5d$ as shown in Figure 5.10. Baturin [10] gives the following expression for the maximum velocity decay for $x > 0.5d$ in terms of the duct velocity:

$$U_m/U_o = 0.72/\sqrt{[(x/d)(x/d + 1)]} \tag{5.25}$$

which is also plotted in Figure 5.10. The spacing, h, however, has a significant effect on the pressure loss coefficient that is based on the duct velocity, as shown in Figure 5.11. For a spacing ratio of $h/d > 0.5, \xi \approx 1.0$, i.e. the baffle plate introduces no additional resistance to the flow and the loss in pressure is only the velocity pressure.

5.4.3 *Diffusers*

Diffusers are ATDs that can supply air jets with different forms and directions and are usually used for ceiling, floor or desk supplies for both air supply and extraction. A supply diffuser can provide a linear jet, a radial jet or jets in more than one direction. Because of the flexibility they offer diffusers are widely used in air distribution systems. There is the slot type that can provide large volumes of air (suitable for large air change rates) with long throws or the round swirl type that can provide rapid diffusion over a short distance. Some of the most common types of diffusers are described here but for additional information and full characteristics readers should refer to the manufacturer's data relevant to the particular diffuser.

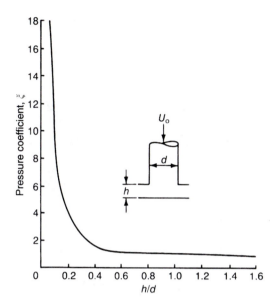

Figure 5.11 Effect of baffle spacing on pressure loss coefficient.

Ceiling diffusers   A ceiling is the most convenient surface for diffusing the initial high momentum of an air jet because it is outside the occupied zone of the room. Ceiling diffusers are probably the most common types of ATDs used for air distribution in commercial buildings. There is a wide variety of ceiling diffusers to suit almost every application. Some of the more widely used types are briefly discussed here.

Circular multi-cone diffusers, Figure 5.12, are used in applications requiring compact diffusers that are capable of supplying large volume-flow rates. These are usually constructed from steel or aluminium spinnings and their core can be adjusted to produce horizontal (along the ceiling) or vertical downward flow patterns. The diffusers can be supplied with dampers for controlling the flow rate. The flow pattern is usually determined by the magnitude of the angle, φ, between the jet outlet and the ceiling. For $\varphi < 30°$ a horizontal flow is produced due to the Coanda effect, but greater values of φ will produce a downward projection as shown in Figure 5.13. For $\varphi < 30°$ and a single cone diffuser, the decay of maximum jet velocity along the ceiling is given by [11]:

$$U_{m}/U_{o} = 2.2\sqrt{\{r_{o}h\cos\phi/[r(r - r_{o})]\}} \tag{5.26}$$

The circular ceiling diffuser can be used for supply or extraction or a combination of both. In the latter case, supply air is discharged through the outer cones and extract air is drawn through the inner cones. However, the extract air flow rate is usually limited to about 80% of the supply air to avoid short-circuiting of supply air to the extract duct.

Circular swirl diffusers with fixed or adjustable vanes are designed to generate radial air jets with a swirling action, Figure 5.14. The jets of air generated by the vanes merge on discharge and produce a swirling jet of air that attaches to the ceiling due to the

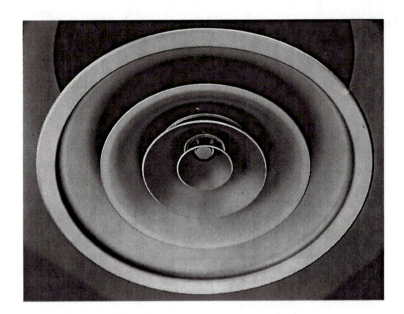

Figure 5.12 A multi-cone ceiling diffuser.

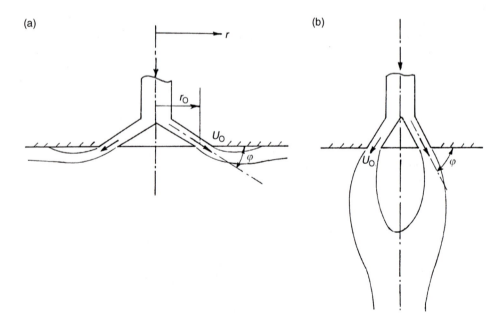

Figure 5.13 Effect of angle on the flow pattern of a circular ceiling diffuser: (a) $\varphi < 30°$ and (b) $\varphi > 30°$.

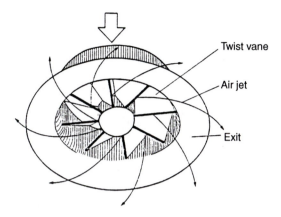

Figure 5.14 A ceiling swirl diffuser.

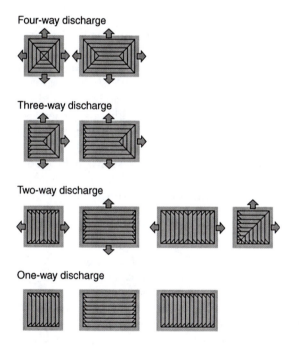

Figure 5.15 Square and rectangular diffuser core types.

Coanda effect. The swirling jet induces large air entrainment causing a rapid diffusion of the jet with room air. This type of ceiling diffuser is suitable for handling large air volumes without the risk of draughts in the occupied zone.

Square and rectangular ceiling diffusers are available in one-, two-, three- and four-way discharge and can be used for supply or extraction. Diffusers are usually available in the same size as ceiling tiles or panels. Most of these devices have adjustable cores to give the system designer flexibility in size and flow pattern selection. Some typical core type diffusers are shown in Figure 5.15.

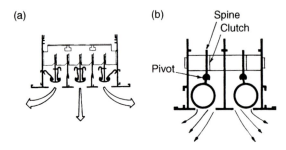

Figure 5.16 Adjustment of flow direction in linear slot diffusers.

Although linear slot diffusers are usually designed for ceiling installations they can also be used for wall, sill or floor supplies. Ceiling slot diffusers normally have adjustable vanes or other control elements for varying the jet direction, but wall, sill and floor diffusers are usually provided with fixed vanes. Figure 5.16(a) shows a slot diffuser that produces a horizontal or vertical jet by vane adjustment and Figure 5.16(b) shows a diffuser that produces similar flow patterns by the movement of a roller. Slot diffusers are available as single or multiple slot outlets and can be used to provide a continuous strip across a ceiling or on the perimeter of a ceiling. They are usually made from extruded aluminium and can be provided with such accessories as plenum boxes, control dampers, flow equalizing grids, blanking-off strips and mitred corners. Slot diffusers are used for air supply and extraction usually without any changes being necessary. The main characteristics of the supply slot diffusers are high air-handling capacity, variable flow pattern capability and good performance at reduced flow rates, which makes them attractive for variable air volume (VAV) systems. In integrated ceiling designs troffer diffusers can be incorporated with luminaires, which is particularly attractive for extract air where part of the lighting load is extracted at source before it enters the air-conditioned space.

Floor diffusers   In high cooling load applications such as computer rooms or high-density occupancy, it is preferable to supply the cool air from the floor so that the heat produced by data processing equipment and/or people can be convected upwards by the supply air. This is more effective in dissipating high-load concentrations than a high-level air supply [12]. Floor ATDs can be in the form of perforated or slotted diffusers producing a flow normal to the plane of the diffuser or a circular swirl diffuser discharging jets of air at an angle (less then 90°) to the diffuser face by inclined slots, Figure 5.17. Often the latter type is preferred because the extensive diffusion of the supply jet with room air as a result of the swirl action produces a rapid decay of jet velocity and temperature difference between the jet air and room air. Figure 5.18 compares the maximum velocity and temperature difference profiles for two circular floor diffusers, one being a plain diffuser and the other a swirl diffuser. The rapid jet diffusion of the swirl diffuser is evident from this Figure 5.18.

Personal diffusers   Personal air supply devices are becoming more widely used because of the flexibility and flow control they offer the occupant. They are mainly used in theatre seats, lecture theatres and office desks. In a theatre seat, the diffuser is situated at the top of the backrest for discharging the air at a small angle towards

(a)   (b)

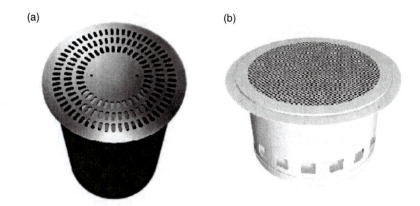

Figure 5.17 Swirl floor diffusers.

(a) $V = 11\ \text{l s}^{-1}$
$\Delta t_o = t_{room} - t_o = 4\ \text{K}$
t_x = temperature at checkpoint x
t_o = supply air temperature

(b)

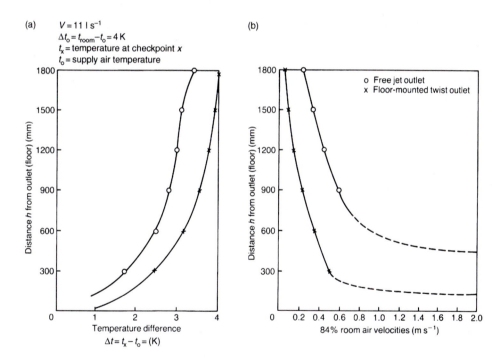

Figure 5.18 Comparison of velocity and temperature decay above a plain circular diffuser and a swirl diffuser [12].

the next rear seat (approximately 0–20° from the vertical), see Figure 5.19. In a lecture theatre desk, a linear diffuser is placed along the front edge of the desk over most of the desk length to produce a jet envelope in front of the seated person's head, i.e. the head is not directly in the path of the jet but within its entrainment zone, see Figure 5.20. The discharge velocity for a backrest or a desk diffuser is usually between

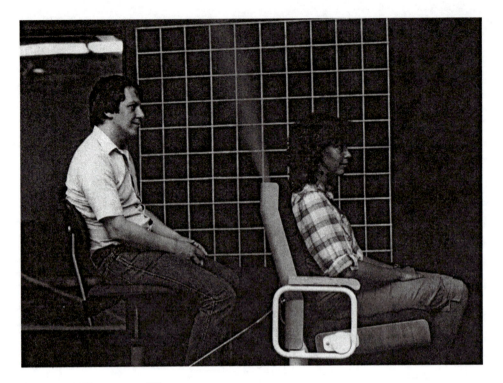

Figure 5.19 Theatre seat diffuser.

1.5 and 2.5 m s^{-1}. Primary conditioned air is normally supplied at a higher velocity into a plenum where induced room air mixes with the primary air before it is discharged from the top of the backrest or the desk to the room. A desk-type personal air supply system, which directs a jet of air towards the seated person's head and provides user control for air speed and temperature, is shown in Figure 5.21. This type offers adjustable jet speed by means of a control disc that changes the swirl angle of the jet.

5.4.4 *Low-velocity terminals*

In displacement air distribution, the air is supplied at low level directly into the occupied zone for minimum mixing with room air. As a result the air must be supplied at low velocity (usually between 0.25 and 0.5 m s^{-1}) and hence a large area of air supply device is needed to provide adequate quantities of air. This is normally achieved by supplying the air through an ATD surface that contains many small holes to produce air at low velocity and turbulence. The air jets supplied from these small holes very quickly merge after leaving the ATD to form a 'stream flow' of air over the floor. The purpose of such terminals is to spread the supply air over the whole floor in order to remove the contaminated room air as it warms up and rises upward. There are many different shapes of low-velocity air terminals, e.g. cylindrical, semi-cylindrical, quarter cylindrical, flat face, etc. Figure 5.22 shows two of these types.

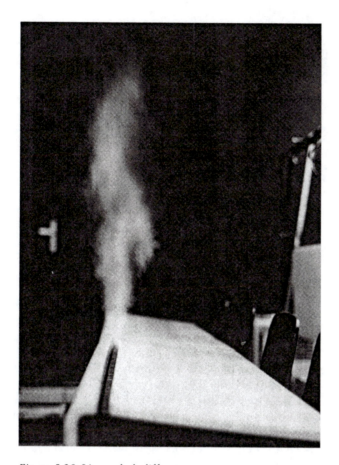

Figure 5.20 Linear desk diffuser.

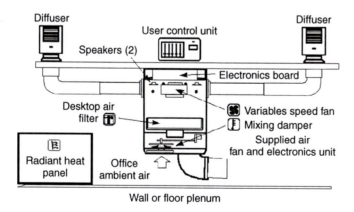

Figure 5.21 Personal swirl diffuser (courtesy of Johnson Controls, Inc.).

Figure 5.22 Types of low-velocity air terminal units (courtesy of Halton Products).

5.4.5   Impinging jet terminals

The impinging jet ventilation concept (see also Section 4.3.3) is based on the principle of supplying air at relatively high velocity downward on to the floor to form a thin layer of fresh air over the floor. The air terminal is normally mounted on a wall or a column at a height of 0.3–1.0 m above the floor. The air leaves the terminal at a velocity of 1–5 m s^{-1} towards the floor. As the jet impinges on the floor it spreads over a large area but the wall jet created is only a few centimetres thick and it travels at higher speeds than in the case of wall displacement systems as it has a higher momentum. This provides a better spread than in the case of wall displacement terminals particularly when floor obstructions are present. Furthermore, lower-supply air temperatures (down to about 16 °C), than those for wall displacement systems, can be used as the jet entrains some room air before it reaches the floor. There are many types of terminal devices that are used in impinging jet ventilation. The range of Air Queen$^{\circledR}$ terminals developed by Air Innovation AB in Sweden is quite exhaustive. Few such devices are shown in Figure 5.23.

5.4.6   Textile terminals

Flexible textile ducts are used for supplying cool air to spaces by diffusing the air through permeable material at a velocity of between 0.07 and 0.18 m s^{-1} with a supply temperature 6–9 K below room temperature. The air flow in the space is influenced by the temperature difference between the air supply and room, ΔT, as shown in Figure 5.24. The air supply ducts are normally hung at a high level and the cool air descends down to the occupied zone by displacing the warm room air. These air terminals are particularly well suited to industrial spaces where large air supply for cooling is needed. It is claimed that up to 300 W m^{-2} of floor area cooling could be achieved without noticeable draught. Applications are found in the food and catering industries, the pharmaceutical industry, the electronic industry, the leisure industry, etc., see Figure 5.25.

Figure 5.23 Some of the impinging jet terminal devices that are in use (courtesy of Air Innovation, AB).

5.4.7   *High-impulse terminals*

This type of air supply terminals consists of a metal duct with small nozzle-shaped holes along the whole length of duct supplying air either horizontally or at an angle of 45° to the horizontal, see Figure 5.26. The diameter of the holes is about 5 or 6 mm supplying air at a velocity between 10 and 15 m s^{-1}. The high impulse of the supply air creates a large entrainment thus maintaining the room air velocity at acceptable levels in the occupied zone. This system can be used for both heating and cooling with maximum heating load of about 180 W m^{-2} and maximum cooling load of about 250 W m^{-2} of floor area.

5.4.8   *Laminar flow panels*

Laminar flow panels are used in clean rooms, laboratories and hospitals where low-turbulence air supply is required. The panels are either positioned on the ceiling to produce a downflow or on a wall to produce a side flow. Although called laminar flow panels, the air flow produced by them is not truly laminar but a low-turbulence flow. This flow is achieved by the large number of perforations in the panel face that produce small air jets that merge together to form a uniform air flow. In clean rooms, the panels are installed side by side to prevent entrainment of supply air with

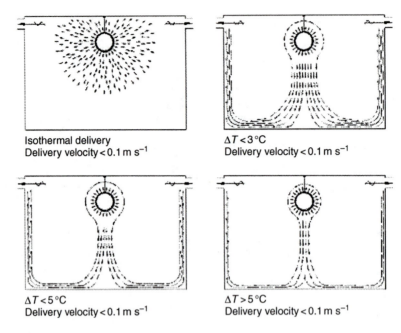

Isothermal delivery
Delivery velocity < 0.1 m s^{-1}

$\Delta T < 3\,°C$
Delivery velocity < 0.1 m s^{-1}

$\Delta T < 5\,°C$
Delivery velocity < 0.1 m s^{-1}

$\Delta T > 5\,°C$
Delivery velocity < 0.1 m s^{-1}

Figure 5.24 Air flow patterns from a textile duct as affected by ΔT.

Figure 5.25 Some applications of textile ventilation ducts (courtesy of KE Fibertec).

room air. In this case displacement ventilation is achieved with low turbulence and the air velocity remains almost constant throughout the room. This velocity may be calculated using equation (5.10). For other less demanding applications laminar flow panels may be mixed with perforated ceiling diffusers that produce conventional air jet supplies. In this case the air supply from the laminar flow panels diffuses with room air as a normal jet.

Laminar flow panels can be supplied with flow control dampers at the inlet to the panel plenum.

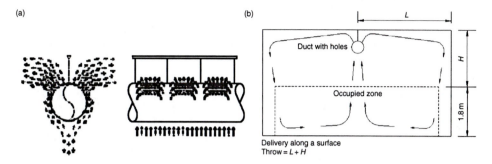

Figure 5.26 Air movement due to high-impulse duct terminals for 45° holes position. (a) Flow around duct and (b) room air flow pattern (courtesy of KE Fibertec).

5.4.9   *External louvres*

External louvres are manufactured from extruded aluminium blades so shaped that ingress of rainwater is minimized. Some types have perforated blades to allow accumulated water to drop onto a sill at the bottom of the louvre and flow outside. Louvres are normally fixed to a wall where air is drawn into a ventilation plant or exhausted from the plant. They should be selected so that large free areas are available with minimum pressure losses. Aesthetic appeal is also taken into consideration when selecting louvres. Positioning of external louvres for air intake requires careful consideration to reduce short-circuiting with exhaust louvres and the ingress of polluted air from outside.

5.5   Selection of air terminal devices

5.5.1   *Supply air terminal devices*

The type of supply ATD required for a particular application is determined by the room load (W m^{-2} of floor area), the desired air flow pattern in the occupied zone, the noise characteristics and the pressure losses, as well as by the aesthetic and architectural requirements. The influence of room load on ATD selection and the air flow patterns produced by different ATDs in different positions in a room are treated in Chapter 6. Acoustic and pressure loss data is usually provided by the ATD supplier. Different spaces require ATD of different noise rating (*NR*). In an office for example, a diffuser of *NR* = 35 can usually be tolerated whereas in a concert hall an *NR* of 15 may be called for. Having decided on the types of ATDs to be used for a particular application and their location in the room (i.e. ceiling, wall, floor, etc.) the exact positioning of each ATD is then required. This is dictated by the aerodynamic and acoustic characteristics of each ATD, which can be extracted from the manufacturer's data sheet for the ATD. However, this data normally relate to ideal conditions of installation and any departure from these conditions will produce a different performance. For example, the data does not normally allow for the presence of a flow control damper, the variation in the length of a supply duct to the plenum chamber, the presence of bends in the supply duct, etc. Furthermore, the effect of texture of room surfaces, particularly the ceiling in the case of ceiling supplies, and the effect of room partitions, furniture and heat sources or sinks are not usually included in the ATD performance data. All of these factors could produce air flow patterns that are different from the manufacturer's published data. Great care must therefore be exercised in deciding the positions of ATDs in a room and wherever possible the manufacturer's installation guidelines should be adhered to. If it is not possible to implement the latter then allowance should be made for the effect of departure from installation guidelines on the performance of the ATD. If a ceiling with obstructions is used as an air diffusion surface then the effect of obstructions must be considered using the procedure outlined in Chapter 4.

The position of a supply ATD in a room will depend on the type selected, the air volume capacity, the throw, drop and spread, the pressure losses across the ATD and the sound level requirement. The air volume flow rate supplied by an ATD, $\dot{V}_s$, depends on the room load, the air supply temperature (determined from air-conditioning cycle analysis), the room temperature and the number of ATDs in the room. This can be calculated using:

$$\dot{V}_s = Q_s/[C_p\rho(t_r - t_s)] \tag{5.27}$$

where Q_s is the sensible thermal load for the ATD to handle (W), and ρ (kg m^{-3}) and C_p (J kg^{-1} K^{-1}) are the supply air density and specific heat respectively. The other factors given previously should be evaluated from the manufacturer's data. These are normally supplied in the form of a nomogram that gives the throw, drop, pressure loss and noise rating for a particular ATD for various flow rates. By knowing the volume of air flow through the diffuser in l s^{-1} (per metre length in the case of a slot diffuser), the nomogram provides the designer with the radius of diffusion (throw) for a specified terminal velocity (0.25, 0.5 or 0.75 m s^{-1}), the Noise Criteria Curve (*NC*) or the Noise Rating (*NR*) level and the pressure loss across the diffuser in pascals. Nomograms are usually prepared from tests carried out on ATDs according to the procedures laid down by a standard, e.g. BS 4773 [1] or ISO 5219 [2]. Corrections to the ATD data for conditions that are different from the standard tests are usually obtained by extrapolation and therefore should only be taken as approximations to the actual performance of the ATD in situ at these conditions.

With the general use of computer software in design, many manufacturers offer computer programs for selecting a suitable ADT that takes into consideration room geometry, thermal load, supply temperature, noise rating, etc. This offers a more convenient and rapid selection of ATDs as well as the choice of selecting different types that may be available, which could provide the required room air distribution.

General guidelines that may be useful in deciding the size and position in the room of supply ATDs commonly used in commercial buildings are given in the following.

Wall-mounted grilles　Wall grilles mounted at high level are selected to discharge the air along the ceiling using the Coanda effect. This is particularly beneficial in the cooling mode as it minimizes the risk of 'dumping', i.e. the projection of a cool jet into the occupied zone. The available ceiling area should be utilized to reduce the throw, thus resulting in smaller grilles and ducting and more efficient air diffusion. The throw is normally based on a jet terminal velocity of 0.25–0.5 m s^{-1} depending on the application. Usually, the lower value is used for heating to obtain low room

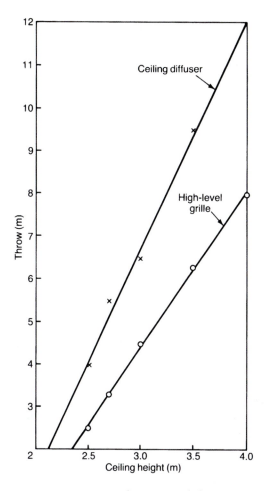

Figure 5.27 Increase of recommended maximum throw with ceiling height.

velocities and the higher value for cooling to give higher room velocities. It is important that the jet does not drop into the occupied zone and for this reason a recommended throw value for high-level grilles depending on the ceiling height is given in Figure 5.27, which is based on data given in [13]. If the required throw exceeds the values given by this figure alternative solutions should be sought, such as using more than one grille on the same wall for the case of low-aspect ratio grilles or putting another grille on the opposite wall for the case of linear grilles. By distributing the air flow in this way the throw can be reduced. If the throw is lower than that given by Figure 5.27 then a smaller grille or a different device needs to be selected.

Ceiling diffusers The ceiling should be divided into convenient strips in the case of slot diffusers or convenient squares in the case of circular or square diffusers, the size of which is based on the maximum throw given by Figure 5.27 according to the height of the ceiling. For a given volume flow rate through a slot diffuser the throw

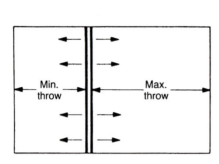
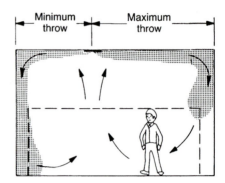

Figure 5.28 Selection of ceiling slot diffusers.

depends on the number of slots selected, i.e. a larger throw with less slots. When the slot diffuser is not in the middle of the ceiling and discharging air in two directions the number of slots is usually determined by the distance of the diffuser from the two walls, known as the minimum or maximum throw, see Figure 5.28. Most diffuser selection software allows for the position of the slot diffuser on the ceiling taking into consideration the air flow rate and the minimum and maximum throws. The number of slots can be selected to produce the required throw, but the optimum number should be a compromise between economic consideration (i.e. minimum number of slots) and comfort consideration (i.e. maximum number of slots). If the optimum number of slots is less than one then a single slot diffuser is selected with the active length reduced by a blank slot. A slot number that is greater than that given in the nomogram or the diffuser selection software, means that the diffuser is not suitable for the application being considered and an alternative solution should be found.

Similarly, for a circular, square or four-way rectangular ceiling diffuser the ceiling is divided into equal square areas with each area serviced by one diffuser, see Figure 5.29. The throw in this case should ideally be equal to the distance from a wall or half the distance to the next adjacent diffuser. If the diffuser is not in the middle of a square area of the ceiling then the procedure outlined for a slot diffuser should also be applied for these diffusers. If the minimum throw required is larger than the throw provided by the smallest size diffuser available, then insufficient jet mixing with room air will result and an alternative solution should then be considered. On the other hand, if the throw of the largest available diffuser is greater than the required throw, then an alternative solution is required in this case too.

Unbound grilles and diffusers   When a grille or diffuser is mounted away from a surface such that the Coanda effect will not be present, the velocity decay of the jet is faster, i.e. the throw is reduced. In this case the throw correction factors given in Table 5.6 should be applied to the device data for wall bound.

The chart in Figure 5.30(a) may be used as a guide for the initial selection of ATDs given the air flow rate per square metre of floor area. Figure 5.30(b) shows a comparison of the relative cost of different ATDs for the same air flow rate to a given room. The cost does not include ducting or flow control accessories. For example,

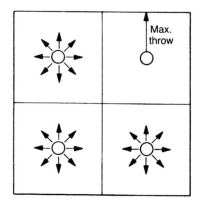

Figure 5.29 Selection of circular ceiling diffusers.

Table 5.6 Throw correction factors for an unbound ATD [13]

Type of ATD	Distance from ceiling (mm)			
	<300	300–600	600–1000	>1000
Low-aspect ratio grille	1.0	0.7	0.7	0.7
Linear grille	1.0	0.8	0.7	0.7
Diffusers	1.0	0.9	0.8	0.7

a slot diffuser could cost between 2.5 and 7.0 times more than a grille with straight adjustable vanes.

Floor grilles and diffusers Floor grilles are usually of the perforated- or slotted-plate type with a free area of 10–20%. The small individual jets issuing from the holes do not entrain room air but rather they entrain other surrounding jets resulting in a rapid reduction of air speed with height above the grille. However, floor diffusers are usually of the circular type that produce free or swirl jets depending on whether the supply slots are straight or twisted. The performances of these two types of diffusers are compared in Section 5.4.3. In positioning floor grilles or diffusers, attention should be given to the areas in the room where large thermal loads are likely to be placed, where large concentration of ATDs are required to dissipate the heat more effectively. At the same time, depending on the type and or size of diffuser this should be placed at a distance of between 0.8 and 1.5 m away from a sedentary occupant to reduce the risk of draught [14]. Air supply temperature should not be less than 18 °C for cooling and not greater than 26 °C for heating.

Low-velocity terminals These ATDs consist of large perforated plates that can have many different shapes and air is supplied uniformly across the plate face. Depending on the application, common ATD shapes are: flat, rectangular, quadrant, semi-circular, circular. In each case the slightly cooler air descends onto the floor as it leaves the ATD and pervades the region close to the ATD. In selecting these ATDs, it is important to

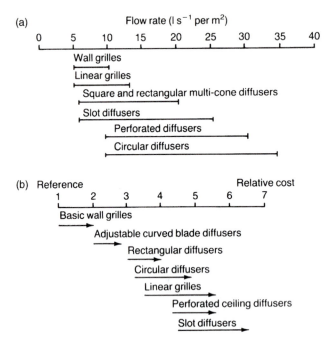

Figure 5.30 Comparison charts for some ATD's. (a) Flow rate range per m² floor area and (b) relative cost of ATDs.

know the maximum velocity at a certain distance from the terminal in order to ensure draught-free air distribution, particularly at ankle height. The radius of diffusion to a velocity of either 0.2 or 0.35 is determined from manufacturer's data for this purpose. This is illustrated in Figure 5.31 by the value of L_v in longitudinal direction and B_v in the lateral direction for semi-circular and circular low-velocity terminals.

Impinging jet terminals   As was shown earlier there are many different designs of ATDs that produce an impinging jet onto the floor. The choice of an ATD for a particular application will depend on:

- direction of air supply, i.e. whether the supply duct is in the ceiling or under the floor
- the room surface or feature to support the ATD
- the air supply rate required by the ATD
- the noise level that can be tolerated
- aesthetic considerations as these ATDs have the advantage of being able to integrate with the architectural or interior design features of the room.

5.5.2 *Extract air terminal devices*

The selection of extract ATDs is usually based on flow rate, pressure losses and noise level as well as aesthetic and architectural considerations. In general, the mass flow rate handled by an extract device may not be the same as that supplied to the space as some

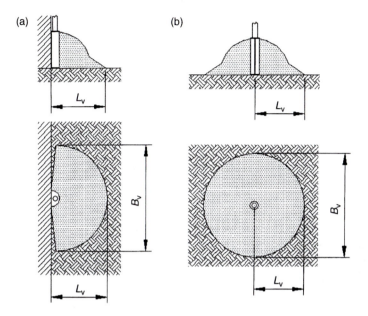

Figure 5.31 Radius of diffusion for: (a) semi-circular low-velocity terminal and (b) circular low-velocity terminal.

air may be exhausted outside through another device or the extract air is kept below the supply rate to maintain a certain positive pressure level in the room. Even when the same mass of air supplied is extracted, the volume flow rate through the extract device may be different from that supplied as a result of the temperature difference between supply and extraction. The ratio of extract air flow rate, $\dot{V}_e$ (m^3 s^{-1}), to supply flow rate, $\dot{V}_s$ (m^3 s^{-1}), is given by:

$$\dot{V}_e/\dot{V}_s = (t_r + 273)/(t_s + 273) \tag{5.28}$$

where t_r is the extract air temperature (i.e. room temperature) and t_s is the supply air temperature, both in °C. The extract volume is larger than the supply volume when cool air is supplied to the room and smaller when warm air is supplied. If pressurization or depressurization of the room is required then this can be achieved by positioning a damper behind the extract device or in the extract duct. Positive or negative room pressure may be achieved through damper adjustment.

Extract devices are usually selected for minimum pressure losses and sound-level requirement for the room. The pressure losses through the device and damper if available should be included in the sizing of extract ducting. Pressure loss and noise data are normally provided by the device manufacturer.

In general, the location of an extract ATD does not significantly influence the air movement in the room except when the inlets are positioned in the occupied zone. In the latter case, the distance from the inlet over which the comfortable velocities occur should be calculated using the method outlined in Section 5.3.4 and the core velocities recommended by ASHRAE [3] and given in Table 5.7.

Table 5.7 Recommended extract velocities based on core area of inlet [3]

Location of inlet	Velocity (m s^{-1})
Above occupied zone	>4
Within occupied zone, not near seats	3.0–4.0
Within occupied zone, near seats	2.0–3.0
Door or wall louvres	1.0–1.5
Undercut door area	1.0–1.5

There are other factors that should be taken into consideration in deciding the position of extract devices such as:

1   avoiding the possibility of short-circuiting the supply air;
2   the extract device should not be located in a stagnant zone, which is at a different temperature from the room (i.e. cold in the heating season and warm in the cooling season), as good temperature control becomes difficult to achieve;
3   in large spaces where temperature gradients can be present a number of extract ATDs should be used;
4   extracts should be located in positions where gaseous contaminants and dust concentration levels are high to reduce the risk of dispersion into the occupied zone.

References

1. BS 4773 (1991) Testing and rating of air terminal devices for air distribution systems – Part 1: Aerodynamic testing, British Standards Institution, London.
2. ISO 5219 (1984) Air distribution and air diffusion – laboratory aerodynamic testing and rating of air terminal devices, International Standards Organisation, Geneva.
3. *ASHRAE Handbook of Fundamentals* (2001) Ch. 32: Space air diffusion, American Society of Heating, Refrigeration and Air-Conditioning Engineers, Atlanta, GA.
4. Nevins, R. G. (1976) *Air Diffusion Dynamics – Theory, Design and Application*, Business News Publishing, Birmingham, MI.
5. Alden, J. L. and Kane, J. M. (1982) *Design of Industrial Ventilation Systems*, 5th edn, Industrial Press, New York.
6. BSRIA (1985) Design guidelines for exhaust hoods, *BSRIA Tech. Note 3/85*, Building Services Research and Information Association, Bracknell.
7. Goodfellow, H. and Tähti, E. (eds) (2001) *Industrial Ventilation – Design Guidebook*, Academic Press, San Diego, California, USA.
8. *ASHRAE Handbook of HVAC Applications* (1999) Ch. 29: Industrial local exhaust systems, American Society of Heating, Refrigeration and Air-Conditioning Engineers, Atlanta, GA.
9. Hayashi, T. *et al.* (1987) *Industrial Ventilation and Air Conditioning*, CRC Press, Boca Raton, FL.
10. Baturin, V. V. (1972) *Fundamentals of Industrial Ventilation*, Pergamon, Oxford.
11. Becher, P. (1966) Air distribution in ventilated rooms, *JIHVE*, 34, 219–27.
12. Sodec, F. (1986) Air distribution systems, Report no. 3554 E, Krantz GmbH, Germany.
13. Technical Guide (1983) Section A – Air diffusion manual, no. UDC 697.9, Waterloo Grille Company Limited, Benfleet, Essex.
14. Sodec, F. and Craig, R. (1991) Underfloor air supply system, Report no. 3787 E, Krantz GmbH, Germany.

6 Design of room air distribution systems

6.1 Introduction

In Chapter 4, the principles of diffusion of various types of air jet that have applications in the ventilation of buildings were presented and the expressions and data required for their calculation were also given. In Chapter 5, the aerodynamic characteristics of air terminal devices (ATDs) which are commonly used for room air distribution and their selection procedures were discussed. It was found that the diffusion of an air jet downstream of the ATD is strongly influenced by the geometry of the ATD as well as other factors relating to the air diffusion boundaries. Although it is essential to know the aerodynamic characteristics of a ventilation air jet for the purpose of design, these alone are insufficient to ensure a comfortable environment in the occupied zone of a room or other enclosures. This is because most available air jet data relate only to the flow within the jet envelope and up to the throw length (refer to Chapter 5 for definition) which is often outside the occupied zone anyway. Although the air movement in the occupied zone is influenced by the initial diffusion of the supply air jet, it is also greatly affected by the way the jet penetrates the occupied zone as well as the natural convective currents produced by heat sources and sinks in the room.

As discussed in Chapter 1, the thermal condition in the occupied zone is determined by the combined effect of the air temperature, mean radiant temperature, air velocity and turbulence intensity as well as relative humidity. To assess thermal comfort, all these variables need to be evaluated individually throughout the occupied zone and then used in a thermal comfort equation, such as Fanger's equation, to predict the comfort level in the zone. The evaluation of these comfort parameters in a room is the main theme of this chapter. Accurate prediction of room thermal environment is not only necessary for providing thermally acceptable conditions for the occupants, but is also required to optimize the energy utilization in each zone of the building and the sizing of the heating or air-conditioning plant.

6.2 Model investigations

The prediction of air movement in enclosures has, until recently, been based totally on the measurements of air velocity and temperatures in a physical model of the building in a laboratory for pre-design evaluations, or actual site measurements for post-design investigations. Although numerical predictions using computational fluid dynamics (CFD) are becoming more widely acceptable in design (see Chapter 8), experimental data obtained from physical model studies are still considered by some designers to be the most reliable source for the design of air distribution systems. A model investigation, if properly conducted, should provide information on:

1   the quantity of air supply required;
2   the most suitable position in the room of the supply and extract ATDs;
3   the best ATD type for the room;
4   the optimum supply air velocity and temperature;
5   the air flow pattern in the room;
6   the distribution of air velocities, turbulence intensities and air temperatures in the occupied zone and
7   the distribution of contaminants in the room.

The next section describes the main parameters which should be considered when conducting a physical-model investigation or when interpreting the model data for full-scale design.

6.2.1   Parameters for model investigations

Before conducting an experimental investigation involving a physical model it is essential that the fundamental similarity laws relevant to the case under investigation be established and implemented during the experiments. Any deviation from the basic laws of similarity pertaining to the physics of the problem could diminish the significance of the results obtained from the model tests. Szücs [1] defines the similarity in the behaviour of systems, i.e. similitude, as 'two systems (phenomena) are similar if their corresponding characteristics (features, parameters) are connected by bi-unique (one-to-one) mappings (representations)'. This statement links any physical quantity in two systems by a transformation function which is unique to it and therefore a complete similitude may require a number of different transformation functions.

For the study of non-isothermal turbulent diffusion of air jets in enclosures three basic similarity criteria must be adhered to, namely *geometric, kinematic* and *thermal similarities* [1–4]. The significance of each of these three criteria in modelling room air movement is discussed below.

Geometric similarity   Two shapes are called geometrically similar if any part of one shape can be bi-uniquely mapped on to a corresponding point of the other shape [1]. If **r** is a radius vector fixing the position of a point P on one shape then the radius vector **r**$'$ of the transformed point P' on a transformed shape is given as:

$$\mathbf{r}' = T\mathbf{r} \tag{6.1}$$

where T is the transformation matrix. In general a transformation matrix is defined as:

$$T = \begin{bmatrix} C_{11} & C_{12} & C_{13} \\ C_{21} & C_{22} & C_{23} \\ C_{31} & C_{32} & C_{33} \end{bmatrix} \tag{6.2}$$

where C_{ij} are the matrix coefficients ($i = 1, 2, 3$ and $j = 1, 2, 3$).

True geometric similarity can only be achieved when the transformation matrix is a diagonal matrix, i.e.:

$$T = \begin{bmatrix} C & 0 & 0 \\ 0 & C & 0 \\ 0 & 0 & C \end{bmatrix} \tag{6.3}$$

where the matrix coefficient C is the scaling factor which must be the same in the three directions x, y and z.

In room air movement modelling, the linear dimensions of the prototype are scaled down using a suitable scaling factor which is normally ≤ 1. This applies to every unit within the prototype. In particular, the sizes of the supply and extract air terminal devices must be scaled down by the same scaling factor. Since the surface roughness could have a significant effect on the development of the boundary layers on the internal room surfaces and hence the heat convection to these surfaces, care should be taken to apply the same scaling factors to the texture of the internal surfaces too.

Kinematic similarity Kinematic similarity is achieved when the ratios of all the corresponding fluid velocities and accelerations in the geometrically similar model and prototype are equal. This normally requires the equality of the ratios of all the forces causing the fluid motion, i.e. the existence of dynamic similarity. Since the flow in any enclosure is described by the Navier–Stokes equation, which represents the sum of the inertia, viscous, pressure and buoyancy forces acting on the fluid, then to determine the conditions for dynamic and therefore kinematic similarity, it is necessary to non-dimensionalize this equation. The Navier–Stokes equation for steady, incompressible and turbulent flow is given by:

$$\frac{\partial}{\partial x_j}(\rho U_i U_j) = -\frac{\partial p}{\partial x_i} + \frac{\partial}{\partial x_j}(-\rho \overline{u_i u_j}) + g_i(\rho - \rho_r) \tag{6.4}$$

where U_i and u_i represent the time-mean and fluctuating velocity components in the x_i direction respectively, p is the pressure, ρ is the fluid density, ρ_r is a reference density and g_i is the gravitational acceleration. If Boussinesq's eddy-viscosity concept is applied, the Reynolds stress term on the right-hand side of equation (6.4) becomes:

$$-\rho \overline{u_i u_j} = \mu_t \left(\frac{\partial U_i}{\partial x_j} + \frac{\partial U_j}{\partial x_i} \right) - \frac{2}{3}\rho k \delta_{ij} \tag{6.5}$$

where μ_t is the turbulent viscosity and k is the turbulent kinetic energy which is given as:

$$k = 0.5\overline{u_i u_j} \tag{6.6}$$

and the Kronecker delta is:

$$\delta_{ij} = \begin{bmatrix} 1 & 0 & 0 \\ 0 & 1 & 0 \\ 0 & 0 & 1 \end{bmatrix}$$

Substituting equation (6.5) in (6.4) and representing $g_i(\rho - \rho_r)$ by $g\beta\rho_r \Delta T$ the following equation results:

$$\frac{\partial}{\partial x_j}(\rho U_i U_j) = -\frac{\partial p}{\partial x_i} + \frac{\partial}{\partial x_j}\left[\mu_t \left(\frac{\partial U_i}{\partial x_j} + \frac{\partial U_j}{\partial x_i} \right) - \frac{2}{3}\rho k \delta_{ij} \right] + g\beta\rho_r \Delta T \tag{6.7}$$

where the coefficient of cubic expansion $\beta \approx 1/T_r$ and $\Delta T(= T - T_r)$ is the difference between the temperature at a point and a reference temperature, T_r.

Equation (6.7) can be non-dimensionalized by dividing the variables by the corresponding reference values, i.e. x_o, U_o, ρ_o, ΔT_o and p_o which may be written as $\rho_o U_o^2$. Denoting the dimensionless variables by (*), equation (6.7) becomes:

$$\left(\frac{\rho_o U_o^2}{x_o}\right) \frac{\partial}{\partial x_j^*}(\rho^* U_i^* U_j^*) = -\left(\frac{\rho_o U_o^2}{x_o}\right) \frac{\partial p^*}{\partial x_j^*} + \left(\frac{\rho_o U_o^2}{x_o}\right)$$

$$\times \left[\frac{\mu_o}{\rho_o x_o U_o} \mu_t^* \left(\frac{\partial U_i^*}{\partial x_j^*} + \frac{\partial U_j^*}{\partial x_i^*}\right) - \frac{2}{3}\rho^* k^* \delta_{ij}\right]$$

$$+ g_i \beta \rho_o \Delta T_o \Delta T^* \tag{6.8}$$

where $T^* = \Delta T/\Delta T_o$ and $\Delta T_o = T - T_o$.

Dividing both sides of equation (6.8) by $\rho U_o^2/x_o$, we obtain:

$$\frac{\partial}{\partial x_j^*}(\rho^* U_i^* U_j^*) = -\frac{\partial p^*}{\partial x_j^*} + \left(\frac{\mu_o}{\rho_o x_o U_o}\right)\mu_t^* \left(\frac{\partial U_i^*}{\partial x_j^*} + \frac{\partial U_j^*}{\partial x_i^*}\right)$$

$$- \frac{2}{3}\rho^* k^* \delta_{ij} + \left(\frac{g_i \beta x_o \Delta T_o}{U_o^2}\right)\Delta T^* \tag{6.9}$$

In addition to the dimensionless variables this equation contains two dimensionless parameters on the right-hand side. The first is the reciprocal of the Reynolds number, i.e.:

$$Re = \rho_o x_o U_o/\mu_o \tag{6.10}$$

which represents the ratio of inertia force to the viscous force. The second parameter in equation (6.9) is the Archimedes number, i.e.:

$$Ar = g\beta x_o \Delta T_o/U_o^2 \tag{6.11}$$

which represents the ratio of the buoyancy force to the inertia force.

The solution of the dimensionless Navier–Stokes equation (6.9) is clearly dependent upon the values of Re and Ar and to achieve kinematic similarity the values of these two numbers in the model and prototype must be equal.

Thermal similarity   Thermal similarity is achieved when the ratio of the difference in temperature between any two points in the model to the difference in temperature between corresponding points in the prototype is a constant value. This condition can only be achieved if the three modes of heat transfer by conduction, convection and radiation in the model and prototype are identical, in addition to the existence of geometric and kinematic similarities. Neglecting heat transfer to the fluid (air) by radiation, the time-mean energy equation is given as:

$$\frac{\partial}{\partial x_j}(\rho U_j T) = \frac{\partial}{\partial x_i}(-\rho \overline{u_i T'}) \tag{6.12}$$

Representing the turbulent heat flux term by (see also Chapter 8):

$$-\rho\overline{u_i T'} = \Gamma_t \frac{\partial T}{\partial x_i} \tag{6.13}$$

where $\Gamma_t = \mu_t/\sigma_t$ is the turbulent diffusion coefficient and σ_t is the turbulent Prandtl (Schmidt) number, equation (6.12) becomes:

$$\frac{\partial}{\partial x_j}(\rho U_j T) = \frac{\partial}{\partial x_i}\left(\Gamma_t \frac{\partial T}{\partial x_i}\right) \tag{6.14}$$

In dimensionless representation this equation can be written as:

$$\left(\frac{\rho_0 U_0 T_0}{x_0}\right)\frac{\partial}{\partial x_j^*}(\rho^* U_j^* T^*) = \left(\frac{\mu_0 T_0}{\sigma_0 x_0^2}\right)\frac{\partial}{\partial x_i^*}\left(\Gamma_t^* \frac{\partial T^*}{\partial x_i^*}\right)$$

Substituting for σ_0 by $\mu_0 C_{p_0}/\lambda_0$ where C_{p_0} and λ_0 are the specific heat at constant pressure and thermal conductivity respectively, and rearranging the equation gives:

$$\frac{\partial}{\partial x_j^*}(\rho^* U_j^* T^*) = \left(\frac{\lambda_0}{C_{p_0}\rho_0 x_0 U_0}\right)\frac{\partial}{\partial x_i^*}\left(\Gamma_r \frac{\partial T^*}{\partial x_i^*}\right) \tag{6.15}$$

The dimensionless parameter on the right-hand side of this equation is the reciprocal of the Peclet number, Pe, i.e.:

$$Pe = C_{p_0}\rho_0 x_0 U_0/\lambda_0 \tag{6.16}$$

which can also be written as:

$$Pe = Pr\,Re \tag{6.17}$$

where

$$Pr = \mu_0 C_{p_0}/\lambda_0 = \text{Prandtl number} \tag{6.18}$$

Therefore, to achieve thermal similarity between the model and prototype the Peclet numbers for both must be equal.

Boundary conditions To achieve similarity of the velocity and temperature fields between the model and prototype, similarity of the boundary conditions must be present in addition to the equality of the dimensionless numbers Re, Pr and Ar which is necessary for kinematic and thermal similarity of the two flows. Similarity of the boundary conditions is attained when there is a similarity of the geometric, hydro-dynamic and thermal conditions at the solid boundaries of the model and prototype. The requirement of geometrically similar boundaries is satisfied if the model and pro-totype are geometrically similar, which is an essential requirement for flow modelling. Hydraulic similarity of the boundaries can be achieved by accurate scaling of the flow boundaries (e.g. air supply and extract openings) including surface roughness. This would ensure similar air flow patterns and turbulence levels at the boundaries of the model and prototype. However, thermal similarity of the boundary conditions is not as easily attained as the two other conditions as explained below.

To achieve similarity of the fluid temperature distribution in the model and prototype the equality of dimensionless heat flux at the boundary is required. However, this may not produce similarity of temperature distribution close to the boundaries. Generally, the heat transfer from the internal boundaries, Q_t, occurs by conduction, convection and radiation, i.e. $Q_t = Q_c + Q_v + Q_r$ where subscripts c, v and r refer to conduction, convection and radiation respectively. However, in a room environment the heat transfer to the air by conduction is negligible except very close to solid boundaries. The influence of heat convection and radiation from internal boundaries on the thermal similarity of two rooms is now considered.

In convection, the heat transfer between a solid boundary and air occurs across the thermal boundary layer which may be turbulent in most room air movement applications. However, very close to the boundary, a very thin fluid layer called the 'laminar sublayer' exists across which heat is 'conducted' from the surface to the turbulent fluid layer adjacent to it. The heat flow across the laminar sublayer may be approximated by conduction through a fluid. Thus:

$$q_c = Q_c/A = \lambda(\partial T/\partial x_j)_{x_j=0} \tag{6.19}$$

where A is the surface area, λ is the thermal conductivity of air ($\mathrm{W\,m^{-1}\,K^{-1}}$) and $\partial T/\partial x_j$ is the temperature gradient normal to the surface. Equation (6.19) may be non-dimensionalized by defining the reference values T_0, x_0, λ_0 and a reference convective heat flux as:

$$q_0 = \rho_0 C_{p_0} U_0 T_0$$

Hence:

$$(\rho_0 C_{p_0} U_0 T_0)q_c^* = (\lambda_0 T_0/x_0)\lambda^*(\partial T^*/\partial x_j^*)_{x_j^*=0}$$

On rearranging, the equation becomes:

$$q_c^* = (\lambda_0/\rho_0 C_{p_0} U_0 x_0)\lambda^*(\partial T^*/\partial x_j^*)_{x_j^*=0} \tag{6.20}$$

The quantity inside the parentheses on the right-hand side represents the reciprocal of $Pr\ (= \mu_0 C_{p_0}/\lambda_0)$ and $Re\ (= \rho_0 U_0 x_0/\mu_0)$. Hence, equation (6.20) may be written as:

$$q_c^* = (\lambda^*/Pe)(\partial T^*/\partial x_j^*)_{x_j^*=0} \tag{6.21}$$

where $Pe = Pr\,Re$.

Equation (6.21) implies that for similarity of the convective heat transfer from the internal boundaries of the model and prototype, Pe must be equal for the two enclosures. The same condition was also required when the similarity of the thermal fields was considered earlier (equation (6.15)). For that purpose field temperature similarity was, a priori, also required in addition to the equality of Pe.

In radiation heat transfer, the net radiation energy transfer for a surface i in an enclosure of N isothermal grey surfaces, is given as [5]:

$$Q_{ri}/(\varepsilon_i A_i) = \sigma T_i^4 - \sum_{m=1}^{N} F_{i-m}\sigma T_m^4 + \sum_{m=1}^{N} F_{i-m}[(1-\epsilon_m)/\epsilon_m]Q_m/A_m \tag{6.22}$$

where F_{i-m} is the shape factor for surfaces i and m, T_i and T_m are the absolute temperatures of surfaces i and m, ϵ_i and ϵ_m are the emissivities of surfaces i and m, A_i and A_m are the areas of surfaces i and m, Q_m is the heat flux of surface m and σ is the Stefan–Boltzmann constant (i.e. $\sigma = 5.67 \times 10^{-8}\,\mathrm{W\,m^{-2}\,K^{-4}}$). The first term on the right-hand side of equation (6.22) represents the energy emitted by the surface i, the second term is the energy absorbed by the surface i and the third term represents the energy reflected by the other surfaces m. Most room surfaces have an approximately equal emissivity of $\epsilon \approx 0.9$ [6] and if the reflectivity $(1 - \epsilon_m)$ of the room surfaces is ignored (small in comparison with ϵ_m), equation (6.22) reduces to:

$$q_{ri} = Q_{ri}/A_i \approx \epsilon\sigma \sum_{m=1}^{N} F_{i-m}(T_i^4 - T_m^4) \tag{6.23}$$

By non-dimensionalizing equation (6.23) using F_o, T_o and q_o $(=\rho_o C_{p_o} U_o T_o)$ we obtain:

$$(\rho_o C_{p_o} U_o T_o)q_{ri}^* = (F_o T_o^4)\epsilon\sigma \sum_{m=1}^{N} F_{i-m}^*(T_i^{*4} - T_m^{*4})$$

Hence:

$$q_{ri}^* = [F_o T_o^4/(\rho_o C_{p_o} U_o T_o)]\epsilon\sigma \sum_{m=1}^{N} F_{i-m}^*(T_i^{*4} - T_m^{*4}) \tag{6.24}$$

For a gas the density is inversely proportional to the absolute temperature, i.e. $\rho_o \propto 1/T_o$. The parameter on the right-hand side of equation (6.24) then reduces to:

$$T_o^4/U_o = \text{constant} \tag{6.25}$$

For similarity of radiation heat transfer at the boundaries of the model and prototype, equation (6.25) must have the same value in the model and prototype.

6.2.2 Reduced-scale modelling

Internal flow For a complete similarity of air flow and heat transfer in both model and prototype it was shown in the preceding sections that the following conditions must be satisfied:

1 For a predominantly convective heat transfer in the room the values of *Pr*, *Re* and *Ar* must be the same in the model and prototype.
2 Where radiant heat transfer at the boundaries is significant, then in addition to the conditions in 1, the parameter T_o^4/U_o must also be equivalent in the model and prototype, where T_o is a reference temperature (K), e.g. supply air temperature, and U_o is a reference velocity, e.g. supply velocity.

The requirement of an equal Prandtl number $(Pr = \mu C_p/\lambda)$ for two different fluids which would also satisfy other modelling parameters is not possible to achieve in practice as was concluded by Nevrala [7]. Water is sometimes used for obtaining high

Re flow in model studies under isothermal conditions but it is not suitable for non-isothermal flows because the values of *Pr* for water greatly decrease with increase in temperature. For these reasons air is still used for modelling non-isothermal flows in enclosures, in which case equality of *Pr* is assumed as *Pr* for air is about 0.7 for the range of temperatures normally used in modelling air flow in rooms.

The other modelling requirement is the equality of *Re* in the model and prototype, i.e.:

$$(\rho_0 U_0 x_0/\mu_0)_m = (\rho_0 U_0 x_0/\mu_0)_p$$

With the same fluid (air) used in the model and prototype this gives:

$$U_{om}/U_{op} = x_{op}/x_{om} = S \tag{6.26}$$

where *S* is the scale factor and U_0 represents the air supply velocity. The other requirement for kinematic similarity is the equality of *Ar*. Thus:

$$(g\beta x_0 \Delta T_0/U_0^2)_m = (g\beta x_0 \Delta T_0/U_0^2)_p$$

If similar thermal conditions existed in the model and prototype it may be assumed that $\beta_m \approx \beta_p$ and $\Delta T_{om} \approx \Delta T_{op}$. Hence the above equation reduces to:

$$U_{om}/U_{op} = 1/\sqrt{S} \tag{6.27}$$

From equations (6.26) and (6.27) it is clear that the requirements for the equality of *Re* are quite different from the requirement for the equality of *Ar* and the two equalities can never be achieved concurrently in a model study. However, in practice the air movement in rooms ventilated with high momentum jets (e.g. in mixing ventilation) is predominantly turbulent in which case the Reynolds number of the supply air jet is greater than 3×10^4. In this case the flow in the jet is not only turbulent but large-scale turbulent eddies also form in the occupied zone by the diffusion of the jet and from convective currents produced by heat sources. Under such conditions the effect of *Re* on the similarity of air flow patterns in the model and prototype is small in comparison with the influence of *Ar*. The experimental results of Müllejans [2] based on tests carried out in three geometrically similar but different scale rooms (in the ratio 1 : 3 : 9) under isothermal and non-isothermal conditions showed that the flow patterns were principally determined by the value of *Ar*. Müllejans could not detect an effect of *Re* in the air movement patterns over a wide range of *Re* for the three test rooms. It might be concluded, therefore, that kinematic and thermal similarity between the model and most mechanically ventilated rooms can be achieved if *Ar* is kept constant. In general this means that:

$$[gx_0 \Delta T_0/(T_0 U_0^2)]_m = [gx_0 \Delta T_0/(T_0 U_0^2)]_p$$

i.e.:

$$S = (U_{op}/U_{om})^2 (T_{op}/T_{om})(\Delta T_{om}/\Delta T_{op}) \tag{6.28}$$

Equation (6.28) can be used to calculate the scale factor of a model where the heat transfer is predominantly by convection. Equation (6.27) shows that the air velocity

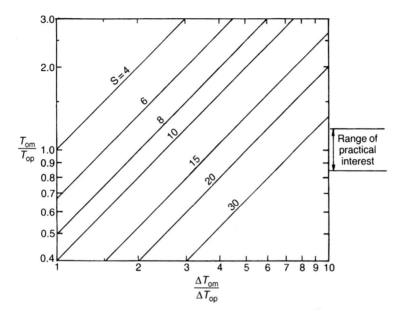

Figure 6.1 Effect of scale factor on the temperature ratios for convective heat transfer $(U_{op}/U_{om} = 2)$.

in the model is always lower than in the prototype if $S > 1$. The mean velocity in the occupied zone is usually in the range 0.2–$0.3 \, \text{m s}^{-1}$ [8, 9]. With such low air velocities special types of hot-film/wire anemometers are required to obtain reliable measurements (see Chapter 9) but even with these instruments velocities below about $0.1 \, \text{m s}^{-1}$ cannot be reliably measured. The ratio of the velocities in the prototype and model, U_{op}/U_{om}, is therefore limited to about two. Using this value in equation (6.28), the scale factor for convective heating or cooling is given as:

$$S = 4(T_{op}/T_{om})(\Delta T_{om}/\Delta T_{op}) \tag{6.29}$$

Figure 6.1 is a log–log plot of equation (6.29) which may be used to estimate the optimum scale factor for modelling convective heat transfer in a room or an enclosure. By limiting the air supply temperature in the model and prototype to 275 and 325 K (i.e. the range of operation of most instruments) the range of scale factors of practical interests are as indicated in Figure 6.1.

Using computational fluid dynamics and measurements in a room ventilated using a continuous slot, Awbi and Nemri [10] have shown that for non-isothermal conditions, it is more important that the Archimedes number, Ar_j, in the model room and the prototype are kept equal than the Reynolds number, Re_j, being equal. This is particularly true when $Ar_j > 0.001$, where subscript j refers to the jet, i.e. equation (6.11) is written as:

$$Ar_j = g\beta x_o \Delta T_o / U_o^2 \tag{6.30}$$

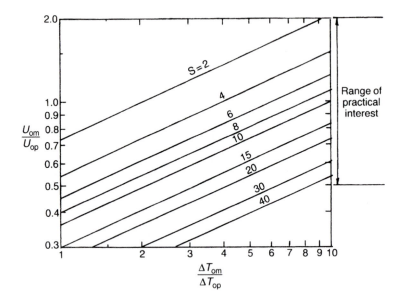

Figure 6.2 Effect of scale factor on the supply velocity and temperature ratios for convective and radiant heat transfer.

where x_o is the height of the supply slot or $\sqrt{A_o}$ where A_o is the supply area (m); ΔT_o is the temperature different between the room and supply air (K) and U_o is the supply velocity (m s^{-1}).

If radiant heat transfer in the room to be modelled is significant in comparison with convective heat transfer (as with radiant or radiant/convective heating or with cooling involving large solar gains) then similarity of radiation heat transfer at the boundaries is required. In this case equation (6.25) is combined with equation (6.28) to produce an expression for the scale factor, thus:

$$S = (U_{op}/U_{om})^{9/4}(\Delta T_{om}/T_{op}) \tag{6.31}$$

or

$$S = (T_{op}/T_{om})^9(\Delta T_{om}/\Delta T_{op}) \tag{6.32}$$

Equations (6.31) and (6.32) are plotted in Figures 6.2 and 6.3 from which the temperature and/or velocity ratios of the model and prototype may be determined for a given scale factor. Alternatively, if these ratios are known the appropriate scale factor may be determined. The ranges of U_{om}/U_{op} and T_{om}/T_{op} of practical interest are also indicated on each figure.

Once the most suitable scale factor has been determined from equations (6.28), (6.31) or (6.32), or from Figures (6.1), (6.2) or (6.3), it is then required to determine the heating/cooling load for the model. The total sensible heat supplied or extracted for a ventilated room can be expressed by:

$$Q = \rho_o C_{p_o} x_o^2 U_o \Delta T_o \tag{6.33}$$

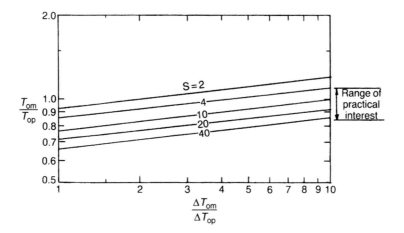

Figure 6.3 Effect of scale factor on the temperature ratios for convective and radiant heat transfer.

where x_o^2 represents the area of the outlet, U_o is the supply velocity and ΔT_o is the difference in temperature between the supply and extract air. Hence, the ratio of total heat of the model and prototype is:

$$Q_m/Q_o = (\rho_o C_{po} x_o^2 U_o \Delta T_o)_m / (\rho_o C_{p_o} x_o^2 U_o \Delta T_o)_p$$

$$= (1/S)^2 (U_{om}/U_{op})(T_{op}/T_{om})(\Delta T_{om}/\Delta T_{op}) \tag{6.34}$$

assuming that $\rho_{om}/\rho_{op} = T_{op}/T_{om}$. Knowing the scale factor, velocity ratio, temperature ratios and prototype heat load the model load may be calculated from equation (6.34) to achieve kinematic and thermal similarities for turbulent flow in rooms. In practice, the scale factor is usually limited to about 20. However, attention should be paid to the design of a flexible air distribution system for a model. This is usually required because of the complexity of the air movement and heat transfer in ventilated enclosures which has been conveniently simplified in deriving the similarity equations given in this section.

Steady-flow modelling Most of the model studies involving room air movement are conducted under steady-state conditions, as these are sufficient for most design considerations and, furthermore, much simpler to perform. Some examples and results obtained using small-scale modelling will be given below.

The effect of Archimedes number on the air movement in three geometrically similar model rooms that were investigated by Müllejans [2] is illustrated in Figure 6.4(a) and (b). The length $(L) \times$ height $(H) \times$ width (B) of the three test rooms were $4.75\,\text{m} \times 2.95\,\text{m} \times 2.88\,\text{m}$, $1.6\,\text{m} \times 1.0\,\text{m} \times 1.0\,\text{m}$ and $0.531\,\text{m} \times 0.333\,\text{m} \times 0.227\,\text{m}$ giving scale factors of 1, 3 and 9, respectively. The air was supplied from a wall at high level using a rectangular opening. A logarithmic scale of Ar is shown as the abscissa in the figures. This is defined as:

$$Ar = g D_h \Delta T_o / (T_m U_r^2) \tag{6.35}$$

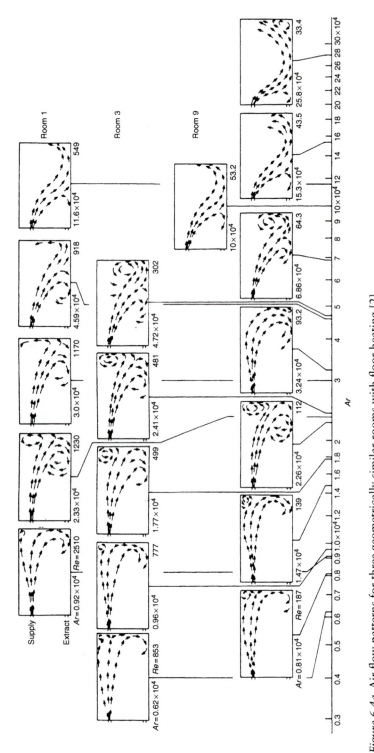

Figure 6.4a Air flow patterns for three geometrically similar rooms with floor heating [2].

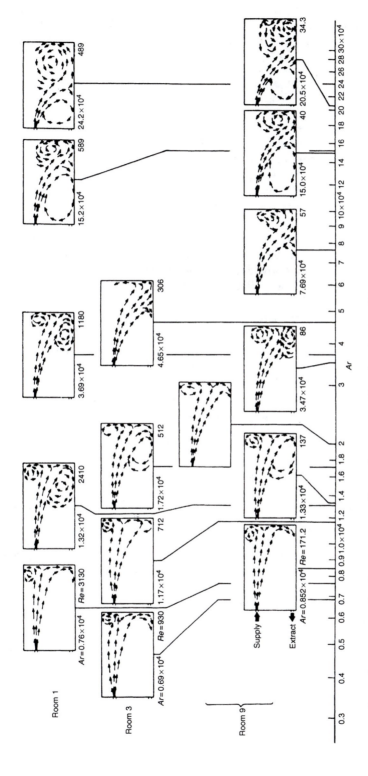

Figure 6.4b Air flow patterns for three geometrically similar rooms with ceiling heating [2].

where D_h = hydraulic diameter of room = $2BH/(B+H)$ (m); U_r = equivalent room velocity = flow rate (m³ s⁻¹)/cross-sectional area (m²) = $\dot{V}/(BH)$ (m s⁻¹); T_m = mean room temperature = $0.5\,(T_o + T_m)$ (K); T_o = supply air temperature (°C); T_w = heated wall temperature (°C) and $\Delta T_o = T_w - T_o$ (K).

The Reynolds number for the room is defined as:

$$Re = D_h U_r / \nu \tag{6.36}$$

where ν = kinematic viscosity of air (m² s⁻¹).

In Müllejans' experiments, the room load is applied by heating one of the surfaces and the cool air is supplied from one wall at high level to offset this load. In Figure 6.4(a) only the floor is heated and in Figure 6.4(b) only the ceiling is heated to represent normal internal heat gain from occupants and equipment. The figures were produced from velocity measurements and smoke flow patterns in the three test models. The flow patterns of almost equal Ar but different Re clearly show the independence of the air movement of Re even though the range of Re spans the low-turbulence region (i.e. $Re < 100$) which is sometimes erroneously referred to as the laminar flow region. The effect of Ar on the air flow pattern is clearly predominant. With floor heating at low Ar, the jet attaches to the ceiling by the combined effect of Coanda and convective currents rising from the floor. As Ar increases, the downward buoyancy force acting on the cool air jet predominates over the Coanda force and causes the jet to deflect downwards until 'dumping' (i.e. deflection of jet into the occupied zone) takes place at high Ar values. With a heated ceiling, attachment of the jet to the ceiling does not occur even at low Ar values because of the convective currents produced by the hot ceiling counteracting the Coanda force.

Müllejans observed similar flow patterns for each Ar range regardless of whether Ar was varied largely by changing the air flow rate to the room, i.e. U_r, or the heated wall temperature T_w. This is an interesting observation because a large change in T_w causes a significant change in the boundary conditions as well as the kinematic conditions caused by the resulting change in Ar. An increase in the heated wall temperature also causes an increase in the temperature of other room surfaces as a result of radiation heat exchange between the surfaces. This increase in room surface temperature increases the convective heat transfer to the room air thus producing an effect on air movement which is similar to a room load increase. The effect of radiation on the air movement was studied by Müllejans by covering a heated ceiling with a reflective aluminium foil. The effect of buoyancy on the downward deflection of the jet was markedly reduced by the reduction in the radiant heat exchange due to the foil although Ar was the same in both cases.

The convective heat transfer data for the three test rooms were correlated by the single expression:

$$Nu = C Re^{m+2n} Ar^n [D_h^2/(bh)]^k \tag{6.37}$$

where Nu = Nusselt's number = $h_c D_h/\lambda$; h_c = mean surface heat transfer coefficient (W m⁻² K⁻¹) = $q/(T_w - T_o)$ where q is the surface heat flux (W m⁻²) and Ar and Re are the Archimedes and Reynolds numbers defined by equations (6.35) and (6.36) respectively. The constant C and exponents m, n and k have the values given in Table 6.1 for the heating of the ceiling, floor, side wall and the wall facing the supply opening. An index for Re of $m + 2n \geq 1.1$ does not represent a turbulent convective heat transfer

Table 6.1 Indices for equation (6.37)

Heated surface	C	m	n	k
Ceiling	0.032	1.10	0	0.12
Floor	0.034	0.95	0.08	0.06
Sidewall	0.030	0.98	0.70	0.04
Wall facing supply opening	0.013	1.00	0.10	0.08

process, as this normally has an index of about 0.8. An index greater than 0.8 is often associated with either a transition boundary layer or a surface-bound vorticity flow. The low index for Ar suggests the relatively small influence of Ar on the surface heating rate. A similar small influence is found in the dimensionless geometric parameter $D_h^2/(bh)$ where b and h represent the width and height of the supply slot respectively. Equation (6.37) was based on experimental data for $b < B$ and applies to values of $(bh/D_h^2)Ar \geq 40$. The equation produces a value of h_c increasing from about 0.4 at 2 air changes per hour (ach) to about 6.0 at 20 ach. The product $(bh/D_h^2)Ar$ takes into consideration the areas of the supply opening and the room and is sometimes referred to as the corrected or modified Archimedes number Ar'.

Lee and Awbi [11] and Lee [12] studied the effects of internal room partitions on ventilation performance in terms of room air change efficiency and ventilation effectiveness. A model test room 1.6 m (L) × 0.8 m (W) × 0.7 m (H) was used and the physical test conditions were simulated numerically by using CFD under isothermal and non-isothermal conditions. Different partition configurations, including its location, height and gap underneath as well as the contaminant source location, were examined using mixing and displacement ventilation. The internal partition was placed in the room for different test configurations, e.g. its location (40–60% L), its height (60–80% H), and its gap underneath (0–10% H). All the model tests were carried out for a jet Reynolds number of 9.2×10^3, which is typical for room ventilation, and different Archimedes number. Because of the small scale model used, a higher supply velocity than what is normally experienced in actual rooms was used. A tracer gas (CO_2) was injected on one side of the partition and the concentration at key points in the model was measured using a gas analyzer. The model and measuring system used is shown in Figure 6.5.

Figure 6.6(a) and (b) summarizes the effect of partition gap on the ventilation performance when the air is supplied and extracted from the ceiling (Air supply A in Figure 6.5(a)). The internal partition tested was placed in the middle of the model room at a height of 80% of room height (H). The subplot (a) shows that the 5% H gap above the floor is an optimum gap for achieving better air change effectiveness in the model room. The ventilation effectiveness in the subplot (b), however, is influenced differently depending on the contaminant source location. With the contaminant source in the supply zone, the partition gap has insignificant influence on the ventilation effectiveness, as the generated contaminant is mixed with the ventilation air before the internal partition starts to influence the airflow pattern. In cases of the contaminant source in the exhaust zone, the partition with 5% H gap produces better ventilation effectiveness. It implies that the ventilation performance can be improved simply by putting a gap underneath the partition. A further study using the same model room investigated the optimal height of the partition gap, different cooling loads and different locations of the air supply position and is described in [12].

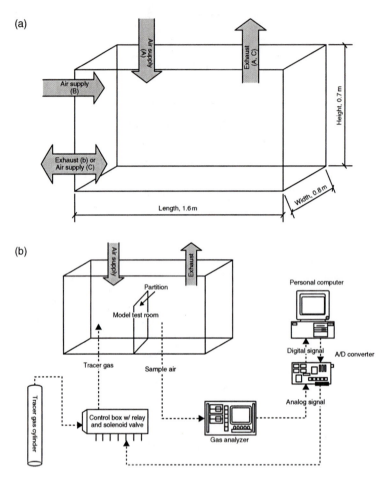

Figure 6.5 Model test room and methods of air supply and exhaust used; (a) model dimensions and positions of air supply/exhaust and (b) tracer gas injection and data logging.

Transient flow modelling   For modelling transient flows such as those occurring in naturally ventilated buildings, saline water solutions of different density distribution can be used (salt box modelling). A scale model of the significant part of the building is usually immersed in a larger tank of water. Buoyancy forces due to heat sources, etc. are simulated by pumping in salt water solution that is denser than the water in the tank whereas heat sinks, i.e. cool surfaces or cool air, are represented by injecting lower density liquid such as alcohol. In these situations, the model is suspended in the water tank upside down to allow gravity effect on the difference in densities between the water in the tank and the salt water or alcohol. Because water is used as the fluid, large Reynolds numbers as well as large Archimedes numbers can be achieved using this technique if sufficient density difference is produced.

In most cases the modelling is carried out over short time periods to reduce the dilution of the tank water with other liquids. Using dyes of different colours the processes

(a)

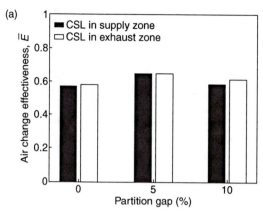

(b)

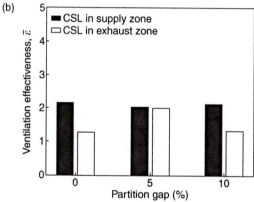

Figure 6.6 Effect of partition gap on the ventilation performance with a partition (80% *H*) located at 50% *L*: (a) room air change effectiveness and (b) ventilation effectiveness.

Notes

(i) CSL = contaminant source location.
(ii) The air change effectiveness here is as defined by equation (2.31).
(iii) The ventilation effectiveness is as defined by equation (2.33) but instead of the mean concentration in the occupied zone the mean for the room is used.

are usually recorded on video and the images are subsequently digitized to produce flow velocities. However, it is also possible to conduct the experiments under steady-state condition by continuously feeding fresh water to the tank and draining water present in the tank. Further details of this modeling technique can be found in [13–15].

External flow Wind flow over buildings and other structures is traditionally studied to determine wind-loading or the influence of wind gust on the frequency response of the structure. This requires modelling of the wind speed, wind profile and the turbulence characteristics (intensity and scales) of the natural wind in a special wind tunnel, usually called a boundary layer or an environmental wind tunnel. An important feature of such a wind tunnel is the representation of the atmospheric boundary layer and the turbulence characteristics of natural wind, but very often it is difficult or almost impossible to achieve a realistic representation of the natural wind, particularly its turbulence characteristics. Information on the dynamic characteristics of natural wind is dealt with in specialist texts, e.g. [16–18]. For most ventilation studies, however, the mean flow is more relevant for understanding the airflow into a building than the transient effects due to wind turbulence. A wind tunnel that is capable of representing a realistic wind profile, wind velocity and turbulence intensity is probably all that is required for ventilation studies. However, wind tunnels are not useful in studying buoyancy-induced ventilation.

In using boundary layer wind tunnels for the testing of building models for natural ventilation, attention should be given to the wind profile across the working section, the scale factor of the model and the blockage caused by the model. The mean wind speed is normally represented by the power law:

$$\frac{V}{V_H} = \left(\frac{z}{H}\right)^a \tag{6.38}$$

where V = the wind speed at a height z (m s^{-1}); V_H = the wind speed at height H (the edge of the boundary layer) (m s^{-1}) and a = the profile index that has to be achieved in the wind tunnel. The index a is dependent on the building terrain and the exposure of the building to the wind. The required value of a is usually attained by using flow conditioners (e.g. honeycombs, gauze screens, etc.) to achieve a uniform velocity distribution initially followed by floor roughness elements (turbulence generators such as small blocks, coffee cups, etc.) upstream of the working section of the wind tunnel to generate large turbulence scales.

The other factor to consider is the model scale factor and its relation to the height of the velocity profile. If the wind-tunnel boundary layer thickness spans most of the working section then one can theoretically use a large model which will still be within the boundary layer profile since the exponent a is independent of the length scale. However, care should be taken of the model blockage (usually expressed as the cross-sectional area of the model to that of the working section) particularly as large wakes are normally generated by flow separation from a building model. It is recommended that this should be below 10% otherwise correction to the measurement should be made [19]. Furthermore, the position of the model in the test section of the wind tunnel is important and normally the extent of the wake of a building in the flow direction is between 6 and 10 building heights [16], and the model should therefore be placed at least this distance away from the end of the test section. In addition, the position of the model should be within the fully developed wind profile.

When pressure measurements are made to obtain pressure coefficients then it is important to define the corresponding reference wind velocity. Defining the pressure coefficient, C_p, by:

$$C_p = \frac{\Delta p}{\frac{1}{2}\rho V_r^2} \tag{6.39}$$

where Δp = the pressure difference between a point on the model and that at a reference point in the wind (Pa); V_r = reference wind speed, normally at a height corresponding to the height of the building or the height of a certain opening (m s^{-1}) and ρ = air density (kg m^{-3}).

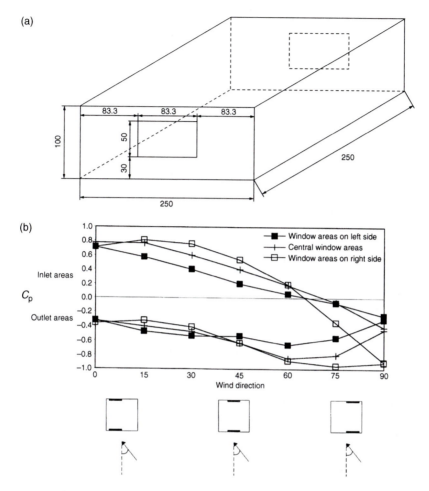

Figure 6.7 Wind-tunnel measurement of pressure coefficients at ventilation openings: (a) model and position of openings: and (b) mean pressure coefficient at the inlet and outlet openings for different wind direction.

Scale models in wind tunnels have been used to measure the wind-induced velocities inside buildings in addition to pressure distribution around buildings required for wind-loading purposes. Data is available from parametric studies using generic building models to assess the effects of building geometry, building orientation, position, size and number of ventilation openings, etc. on the ventilation characteristics of these models. The data is either in the form of pressure distribution around the building envelope and inside the building or internal air velocity measurements at a certain height in the occupied zone. A survey of some of these studies is given by Ernest [20]. Figure 6.7 (a) shows a 1 : 30 scale model (250 mm × 250 mm × 100 mm) for wind-tunnel measurements of the mean pressure coefficients, C_p, at the inlet and outlet openings at different positions of the inlet and outlet openings on the windward and leeward faces of the model [20]. For the purpose of pressure measurement at the positions of the ventilation openings, the model openings were closed and the pressure was measured at a number of points distributed over the positions of the inlet and outlet openings. The mean pressure coefficients for the three positions of inlet and outlet openings are shown in Figure 6.7(b) for wind incidence angles from 0 to 90°.

6.2.3   *Full-scale laboratory investigations*

It was shown in the previous section that air movement tests with reduced-scale models require correct scaling of the supply air flow rate, supply and extract temperature, boundary temperatures and heat sources, in addition to the geometric scaling of the model itself. Sometimes, difficulty may arise in the scaling of different variables in the ratios dictated by similarity laws, as the required scaling of each variable may not be possible to implement concurrently. In such circumstances, some judgment may have to be exercised to select the most significant variables that should be correctly scaled in preference to others. This choice normally depends on the particular case to be modelled and a reference to the similarity conditions discussed earlier should provide the necessary guidance in reaching the correct choice of variables that ought to be scaled correctly.

To overcome the problems associated with scaling a full-scale physical model is built which represents a module of the building being investigated. In the case of a building comprising a number of rooms a representative room is modelled, and in the case of an open-plan building, the model is selected on the basis of the proposed air distribution system layout. In the latter case, attention should be paid to represent the aerodynamic boundaries of the space. The aim of full-scale model investigations is the evaluation of air temperatures, radiant temperatures and air velocities in the occupied zone as well as the assessment of air flow patterns in the room. Using data from such tests with a suitable thermal comfort criterion it would then be possible to evaluate the comfort level in the occupied zone. Room air movement investigations involving full-scale models are widely reported in the literature and, in the remainder of this section, the results of some investigations are presented and discussed. Some of these findings will form the basis of the design procedures described in Section 6.4.

Air movement analysis procedure

In North America three major full-scale investigations were carried out by Koestel and coworkers at the Case Institute of Technology [21, 22], by Straub and his colleagues at the University of Illinois [23] and by Nevins and others at Kansas State University [24, 25] over a period of some 20 years. The tests of Koestel *et al.* were performed in a full-size model of a living room with the external walls exposed to winter and summer temperatures using a high-level side-wall grille, baseboard diffusers or circular ceiling diffusers as the air supply terminals. The maximum air velocity in the occupied zone is plotted in Figure 6.8 (a, b and c) versus the air change rate in the room for each supply mode and for cooling and heating. The percentages of maximum objectionable votes estimated using the draught temperature concept, described later in this chapter, are shown in Figure 6.9 (a, b and c) for each air supply. Koestel and colleagues considered a figure of 10% of the subjects objecting as acceptable. For the side-wall supply, Figures 6.8(a) and 6.9(a), it was found that the air velocities in the occupied

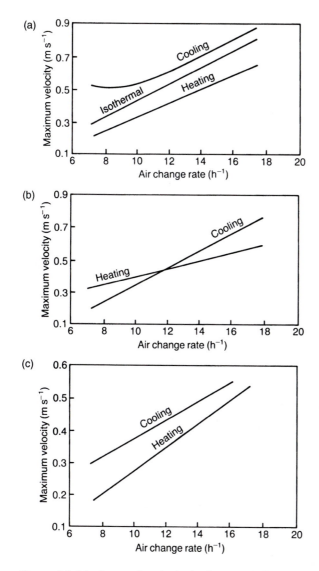

Figure 6.8 Maximum air velocity in the occupied zone. (a) Side-wall grille; (b) baseboard diffusers and (c) circular ceiling diffusers.

zone were highest during cooling and lowest during heating, with intermediate values for isothermal flow. The highest velocities were recorded in the upper region of the occupied zone with values exceeding $0.2\,\mathrm{m\,s^{-1}}$ for heating and $0.5\,\mathrm{m\,s^{-1}}$ for cooling. More than 30% objectionability was recorded during cooling. The air temperature difference between floor and ceiling was less than $1.7\,\mathrm{K}$.

With baseboard diffusers, the maximum air velocities in the occupied zone were lower than the previous case, Figure 6.8(b), but high local velocities occurred close to the air supply wall. The percentage objectionability, Figure 6.9(b), was similar to

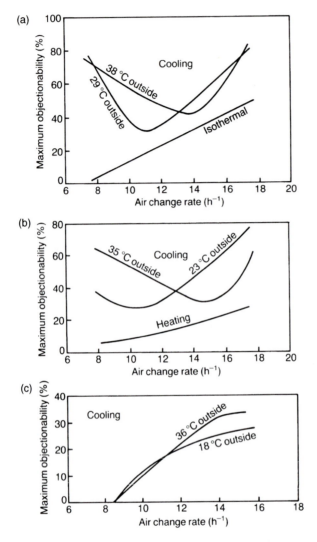

Figure 6.9 Maximum objectionability of draughts. (a) Side-wall grille; (b) baseboard diffusers and (c) circular ceiling diffusers.

the case of side-wall supply, i.e. over 30% with cooling but lower with heating. The vertical temperature difference in this case was very low.

The best results were obtained with circular ceiling supplies, Figures 6.8(c) and 6.9(c), where good diffusion was achieved in the occupied zone. The only draw-back was that the maximum air change rate had to be limited to 10 ach to meet the 10% objectionability criterion. This was due to the draught occurring between two diffusers at a height of 1.8 m. This problem could have been overcome if a greater diffuser separation distance was used by using a larger-size test room.

A comprehensive laboratory study of room air movement was conducted at the University of Illinois by H. E. Straub [23] and others. They used a research facility consisting of a testing room of dimensions 3.96 m × 5.49 m × 2.59 m ceiling height enclosed by a larger room to achieve controlled conditions inside and outside the testing room. Measurements of air temperature and velocity were taken at heights of 0.1, 0.76, 1.52, 1.98, 2.29 and 2.49 m at 33 points in each horizontal plane. External summer and winter temperatures were simulated by controlling the temperature in the space between the two enclosures. Several types of air terminal device were investigated to produce a horizontal and a vertical projection of the supply air, such as floor registers, floor diffusers, sill grilles, side-wall grilles and ceiling diffusers. Several air temperature and air velocity indices were applied to the thermal condition in the occupied zone to evaluate the air diffusion performance of the devices used. From the large amount of data collated of velocity, temperature and smoke flow patterns, a five-step procedure was developed to analyse the air movement in heated and cooled rooms. Figure 6.10 illustrates the five-step analysis procedure as applied to a floor air supply projecting

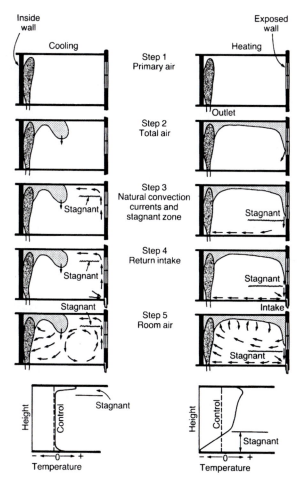

Figure 6.10 Air movement analysis procedure for a floor supply.

upwards close to an internal wall of the room. The five-point procedure given below can also be applied to mixing ventilation with other types and locations of outlets taking into consideration the differences in the flow patterns produced by each outlet.

Primary air   This is the starting point for analysing the air movement. The primary air consists of a mixture of supply air and entrained room air within a velocity envelope greater than $0.75\,\mathrm{m\,s^{-1}}$. This region of the air motion is amenable to analytical treatment using the appropriate jet flow equations presented in Chapter 4.

Total air   The total air is assumed to be the flow in the main body of the supply jet which has a relatively high velocity. The flow in this region is determined by the type of air outlet as well as the room boundary, heating or cooling load, obstructions, furniture, etc. Ideally this region should enclose much of the occupied zone by ensuring the attachment of the jet to the ceiling and walls. Buoyancy forces due to the difference in temperature between the air in this region and the room may prevent an ideal spread of the total air over the room surfaces. A partial analytical treatment of the flow in this region is possible using the data on buoyancy and other jet interference effects given in Chapter 4.

Natural convection currents   These are caused by buoyancy forces resulting from the difference in temperature between the room air and the air in contact with a warm or a cold surface. These currents are normally produced by windows, external walls, occupancy, lighting, computers, IT equipment, etc., and could have a major influence on the air motion particularly in the occupied zone where air velocities are generally low. They could set up circulatory flow within the space which could interfere with the general motion produced by the total air. Downdraught from a cold window is a common occurrence in winter and plumes from heat sources is also very common (see Chapter 4).

Although convective currents enhance the air motion in a room, they can also produce stagnant air zones where mixing of total air with room air is poor, as illustrated in Figure 6.10. These zones can degrade the thermal comfort in the conditioned space and can also increase the plant load. There are no simple analytical solutions for predicting convective currents or stagnant zones in a room and one has to rely on, CFD simulation (see Chapter 8) or flow visualization tests (see Chapter 9) to provide a quantitative and qualitative account of the flow pattern in these regions. A good air distribution design attempts to minimize the effect of convective currents on the general air movement in the space and to avoid the occurrence of stagnant zones.

Return intake   The type and position of a return intake (i.e. extract inlet) has little influence on the air motion in the room (even convective currents) except in the immediate vicinity of the intake. However, this does not suggest insignificant effects of the intakes on other aspects of room air distribution (refer to Chapter 5 for further details).

Room air   This is the bulk of the air motion in the room which also includes the occupied zone. The thermal comfort is determined by the pattern of room air currents. These currents are influenced by a number of factors as discussed earlier. A good design ensures a proper mixing of the primary and total air with room air to counteract convective currents and to prevent stagnant zones. If convective currents are present then some discomfort may occur to the occupants nearest to them, such as window draught or thermal plumes. Furthermore, stagnant zones produce large temperature gradients which can affect the temperature distribution in the conditioned space. Stratified layers of increasing temperature with height are the main characteristic of stagnant zones as shown in Figure 6.10.

Since stagnant zones are primarily produced by convective currents the extent of the region occupied by these zones will depend on the position and value of the heat source or sink producing these currents, i.e. primarily on the room load and its distribution. The thermal conditions in the room air region can only be evaluated by experimental measurements in a physical model or by solving the flow and energy equations numerically using CFD (see Chapter 8).

The Kansas State University team [24] modelled an intermediate floor of a multi-storey office building with a test room of 6.10 m × 3.66 m × 2.74 m ceiling height enclosed by a chamber for controlling the external temperature of the room. Temperatures and velocities were measured in the room at 216 locations using a 600 mm × 600 mm × 600 mm measuring grid, 100 mm from the floor and 300 mm from the walls. Tests were conducted with a high side-wall grille, a sill grille, two- and four-slot ceiling diffusers, circular ceiling diffusers, light troffer diffusers and perforated and louvered square ceiling diffusers. The effective draught temperature criterion [26] was used together with an 80% occupant satisfaction limit [25] to analyse the vast experimental data obtained from 256 tests. The effective draught temperature, θ, is given as:

$$\theta = (T_x - T_r) - 8(v_x - 0.15) \tag{6.40}$$

where T_x = local temperature (°C); T_r = average room temperature (°C) and v_x = local air velocity (m s^{-1}). The draught temperature limit was taken as 1.1 K for warm sensation and −1.7 K for cool sensation and the maximum air velocity was taken as 0.35 m s^{-1}. These values represent the limits for sedentary (office) occupation [26].

Using the effective draught temperature criterion above, Nevins and his coworkers established an analysis procedure based on the measured values of air temperature and velocity at uniformly spaced points throughout the occupied zone or through a vertical, centreline plane through the air supply outlet. The number of points at which the draught temperature, defined by equation (6.40), satisfies the comfort limits above, expressed as a percentage of the total number of points, was defined as the air diffusion performance index (*ADPI*). The Kansas test data have shown the *ADPI* to be a function of the type of air terminal device, the room load, the supply air flow rate and the room geometry for all systems tested except ceiling air distribution using slotted or perforated square panels. For this type of air distribution most of the test data was correlated by the single expression:

$$ADPI = 99.8 - 0.088q \tag{6.41}$$

where q is the total room load in W m^{-2}. The dependency of *ADPI* on a number of factors excludes its use as a universal index for assessing room air movement. Despite this drawback the index is extensively used in North America and it is recommended by the ASHRAE Handbook [26] for evaluating the performance of air terminal devices. Another disadvantage of this method is the uncertainty in defining an acceptable limit of *ADPI* for comfort. A value of 80% is usually considered to be the minimum acceptable [25].

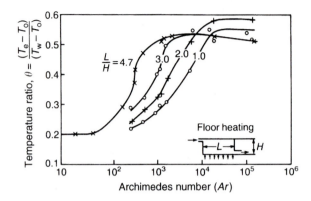

Figure 6.11 Effect of Archimedes number on the room temperature ratio.

Table 6.2 Critical room Archimedes number (Ar_c)

L/H	4.7	3.0	2.0	1.0
Ar_c	2000	3000	10,000	11,000

Effect of room length and size of air outlet

Linke [27] has carried out extensive work on the diffusion of air jets and the air movement in rooms at the Technische Hochschule, Aachen. In particular, Linke investigated the effect of room length on the air movement pattern using smoke in test rooms and aluminium powder in water tunnels. The results of the water-tunnel tests are shown in Figure 6.11. The tunnel had a square cross-section of 100×100 mm and a variable length, L, from 100 to 470 mm giving values of L/H between 1 and 4.7. The water was supplied from the square side wall using a 3.1×100 mm slot at the top or middle of the wall and a circular 5 mm opening at the centre of the wall. The results show that, with symmetrical and asymmetrical supply and over a range of Re ($Re = d_h U_o/v$ where d_h is the hydraulic diameter of the outlet) between 1825 and 12,000, the jet breaks up at a maximum distance of about three room heights, H, from the supply. Linke also studied the effect of room Archimedes number (defined by equation (6.35)) on the non-dimensional temperature ratio, θ, defined as:

$$\theta = (T_e - T_o)/(T_w - T_o) \tag{6.42}$$

where subscript e, o and w refer to extract, supply and heated wall. Figure 6.11 shows plots of the non-dimensional temperature θ with Ar for four rooms of $L/H = 1, 2, 3$ and 4.7 when the floor was uniformly heated. In each case, θ increases with Ar until a maximum value is reached and then remains almost constant with further increase in Ar. This is the room critical Archimedes number, Ar_c, which is given in Table 6.2 for each of the four rooms.

Once θ_{max} is reached the air distribution in the room becomes less effective in removing higher heat loads probably due to the dumping of the air jet at high Ar. Using an average value of $\theta_{max} = 0.54$ for the four rooms, Linke derived an expression for the critical (minimum) room air change rate, N_c, in terms of the heat load and

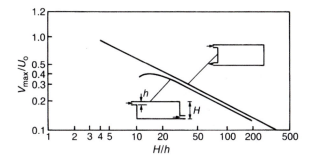

Figure 6.12 The effect of room height/slot height on the maximum velocity in the occupied zone.

Ar_c given by:

$$N_c = 134\sqrt[3]{[D_h q/(L^2 H Ar_c)]} \quad \text{h}^{-1} \tag{6.43}$$

where q is the load in W m^{-2} and the physical properties of air are taken at $22\,^\circ\text{C}$.

Regenscheit [28] investigated experimentally the diffusion of a plane wall jet in a room and determined the effect of the ratio of room height to supply slot height, H/h, on the maximum air velocity in the occupied zone. The results are shown in Figure 6.12 for two extract positions, one at low level on the wall facing the outlet slot and the other on the same wall as the supply also at low level. In both cases the maximum velocity in the occupied zone decreases as the ratio H/h increases but the effect of extract position is marginal.

Effect of air supply method on the room air movement

A comprehensive programme of experimental evaluation of air movement in ventilated rooms was undertaken at the Building Services Research and Information Association (BSRIA) in the UK to develop design procedures for side-wall-mounted grilles, sill-mounted grilles and for circular and linear diffusers mounted on the ceiling. The tests were carried out in a room of dimensions 4.9 m × 3.7 m × 2.75 m ceiling height, constructed to represent a typical office module with one external wall. The external wall in the test room was enclosed with a compartment fitted with heating and cooling appliances to simulate external temperatures. The cooling load was simulated by electrically heated carpets providing a uniform load distribution over the floor. Air velocities and temperatures were measured at 539 and 294 positions, respectively, throughout the room. The test facility is described by Jackman [29]. For each type of supply terminal the air movement in the occupied zone, based on the velocity and temperature distribution within the occupied zone, was correlated with the momentum of the supply air, room heat load, and room geometry. A summary of the main findings of this work is given below and the design procedures developed are presented in Section 6.4. The results of other studies at BSRIA and elsewhere are also presented to cover other air supply methods used in practice.

High-level side-wall supply   The supply opening was positioned at the centre of the side wall (3.7 m×2.75 m) close to the ceiling and tests were conducted under isothermal and cooling conditions. The supply openings used were 150 mm diameter circular; 150 mm × 150 mm, 230 mm × 230 mm and 305 mm × 305 mm square; and 610 mm × 150 mm and 1070 mm × 75 mm rectangular. As a result of this work, Jackman [29] proposed a change in the previously used throw criterion of 0.25 m s^{-1} at 0.75 L (where L is the room length in the jet flow direction) to 0.5 m s^{-1} at 0.75 L, as the lower throw velocity produced very low air movement in the occupied zone. Using a throw velocity of 0.5 m s^{-1}, Jackman derived an empirical equation based on experimental results for estimating the mean air speed in the occupied zone v_r, given as:

$$v_r = 0.73\sqrt{[M_o/(BH)]} \tag{6.44}$$

where B and H are the width and height of the room and M_o is the momentum of the supply air, given by:

$$M_o = \rho A_o U_o^2 \tag{6.45}$$

where A_o is the effective area of the supply opening and U_o is the supply velocity. Equation (6.44) and Jackman's data are plotted in Figure 6.13 and, as shown, the equation correlates the experimental results satisfactorily up to a momentum of 1.2 N, but beyond that the equation overestimates the room velocity. This deviation was attributed to ignoring velocities greater than 0.5 m s^{-1} in calculating the average room velocity and such high air velocities could occur with a large supply momentum at a number of locations in the occupied zone.

By combining equations (6.44) and (6.45) and using a standard air density ($\rho = 1.2$ kg m^{-3}) it can be shown that the average room velocity is given by:

$$v_r = 0.80\sqrt{[A_o/(BH)]}\,U_o \tag{6.46}$$

where $A_o/(BH)$ is the ratio of supply opening area to the area of the wall containing the opening.

The trajectory of a non-isothermal jet projected close to the ceiling was also investigated by Jackman. He found it to be influenced by the Archimedes number, the distance of the supply opening from the ceiling and the aspect ratio of the opening. The attraction of the jet towards the ceiling due to the Coanda effect was found to be greater

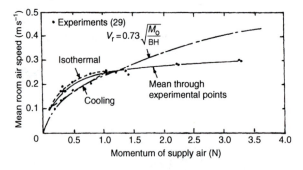

Figure 6.13 Mean room air speed for a sidewall supply.

for high aspect ratio openings and for openings closer to the ceiling. In order to correlate the experimental data, Jackman modified Müllejans' [2] Archimedes number Ar' [$= (bh/D_h^2)Ar$] further to include these parameters and obtained the following Archimedes expression:

$$Ar'' = (2d/b)(h/b)Ar'$$
$$= (2d/b)(h/b)(bh/D_h)Ar \qquad (6.47)$$
$$= [2dhA_o/(bD_h)^2]Ar$$

where b and h are the width and height of the supply opening, d is the distance of the upper edge of the opening from the ceiling, D_h is the hydraulic diameter of the room and Ar is defined by equation (6.35). Using the modified Archimedes number, Ar'', the vertical displacement of the jet centreline from the opening centreline is given by:

$$y/\sqrt{A_o} = \{0.04\,x^3/[BH(B+H)]\}Ar'' \qquad (6.48)$$

Assuming a maximum design value of $Ar = 10^4$, which was based on Linke's [27] results (see Figure 6.11), Jackman rearranged equation (6.47) to yield an expression relating to the drop and aspect ratio of the supply opening. Thus:

$$r = 34.2\{(A_od/y)[L^3(B+H)/(BH)^3]\}^{2/3} \qquad (6.49)$$

where $r = b/h$, which is the aspect ratio of the supply opening. Equations (6.44), (6.48) and (6.49) are used in the design procedure for side-wall supplies described in Section 6.4.

Sill supply   Jackman [30] used the test room described earlier to investigate the room air movement produced by air supplied from sill-mounted grilles under cooling and heating condition. The sill plenum positioned below the window was used to supply air vertically upwards from a single grille of core dimensions 1070 mm × 75 mm and two grilles each of dimensions 1070 mm × 65 mm. Measurements of air velocity and temperature were carried out throughout the test room in the same way as the tests with side-wall supply. The average air speed in the occupied zone was also correlated with the momentum of the supply air and the results are shown in Figure 6.14. The cooling tests produced about 20% higher velocities than the heating tests. The following equation was derived from the velocity measurements:

$$v_r = 0.37[M_o/(BH)]^{0.33} \qquad (6.50)$$

Substituting equation (6.45) for M_o this becomes:

$$v_r = 0.39[A_o/(BH)]^{0.33}U_o^{0.66} \qquad (6.51)$$

The experimental points show higher values of v_r than the line representing equation (6.50) because, for these points, velocities below $0.1\,\mathrm{m\,s^{-1}}$ were not included in the calculation of v_r. The equation (6.44) representing the results of the side-wall supplies, which is also plotted in Figure 6.14, gives higher mean room speeds than equation (6.50). This is because a sill supply makes better use of the room surfaces for diffusing the jet than a side-wall supply and it is preferable if low room velocities are sought.

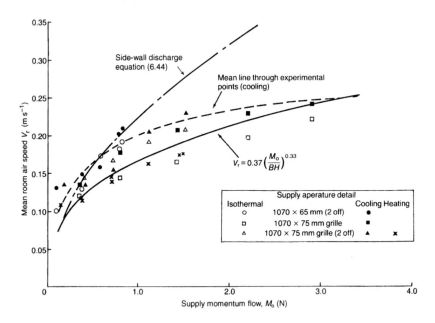

Figure 6.14 Mean room air speed for a sill supply [30].

The trajectory of the ceiling jet is given by:

$$y = 0.00196 \left[\frac{\Delta t_c x^3}{\sqrt{(A_o) U_c^2}} \right] \tag{6.52}$$

where Δt_c is the temperature difference between the jet at the ceiling/wall corner above the sill and extract air and U_c is the maximum jet velocity at the same position. Values of t_c and U_c can be calculated from the equations below which were derived from Jackman's test results:

$$\Delta t_c / \Delta t_o = 1.38 / (H'/b)^{0.25} \tag{6.53}$$

and

$$U_c / U_o = \sqrt{[5.4/(H'/b)]} + 2.15 \, Ar(H'/b - 5.4) \tag{6.54}$$

where Δt_o is the difference in temperature between supply and return air, H' is the height of the ceiling above the sill opening, b is the effective width of the opening and Ar is the Archimedes number for the supply jet given as:

$$Ar = g\beta \Delta t_o b / U_o^2 \tag{6.55}$$

The maximum velocity of the jet across the ceiling, U_m, is given by:

$$U_c / U_m = 1 + 0.042(B/b)[x_c/\sqrt{(A_o)} - 5] \tag{6.56}$$

where b is the length of the supply opening and x_c is the distance along the ceiling measured from the wall/ceiling corner.

Equations (6.50), and (6.52) to (6.56) were used by Jackman to produce a design procedure for sill supplies based on nomograms and charts.

Ceiling supply The air movement produced by circular and linear ceiling diffusers was investigated by Jackman [31] using the same test facility as for the side-wall and sill supplies described earlier. As a result of these tests a design procedure was developed for ceiling supplies using tabulated data. Tests were carried out with three different circular diffusers centrally mounted on the ceiling, namely two single-baffle plate diffusers and a multi-cone diffuser, all with a neck diameter of 150 mm. Further tests were performed with a two-slot linear diffuser (20 mm slot) mounted either centrally in the ceiling with a two-way discharge or close to the external wall with a one-way discharge along the ceiling.

Typical flow patterns using smoke and measured air velocities are shown in Figures 6.15 and 6.16 for the circular and linear diffusers, respectively. The three circular diffusers produced similar patterns. The flow patterns are symmetrical with isothermal or cooling with a uniformly distributed load on the floor, but become asymmetrical when a cold window is present in the heating mode or a hot window in the cooling mode. The downdraught produced by a cold window is clearly illustrated in Figure 6.15(b) where a cool air stream flows close to the floor. Discomfort to the occupants will be expected in this instant unless measures are taken to counteract the downdraught from the window. Figure 6.15(c) shows the opposite effect where an upward convective current is produced by the hot window interacting with the ceiling jet and the resulting flow entering the upper region of the occupied zone at high air

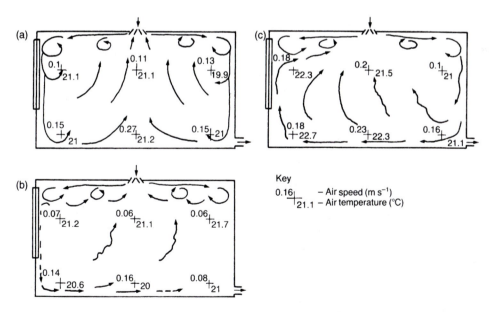

Figure 6.15 Air flow patterns with circular ceiling diffusers. (a) Isothermal and cool air supply with uniformly distributed floor heating; (b) warm air supply with cold window and (c) cool air supply with heated window [31].

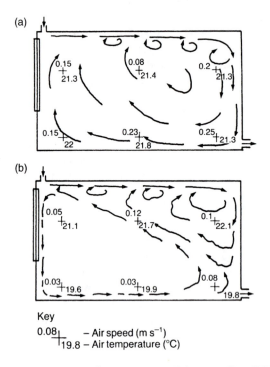

(a)

(b)

Key

0.08 — Air speed (m s^{-1})
19.8 — Air temperature (°C)

Figure 6.16 Air flow patterns with linear ceiling diffusers. (a) Isothermal and cool air supply and (b) warm air supply with cold window [31].

speed. The flow patterns produced by a linear diffuser on the centre of the ceiling were similar to the circular diffuser patterns shown in Figure 6.15. However, different flow patterns were produced with the linear diffuser at the ceiling end close to the window as shown in Figure 6.16. A good mixing of primary air with room air is evident in Figure 6.16(a) for isothermal and cool air supplies, but two triangular circulation zones were produced under heating with a cold window as a result of the window downdraught, Figure 6.16(b). Mixing of primary air with room air is very poor in this case and discomfort to the occupants will be expected as a result of a cool air stream at floor level and the large vertical temperature gradient in the occupied zone.

The variation of mean room air speed with the supply air momentum is shown in Figures 6.17 and 6.18 for the circular and linear diffusers respectively. For the same supply momentum, the circular diffusers produced higher room speeds particularly for large momenta where up to 40% higher room speeds occurred. Furthermore, lower room speeds were produced when the linear diffuser was located in the middle than at one end of the ceiling. The average results for isothermal conditions were represented by the following empirical expressions:

linear diffuser:

$$v_r = 1.08 \left\{ M_o L / [B(L^2 + H^2)] \right\}^{0.5} \tag{6.57}$$

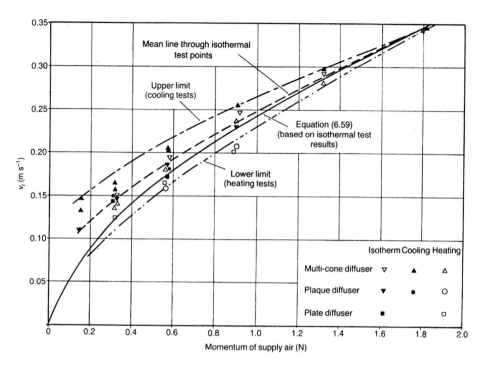

Figure 6.17 Mean room air speed for circular ceiling diffusers [31].

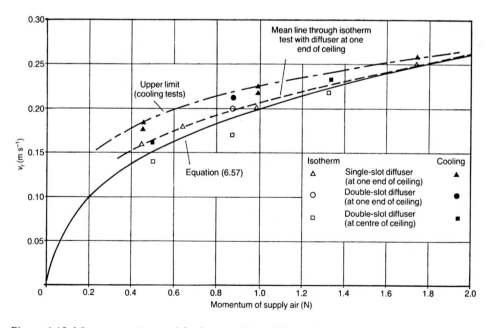

Figure 6.18 Mean room air speed for linear ceiling diffusers [31].

or

$$v_r = 0.25 \, [LT/(L^2 + H^2)]^{0.5} \tag{6.58}$$

circular diffuser:

$$v_r = 0.765 \, [M_o/(0.25 \, LB + H^2)]^{0.5} \tag{6.59}$$

or

$$v_r = 0.381 \, T/[0.25 \, LB + H^2]^{0.5} \tag{6.60}$$

where L is the room length, T is the throw length and B is the room width.

Equations (6.57) and (6.59) are plotted in Figures 6.18 and 6.17 respectively for the test room used by Jackman assuming a throw of $T = 0.75 \, L$.

The results for ceiling supplies presented so far are only valid for supply air temperature differences, Δt_o, of 5–10 K from the mean room air temperature. For such temperature differentials the air movement in the occupied zone is primarily influenced by the momentum of the supply air as shown earlier. However, with higher room loads, i.e. a higher temperature differential, the effect of convective currents, caused by the greater difference in temperature, on the air movement becomes significant and should be combined with the movement due to the supply momentum. Holmes and Caygill [32] extended Jackman's work on ceiling supplies with cooling to cover a temperature differential of up to 15 K. They found that the parameter $M_o/(QH)$, where Q is the room load (kW) and H the ceiling height (m), has a significant effect on the mean room air speed and the temperature gradient within the occupied zone. Flow visualization tests indicated that when $M_o/(QH) < 0.07$ the convective currents produced by internal heat loads dominate the room air movement and the temperature gradient in the occupied zone, Δt_r, increase as a result.

Taking into consideration the effect of the parameter $M_o/(QH)$, Holmes and Caygill obtained expressions for calculating the mean room air speed and the maximum temperature gradient in the occupied zone. For a room with a centrally mounted ceiling diffuser, the room speed, v_r, is given by:

$$\frac{v_r^2}{a^2 M_o} \left(\frac{L^2}{4} + H^2 \right) = 0.22 \left[0.19 \left(\frac{QH}{M_o} \right)^2 + 1 \right]^{1/2} \tag{6.61}$$

where a is an empirical constant having a value of 1.0 for linear diffusers and 1.2 for circular diffusers. The temperature gradient, Δt_r, is given as:

$$\Delta t_r = 4.41 \, [M_o/(LBH)][QH/M_o]^2 \tag{6.62}$$

In equations (6.61) and (6.62) L is the room length, B is the room width and H is the ceiling height.

In general, Holmes and Caygill found that circular diffusers produce the most satisfactory air distribution but linear diffusers were more desirable with high room loads because of their lower room speed characteristic.

Displacement ventilation   The principle of displacement ventilation is the scavenging of room air by the supply air. Depending on the application, displacement ventilation can normally be achieved in any of four ways: side-wall supply (cross-flow), floor supply (upward flow), ceiling supply (downward flow), and low-level over-floor supply. The latter method is the most common although it is not a truly displacement flow because of the presence of a lower displacement zone and an upper mixing zone, as explained in Section 4.6.5. The other three systems provide true displacement when a whole surface is used as an air outlet, however a downward system produces air currents in the opposite direction to the natural ventilation currents produced by heat sources at low level, which could result in an unstable air flow pattern. In an upward system the air is supplied in the same direction as the natural currents, although a limitation on the magnitude of supply air velocity is normally imposed to avoid draughts in the occupied zone.

Linke [33] studied upward and downward flow systems in lecture theatres and established critical Archimedes number values for these two types of displacement ventilation methods. Defining the Archimedes number for the room in terms of its height, H, which is the most significant dimension in displacement ventilation, then:

$$Ar = g\beta H \Delta T_0 / v_r^2 \tag{6.63}$$

where ΔT_0 is the temperature difference between the room air and supply air, v_r is the mean room air speed which is defined by:

$$v_r = \dot{V}/A \tag{6.64}$$

$\dot{V}$ being the volume of air supply ($m^3\,s^{-1}$) and A the floor (ceiling) area. Writing $\dot{V}$ in terms of the room air change rate per hour, N, i.e.:

$$\dot{V} = NV/3600$$

equation (6.64) becomes:

$$v_r = NH/3600 \tag{6.65}$$

where $V = AH$.

On substituting equation (6.65) into equation (6.63), we obtain a definition for Ar in terms of N. Thus:

$$Ar = (3600)^2 g\beta \Delta T_0 / (N^2 H) \tag{6.66}$$

In Figure 6.19, this equation is used to show the relationship between N and Ar for different values of $\Delta T_0/H$.

The room cooling load, q ($W\,m^{-2}$), is given by:

$$q = Q/A = C_p \dot{V} \Delta T_0 / A$$

i.e.:

$$q = (\rho C_p / 3600) NH \Delta T_0 \tag{6.67}$$

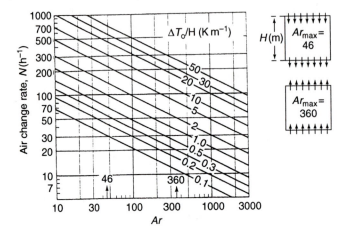

Figure 6.19 Effect of Archimedes number on the air change rate of upward and downward displacement systems.

On substituting for ΔT_o from this equation into equation (6.66), the expression for Ar becomes:

$$Ar = (3600)^3 g\beta q/(\rho C_p N^3 H^2) \tag{6.68}$$

Substituting standard values for g, ρ and C_p in equation (6.68) and rearranging, the air change rate is given by:

$$N = 108.93\sqrt[3]{[q/(H^2 Ar)]} \tag{6.69}$$

According to Linke [33] a stable air movement pattern is achieved in a downward flow system when $Ar \leq 46$ and in an upward system when $Ar \leq 360$ (see Figure 6.19). Using these limits in equation (6.69), the critical (minimum) air change rate for the two displacement systems is given by:

for a downward system:

$$N_c = 30.4\sqrt[3]{(q/H^2)} \tag{6.70}$$

for an upward system:

$$N_c = 15.3\sqrt[3]{(q/H^2)} \tag{6.71}$$

These two equations show that for a stable room air movement, an upward displacement system requires half the air change rate of a downward displacement system for removing the same room load. Because an upward system supplies the air directly into the occupied zone only the load in the occupied zone need be considered in equation (6.71), i.e. convective load due to lighting is not included. Consequently,

Design of room air distribution systems   257

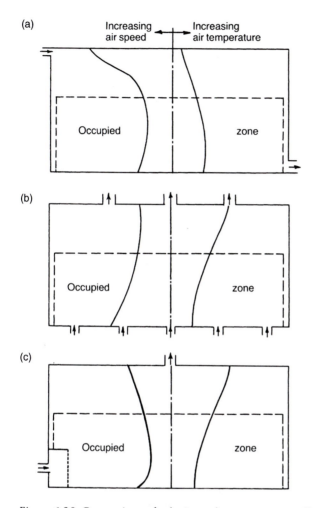

Figure 6.20 Comparison of velocity and temperature profiles for mixing and displacement ventilation (cooling). (a) Mixing ventilation; (b) floor displacement ventilation and (c) low-level wall displacement.

an upward system is most suitable for rooms with large heat loads (up to about 700 W m^{-2} [34]) such as in high occupancy density (theatres, auditoria, etc.), computer rooms, industrial processing enclosures, etc. Figure 6.20 shows comparisons of the expected temperature and room air speed distribution in a room ventilated by mixing, floor displacement and low-level wall displacement ventilation.

   The use of low-level wall supply has become very common in Europe during the last two decades. The advantages of this method over other displacement methods is that only few air terminal units are needed to supply cool air at low velocity (<0.5 m s^{-1}) over the floor and allow room heat sources to create the upward buoyancy. This method of air supply inevitably creates vertical temperature gradient and it is therefore important that this is maintained within the limits acceptable for comfort, e.g. <3 K between head and feet level for sedentary position (i.e. between 0.1

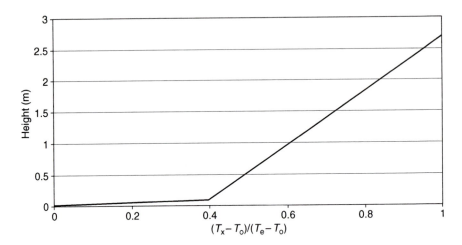

Figure 6.21 Typical variation of room air temperature with height for low-level air supply.

and 1.1 m above floor) [35]. Experimental measurements have shown that there is a linear air temperature variation with room height from a point just above floor level [36, 37]. The design of a low-level displacement system involves the prediction of temperature gradient so that the limit required for comfort may be achieved in the occupied zone. The linear temperature gradient found from experiments has been used to establish a method of determining temperature gradient in office rooms of maximum ceiling height of 4 m for low-level displacement flow [38]. The basic assumptions used are

- the existence of a linear temperature gradient from 0.1 m above the floor to ceiling;
- the extract air temperature is the same as that at ceiling height;
- the air temperature difference between 0.1 m above floor and the supply $(T_f - T_o)$ is 0.4 times the overall temperature difference between extract and supply $(T_e - T_o)$.

These assumptions lead to a non-dimensional temperature expression (see also Figure 6.21) that may be used for design purposes and given by:

$$\theta = (T_x - T_o)/(T_e - T_o) \tag{6.72}$$

Here, T_o and T_e (°C) are the supply and extract temperatures, T_x (°C) is the room temperature at height y (m). The air temperature gradient, G_T (K m^{-1}), is given by:

$$G_T = 0.6(T_e - T_o)/(H - 0.1) \tag{6.73}$$

where H is the ceiling height (m) and T_f (°C) is the temperature at 0.1 m above the floor.

Ceiling cooling devices   Experience with displacement ventilation has shown that standard low-level supply systems have a maximum cooling capacity of 40 W m^{-2} of floor area. In situations that demand a greater cooling capacity the standard displacement system will not be adequate for maintaining a displacement flow in the

(a)

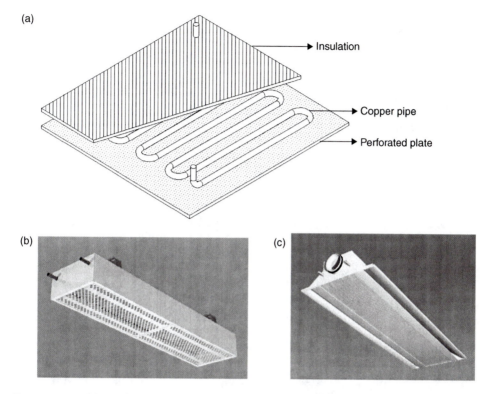

Insulation

Copper pipe

Perforated plate

(b)   (c)

Figure 6.22 Additional cooling devices used with low-level displacement ventilation. (a) Cold ceiling panel; (b) passive cold ceiling beam and (c) active cold ceiling beam. (b) and (c) are Courtesy of ABB.

lower zone, which is necessary for achieving good thermal comfort and air quality. Increased cooling capacity may be achieved through the use of either cold ceiling panels (up to $100 \, \mathrm{W \, m^{-2}}$ floor area), passive chilled beams (additional $250 \, \mathrm{W}$ per linear metre length), or active cooling beams ($350 \, \mathrm{W \, m^{-1}}$ length) with a displacement system.

Cold ceiling panels consist of a serpentine pipe carrying cold water ($\approx 14 \, °\mathrm{C}$ to avoid water vapour condensation on the panel surface) attached to perforated flat plate backed by insulation, see Figure 6.22(a). Cooling is provided by convection of cool air downward and radiation from the panel surface. The convective flow will generate a mixing zone directly below the ceiling panel, the extent of which is dependent on the temperature difference between panel surface temperature and local air temperature. Because a significant area of a suspended ceiling contains these panels the radiation heat output from the panels is a significant component of the total cooling capacity of the panels.

Cold beams consist of copper pipes covered with fins and baffle plates below the fins to direct the air flow downward, see Figure 6.22(b). Most of the beam cooling is by convection, with radiation being a small component of the total. These are called passive (or static) cold beams or chilled beams as the cooling is by natural convection.

An active cold beam on the other hand, contains an air duct that blows air over the finned tubes downward, see Figure 6.22(c).

The effect of cold ceiling panels or beams is to suppress the vertical temperature gradient in the upper zone of the room with most of the gradient remaining within the occupied zone. Cold beams provide the most cooling but the temperature gradient is suppressed further as a result of the breakdown of displacement flow pattern in the room and the resulting flow becomes akin to a mixing system. The air quality in the occupied zone consequently deteriorates in comparison with the basic displacement system. Additional information on the performance of cold panels and beams may be found in [39–41].

Variable air volume In a constant air volume (CAV) system room load variation is dealt with by varying the temperature of the air supply to the room by reheating or by mixing two air streams which could involve energy wastage. Depending on the complexity of the design, a number of CAV systems are available to achieve temperature control with varying degrees of energy penalty. These are outside the scope of this book and interested readers should consult an air-conditioning textbook. In the variable air volume (VAV) systems, changes in room load are accommodated by controlling the volume of air supply to the room without changing the supply temperature until the minimum permissible air supply is reached. The main advantage of a VAV system over a CAV system is in the operating cost due either to the reduction in reheat and mixing or to a reduction in fan power or a combination of these. It should be noted that fan power is directly proportional to the cube of the volume flow rate and substantial saving in energy can be achieved at low thermal loads by using a VAV system. However, the disadvantages of the VAV system may be unstable operation caused by variation in duct pressure, acoustic problems as a result of flow throttling, temperature stratification with heating and the possibility of dumping (separation of air jet from the ceiling) under cooling with reduced air flow rates. These problems can, however, be overcome with a good design practice.

Volume flow control is normally achieved by one of the following methods:

1 control of a damper in a box upstream of a conventional air terminal device actuated by a room thermostat;
2 control of the effective area of the air terminal device by room thermostat although the volume flow rate may still be controlled separately;
3 pulsating control of volume flow rate by means of a control box upstream of a conventional air terminal device.

In all previous treatments of air movement in rooms a steady-state air supply has been assumed, but any of these methods of VAV air flow control can produce deviations from the steady-state approach. Holmes [42] carried out tests in the BSRIA test room described earlier with dynamic room loads and air supplies. Hourly measurement of mean room air speeds under dynamic conditions were in good agreement with those calculated using equation (6.61) which is based on steady-state results. This was a confirmation that room air movement with a VAV system can be treated using equations derived from CAV ceiling supply systems. However, a VAV system can operate over a wide range of air flow rates and the effect of variation in the momentum of the supply air as well as room load must be considered in the design. A design procedure for VAV air distribution is given in Section 6.4.3.

Comparison between high-level and low-level supplies

The effect of the position of supply air device on the comfort level in a room has been investigated in [43, 44] in a study aimed at assessing the effectiveness of a warm-air heating system that utilizes a low-grade energy source with supply air temperatures considerably below those provided by conventional warm-air systems. The investigation was carried out in a purpose-built test chamber of internal dimensions 5.5 m × 3.5 m × 2.5 m ceiling height surrounded at two perpendicular walls and the floor by another enclosure which was maintained at a UK winter design temperature of $-1\,°C$. Measurements of air velocity and air temperature were performed using a computer-controlled measuring and data logging system at 385 positions in the occupied zone, covering heights of 0.15, 0.5, 0.85, 1.2 and 1.8 m above the floor, and the mean radiant temperature was measure at a height of 1.2 m throughout the chamber. Three methods of supplying the air were investigated: direct supply into the occupied zone at low level; indirect supply outside the occupied zone at high level; and air diffusion across the walls at low level using baffle plates. A total of five different ATDs were used at low level and four ATDs at high level. Fanger's thermal comfort criterion (see Chapter 1) was used to assess the comfort level in the occupied zone. Further details of the test room and experimental measurements are given in [44].

The difference in the average room air temperature between head height and ankle height $(t_{1.8} - t_{0.15})$ is shown in Figure 6.23 plotted against air change rate. This figure

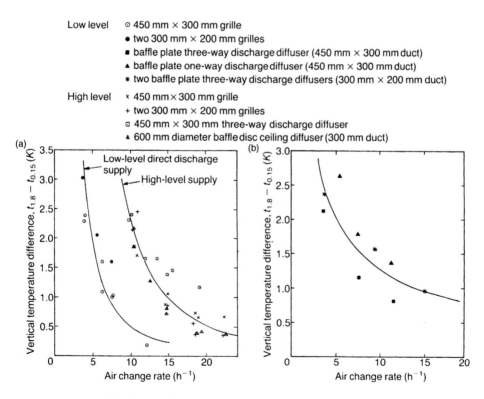

Figure 6.23 Vertical gradient of mean air temperature in a heated room. (a) Low-level and high-level supplies and (b) low-level baffle-plate diffusers.

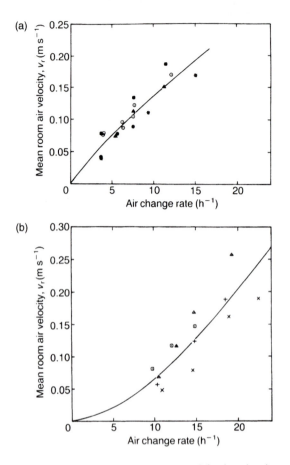

Figure 6.24 (a) Mean room air speed for low-level warm air supplies and (b) mean room air speed for high-level warm air supplies.

shows that the vertical temperature gradient in the occupied zone is dependent on whether the air is supplied at low level or at high level, but almost independent of the number and type of supply devices used in each case. The high-level supply mode produced a larger temperature gradient than the low-level direct discharge mode and in both cases the gradient decreased with increasing air change rate. Figure 6.24 shows the mean room air speed, v_r, versus the air change rate. The low-level supplies produced higher v_r than the high-level supplies but in each case the results were almost independent of the type of supply device used.

The supply and extract air temperatures that provide thermal environments in the occupied zone with predicted percentage of dissatisfied (*PPD*) less than the recommended value of 10%, ISO 7730 [35], are shown plotted versus air change rate in Figure 6.25. For each of the three supply modes, the range of air change rate for comfortable conditions is indicated in this figure. To cover the full range of 4–23 ach, it would be necessary to use the low-level supply mode for air change rates below 10 ach and the high-level supply mode for higher air change rates. The low-level wall

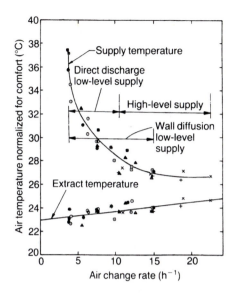

Figure 6.25 Supply and extract air temperatures.

air diffusion supply could provide a comfortable environment for air change rates between 4 and 15 ach. Figure 6.25 also shows that for a winter design temperature of −1 °C, a high-level supply is necessary to provide an acceptable condition if the supply temperature is between 26 and 28 °C. For a supply temperature above 28 °C, the high-level supply will not provide an acceptable thermal environment and therefore a low-level supply is required. If low-level supplies are used the wall diffusion using baffle plates produces a wide range of air change rates and supply temperatures over which comfort can be achieved.

6.3   Case studies

Detailed evaluation of air movement in actual buildings is normally only carried out for the purpose of commissioning, for substantiating the design calculations, or for troubleshooting to identify the cause of problems associated with the air distribution system. Field investigations can also provide facility managers with useful information for plant control settings. Because of the high internal loads and the large volumes involved, the design of air distribution systems for large enclosures normally represents a challenge to the ventilation system designer. Whereas the air flow patterns in most buildings can be adequately modelled in the laboratory using a full-scale mock-up, the modelling of large enclosures, e.g. auditoria, cannot realistically be achieved with a model size greater than one-sixth of the full scale [9] and sometimes scale ratios as small as 1 : 20 are used [45]. With such scale ratios kinematic similarity, thermal similarity and similarity of the boundary conditions in the model and prototype are extremely difficult to achieve. Although model tests can still provide useful information on the air movement for optimizing the air distribution design, precise comfort assessment can only be achieved with full-scale measurements under normal load conditions. For this reason, most reported field investigations have been carried out on large spaces [46].

In this section three types of mechanically ventilated large spaces where sufficiently detailed measurement are available will be described.

6.3.1   Lecture theatre

A widely quoted field investigation is that by Linke [33] on a 500 seat lecture theatre at the Technical University in Aachen, Germany. In the air distribution design of large spaces, usually, either a downward or an upward displacement ventilation system is used and Linke conducted an investigation to compare the air movement patterns and temperature distribution produced by two such systems. Cool air was supplied at the rate of $4.70 \, m^3 \, s^{-1}$ (9.1 ach) from openings distributed over the ceiling and extracted beneath the seats for the downward flow and supplied from beneath the seats and extracted from the ceiling for the upward flow system. This produced an air supply rate of $9.4 \, l \, s^{-1}$ per person with full occupancy. Figure 6.26(a–e) shows the measured air flow patterns and the air temperature along the theatre above the seated occupants and also the supply and extract temperatures for full occupancy, uniformly distributed half-occupancy, concentrated half- and quarter-occupancy and unoccupied theatre. The floor supply with full and uniformly distributed half-occupancy (Figures 6.26(a1) and (b1)) produced a stable and a uniform air flow pattern throughout the theatre with the flow direction at low level from the front to the back. Extensive circulatory air motion only occurred when the occupancy was concentrated at the front of the theatre (Figure 6.26(c)) and this was produced by the natural convective currents rising from the occupants themselves. In contrast, the ceiling supply produced turbulent and unstable air movement patterns with unpredictable flow directions at low level caused by local circulatory motion. For more uniform air distribution with ceiling supplies a larger air change rate would have been necessary as equations (6.70) and (6.71) suggest.

The vertical air temperature gradients for floor and ceiling supplies are compared in Figure 6.27 for full and half-occupancy. With a floor supply the steepest temperature gradient (increasing with height) occurs over a distance 1.2 m from the floor, i.e. from foot to head, which has a range of 3–8 K depending on the load. With a ceiling supply, there is little temperature gradient up to 1.2 m above the floor but the steepest gradient (decreasing with height) occurs above head height, giving a value between 4 and 8 K. Temperature measurements in the seat region and shown in Figure 6.28(a) and (b), confirm the presence of a large gradient with a floor supply and a small gradient with a ceiling supply [33]. The steep temperature gradient in the seat region can be greatly reduced by supplying the air from an opening in the back rest of the seat instead of the floor as shown in Figure 6.29(a) and (b) from measurements by Sodec [34]. This flow arrangement also produces lower air speeds around the occupant, particularly at ankle level, where high speeds that are usually associated with floor supplies can produce discomfort especially as the temperature is low (i.e. almost supply temperature) at this height.

6.3.2   Concert hall

Bouwman and van Gunst [47] investigated the air movement in the de Doelen concert hall in Rotterdam which was ventilated by ceiling-mounted circular air diffusers. A plan and a side elevation of the hall are shown in Figure 6.30. The maximum

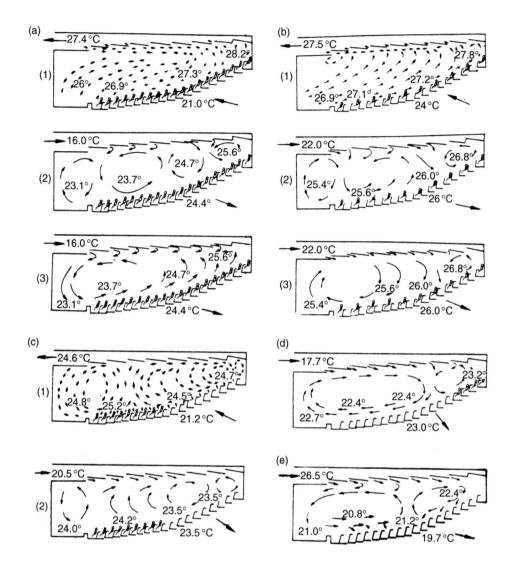

Figure 6.26 Air flow patterns and temperature produced by upward and downward ventilation in a lecture theatre. (a) 100% occupancy; (b) 50% occupancy distributed evenly; (c) 50% occupancy in front half of auditorium; (d) 25% occupancy at back of auditorium and (e) unoccupied auditorium [33].

dimensions of the hall were 60 m long × 40 m wide × 18 m high with a maximum seating capacity of 2500. Measurements of air speeds and temperatures were carried out in: (i) a 1:10 model under isothermal conditions, (ii) a full-scale mock-up 13 m long × 9 m wide × 7 m high representing a central part of the hall under cooling and (iii) in the hall itself with simulated occupancy and a total load of 200 kW. The hall was provided with two air supply systems, one for each half, which could be independently controlled to give a maximum air flow rate of 11 l s^{-1} per person. The

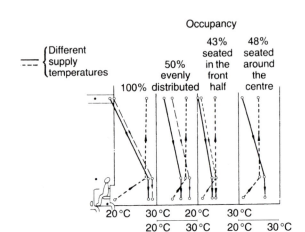

Figure 6.27 Vertical temperature gradients for downward and upward ventilation [33].

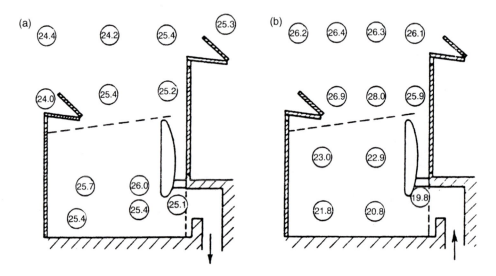

Figure 6.28 Temperature distribution (°C) in a seat region with downward and upward air supply. (a) Downward supply and (b) upward supply [33].

flow patterns produced by the field tests are shown in Figure 6.31. With no mechanical ventilation (Figure 6.31(a)) the air motion in the hall is generated by convective currents from occupancy alone and similar but larger velocities were observed with full ventilation supplying air at 17 °C for the front half and at 19 °C for the rear half (Figure 6.31(b)). By reversing the temperature of the two air supplies so that a higher-temperature air is supplied at the front half of the hall, the flow pattern was completely reversed, (Figure 6.31(c)), in which case the occupants would be exposed to air currents at the back of the neck, which is not a desirable situation. When the

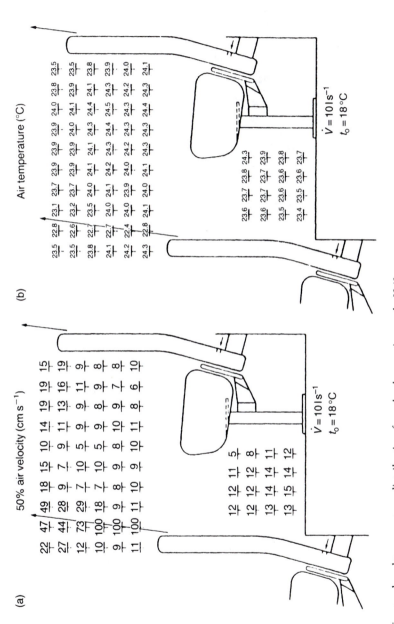

Figure 6.29 Air speed and temperature distribution for a back rest air supply [34].

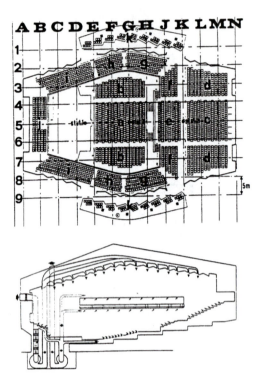

Figure 6.30 A plan and an elevation of the de Doellen Concert Hall.

supply temperature of the two systems was the same (22 °C) but only 50% of the air was supplied at the rear half, an unstable air flow pattern was produced as shown in Figure 6.31(d). A more stable flow pattern was produced when the occupancy was reduced to 70% and the supply temperature of the rear half of the hall increased to 24 °C, see Figure 6.31(e). Although this is a stable flow dominated by two large vortices each controlling the flow in one-half of the hall, the flow direction in the rear half at occupancy height is towards the back of the neck, which is undesirable. Measured air speeds at the front, middle and rear of the hall for different ventilation rates are given in Table 6.3. As would be expected the speed increases with ventilation rates and occupancy and is always lower at the rear of the hall. The vertical distribution of air speed at the middle of the hall with full occupancy and 75% front and 50% rear air supplies is shown in Figure 6.32. The figure shows an increase in speed with height in the occupied region.

6.3.3 *Sports arena*

A scale model and full-scale investigation of the air movement and temperature distribution in a Sumo wrestling arena was carried out by Togari and Hayakawa [45]. The 11,000 seat Shinkokugikan wrestling arena in Tokyo, which was completed in 1986, has a square plan of 80 m × 80 m and a maximum ceiling height of 40 m,

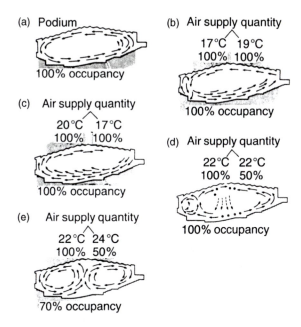

(a) Podium

100% occupancy

(b) Air supply quantity

17°C 19°C
100% 100%

100% occupancy

(c) Air supply quantity

20°C 17°C
100% 100%

100% occupancy

(d) Air supply quantity

22°C 22°C
100% 50%

100% occupancy

(e) Air supply quantity

22°C 24°C
100% 50%

70% occupancy

Figure 6.31 Air movement patterns in the de Doellen Concert Hall. (a) Air movement
pattern due to convection currents from audience; (b) as (a) but with full air
supply at 17 °C and 19 °C to front and back of hall respectively; (c) as (b) but
with temperatures of air supply reversed producing a change of direction in
the air movement pattern; (d) opposing circulation achieved by closing some
of the diffusers in the back half of the auditorium – some unstable downward
flow occurs and (e) stable opposing circulations produced as for (d) with the
aid of convection currents [47].

Table 6.3 Air speed measurements in the de Doellen Concert Hall [47]

Occupancy (%) (2500 max)	Ventilation rate (%)		Air speed (m s^{-1}) Position in hall		
	Front	Rear	Front	Centre	Rear
100	0	0	0.260	0.290	0.120
0	75	50	0.200	0.190	0.230
100	100	100	0.345	0.375	0.200
100	75	50	0.330	0.390	0.180

Figure 6.33. The three (spectator) floor arena has five main occupancy zones including
the square wrestling 'Dohyo' ring in the centre of the ground floor. These zones were
supplied with conditioned air (cooled during occupancy even in winter) from six main
ductworks using four types of outlet. Swirl-type diffusers (A-1) were mounted in the
ceiling at the rear of the first (spectator) floor to supply air to the first floor, as shown
in Figure 6.34. Slot diffusers (S-1) positioned at the outer perimeter of the second floor
seating gallery were used to supply air to the first floor and the wrestling ring, and slot
diffusers (S-2) were used to supply air to the third floor. Long throw nozzles (N-2)

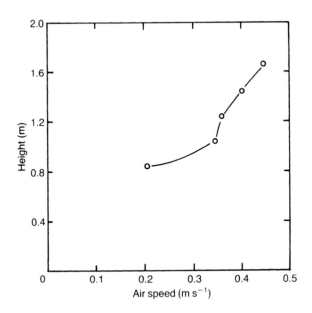

Figure 6.32 Vertical distribution of air speed for downward ventilation in a concert hall [47].

Figure 6.33 The Shinkokugikan Sumo Wrestling Arena in Tokyo.

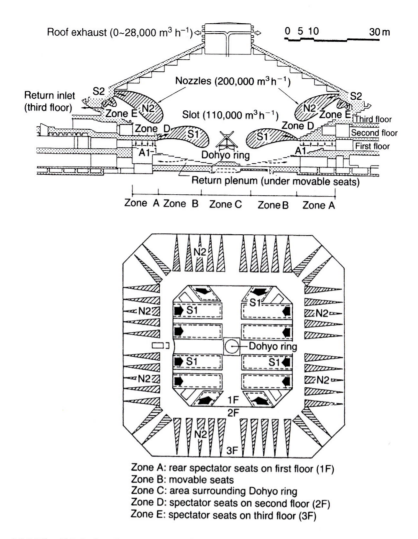

Roof exhaust (0~28,000 m³ h⁻¹)

0 5 10   30 m

Nozzles (200,000 m³ h⁻¹)

Return inlet
(third floor)
S2
Zone E   N2
Slot (110,000 m³ h⁻¹)
Zone D
S1
A1
Dohyo ring
S2
N2   Zone E   Third floor
Zone D   Second floor
First floor
S1
A1
Return plenum (under movable seats)

Zone A Zone B   Zone C   Zone B   Zone A

N2
S1   S1
N2   N2
Dohyo ring
S1   S1
N2   N2
1F
2F
N2
3F

Zone A: rear spectator seats on first floor (1F)
Zone B: movable seats
Zone C: area surrounding Dohyo ring
Zone D: spectator seats on second floor (2F)
Zone E: spectator seats on third floor (3F)

Figure 6.34 The Shinkokugikan Arena air distribution [45].

positioned on the two side walls at high level above the third floor were used to project the air over the second and third floors. The slot diffusers S-2 and nozzles N-2 were supplied by a single duct. Because of the large radiant heat received at the wrestling ring from spotlights (168 kW total lighting load) the air temperature required at the ring was 25.5 °C or lower (i.e. about 1 K lower than the galleries) and the air speed was 0.5 m s⁻¹. The total air supply rate from the four diffuser types was 127 m³ s⁻¹ in the ratios A-1 = 32.6%, S-1 = 23.9%, S-2 = 6.5% and N-2 = 37.0%. About 90% of the air extracted was taken from the movable seat area on the first floor and 10% was exhausted by fans at the ridge of the building.

A 1:20 scale model with dimensions 4.0 m × 4.0 m × 1.9 m ceiling height was constructed to provide design data on the air movement and the temperature distribution and also to devise a temperature control strategy for the six supply ducts. In addition to temperature and air velocity, tracer gas concentration measurements and smoke plumes were also used to provide information on the air mixing in the arena. The model tests were carried out at the same Archimedes number (referred in this case to the outlets) as in the full scale and the heat loads were simulated by electrically heating the internal surfaces and the seating galleries. To achieve equality of Ar in the model and full scale the following parameters were used:

Scale factor $(S) = 20$
Ratio of temperature difference $(\Delta T_{om}/\Delta T_{op}) = 2.0$
Ratio of air supply velocity $(U_{om}/U_{op}) = 0.32$

These ratios of velocity and temperature difference were used to normalize the air velocities and temperatures measured in the model to obtain the full-scale equivalent. Typical normalized temperature and velocity distributions for maximum cooling

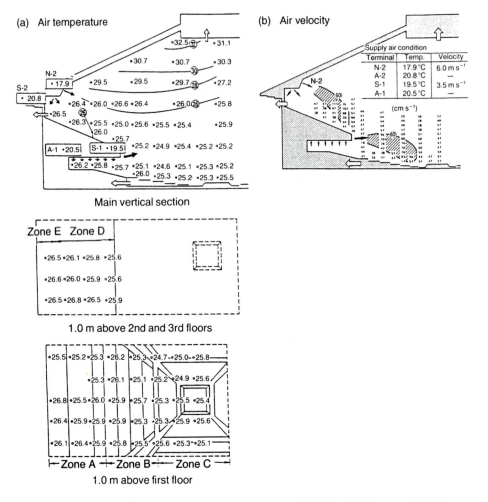

Figure 6.35 Air temperature and velocity distribution in the Shinkokugikan model [45].

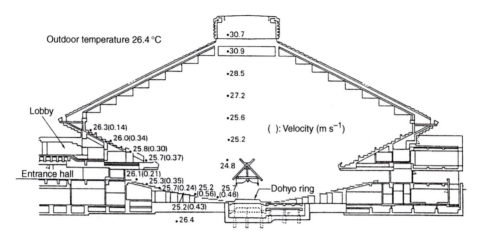

Outdoor temperature 26.4 °C

•30.7
•30.9
•28.5
•27.2
•25.6

Lobby
26.3(0.14)
26.0(0.34)
25.8(0.30)
25.7(0.37)
•25.2
( ): Velocity (m s⁻¹)

24.8
Entrance hall
26.1(0.21)
25.3(0.35)
25.7(0.24) 25.2 25.7
(0.56) (0.46)
—Dohyo ring
25.2(0.43)
•26.4

Figure 6.36 Air temperature and velocity distribution in the Shinkokugikan wrestling arena [45].

are shown in Figure 6.35(a) and (b). The average temperature in the wrestling ring is 25.5 °C and in the spectator galleries the temperature range is 25.5–26.5 °C. Above the occupied regions the temperature increased with height reaching an extract temperature of about 31 °C at the roof.

Air temperature and velocity distribution measured in the actual building after its construction under summer cooling conditions are shown in Figure 6.36. These values compare favourably with those obtained from the test model shown in Figure 6.35. To evaluate the mixing characteristics of the four different supply outlets tests were carried out on the scale model using ethylene as a tracer gas. Figure 6.37(a–d) shows the distribution of the concentration of the gas in the arena as a ratio of its concentration at the air supply terminals. In Figures 6.37(a) and (c) the tracer gas was supplied from the swirl diffusers (A-1) and the nozzle diffusers (N-2) with a resulting high concentration ratios measured in the two zones served by these diffusers, thus confirming the design expectation. With the gas supplied through slot diffusers (S-1) and (S-2), Figures 6.37(b) and (d), the concentration ratios in the zones served by these diffusers is considerably lower than in the previous two zones indicating a wider spread of the air supplied from these outlets into the arena.

It is interesting to note the distinct zones of the arena which can only be adequately ventilated by proper distribution of the supply air to each zone. That is, a high-level supply from the slot diffusers (S-2) or the nozzle diffusers (N-2) only may produce an acceptable environment on the second and third floors of the arena but conditions in the lower regions will be quite unacceptable, see Figures 6.37(c) and (d).

The Shinkokugikan project is a good example where a reduced scale model investigation when properly scaled (see Section 6.2.1) and conducted can provide extremely useful and accurate predictions of the thermal climate in the actual building. The model data can then be used in optimizing the air distribution systems and in the case of the Shinkokugikan arena also provided the designers with valuable information which was used in the control of air temperature in the three supply ducts, i.e. different supply temperatures were used for each duct.

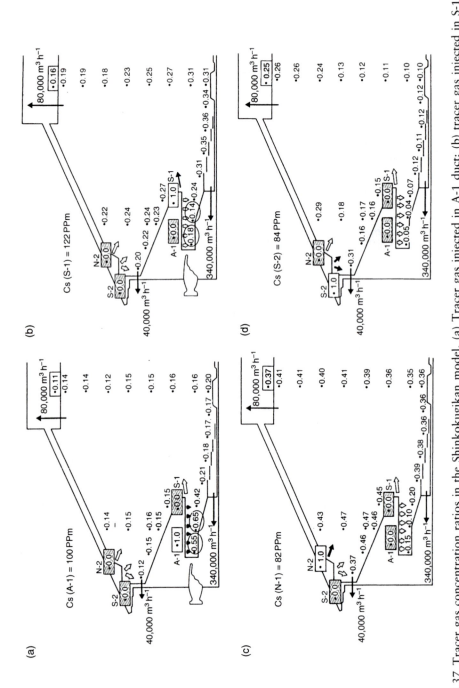

Figure 6.37 Tracer gas concentration ratios in the Shinkokugikan model. (a) Tracer gas injected in A-1 duct; (b) tracer gas injected in S-1 duct; (c) tracer gas injected in N-2 duct and (d) tracer gas injected in S-2 duct [45].

6.4 Design procedures

Based on the technical information presented in the previous sections on reduced-scale model and full-scale investigations of the air movements in enclosures, design procedures are presented in this section for different air supply methods. When adhered to, these procedures should produce satisfactory room air distribution under normal occupancy and usage and with conventional building designs. Under other circumstances it is recommended that investigations with a physical model (either reduced-scale or full-scale or both) and/or CFD modelling of the flow (see Chapter 8) be carried out for assessing the air distribution design.

6.4.1 Side-wall supply

Using the test results for high-level side-wall supplies described in Section 6.2.3, Jackman [29] developed a design procedure based on graphs and nomograms to estimate the air supply rate, the temperature difference between supply and room and the dimensions of supply openings. It is claimed [29] that following this design procedure unacceptably high velocities in a room supplied with air from a side-wall at high level can be avoided. Jackman's design procedure is presented here in a form which is more suitable for use in a computer program instead of reading data from nomograms. The new method, which is based on the use of algebraic equations, is particularly suitable for more complex design procedures which would otherwise involve the use of a number of nomograms, charts and tables such as those for sill or ceiling supplies described later. The procedures should be applied using the following steps.

1 Calculate the minimum room air change rate using equation (6.43) and a value of $Ar_c = 10^4$, i.e.:

$$N = 7.84\sqrt[3]{qB/[(B+H)L^2]} \quad \text{h}^{-1}$$

where L, B and H are the length, width and height of the room and q is the room load in W m^2.

2 Calculate the supply volume flow rate from:

$$\dot{V} = NBLH/3600 \quad \text{m}^3\,\text{s}^{-1}$$

3 Calculate the difference between the supply and extract air temperatures using:

$$\Delta t_o = 3q/(NH) \quad \text{K}$$

4 Calculate the supply air velocity, U_o, for a given effective area, A_o, of the supply opening (initially assumed) using:

$$U_o = \dot{V}/A_o \quad \text{m s}^{-1}$$

where $\dot{V}$ is the volume flow rate calculated in step 2.

5 Calculate the throw length, T, of the jet using:

$$T = 11.6\sqrt{(M_o/\rho)} = 11.6\, U_o\sqrt{A_o} \quad \text{m}$$

which should be approximately $0.75\,L$. If $T \neq 0.75\,L$ then select another effective area A_o and repeat the calculations until $T \approx 0.75\,L$.

6   Calculate the momentum of the supply air using equation (6.45):

$$M_o = \rho A_o U_o^2 \quad N$$

7   Calculate the mean room speed using equation (6.44):

$$v_r = 0.73\sqrt{[M_o/(BH)]}$$

If $v_r < 0.1$ or $v_r > 0.3\,\mathrm{m\,s^{-1}}$ then select different air terminal devices and repeat steps 4–6.

8   Calculate the maximum tolerable drop of the jet centreline, y, using:

$$y = H - (Z + d + 0.5h + \delta)$$

where Z is the height of the occupied zone (i.e. 1.8 m for standing or 1.2 m for seated occupancy), d is the distance between the top of the supply opening and the ceiling, h is the vertical dimension of the core of the supply opening (to be assumed initially) and $\delta = 0.075\,L$ is the jet spread.

9   Calculate the aspect ratio, r, of the supply opening using equation (6.49):

$$r = 34.2 \left[\frac{A_o d}{y} \left(\frac{L^3 (B+H)}{(BH)^3} \right) \right]^{2/3}$$

10 Select an air terminal device with an aspect ratio $\geq r$ and an effective area A_o. Check that $h \geq$ vertical dimension of ATD otherwise repeat steps 8 and 9.

If $d > 0.5b$, where b is the length of ATD, then a cold jet may not attach to the ceiling and steps 8 and 9 are invalid and the jet behaves as a free jet.

Example 6.1 It is proposed to air-condition a room 8 m × 4 m × 3.5 m ceiling height and having a heat gain of 40 W m^{-2}, by supplying cool air from a grille mounted on a small side wall 200 mm below the ceiling. Select a suitable grille and assess the thermal environment in the occupied zone, assumed 1.8 m high.

1   The required air change rate is:

$$N = 7.84\sqrt[3]{[4 \times 40/(4 + 3.5) \times 64]}$$
$$= 5.43 \quad \mathrm{h^{-1}}$$

2   Volume flow rate:

$$\dot{V} = 5.43 \times 4 \times 8 \times 3.5/3600$$
$$= 0.169 \quad \mathrm{m^3\,s^{-1}}$$

3   The temperature difference:

$$\Delta t_o = 3 \times 40/(5.43 \times 3.5) = 6.3 \quad K$$

4   Assuming a grille effective area of 0.08 m^2 the supply velocity is $U_o = 0.169/0.08 = 2.11\,\mathrm{m\,s^{-1}}$.

5 The throw lengths $T = 11.6 \times 2.11 \times \sqrt{0.08} = 6.92$ m, which is $0.86\,L$ (i.e. greater than $0.75\,L$); hence select a greater A_o and repeat calculations starting at step 4. Let $A_o = 0.10\,\text{m}^2$. Hence:

$$U_o = 0.169/0.1 = 1.69 \quad \text{m s}^{-1}$$

$$T = 11.6 \times 1.69\sqrt{0.1} = 6.2 \quad \text{m}$$

which is $0.77\,L$, i.e. satisfactory.

6 The supply momentum:

$$M_o = 1.22 \times 0.1 \times (1.69)^2 = 0.35 \quad \text{N}$$

7 The mean room speed:

$$v_r = 0.73\sqrt{[0.35/(4 \times 3.5)]}$$
$$= 0.12 \quad \text{m s}^{-1}$$

which is acceptable.

8 Calculate the drop:

$$y = 3.5 - (1.8 + 0.2 + 0.1 + 0.6)$$
$$= 0.8 \quad \text{m}$$

9 Calculate the aspect ratio of opening:

$$r = 34.2[(0.1 \times 0.2/0.8)(512 \times 7.5/2744)]^{2/3}$$

giving $r = 3.66$.

A suitable grille size would be 800 mm × 200 mm, i.e. aspect ratio of 4 and a vertical dimension which is that assumed in step 8.

Furthermore, the distance of the grille from the ceiling, $d = 0.2$ m, is less than half the grille width 0.4 m, i.e. dumping is unlikely to occur.

If a much larger room is being considered then it should be divided into smaller zones of equal areas and the above procedure applied to each zone.

6.4.2 Sill supply

A design procedure was developed by Jackman [30] to enable the determination of air distribution parameters required to produce specified conditions in the occupied zone of a room heated or cooled by air supplied upwards from a sill grille below a window. This procedure was based on the use of cumbersome nomograms and graphs, but the method described here uses empirical expressions derived from Jackman's test data (see Section 6.2.3). This is found to be more convenient for design analysis particularly when using a computer program. A ceiling free from obstructions is assumed throughout. For a given room load, q, and supply/extract air temperature difference, Δt_o, the procedure is applied as follows.

1 Calculate the volume of air supply rate using:

$$\dot{V} = qLB/(\rho C_p \Delta t_o)$$

2   Choose an acceptable mean room speed, v_r, and calculate the supply air velocity using equation (6.50) and $M_o = \rho U_o \dot{V}$. Thus:

$$U_o = 19.75 \, BH \, v_r^3/(\rho \dot{V})$$

3   Calculate the effective grille area from:

$$A_o = \dot{V}/U_o$$

and assuming a grille length, b, determine the effective width of the grille, h.

4   Calculate the Archimedes number of the supply air:

$$Ar = g\beta \Delta t_o h/U_o^2$$

Care should be taken in giving the correct sign of Ar, i.e. Δt_o is negative for cooling and positive for heating.

5   The maximum jet velocity at the wall/ceiling corner, U_c, is determined from equation (6.54), i.e.:

$$U_c/U_o = \sqrt{[5.4/(H'/h)]} + 2.15 \, Ar(H'/h - 5.4)$$

where H' is the vertical distance from the grille to ceiling.

6   Calculate the air temperature difference between the wall/ceiling corner and extract from equation (6.53):

$$\Delta t_c/\Delta t_o = 1.38/(H'/h)^{0.25}$$

7   Calculate the maximum velocity of the jet on the ceiling U_m at $0.75 \times$ room length from equation (6.56), i.e.:

$$U_c/U_m = 1 + 0.042(B/b)(0.75 \, L/\sqrt{A_o} - 5)$$

where b is the grille length.

8   Calculate the distance below the ceiling at which U_m in step 7 occurs (drop) using equation (6.52):

$$y = 8.27 \times 10^{-4} \Delta t_c L^3 / \left[\sqrt{(A_o)} U_c^2 \right]$$

It is recommended that a value of $U_m = 0.5 \, \mathrm{m\,s^{-1}}$ should be obtained at $0.75 \, L$ and if the value calculated in step 7 is considerably different then other grille dimensions should be selected and steps 4–8 repeated. For specified grille dimensions the same steps can be used to assess the room environment but in this case step 3 precedes step 2, i.e. the supply velocity is first calculated and then used to determine the mean room speed.

Example 6.2 The room in Example 6.1 is to be air-conditioned by supplying air at a temperature difference $-3.0 \, \mathrm{K}$ from a linear grille mounted on a 1 m high sill along a wall. Assess the air movement in the room assuming a mean room speed of $0.2 \, \mathrm{m\,s^{-1}}$.

1   The volume flow rate is:

$$\dot{V} = 40 \times 32/(1.2 \times 1000 \times 3)$$
$$= 0.356 \quad \mathrm{m^3\,s^{-1}}$$

2   The supply air velocity is:

$$U_o = 19.75 \times 4 \times 3.5 \times (0.2)^3 \div (1.2 \times 0.356)$$
$$= 5.18 \quad \mathrm{m\,s}^{-1}$$

3   The effective area of the supply grille is:

$$A_o = 0.356/5.18 = 0.069 \quad \mathrm{m}^2$$

Assuming a grille length, b, of 1.2 m, the effective width of the grille is:

$$b = 0.069/1.2 = 0.0575 \,\mathrm{m} \text{ or } 57.5 \,\mathrm{mm}$$

4   The Archimedes number is given as:

$$Ar = 9.806 \times (-3) \times 0.0575/[295 \times (5.18)^2]$$
$$= 2.14 \times 10^{-4}$$

5   The velocity at the wall/ceiling corner is given by:

$$H'/b = 2.5/0.0575 = 43.5$$
$$U_c/5.18 = \sqrt{(5.4/43.5)} - 2.15 \times 2.14 \times 10^{-4} \, (43.5 - 5.4)$$
$$= 0.335$$

Therefore, $U_c = 1.73 \,\mathrm{m\,s}^{-1}$.

6   Calculate the temperature difference Δt_c:

$$\Delta t_c/(-3) = 1.38/(43.5)^{0.25} = 0.54$$
$$\Delta t_c = -1.6 \quad \mathrm{K}$$

7   Calculate the maximum ceiling velocity at $0.75\,L$:

$$1.73/U_m = 1 + (0.042 \times 4/1.2)(0.75 \times 8/\sqrt{(0.069)} - 5)$$
$$= 3.5$$
$$U_m = 0.49 \quad \mathrm{m\,s}^{-1}$$

i.e. the throw is at $0.75\,L$ which is a good design.

8   The drop of ceiling jet is:

$$y = 8.27 \times 10^{-4} \times 1.6 \times 512/[\sqrt{(0.069)} \times (1.73)^2]$$
$$= 0.86 \quad \mathrm{m}$$

Measured from the floor the drop is at $3.5 - 0.86 = 2.64\,\mathrm{m}$, i.e. above the occupied zone.

If room heating is considered with the same load and temperature difference as above then steps 1 to 4 will be unchanged and the procedure will start at 5 as follows.

5    $U_c/5.18 = \sqrt{(5.4/43.5)} + 2.15 \times 2.14 \times 10^{-4} \, (43.5 - 5.4)$

         $= 0.37$

         $U_c = 1.92 \quad \mathrm{m\,s^{-1}}$

6    $\Delta t_c = 1.6 \quad \mathrm{K}$

7    $1.92/U_m = 3.5$

         $U_m = 0.55 \quad \mathrm{m\,s^{-1}}$

This is a slightly higher velocity than in the cooling mode because the buoyancy force in this case increases the jet velocity.

8    $y = 8.27 \times 10^{-4} \times 1.6 \times 512/\left[\sqrt{(0.069)} \times (1.92)^2\right]$

         $= 0.70 \quad \mathrm{m}$

As would be expected the drop of a hot jet is lower than for a cold jet.

6.4.3   Ceiling supply

A design procedure for circular and linear ceiling diffusers was developed by Jackman [31] using tabulated data derived from laboratory test results. These design tables enable the selection of a suitable diffuser depending on the room geometry and thermal load. As in the case of the two design procedures described earlier, the design method which will be described here is based on empirical expressions derived from Jackman's experimental data.

Circular ceiling diffusers

The following design procedure relates to a single circular diffuser mounted at the centre of a square or nearly square smooth ceiling. However, the procedure may also be applied to large rooms which can be divided into equal square zones each served by a single diffuser. If the ceiling or zone is not square then the aspect ratio should be limited to 1.5. Noise data are not considered here and the diffuser manufacturer's selection data should be consulted for this purpose.

1   Calculate the mean room speed using:

$$v_r = 0.143 \, L \, \sqrt{(L^2/4 + H^2)} \quad \mathrm{m\,s^{-1}}$$

     where L is the length or width of the room, whichever is smaller, or the distance between two diffusers mounted on the same ceiling.

2   Given a supply/extract air temperature difference calculate the volume of air supply rate using:

$$\dot{V} = qLB/(\rho C_p \Delta t_o) \quad \mathrm{m^3\,s^{-1}}$$

3   Calculate the effective area of the diffuser from:

$$A_o = 0.0484 \rho (\dot{V}/v_r) \quad \mathrm{m^2}$$

     If the diffuser outlet diameter, d_o, is known then the effective width of the slot can also be calculated from $h = A_o/(\pi d_o) \, \mathrm{m}$.

4   Calculate the air supply velocity:

$$U_o = \dot{V}/A_o \quad \mathrm{m\,s^{-1}}$$

5   Calculate the throw to a velocity of $0.5\,\mathrm{m\,s^{-1}}$ using:

$$T = 4\,K_v^2 U_o^2 h \quad \mathrm{m}$$

For a circular cone diffuser the throw constant $K_v \approx 1.05$. The throw length T should be approximately equal to $0.375\,L$. Alternatively, the throw may be obtained from the diffuser data sheet using the volume flow rate and the effective area calculated in steps 2 and 3.

6   If $M_o/(qLBH) < 2.6 \times 10^{-4}$ then calculate a new value of v_r using equation (6.61) – see the section on ceiling supply with low flow rates.

Linear ceiling diffusers

The design procedure described below applies to: (i) a single linear diffuser mounted at the centre of the ceiling discharging equal volumes of air in two horizontal directions and (ii) a single linear diffuser mounted at one end of the ceiling discharging air horizontally in one direction.

As in the circular diffuser design, a large ceiling may be subdivided into equal areas each provided with a linear diffuser.

1   Calculate the mean room speed using:

$$v_r = 0.217\,L/\sqrt{(L^2 + H^2)} \quad \mathrm{m\,s^{-1}}$$

where L is the room length for an end-mounted diffuser and half the room length for a centrally mounted diffuser.

2   Calculate the volume of air supply rate per metre length using:

$$\dot{V} = qL/(\rho C_p \Delta t_o) \quad \mathrm{m^3\,s^{-1}\,m^{-1}}$$

L is taken as in step 1.

3   Calculate the effective slot width from:

$$h = 0.0484 \rho B (\dot{V}/v_r)^2 \quad \mathrm{m}$$

For a centrally mounted two-way discharge diffuser the width of each slot is $0.5\,h$.

4   Calculate the air supply velocity:

$$U_o = \dot{V}/h \quad \mathrm{m\,s^{-1}}$$

5   Calculate the throw to a velocity of $0.5\,\mathrm{m\,s^{-1}}$ using:

$$T = 4\,K_v^2 U_o^2 h \quad \mathrm{m}$$

where $K_v \approx 2.35$ (see Table 5.1). The throw length should also be equal to $0.75\,L$. Alternatively, T may be determined from the diffuser data sheet.

6   If $M_o/(qLBH) < 2.6 \times 10^{-4}$ then calculate a new value of v_r using equation (6.61).

Example 6.3 A room of dimensions 4 m × 4 m × 3 m ceiling height with a thermal load of 50 W m^{-2}, is to be supplied with cool air from a circular diffuser mounted at the centre of the ceiling. For a supply/extract air temperature difference of −5 K, select a suitable size diffuser and assess the thermal environment in the room.

1   The mean room speed is:

$$v_r = 0.143 \times 4/\sqrt{(4+9)} = 0.159 \quad \text{m s}^{-1}$$

which is satisfactory.

2   The volume of air supply:

$$\dot{V} = 50 \times 4 \times 4/(1.2 \times 1000 \times 5)$$

$$= 0.133 \quad \text{m}^3 \text{s}^{-1}$$

3   The effective area of the diffuser:

$$A_o = 0.0484 \times 1.2(0.133/0.159)^2$$

$$= 0.0408 \quad \text{m}^2$$

If a diffuser outlet diameter of 0.4 m is assumed then the effective slot width, $b = 0.0408/(0.4\pi) = 0.0325\,\text{m} = 32.5\,\text{mm}$.

4   The supply velocity is:

$$U_o = 0.133/0.0408 = 3.25 \quad \text{m s}^{-1}$$

5   Assuming a throw constant, $K_v = 1.05$, the throw is:

$$T = 4 \times (1.05)^2 \times (3.26)^2 \times 0.0325$$

$$= 1.52 \quad \text{m}$$

which is 0.38 L, i.e. satisfactory.

6   Calculate:

$$M_o/(QH) = \rho A_o U_o^2/(QH)$$

$$= 1.2 \times 0.0408 \times (3.26)^2/(800 \times 3)$$

$$= 2.17 \times 10^{-4}$$

which is slightly less than the threshold for correcting v_r.
As a further check equation (6.61) is used to calculate a new value of v_r Thus:

$$[v_r^2/(1.44 \times 0.52)](4+9) = 0.22\,[0.19(4.61)^2 + 1]^{1/2}$$

giving $v_r = 0.169\,\text{m s}^{-1}$ which is only 0.01 m s^{-1} higher than the previous value.
     The maximum temperature gradient in the occupied zone may also be calculated using equation (6.62). Thus:

$$\Delta t_r = [4.4 \times 0.52/(4 \times 4 \times 3)](4.61)^2$$

$$= 1.0 \quad \text{K}$$

***Example* 6.4** A room of dimensions 4 in $\times$ 4 m $\times$ 3 m ceiling height with a thermal load of 50 W m^{-2}, is to be air-conditioned by supplying cool air from a linear diffuser across the ceiling at one end of the room, discharging air in one direction across the ceiling. A suitable linear diffuser is to be selected for this purpose.

1   The mean room speed is:

$$v_r = 0.217 \times 4/\sqrt{(16+9)}$$

$$= 0.174 \quad ms^{-1}$$

which is satisfactory.

2   The volume flow rate per metre length is:

$$\dot{V} = 50 \times 4/(1.2 \times 1000 \times 5)$$

$$= 0.0333 \quad m^3 s^{-1} m^{-1}$$

3   The effective slot width:

$$h = 0.0484 \times 1.2 \times 4 \times (0.0333/0.174)^2$$

$$= 0.00852\, m = 8.52 \quad mm$$

4   The air supply velocity:

$$U_o = 0.333/0.00852 = 3.91 \quad m s^{-1}$$

5   The throw length assuming a throw constant of 2.35 is:

$$T = 4 \times (2.35)^2 \times (3.91)^2 \times 0.00852$$

$$= 2.88 \quad m$$

i.e. 0.72 L which is acceptable.

6   Calculate:

$$M_o/(QH) = A_o U_o^2/(QH)$$

$$= 1.2 \times 0.00852(3.91)^2/(800 \times 3)$$

$$= 2.61 \times 10^{-4}$$

i.e. equal to the limiting value. Hence v_r is correct.

Ceiling supply with low flow rates

The room speed obtained using the above design procedures for ceiling supplies is generally applicable for moderate or low room thermal loads, i.e. $M_o/(QH) > 0.26$ where Q is the load (kW) and M_o is the supply air momentum (N) [32]. For $M_o/(QH) < 0.26$ the procedure described earlier will produce low room speeds, v_r, and the correct room speed should be calculated using equation (6.61). In this case the vertical temperature

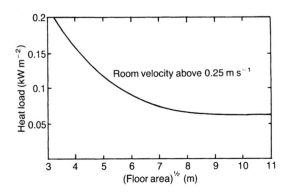

Figure 6.38 Design limits for circular ceiling diffusers [32].

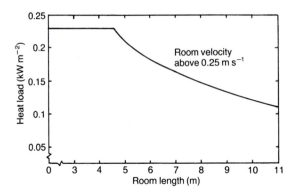

Figure 6.39 Design limits for linear ceiling diffusers [32].

gradient, Δt_r, in the occupied zone becomes more significant and equation (6.62) is used to calculate its value. The momentum of the supply jets is obtained from:

$$M_o = \rho A_o U_o^2$$

When thermal loads are concentrated over a small floor area, room speeds in excess of $0.25\,\mathrm{m\,s^{-1}}$ may occur which is not usually tolerated. Figures 6.38 and 6.39, which were obtained from [32], show the limiting curves for a room speed of $0.25\,\mathrm{m\,s^{-1}}$ in the case of a circular and a linear diffuser respectively. To achieve good air distribution, the load in $\mathrm{W\,m^{-2}}$ should be lower than the values given in Figures 6.38 and 6.39 for a given floor area or room length.

For a large ceiling that requires subdivision into smaller zones it is recommended that the area or length of each zone be equal to or less than that given by Figures 6.38 or 6.39 so that high room speeds can be avoided.

Example 6.5 The load in Example 6.3 is increased to $150\,\mathrm{W\,m^{-2}}$ and to maintain a constant air supply rate the temperature differential is increased in the same ratio as the loads in the two cases. Assess the room condition.

Referring to Figure 6.38, a room load of $150\,\mathrm{W\,m^{-2}}$ is just below the limiting line and it is possible that a high room speed will be obtained with a circular diffuser.

1 The mean room speed is:

$$v_{\mathrm{r}} = 0.143 \times 4/\sqrt{(4+9)} = 0.159 \quad \mathrm{m\,s^{-1}}$$

2 The volume of air supply is:

$$\dot{V} = 150 \times 4 \times 4/(1.26) \times 1000 \times 15$$
$$= 0.127 \quad \mathrm{m^3\,s^{-1}}$$

3 The effective area of diffuser is:

$$A_{\mathrm{o}} = 0.0484 \times 1.26(0.127/0.159)^2$$
$$= 0.0389 \quad \mathrm{m^2}$$

For a 0.4 m diameter circular diffuser, the effective slot width is:

$$b = 0.0389/(0.4\pi) = 0.031 \quad \mathrm{m\ or\ 31} \quad \mathrm{mm}$$

4 The supply velocity is:

$$U_{\mathrm{o}} = 0.127/0.0389 = 3.26 \quad \mathrm{m\,s^{-1}}$$

5 The throw is:

$$T = 4 \times (1.05)^2 \times (3.26)^2 \times 0.031$$
$$= 1.46 \quad \mathrm{m}$$

which is $0.36\,L$, i.e. satisfactory.

6 The parameter:

$$M_{\mathrm{o}}/(QH) = \rho A_{\mathrm{o}} U_{\mathrm{o}}^2/(QH)$$
$$= 1.26 \times 0.0389 \times (3.26)^2/(2400 \times 3)$$
$$= 7.23 \times 10^{-5}$$

This is considerably less than the limiting value and hence a new v_{r} should be calculated using equation (6.61). Thus:

$$[v_{\mathrm{r}}^2/(1.44 \times 0.52)](4+9) = 0.22\,[0.19(1/0.0723)^2 + 1]^{1/2}$$

giving $v_{\mathrm{r}} = 0.278\,\mathrm{m\,s^{-1}}$, which is too high. Therefore, a linear diffuser should be considered as it is expected to produce a lower v_{r} (see Figure 6.39).

The maximum vertical temperature gradient in the occupied zone is:

$$\Delta t_{\mathrm{r}} = [4.41 \times 0.52/(4 \times 4 \times 3)]\,(1/0.0723)^2$$
$$= 9.1 \quad \mathrm{K}$$

which is also unacceptable. Therefore, a linear diffuser should be considered.

Example 6.6 Recalculate the design in Example 6.5 with a linear diffuser mounted on the ceiling at one end of the room.

1   $v_r = 0.217 \times 4/\sqrt{(16+9)}$
    $= 0.174 \quad \mathrm{m\,s^{-1}}$

2   $\dot{V} = 150 \times 4/(1.26 \times 1000 \times 15)$
    $= 0.0317 \quad \mathrm{m^3\,s^{-1}\,m^{-1}}$

3   $h = 0.0484 \times 1.26 \times 4(0.0317/0.174)^2$
    $= 0.0081 \quad \mathrm{m\ or\ 8.1 \quad mm}$

4   $U_o = 0.0317/0.0081 = 3.90 \quad \mathrm{m\,s^{-1}}$

5   $T = 4 \times (2.35)^2 \times (3.9)^2 \times 0.0081$
    $= 2.72 \quad \mathrm{m}$
    i.e. $0.68\,L$ which is somewhat short but may be acceptable.

6   $M_o/(QH) = 1.26 \times 0.0081 \times 4 \times (3.9)^2/(2400 \times 3)$
    $= 8.67 \times 10^{-5}$
    Calculate v_r using equation (6.55):

    $[v_r^2/(1 \times 0.62)](4+9) = 0.22[0.19(11.53)^2 + 1]^{1/2}$

    giving $v_r = 0.232\,\mathrm{m\,s^{-1}}$, i.e. acceptable.
    The temperature gradient is:

    $\Delta t_r = [4.41 \times 0.6/(4 \times 4 \times 3)](11.53)^2$
    $= 7.6 \quad \mathrm{K}$

This is still a large value but 1.5 K lower than with a circular diffuser.

Variable air volume supply

A design procedure for a variable air volume system based on Holmes' experimental [42] results is given below. The limiting factors that should be considered are: (i) a maximum and a minimum room speed, v_r (e.g. 0.25 and 0.1 $\mathrm{m\,s^{-1}}$ respectively); (ii) a maximum local velocity in the occupied zone, U_m (e.g. 0.35 $\mathrm{m\,s^{-1}}$); (iii) a maximum room load limit obtained from $M_o/(QH) = 0.07$ and (iv) a maximum noise rating.
The room heat load is given by:

$$Q = \rho C_p \dot{V} \Delta t_o \tag{6.74}$$

where $\dot{V}$ is the air volume flow rate ($\mathrm{m\,s^{-1}}$) and Δt_o is the difference between supply air and return air temperature (K).
The momentum of air supply is given as:

$$M_o = \rho \dot{V}^2/A_o \tag{6.75}$$

where A_o is the effective area of the air supply terminal. By defining:

$$\tau = \sqrt{(A_o)}\Delta t_o \tag{6.76}$$

equation (6.75) becomes:

$$M_o = \rho \dot{V}^2 \Delta t_o^2 / \tau^2 \tag{6.77}$$

Substituting for Δt_o from equation (6.74) into (6.77):

$$M_o = Q^2 / (\rho C_p^2 \tau^2) \tag{6.78}$$

This equation relates the momentum of the supply air jet to the room load. The maximum heat load for a ceiling supply was found by Holmes and Caygill [32] to be given by:

$$M_o/(QH) = 0.07 \tag{6.79}$$

where M_o has the unit (N), Q (kW) and H (m).

Combining equations (6.78) and (6.79) and using standard values of ρ and C_p for air, Holmes [42] obtained the following expression for the maximum room load:

$$Q = 0.088\,\tau^2 H \tag{6.80}$$

Equation (6.80) represents one of the extreme limits of a VAV design, i.e. the maximum load that can be handled by the system.

Using velocity decay equations for circular and linear ceiling diffusers and equations (6.74) and (6.75), Holmes obtained the following expressions for calculating the maximum local air velocity, v_m, in the occupied zone (a height of 1.8 in):

For circular diffusers:

$$v_m = 0.9Q/[\tau(0.5L + H - 1.8)] \tag{6.81}$$

For continuous linear slot diffusers:

$$v_m = 1.06Q/\{\tau\sqrt{[L(0.5L + H - 1.8)]}\} \tag{6.82}$$

For intermittent slot diffusers:

$$v_m = 0.88Q/\{\tau\sqrt{[rL(0.5L + H - 1.8)]}\} \tag{6.83}$$

where H is the ceiling height and r is the ratio of active length of diffuser to room length. Equations (6.81–6.83) assume a centrally mounted diffuser in a square ceiling of area L^2. If the ceiling is not square then for a circular diffuser L is taken as $\sqrt{(BL)}$ where B is the width of the room, and for a slot diffuser the room dimension normal to the slot is taken as L. Usually the maximum local velocity that is tolerated is 0.35 in m s^{-1}.

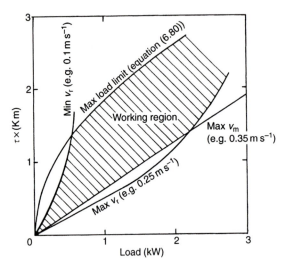

Figure 6.40 Typical VAV design limits.

Equations (6.81–6.83) are used to determine the room load corresponding to a maximum tolerable air velocity in the occupied zone. Typical curves representing the limits of τ given by the maximum load requirement, equation (6.80), and the maximum draught velocity requirement, equation (6.81), are shown in Figure 6.40.

Another criterion that should be considered in any air distribution design is the mean room speed in the occupied zone, v_r. By substituting equation (6.78) into equation (6.61) the mean room air speed can be written as:

$$v_r = 0.42aQ[0.3(H\tau^2/Q)^2]^{1/4}/[\tau\sqrt{(0.25L^2 + H^2)}] \qquad (6.84)$$

By choosing an appropriate minimum and maximum value for v_r (say 0.1 and $0.25\,\text{m}\,\text{s}^{-1}$) two curves can be plotted on Figure 6.40 to represent the two extremes of room speed. Thus the figure represents the four limits required for VAV design and the conditions within the shaded area should provide an acceptable design.

6.4.4 Low-level displacement supply

Low-level displacement ventilation is driven by buoyancy forces in plumes from heat sources, consequently the design calculations are more complex than for the mixing systems described earlier. Additional factors that need to be considered in this case are:

- Temperature gradient
- Position of neutral height
- Contaminant gradient.

Quantifying these factors requires complex analysis, see Chapter 4. However, a general design procedure that is based on that described in Section 6.2.3 can be used for rooms with ceiling height up to 4 m if a more elaborate method is not pursued. Using equation (6.73) with acceptable value of a temperature gradient for the room, G_T, the difference between the extract air and supply air temperatures are calculated:

$$(T_e - T_o) = G_T(H - 0.1)/0.6 \qquad (6.85)$$

A value for G_T based on a vertical temperature difference of 3 K in the occupied zone ($H = 1.1$ m) of $3 \, \text{K m}^{-1}$, would produce a temperature difference $(T_e - T_o) = 5$ K. Using this value of temperature difference and knowing the room load, the air flow rate, $\dot{V}$, may be calculated using:

$$Q = \rho C_p \dot{V}(T_e - T_o) \qquad (6.86)$$

where Q is the room load (W), ρ is the density (kg m^{-3}), and $\dot{V}$ is the volume flow rate ($\text{m}^3 \, \text{s}^{-1}$). Recommended values for the air supply temperature, T_o, are:

sedentary occupation $18 \sim 20 \, °\text{C}$

active occupation $\quad 17 \sim 19 \, °\text{C}$

industrial occupation $15 \sim 17 \, °\text{C}$

This design method is referred to as the *thermal comfort method*. Another design method that is also used for calculating the flow rate is based on the neutral height in the room and is referred to as the *air quality method*. The latter is based on the flow produced by all the plumes in the room which is calculated using the method described in Section 4.6.

Once the air flow rate and supply temperature is determined using one of the above methods, a suitable air terminal device can then be selected. For large rooms, it is recommended that several small devices are used rather than one large unit to provide more uniform air distribution. In positioning the air supply units, consideration should be given to room layout, seating arrangement, room partitions and floor obstructions, etc.

Example 6.7 A lecture room 20 m × 10 m × 4 m ceiling height accommodates 100 sedentary occupants. Lighting on the ceiling produces a total load of $20 \, \text{W m}^{-2}$. Calculate the design supply air flow rate for achieving the following conditions in the room: (i) temperature gradient of $1.5 \, \text{K m}^{-1}$ and (ii) stratification (neutral) height of 1.5 m.

(i) Assuming that each occupant produces 100 W then the occupancy load $= 100 \times 100 = 10 \, \text{kW}$.
   Lighting load $= 0.3 \times 20 \times 10 \times 20 = 1.2 \, \text{kW}$.
   Total room load $= 10.0 + 1.2 = 11.2 \, \text{kW}$.
   Equation (6.85) is used to calculate the temperature difference between extract and supply air:

$$T_e - T_o = G_T(H - 0.1)/0.6 = 1.5(4.0 - 0.1)/0.6 = 9.75 \, \text{K}.$$

The air flow rate is then calculated from equation (6.86):

$$\dot{V} = \frac{Q}{\rho C_p (T_e - T_o)} = \frac{11.2}{1.2 \times 1.0 \times 9.75} = 0.957 \quad \text{m}^3\,\text{s}^{-1}$$

i.e. approximately $10\,\text{l}\,\text{s}^{-1}$ per person.

(ii) Using the plume flow equation (4.73) for a point source and assuming an equivalent diameter of an occupant to be 0.4 m and the height of a seated occupant above the floor is 1.1 m gives:

Flow rate in plume person: $q_y = 5.5 \times 10^{-3}\, P_c^{1/3} (y + y_o)^{5/3}$ (4.73)

The distance of the virtual source from the seated occupant is given by equation (4.78):

$$y_o = 1.7 - 2.1\,d = 1.7 - 2.1 \times 0.4 = 0.86\,\text{m}.$$

Hence, $q_v = 5.5 \times 10^{-3}\,(100)^{1/3}\,(1.5 - 1.1 + 0.86)^{5/3} = 0.037\,\text{m}^3\,\text{s}^{-1}$. The total flow from all the occupants $= 100 \times 0.037 = 3.70\,\text{m}^3\,\text{s}^{-1}$.

The plume flow from a horizontal surface is calculated using equation (4.72):

$$q_y = 0.5 \times 10^{-3}\,l^{2/3}\,P_c^{1/3}\,(y + w)^{5/3}$$

$$= 0.5 \times 10^{-3} \times (20)^{2/3} \times (1200)^{1/3} \times (1.5 + 10)^{5/3}$$

$$= 2.280 \quad \text{m}^3\,\text{s}^{-1}$$

Total air flow rate $= 3.7 + 2.28 = 5.98\,\text{m}^3\,\text{s}^{-1}$ or about $60\,\text{l}\,\text{s}^{-1}$ per person.

  This flow rate is about 6 times that calculated using the thermal comfort method. The large difference is due to the fundamental difference between the two methods and some of it may also be due to the assumption that the lighting load is distributed over the floor whereas in reality some of this load will be on the walls.

6.4.5   Downward and upward ventilation

It was shown in Section 6.2.3 that, for a given air change rate, upward ventilation produces a more stable air distribution than downward ventilation. Equations (6.70) and (6.71) show that for a stable air distribution the downward ventilation requires twice the air change rate of the upward ventilation. However, each method has its own merits and they are both used in practice. The main factors which should be considered in the design of an air distribution system that is based on each of these methods are given here.

Downward ventilation

This type of ventilation is common in buildings with local concentration of thermal loads, such as process work, where floor-mounted air supply outlets are not desirable. The minimum air change rate required can be determined using equation (6.70) which is:

$$N_c = 30.4 \ \sqrt[3]{(q/H^2)} \quad \text{h}^{-1}$$

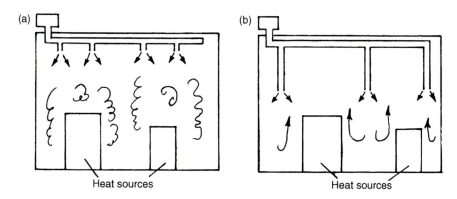

Figure 6.41 Downward displacement ventilation for high concentration thermal loads. (a) Poor air distribution and (b) good air distribution.

Table 6.4 Classes for cleanrooms and clean zones

Class (ISO)	Concentration limits (particles per m³)					
	Particle size (μm)					
	0.1	0.2	0.3	0.5	1.0	5.0
1	10	2				
2	100	24	10	4		
3	1,000	237	102	35	8	
4	10,000	2,370	1,020	352	83	
5	100,000	23,700	10,200	3,520	832	29
6	1,000,000	237,000	102,000	35,200	8,320	293
7				352,000	83,200	2,930
8				3,520,000	832,000	29,300
9				35,200,000	8,320,000	293,000

where q is the thermal load (W m^{-2}) and H is the ceiling height (m). If high air velocities are tolerated in the occupied zone then, wherever practical, the height of the supply outlet should be selected as low as possible. In addition, and to minimize turbulence caused by convective currents, the supply outlets should not be placed directly above a high thermal load source but in between sources, see Figure 6.41.

Another widely used downward ventilation system is in clean room air distribution. The main emphasis in clean room design is low concentration of airborne particles and low air turbulence in the room. Clean rooms have applications in the semiconductor industry, pharmaceutical industry, medical supplies industry, hospitals, food industry, etc. They are classed according to the number of particles of a given size that are tolerated per cubic metre of air, see Table 6.4 [48, 49].

That is, a Class 4 clean room will allow 1020 particles of diameter 0.3 μm in one cubic metre of air or 8 particles of diameter 1.0 μm, and a Class 9 will tolerate 35,200,000 particles of size 0.5 μm in a cubic metre, etc. In a Class ≤ 5 air is supplied to a ceiling void through a bank of high-efficiency particulate air filters (HEPA filters) and discharged through perforated ceiling plates (Figure 6.42(a)). In a

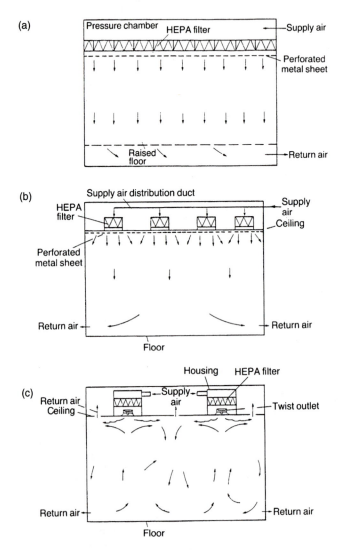

Figure 6.42 (a) Air supply for a clean room at Class ≤ 5. (b) Air supply for a clean room at Class 6. (c) Air supply for a clean room of class ≥ 7.

Class 6 clean room, the air is supplied through individual HEPA filter hoods spread over the ceiling above the perforated plate, as shown in Figure 6.42(b). A perforated ceiling is not required for a clean room of Class 7 and over but air is supplied from individual ceiling-mounted outlets fitted with HEPA filters, as shown in Figure 6.42(c). To prevent airborne particles settling on room and equipment surfaces an air velocity of between 0.35 and 0.45 m s^{-1} is usually required throughout the room. To achieve such a high room velocity, air change rates in excess of 300 h^{-1} are required. At such high flow rates the difference between supply and extract air temperature, Δt_o, is small and therefore the effect of buoyancy on the air distribution is insignificant except near localized heat sources.

Upward ventilation

Upward ventilation is most suitable for rooms with high thermal loads as it provides more stable air distribution and requires about half the air change rate as downward ventilation. Air may be supplied from floor-mounted outlets as in offices and data processing rooms, or from desk or seat outlets as in lecture theatres and assembly rooms. The minimum air change rate for a satisfactory air distribution is given by equation (6.71). Thus:

$$N_c = 15.3\sqrt[3]{(q/H^2)} \quad \text{h}^{-1}$$

Floor displacement ventilation using perforated or slotted grilles with low supply velocity are suitable for room loads of up to about 50 W m^{-2}. Other floor supplies may be suitable for room loads in excess of 140 W m^{-2} [34] which normally require air change rates more than 40 h^{-1}. Because the air is supplied directly into the occupied zone steps must be taken in the design to avoid draughts at low level and this requires careful selection and positioning of the air terminal devices and deciding on the most suitable supply air velocity for each device. The floor outlets which are commonly used are perforated or slotted floor grilles, circular outlets producing a free jet and circular swirl diffusers producing a swirling jet. Typical supply velocities for a slotted grille will be 0.25–0.7 m s^{-1} and for the two circular outlets will be in the range 2 to 4 m s^{-1}. Typical air flow rates supplied by floor outlets are [34]:

Slotted grille	28 to 40 l s^{-1}
Circular (free jet)	8.5 to 14 l s^{-1}
Circular (swirl jet)	8.5 to 50 l s^{-1}

A typical supply air temperature from a floor outlet would be 18 °C and a supply to extract air temperature difference in the range 8–12 K. Although an extract temperature of 26–30 °C is normally associated with a floor air distribution system, this does not imply room air temperatures of the same values. In other air distribution systems the extract air temperature is usually very close to the average air temperature in the occupied zone, but with an upward system the air temperature increases with height as the air 'picks up' the heat in its ascending motion (see also Section 4.6.5). The extract in an upward system is normally at high level (such as through troffer diffusers) with an extract temperature a few degrees above the temperature in the occupied zone.

Attention must be given to the location of floor outlets in the room particularly with regard to occupant comfort and the presence of large heat sources. To avoid uncomfortable draughts to occupants, the minimum distance between the outlet and an occupant should be 1 m in the case of the circular outlets and 1.5 m in the case of other floor outlets [34]. To reduce the effects of warm convective currents on the occupants, the air outlets should be mounted as close as possible to high heat sources in the room. The flow rates due to convective currents produced by internal heat sources can be determined from the plume equations given in Section 4.6.

Upward ventilation in theatres and conference rooms usually means supplying the air from seat and desk outlets respectively, i.e. a fixed seating arrangement will be required. In desk or seat supplies, an induction air system is used where induced room air is mixed with primary (conditioned) air to form the supply air. The induced air volume flow rate is normally between about 40 and 60% of the primary air flow.

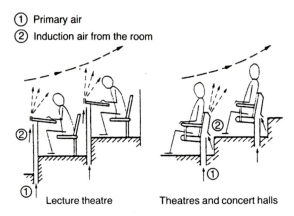

① Primary air
② Induction air from the room

Lecture theatre   Theatres and concert halls

Figure 6.43 Typical desk and seat air supply arrangements.

A desk outlet is either a slot at the front edge of the desk supplying air in a fixed upward direction (Figure 5.20) or a circular swivel outlet mounted on the desk which can vary the supply air direction to suit the user (Figure 5.21). With a supply outlet at the top of a seat back rest (Figure 5.19) the air is discharged at an angle of 0 to 20° from the vertical. Figure 6.43 shows a desk and a seat back rest supply commonly used in lecture theatres and assembly rooms respectively. In either case, the head of the individual should not be in the path of the jet but within the entrainment zone of the jet. A typical supply velocity in each case would be in the range $1.5–2.5 \, \mathrm{m \, s^{-1}}$. These types of air distribution have the following features.

1   The cool supply air is discharged into the upper half of the occupied zone whereas the lower part is served by induction of room air, i.e. draught at ankle level is eliminated.
2   The fresh air (primary air) is supplied directly to the occupant before mixing with room air as in the case of high-level supplies.
3   A small temperature gradient is achieved in the occupied zone (typically 1 K).
4   A uniform air distribution is provided by supplying air to each occupant.
5   The floor void can be used as a common supply plenum thus reducing duct noise.

Typical air velocity and temperature distribution around a seat with $10 \, \mathrm{l \, s^{-1}}$ primary air supplied at 18 °C from a back rest outlet is shown in Figure 6.29. It can be seen that the air velocities at ankle level are below $0.15 \, \mathrm{in \, m \, s^{-1}}$ and at head level below $0.2 \, \mathrm{m \, s^{-1}}$ which should provide a comfortable environment for a sedentary person if these velocities are coupled with the air temperatures also given in the figure. Figure 6.44 shows typical velocity profiles for a linear desk grille supplying $7 \, \mathrm{l \, s^{-1}}$ of primary air.
    The advantages of a desk or a seat air supply system over a high-level ceiling or wall supply are as follows.

1   The outdoor air is supplied directly to the occupants and not extensively diluted by room air.
2   A greater temperature differential between supply and extract air, and hence lower air supply rates.

3   Because higher primary air temperatures are used (18 °C as compared with 10 to 16 °C in high-level supplies) the energy consumption of the refrigeration plant is lower.
4   Lower capital cost due to the above features.

Desk air supply in offices must be supplemented with floor air supplies in order to offset the room load since the air supply rate from a desk outlet is normally limited to about $14 \mathrm{l\,s^{-1}}$. A typical upward ventilation arrangement is shown in Figure 6.45. In offices, whether linear slots or circular outlets are used, some flexibility is usually offered to the individual by the ability to adjust the velocity and direction of the supply air. Apart from the advantage of providing a personal ventilation capability, upward ventilation systems in offices do not have the economic advantage of the assembly room systems over the ceiling or wall supply systems. This is because the temperature differential between supply and extract in the office system is limited for a thermal

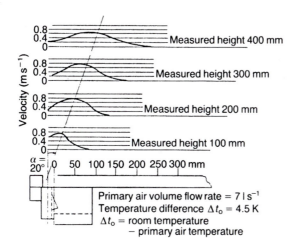

Figure 6.44 Velocity profiles of an air jet from a linear desk grille [34].

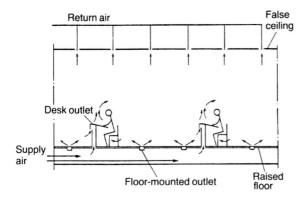

Figure 6.45 Typical upward ventilation for an office.

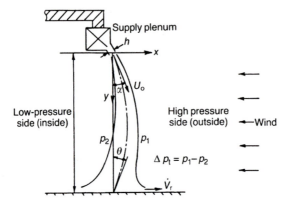

Figure 6.46 Schematic of an air curtain.

comfort reason. Hence, no great reduction in air flow rates can normally be achieved by employing upward ventilation in offices.

Another type of floor displacement ventilation that uses wall-mounted low velocity air terminals is described in 'Displacement ventilation' of Section 6.2.3. In this system the air is supplied from a wall unit over the floor which then rises upwards as it heats up by the room load.

6.4.6   Air curtains

An air curtain is a jet of air blown across a doorway or an entrance to prevent or reduce the ingress of air from outside the building or between two zones of a building. The reduction in air exchange across an unprotected entrance may be for the purpose of reducing the demand on heating or cooling systems or reducing the transfer of contaminants through the entrance. The principle of an air curtain is to supply a jet over the whole opening with sufficient momentum to counter the pressure force across it due to wind, buoyancy and the difference in pressure across the opening. A relationship between the total pressure difference across the opening and the momentum of the jet can be derived by referring to Figure 6.46.

The momentum of the jet is given by:

$$M_o = \rho_o U_o^2 Bh \qquad (6.87)$$

where M_o jet momentum (N); U_o = discharge velocity (m s^{-1}); B = length of supply slot (m); h = effective height of supply slot (m) and ρ_o = density of supply air (kg m^{-3}). If the jet is set at an angle α to the opening and assuming conservation of jet momentum, the change in momentum in a direction normal to the plane of the opening is as follows:

$$\text{Transverse force, } F_M = \text{change in momentum} = 2\rho_o U_o^2 Bh \sin \alpha \qquad (6.88)$$

To prevent air flow across the jet, the momentum force, F_M, must be equal to or greater than the force due to pressure difference across the opening, F_M, which is given as:

$$F_P = BH\Delta p_t \qquad (6.89)$$

where H = height of opening (m); B = width of opening (m) and Δp_t = total pressure difference across the opening (Pa).

If $F_M = F_p$, the following expression for the supply velocity, U_o, is obtained:

$$U_o = \sqrt{\{(H/h)[\Delta p_t/(2\rho_o \sin \alpha)]\}} \tag{6.90}$$

Equation (6.90) shows that for a given value of H, Δp_t and ρ_o, the values of h and U_o depend on the discharge angle, α. For a given discharge angle, the optimum performance of an air curtain is achieved when the point of jet impact is opposite the discharge point and both points are in the plane of the opening. The trajectory of the curtain centreline is a parabola and to achieve the optimum condition just described it must be symmetrical about a horizontal axis, i.e. the impact angle, θ, is equal to the discharge angle, α. The preferred value of α is usually between 12° and 40°.

The thickness of the curtain grows with distance from the supply plenum as the jet entrains air and its velocity decreases, but the momentum remains almost constant except for viscous dissipation due to turbulence. The quantity of air entrained at each side of the curtain can be calculated using the following expression:

$$\dot{V}_e = \dot{V}_o[0.845\sqrt{(\tau H/h + 0.2)} - 0.5] \tag{6.91}$$

where $\dot{V}_e$ = volume of air entrained (m³ s⁻¹); $\dot{V}_e = U_o h B$ is the discharge flow rate (m s⁻¹) and τ = turbulence coefficient which has a value of 0.17 for the concave (low-pressure) side and 0.2 for the convex (high-pressure) side [50]. If the air jet reaching the other side of the doorway impinges on a flat surface (e.g. floor) it splits into two streams each entering one zone. However, in some applications most of the jet is collected by an extract plenum for recirculation. In the former case, some of the jet is returned to the low-pressure zone (inside) at a rate which depends on the impact angle, θ.

Figure 6.47 shows the ratio $\dot{V}_r/\dot{V}$ plotted versus θ, where $\dot{V}_r$ is the flow rate rejected to outside (high-pressure side) and $\dot{V} = \dot{V}_e + \dot{V}_o$. In practice, it is desirable to reject as much of the air to the high-pressure side as possible and this requires a small value of θ. This can be achieved by either designing for a small discharge angle, α and the actual opening height, H, or adopting a design height in excess of the opening height, H_d. An extreme case is to use a design height $H_d = 2H$ to give $\theta = 0$.

The equation for the trajectory of the curtain centreline (see Figure 6.48) is given as:

$$x = y(1 - y/H_d) \tan \alpha \tag{6.92}$$

where x = horizontal distance measured from the supply point (m); y = vertical distance measured from the supply point (m) and H_d = design height of curtain $\geq H$ (m).

The non-dimensional trajectory of the centreline of an air curtain is plotted in Figure 6.48 using equation (6.92). The right ordinate in Figure 6.48 represents a factor which should be used in equation (6.90) to replace 2 as follows:

$$U_o = \sqrt{\{(H/h)[\Delta p_t/(f \rho_o \sin \alpha)]\}} \tag{6.93}$$

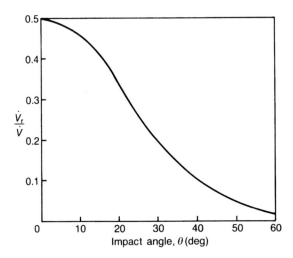

Figure 6.47 Ratio of curtain air rejected to the high pressure side as a function of impact angle [50].

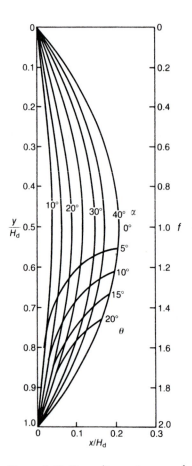

Figure 6.48 Centreline trajectory of an air curtain.

This equation should be used when $H < H_d$ in which case the factor f is obtained from Figure 6.48, given a value for H/H_d. Typical air supply velocities used in practice are $10–12\,\mathrm{m\,s^{-1}}$.

The total pressure difference, Δp_t, is the sum of the pressures due to wind and buoyancy and the pressure difference across the opening as a result of mechanical ventilation, i.e.:

$$\Delta p_t = p_w + p_s + \Delta p_m \tag{6.94}$$

where p_w = wind pressure (Pa); p_s = stack pressure (Pa) and Δp_m = difference in pressure between the two zones (Pa). The values of p_w and p_s can be determined using the procedures described in Chapter 7 and Δp_m can be determined from the difference in pressure between the supply and extract air terminals.

There are two types of air curtain in use: cold- and warm-air curtains. Cold-air curtains use either outside air or a mixture of outside air and internal air extracted at high level or from the extract duct of a ventilation system. In warm-air curtains the air is heated before it is discharged across the opening. The air discharge may be from the top, bottom (floor) or side of the opening. The effectiveness of the air curtain can be improved by extracting the curtain air at the impact point and recirculating it. This also reduces the heating requirement in the case of warm-air curtains. Further information on the optimum air discharge locations for different applications can be found in [51].

Warm-air curtains are used to provide a more comfortable environment for the people using the doorway. Because the air is discharged at high velocities ($\approx 10\,\mathrm{m\,s^{-1}}$) a supply temperature of between 25 and 35 °C is usually used to reduce the cooling effect of curtain air on the skin. The temperature of the air entering the building may be estimated from an energy balance using the discharge air flow rate, the entrained flow rate and the temperatures of the supply and entrained air on both sides of the curtain. The entrained air flow rate is calculated using equation (6.91). The heating load for the air curtain unit depends on the difference in temperature between the discharge air and the air supplied to the plenum as well as the flow rate. However, the net heating load is the difference between the air curtain load and the heat entering the building by ingress of air from the curtain, which may be estimated by finding the impact angle, θ, from Figure 6.48 and using this in Figure 6.47.

6.4.7 General guidelines

The selection of an air distribution system is categorized according to the temperature difference between supply and extract air in Table 6.5. This should provide a quick initial selection of one or more system options outlined in previous sections.

The value of supply–extract air temperature difference is dependent on the room load and the required air change rate for the room. The air change rate, N, is dependent on the application and the upper limit is usually dictated by comfort considerations (i.e. draught) and fan power. The temperature difference, Δt_o, in a ceiling or wall supply is approximately equal to the difference between the air supply and room temperature but this is not the case for a floor supply.

Studies by Fanger *et al.* [8, 52] with sedentary subjects exposed to air velocities in the range of $0.05–0.40\,\mathrm{m\,s^{-1}}$ and turbulence intensity in the range 30–65% have shown that the subjects were more sensitive to draught the higher the turbulence level

Table 6.5 Air distribution methods for different Δt_o [34]

Δt_o(K)	N(h^{-1})	Suitable air distribution systems
≤ 1	$N \leq 20$	All systems
	$20 < N \leq 100$	(i) Downward supply with ceiling swirl diffusers
		(ii) Upward supply with floor swirl diffusers
	$100 < N < 500$	'Laminar' ceiling supply with HEPA filters
1–2	$N \leq 12$	All systems
	$12 < N \leq 25$	(i) Ceiling and wall supplies (mixing ventilation)
		(ii) Floor supplies
	$25 < N \leq 100$	(i) Downward supply with ceiling swirl diffusers
		(ii) Upward supply with floor swirl diffusers
3–4	$N \leq 10$	All systems
	$10 < N \leq 20$	(i) Ceiling and wall supplies (mixing ventilation)
		(ii) Floor supplies
	$20 < N \leq 40$	(i) Downward supply with ceiling swirl diffusers
		(ii) Upward supply with floor swirl diffusers
	$40 < N \leq 100$	Upward supply with floor swirl diffusers
5–6	$N \leq 10$	All systems
	$10 < N < 16$	(i) Ceiling and wall supplies (mixing ventilation)
		(ii) Floor supplies
	$16 < N < 20$	(i) Downward supply with ceiling swirl diffusers
		(ii) Upward supply with floor swirl diffusers
	$20 < N < 80$	Upward supply with floor swirl diffusers
7–10	$N \leq 8$	All systems
	$8 < N \leq 14$	(i) Ceiling and wall supplies (mixing ventilation)
		(ii) Floor supplies
	$14 < N \leq 70$	Upward supply with floor swirl diffusers
11–12	$N \leq 60$	(i) Upward supply with floor swirl diffusers
		(ii) Desk or sear supplies

becomes. Such conditions represent typical air movements in ventilated spaces and were used to derive an empirical expression for the percentage of dissatisfied (*PD*) as a result of draught. Details of the effect of draught and turbulence on human comfort are presented in Chapter 1. These ought to be taken into consideration during the design of room air distribution systems.

But for very few exceptions, the test results presented in this chapter were obtained in empty spaces (without furnishings and people) and the design procedures based on these results will strictly apply to empty spaces only. The presence of furniture and people will influence the air movement in a ventilated room in a way which is directly related to the distribution of obstructions, people and the method of air supply used. The effect of light furniture and people on the maximum velocity in the occupied zone, v_m, for a room supplied by a wall-mounted diffuser is shown in Figure 6.49 which is obtained from [53]. The reduction in v_m due to furniture and people is most prominent at large air change rates.

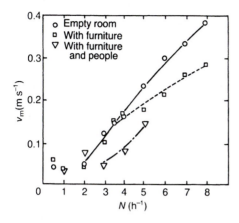

Figure 6.49 Effect of furniture and people on the maximum velocity in the occupied zone for different air change rates – wall supply.

References

1. Szücs, E. (1980) *Similitude and Modelling*, Elsevier, Amsterdam.
2. Müllejans, H. (1966) über die Ähnlichkeit der nicht-isothermen Strömung und den Wandubergang in Räumen mit Strahllüftung. *Forschungsber. des Landes NRW*, no. 1656, Westdeutscher Verlag, Köln. The similarity between non-isothermal flow and heat transfer in mechanically ventilated rooms. *BSRIA Trans.*, 202, Bracknell, UK.
3. Moog, W. and Sodec, F. (1976) Raumströmungsuntersuchungen für das Projekt Stadthalle Aachen – Part 1: Theoretische Betrachtung zur Raum strömung Experimentelle Vorunter-suchungen zur Übertragbarkeit von Raumströmungen. *Heiz. Lüftung Haustech.*, no. 11, 390–400.
4. Rolloos, M. (1977) Possibilities and limitations for the prediction of air flow patterns, temperature and velocities in large halls using scale models. *Meet. of Commission E1 of the International Institute of Refrigeration, Yugoslavia*, pp. 245–56.
5. Janna, W. S. (1986) *Engineering Heat Transfer*, PWS Publishers, Boston, MA.
6. CIBSE Guide C (2001) *New Reference Data, Section 3: Heat transfer*, Chartered Institution of Building Services Engineers, London.
7. Nevrala, D. J. (1979) Modelling of air movement in rooms. PhD thesis, Cranfield Institute of Technology, UK.
8. Hanzawa, H., Melikov, A. K. and Fanger, P. O. (1987) Air flow characteristics in the occupied zone of ventilated spaces. *ASHRAE Trans.*, 93 (1), 524–39.
9. Moog, W. and Sodec, F. (1976) Raumströmungsuntersuchungen für das Projekt Stadhalle Aachen – Part 2: Experimentelle Untersuchungen. *Heiz. Lüftung Haustech.*, no. 12, 442–8.
10. Awbi, H. B. and Nemri, M. M. (1990) Scale effect in room airflow studies. *Energy and Buildings*, 14, 207–10.
11. Lee, H. and Awbi, H. B. (2001) Internal partitioning and air movement in mixing ventilation, *Proc. 4th International Conference on Indoor Air Quality, Ventilation & Energy Conservation in Buildings (IAQVEC 2001)*, Vol. I, 245–52, Changsha, Hunan, China.
12. Lee, H. (2001) The influence of internal partitions on the air movement and contaminant dispersion in mechanically-ventilated rooms, PhD Thesis, University of Reading, UK.

13. Linden, P. F., Lane-Serff, G. F. and Smeed, D. A. (1990) Emptying filling boxes: the fluid mechanics of natural ventilation, *J. Fluid Mech.*, **212**, 309–35.
14. Cooper, P., Mayo. G. A. and Sorensen, P. (1998) Natural displacement ventilation of an enclosure with a distributed buoyancy source applied to one vertical wall, *Proc. Roomvent '98*, Vol. 1, pp. 45–51, Stockholm.
15. Hunt, G. R., Cooper, P. and Linden, P. F. (2000) Thermal stratification produced by plumes and jets in enclosed spaces (ed. H. B. Awbi), *Proc. Roomvent 2000*, Vol. 1, pp. 191–8, Elsevier, Oxford.
16. Lawson, T. V. (1980) *Wind Effect on Buildings*, Applied Science Publishers, London.
17. Aynsley, R. M., Melbourne, W. and Vickery, B. J. (1977) *Architectural Aerodynamics*, Applied Science Publishers, London.
18. Houghton, E. L. and Carruthers, N. N. (1976) *Wind Forces on Buildings and Structures*, Edward Arnold, London.
19. Awbi, H. B. (1978) Wind-tunnel-wall constraint on two-dimensional rectangular-section prisms, *J. Industrial Aerodynamics*, **3**, 285–306.
20. Ernest, D. R. (1991) Predicting wind-induced indoor air motion, occupant comfort, and cooling loads in naturally ventilated buildings, PhD Dissertation, University of California, Berkeley, USA.
21. Koestel, A. and Tuve, G. L. (1955) Performance and evaluation of room air distribution systems. *ASHVE Trans.*, **61**, 533.
22. Reinmann, J. J., Koestel, A. and Tuve, G. L. (1959) Evaluation of three air distribution systems for summer cooling. *ASHRAE Trans.*, **65**, 717.
23. Straub, H. E. (1962) What you should know about room air distribution. *Heat. Piping Air Conditioning*, **34**, 210–20.
24. Miller, P. L. and Nevins, R. G. (1972) An analysis of the performance of room air distribution systems. *ASHRAE Trans.*, **78** (2), 191–8.
25. Nevins, R. G. (1976) *Air Diffusion Dynamics in Theory, Design and Application*, Business News Publishing, Birmingham, MI.
26. *ASHRAE Handbook of Fundamentals* (2001) Ch. 32: Space air diffusion, American Society of Heating, Refrigeration and Air-Conditioning Engineers, Atlanta, GA.
27. Linke, W. (1966) Eigenschaften der Strahllütung. Kältetech. Klim., **18**, 122–6 (Aspects of jet ventilation. *BSRIA Trans.*, 103.).
28. Regenscheit, B. (1959) Die Luftbewegung in Klimatisierten Räumen. Kältetechnik, **11**, 3–11.
29. Jackman, P. J. (1970) Air movement in rooms with side-wall mounted grilles – A design procedure. *BSRIA Lab. Rep.* no. 65, Building Services Research and Information Association, Bracknell, UK.
30. Jackman, P. J. (1971) Air movement in rooms with sill-mounted grilles – A design procedure. *BSRIA Lab. Rep.* no. 71, Building Services Research and Information Association, Bracknell, UK.
31. Jackman, P. J. (1973) Air movement in rooms with ceiling-mounted diffusers (including supplements A and B). *BSRIA Rep.* no. 81, Building Services Research and Information Association, Bracknell, UK.
32. Holmes, M. J. and Caygill, C. (1973) Air movement in rooms with low supply air flow rates. *BSRIA Rep.* no. 83, Building Services Research and Information Association, Bracknell, UK.
33. Linke, W. (1962) Lüftung von oben nach unten oder umgekehrt? (Ventilation from above or below?) Kältetechnik, **14** (5), 142–9.
34. Sodec, F. (1986) Air distribution systems. *Rep.* no. 3554E, Krantz GmbH, Germany.
35. ISO 7730 (1994) Moderate thermal environments – Determination of the PMV and PPD indices and specification of the conditions for thermal comfort, 2nd edn, International Standards Organisation, Geneva.

36. Nielsen, P. V. (1993) *Displacement Ventilation – Theory and Design*, Aalborg University, Denmark, August 1993, ISSN 0902-8002 U9306.

37. Skistad, H. (1994) *Displacement Ventilation*, Wiley.

38. Jackman, P. J. (1990) Displacement ventilation. *BSRIA TM 2/90*, Building Services Research and Information Association, Bracknell, UK.

39. Abbas, T. (1999) Displacement ventilation and static cooling devices. *BSRIA Code of Practice COP 17/99*, Building Services Research and Information Association, Bracknell, UK.

40. Alamdari, F. (1998) Displacement ventilation and cooled ceilings. *Proc. Roomvent' 98*, Vol. 1, pp. 197–204, Stockholm.

41. Fredriksson, J., Sandberg, M. and Moshfeg, B. (2000) Experimental investigation of the velocity field and air flow pattern generated by cooling ceiling beams (ed. H. B. Awbi), *Proc. Roomvent 2000*, Vol. 1, pp. 619–25, Elsevier, Oxford.

42. Holmes, M. J. (1974) Room air distribution with variable air volume supply systems. *BSRIA Rep.* no. 15/107, Building Services Research and Information Association, Bracknell, UK.

43. Awbi, H. B. (1983) Domestic warm air distribution with low-grade heat systems, *Proc. Int. Seminar on Energy Saving in Buildings, The Hague, 14–16 November 1983*, Reidel, Dordrecht, pp. 628–38.

44. Awbi, H. B. and Savin, S. J. (1984) Air distribution methods for domestic warm air heating systems using low grade heat sources – Tests with low level and high level air supply terminals. *Commission of the European Community Rep.* no. EUR 9237 EN.

45. Togari, S. and Hayakawa, S. (1987) Scale model experiment of air distribution in the large space of the Shinkokugikan Sumo wrestling arena, *Proc. Air Distribution in Ventilated Spaces (Roomvent '87)* Stockholm 10–12 June 1987, session 4b.

46. Heiselberg, P., Murakami, S. and Roulet, C.-A. (eds) (1997) Ventilation of large spaces in buildings: Part 3 – Analysis and prediction techniques, Annex 26 Report, International Energy Agency.

47. Bouwman, H. B. and van Gunst, E. (1967) Die Luftbewegung in der grossen Konzerthalle de Doelen in Rotterdam. *Kältetech. Klim.*, **19** (8), 257–63.

48. ASHRAE Handbook HVAC Applications (1999) Ch. 15: *Clean spaces*, American Society of Heating Refrigeration and Air-Conditioning Engineers, Atlanta, GA.

49. BS 5295: Part 1 (1989) *Environmental Cleanliness in Enclosed Spaces*, British Standards Institution, London.

50. Mott, L. F. (1962) Design for protection by air curtain. *Heat. Air Conditioning*, February, pp. 164–6.

51. Danielssen, P. O. (1973) Air curtains and doors. *The Steam Heat. Eng.*, October.

52. Fanger, P. O. *et al.* (1988) Air turbulence and sensation of draught. *Energy Build.*, **12**, 21–39.

53. Nielsen, P. V. (1989) Numerical prediction of air distribution in rooms – Status and potentials. *Building Systems: Room and Air Contaminant Distribution* (ed. L. L. Christianson), American Society of Heating, Refrigeration and Air-Conditioning Engineers, Atlanta, GA, pp. 31–8.

7   Natural, hybrid and low energy ventilation

7.1   Introduction

The air infiltration through a building envelope was dealt with in Chapter 3. Air infiltration is the leakage of air to or from a building due to imperfect construction or porosity of building materials. It occurs in all buildings but at rates that are determined by the design, geometry, construction details, building materials and quality control for the building. Natural ventilation, on the other hand, is the term used to describe the air flow to or from a building through specific openings in the building envelop, such as openable windows, ventilators, ventilation shafts, etc. As in the case of air infiltration, the resulting ventilation process is caused by naturally produced pressures due to wind and stack effects.

Since they are controlled by the (random) climatic conditions surrounding the building, both air infiltration and natural ventilation are adventitious, unless some control is used for the ventilation openings in the latter. Air infiltration is the most uncontrollable form of ventilation and can produce undesirable effects such as increasing the heating (or cooling) loads and draught in cold weather. This can of course be reduced to very low values using airtight construction methods and draught-proofing, which are advantageous from an energy conservation point of view, but could be detrimental to the indoor air quality and the building fabric. A low ventilation rate increases the concentration of contaminants and moisture within the building and could lead to water vapour condensation during cold weather. It is therefore necessary, particularly in modern airtight buildings, to include dedicated ventilation openings in the design of the building fabric and ensure adequate supply of fresh air. Since natural ventilation is mainly climatically driven, some control of the ventilation openings will be required to regulate the air supply to the building. Air flow control is necessary as natural ventilation is not restricted to the provision of fresh air but, in most buildings, it is also a means of controlling overheating in hot seasons. Hence, a naturally ventilated building should have provisions for adequate control of the air flow as this is a constantly varying quantity both due to fluctuating weather conditions and also fresh air demand.

In this chapter, I will be presenting the principles and methodologies used for calculating air flow through ventilation openings due to wind and buoyancy forces, and the methods used for combining natural ventilation with mechanical ventilation to cater for situations of high demand for fresh air flow. This combination is often referred to as 'hybrid ventilation'. A lot of the basic concepts related to wind- and buoyancy-induced air flow have been presented in Chapter 3, but the main focus in this chapter is the application of these concepts. Low-energy ventilation concepts, which rely on solar energy or heat recovery are also discussed as these are gradually being introduced in building design to reduce the energy usage of buildings.

7.2   Wind characteristics

7.2.1   *Climatic scales*

Wind is by far the most significant component of the driving force in natural venti-lation, particularly in hot seasons. The wind flow over the earth's surface is a very complex phenomenon that is governed by a number of variables such as: earth's rota-tion; temperature differences between oceans and land and polar and tropical air; geographical location and landscape. In meteorology, the following scales are used for describing climatic models: *global scale, regional scale, local scale and micro-climate scale*. The *global scale* concerns astronomical factors that relate to the size, shape, and self rotation of earth and its elliptic rotation around the sun and this cov-ers a range of thousands of kilometres. These factors create the diurnal and seasonal variations according to latitude, as well continental variations due to the distribution of land and oceans. The *regional scale* relates to regional climatic features such as geographic landscape (e.g. the influence of mountains, hills, valleys, etc.), proximity to ocean and location of region with respect to zones of general wind circulation. This scale covers wind flows over hundreds of kilometres. The *local scale* relates to local geography or water mass (such as hills, valleys, lakes, large rivers, etc.) and urban-ization (such as heat islands) and how the local climate, including the wind flow, is affected by these factors. This stretches over a distance of about 10 km. The local cli-mate is of course influenced by energy balance at the regional scale. The *microclimate scale* relates to small towns or districts where local features used in their construction could have influence on the wind flow, such as nature of town planning, presence of artificial climate modifiers (e.g. windbreaks, hedges, etc.), the presence of water, etc. This scale covers a few hundred metres and is greatly influenced by man's planning and activities.

7.2.2   *Wind structure*

The natural wind is highly variable, random, turbulent and its speed varies with height above ground due to the existence of atmospheric boundary layer. In addition to the global turbulent nature of the wind, turbulence in the lower region of the atmospheric boundary layer is also generated by ground obstructions as well as thermal currents. The instantaneous wind speed, $V(t)$, varies with time as illustrated in Figure 7.1. As in any turbulent flow, the wind velocity at any point can be written as the sum of a time-mean value and a fluctuating component, i.e.:

$$V(t) = \overline{V} + v(t) \tag{7.1}$$

where $V(t)$ is the instantaneous wind speed, $\overline{V}$ is the time-mean value and $v(t)$ is the fluctuating component. The mean value of $v(t)$ averaged over a reasonable long time period is zero but a measure of the wind turbulence is given by the standard deviation, i.e. the mean square value, $\sqrt{v^2(t)}$. The time-mean wind speed, $\overline{V}$, is given

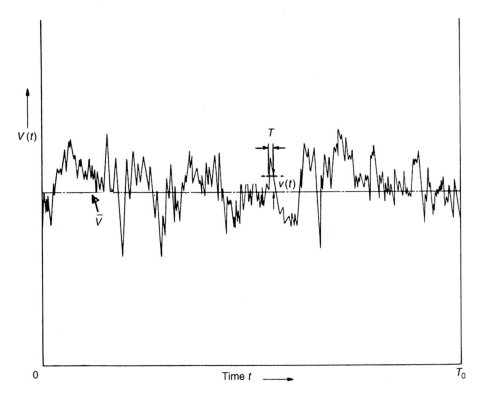

Figure 7.1 Determination of time-mean wind speed for a given time period.

by integrating the instantaneous value over the desired time period, T_0, as given below:

$$\overline{V} = \frac{1}{T_0} \int_0^{T_0} V(t)\mathrm{d}t \qquad (7.2)$$

The quantity $\sqrt{\overline{v^2(t)}}/\overline{V}$ is known as the turbulence intensity *TI*.

For a continuous wind speed record, it is of course possible to average the wind speed over any desired time period, T, and determine a series of values of wind speed, V_T, averaged over that time period (T can be a few seconds or several minutes). A maximum 'mean' wind speed, V_T, for a given period can be defined as the maximum value of the T-second average. This average wind speed represents the T-s gust, e.g. 3-s gust. It is also possible to define extreme values of V_T that have a particular probability of not being exceeded in a given period. Such values are obtained from records of annual maxima from a collection of wind records measured over a number of years (two or more successive years) and applying statistical analysis. As an example, it is possible to specify an extreme value for V_T, $\hat{V}_T$, which is exceeded only once in a period of 50 years that has a probability 0.02 of being exceeded in any one year during that period. Such wind speed values are used to meet certain design criteria that have an acceptable degree of risk of being exceeded. In ventilation design however such extreme wind speeds are rarely used and instead mean values over a relatively long time periods are more relevant.

The value of $\overline{V}$ should ideally be independent of time and the time period, T, but in the lower atmosphere the wind speed is usually averaged over a relatively long period (between 10 min and 2 h). For periods $T < 10$ min, the mean wind speed tends to fluctuate widely from one interval to another due to the influence of gust components. For $T > 2$ h, however, the change in the large-scale wind pressure will cause variation in $\overline{V}$ from one period to the next. Mean wind speeds, $\overline{V}$, available from meteorological stations are for a height of 10 m measured in open country and normally averaged over 1-h period.

Although the mean wind speed increases with height from the ground up to a certain height (boundary layer thickness) that depends on the height and distribution of ground obstructions, turbulence decreases with height. The profile of the time-mean wind speed is logarithmic with respect to the height above ground and various expressions are available for describing the wind speed profile as discussed in Chapter 3. A common expression is that represented by equation (3.6) which is:

$$\frac{\overline{V}}{\overline{V}_r} = c\,H^a \tag{7.3}$$

where $\overline{V}_r$ is time-mean wind speed at a reference height, usually taken 10 m above flat ground level; c is a constant; and a an exponent the values of which are given in Table 3.2.

7.2.3 *Design wind speed*

Wind data for different locations in a country are given in design guides and standards and are also archived by a country's meteorological office. The data is often based on data from continuously recording anemographs exposed at a height of 10 m in open, level terrain. It is presented as mean wind speeds for a certain period (a number of years or decades) that is likely to be exceeded by a specified percentage of time in a year for the location concerned. A country wind speed maps for an hourly mean wind speed that is exceeded for a certain percentage of time in any time period (years or decades) is another way of presenting design wind speeds, e.g. [1, 2] for the UK. This data is used in different applications, such as for the calculation of wind loading on structures, calculation of thermal loads for a building or for ventilation rate calculations, and therefore each application requires a mean wind speed that is relevant to that application. The risk of speed being exceeded annually varies from about 0.01% for wind loading calculations of high risk installations, e.g. nuclear installations [1], to 80% of the summer months for summertime cooling calculations [3]. However, for the design of ventilation systems hourly mean wind speeds corresponding to a height of 10 m are used, which are exceeded for 1, 2.5, 5 and up to 50% of the time in a period of many years according to what the speed is used for, see e.g. [2, 4].

B.S. 5925 [2] presents wind speed maps for the UK in terms of wind speeds exceeded for 50% of the time. This speed is represented by $\overline{V}_{50}$ and it is used in conjunction with the frequency distributions given in Table 7.1 to determine a value for $\overline{V}_r$ exceeded for any chosen proportion of the time at any site in the UK. Values of $\overline{V}_{50}$ given in [2]

Table 7.1 Values of the ratio of mean wind speed exceeded for a given percentage of time to the 50% mean wind speed $\overline{V}_{50}$ for the UK [2]

Percentage	Location	
	Exposed coastal	Sheltered inland
80	0.56	0.46
75	0.64	0.56
70	0.71	0.65
60	0.86	0.83
50	1.00	1.00
40	1.15	1.18
30	1.33	1.39
25	1.42	1.51
20	1.54	1.66
15	1.70	1.80
10	1.84	2.03

vary from $4 \, \mathrm{m \, s^{-1}}$ in the south of England inland to $6.5 \, \mathrm{m \, s^{-1}}$ in the Scottish islands with an average value of $4.5 \, \mathrm{m \, s^{-1}}$ for most inland areas of Great Britain.

7.3 Wind-induced ventilation

Among other things, the wind-induced air flow into a building is affected by the pressure distribution around the building and more specifically at the openings in the building structure (both external and internal). In Chapter 3, the wind pressure, p_w, is defined by (3.5), which is also given below:

$$p_w = 0.5 \, C_p \rho_0 \overline{V}^2 \quad \mathrm{Pa} \tag{7.4}$$

where C_p = static pressure coefficient; $\overline{V}$ = wind speed at datum level, usually height of building or opening $(\mathrm{m \, s^{-1}})$. The static pressure coefficient, C_p, is defined as:

$$C_p = (p - p_0)/0.5\rho_0 \overline{V}^2 \tag{7.5}$$

where p = static pressure at some point on the building (Pa); p_0 = static pressure of the free stream corresponding to $\overline{V}_r$ (Pa); ρ_0 = density of free stream $(\mathrm{kg \, m^{-3}})$; $\overline{V}$ = free stream velocity normally calculated at building height or other reference height $(\mathrm{m \, s^{-1}})$.

The pressure distribution on a building is determined by the orientation of the building towards the prevailing wind as well as its geometry. The windward face (wind facing) is subjected to positive wind pressure coefficients (higher pressure than the static pressure of the wind) as a result of the impact of the wind and its deflection on the surface, whereas the roof and the leeward face (downstream of wind) are subjected to negative pressure coefficients (lower pressure than the static pressure of the wind) because of boundary layer separation from the surfaces at sharp edges joining the roof

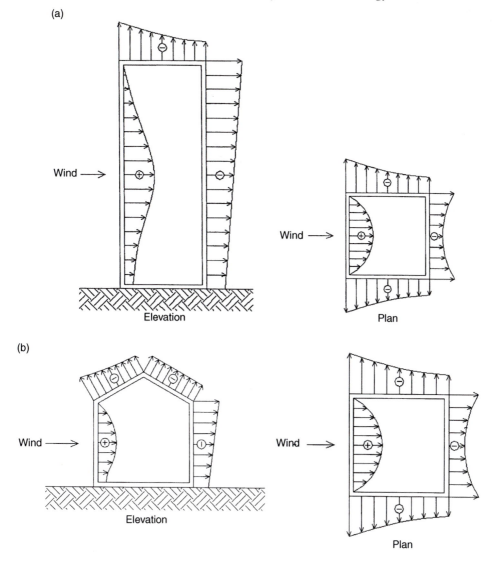

Figure 7.2 Wind pressure coefficient distribution on: (a) flat-roof building and (b) pitched-roof building.

and the windward wall. The pressure coefficients on the side faces of a building can be either positive or negative depending on their inclination with respect to the prevailing wind. Figure 7.2 (a and b) shows typical pressure coefficient distribution on a flat-roof and a pitched-roof building. Values of C_p for different building shapes and aspect ratios (height/width or length/width) expressed as a single value for each face can be found in [2, 5, 6]. However, the data in BS 5925 [2] ignores the significant influence of surrounding obstructions, which act as wind shields. This effect as well as the effect of wind direction on C_p, are given in the data of [6]. Tables 7.2 and 7.3 give average surface values of C_p for a square and a rectangular plan building up to three stories high for three exposures [6].

Table 7.2 Surface averaged pressure coefficients for up to three storeys high, square plan building in urban location [6]

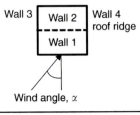

Surface	Wind pressure coefficient, C_p, for wind angle, $\alpha°$							
	0	45	90	135	180	225	270	315
(a) Building exposed (open flat country)								
Wall 1	0.7	0.35	−0.5	−0.4	−0.2	−0.4	−0.5	−0.35
Wall 2	−0.2	−0.4	−0.5	0.35	0.7	0.35	−0.5	−0.4
Wall 3	−0.5	0.35	0.7	0.35	−0.5	−0.4	−0.2	−0.4
Wall 4	−0.5	−0.4	−0.2	−0.4	−0.5	0.35	0.7	0.35
Roof pitch <10°:								
Front	−0.8	−0.7	−0.6	−0.5	−0.4	−0.5	−0.6	−0.7
Rear	−0.4	−0.5	−0.6	−0.7	−0.8	−0.7	−0.6	−0.5
Roof pitch 11 ~ 30°:								
Front	−0.4	−0.5	−0.6	−0.5	−0.4	−0.5	−0.6	−0.7
Rear	−0.4	−0.5	−0.6	−0.5	−0.4	−0.5	−0.6	−0.5
Roof pitch >30°:								
Front	0.3	−0.4	−0.6	−0.4	−0.5	−0.4	−0.6	−0.4
Rear	−0.5	−0.4	−0.6	−0.4	−0.3	−0.4	−0.6	−0.4
(b) Building semi-sheltered (open country with scattered wind breaks lower than height of building)								
Wall 1	0.4	0.1	−0.3	−0.35	−0.2	−0.35	−0.3	−0.1
Wall 2	−0.2	−0.35	−0.3	0.1	0.4	0.1	−0.3	−0.35
Wall 3	−0.3	0.1	0.4	0.1	−0.3	−0.35	−0.2	−0.35
Wall 4	−0.3	−0.35	−0.2	−0.35	−0.3	0.1	0.4	0.1
Roof pitch <10°:								
Front	−0.6	−0.5	−0.4	−0.5	−0.6	−0.5	−0.4	−0.5
Rear	−0.6	−0.5	−0.4	−0.5	−0.6	−0.5	−0.4	−0.5
Roof pitch 11 ~ 30°:								
Front	−0.35	−0.45	−0.55	−0.45	−0.35	−0.45	−0.55	−0.45
Rear	−0.35	−0.45	−0.55	−0.45	−0.35	−0.45	−0.55	−0.45
Roof pitch >30°:								
Front	0.3	−0.5	−0.6	−0.5	−0.5	−0.5	−0.6	−0.5
Rear	−0.5	−0.5	−0.6	−0.5	−0.3	−0.5	−0.6	−0.5
(c) Building in urban location (surrounded by buildings of equal heights)								
Wall 1	0.2	0.05	−0.25	−0.3	−0.25	−0.3	−0.25	−0.05
Wall 2	−0.25	−0.3	−0.25	0.05	0.2	0.05	−0.25	−0.3
Wall 3	−0.25	0.05	0.2	0.05	−0.25	−0.3	−0.25	−0.3
Wall 4	−0.25	−0.3	−0.25	−0.3	−0.25	0.05	0.2	0.05
Roof pitch <10°:								
Front	−0.5	−0.5	−0.4	−0.5	−0.5	−0.5	−0.4	−0.5
Rear	−0.5	−0.5	−0.4	−0.5	−0.5	−0.5	−0.4	−0.5
Roof pitch 11 ~ 30°:								
Front	−0.3	−0.4	−0.5	−0.4	−0.3	−0.4	−0.5	−0.4
Rear	−0.3	−0.4	−0.5	−0.4	−0.3	−0.4	−0.5	−0.4
Roof pitch >30°:								
Front	0.25	−0.3	−0.5	−0.3	−0.4	−0.3	−0.5	−0.3
Rear	−0.4	−0.3	−0.5	−0.3	0.25	−0.3	−0.5	−0.3

Table 7.3 Surface averaged pressure coefficients for up to three storeys high building with plan aspect ratio 2 in urban location [6]

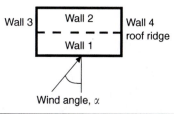

Wind angle, α

Surface	Wind pressure coefficient, C_p, for wind angle, α°							
	0	45	90	135	180	225	270	315
(a) Building exposed (open flat country)								
Wall 1	0.5	0.25	−0.5	−0.8	−0.7	−0.8	−0.5	−0.25
Wall 2	−0.7	−0.8	−0.5	0.25	0.5	0.25	−0.5	−0.8
Wall 3	−0.9	0.2	0.6	0.2	−0.9	−0.6	−0.35	−0.6
Wall 4	−0.9	−0.6	−0.35	−0.6	−0.9	0.2	0.6	0.2
Roof pitch <10°:								
Front	−0.7	−0.7	−0.8	−0.7	−0.7	−0.7	−0.8	−0.7
Rear	−0.7	−0.7	−0.8	−0.7	−0.7	−0.7	−0.8	−0.7
Roof pitch 11 ~ 30°:								
Front	−0.7	−0.7	−0.7	−0.6	−0.5	−0.6	−0.7	−0.7
Rear	−0.5	−0.6	−0.7	−0.7	−0.7	−0.7	−0.7	−0.6
Roof pitch >30°:								
Front	0.25	0	−0.6	−0.9	−0.8	−0.9	−0.6	0
Rear	−0.8	−0.9	−0.6	0	0.25	0	−0.6	−0.9
(b) Building semi-sheltered (open country with scattered wind breaks lower than height of building)								
Wall 1	0.25	0.06	−0.35	−0.6	−0.5	−0.6	−0.35	0.06
Wall 2	−0.5	−0.6	−0.35	0.06	0.25	0.06	−0.35	−0.6
Wall 3	−0.6	0.2	0.4	0.2	−0.6	−0.5	−0.3	−0.5
Wall 4	−0.6	−0.5	−0.3	−0.5	−0.6	0.2	0.4	0.2
Roof pitch <10°:								
Front	−0.6	−0.6	−0.6	−0.6	−0.6	−0.6	−0.6	−0.6
Rear	−0.6	−0.6	−0.6	−0.6	−0.6	−0.6	−0.6	−0.6
Roof pitch 11 ~ 30°:								
Front	−0.6	−0.6	−0.55	−0.55	−0.45	−0.55	−0.55	−0.6
Rear	−0.45	−0.55	−0.55	−0.6	−0.6	−0.6	−0.55	−0.55
Roof pitch >30°:								
Front	0.15	−0.08	−0.4	−0.75	−0.6	−0.75	−0.4	−0.08
Rear	−0.6	−0.75	−0.4	−0.08	−0.15	−0.08	−0.4	−0.75
(c) Building in urban location (surrounded by buildings of equal heights)								
Wall 1	0.06	0.12	−0.2	−0.38	−0.3	−0.38	−0.2	−0.12
Wall 2	−0.3	−0.38	−0.2	0.12	0.06	0.12	−0.2	−0.38
Wall 3	−0.3	0.15	0.18	0.15	−0.3	−0.32	−0.2	−0.32
Wall 4	−0.3	−0.32	−0.2	−0.32	−0.3	0.15	0.18	0.15
Roof pitch <10°:								
Front	−0.49	−0.46	−0.41	−0.46	−0.49	−0.46	−0.41	−0.46
Rear	−0.49	−0.46	−0.41	−0.46	−0.49	−0.46	−0.41	−0.46
Roof pitch 11 ~ 30°:								
Front	−0.49	−0.46	−0.41	−0.46	−0.4	−0.46	−0.41	−0.46
Rear	−0.4	−0.46	−0.41	−0.46	−0.49	−0.46	−0.41	−0.46
Roof pitch >30°:								
Front	0.06	−0.15	−0.23	−0.6	−0.42	−0.6	−0.23	−0.15
Rear	−0.42	−0.6	−0.23	−0.15	0.06	−0.15	−0.23	−0.6

Algorithms have also been developed for selecting C_p values for different building shapes to provide more convenient access to the large databases of C_p that have been derived from wind tunnel tests. Two such algorithms are described in references [7, 8].

In addition to the external pressure at an opening, the flow through an opening is also affected by the internal pressure, which can be influenced by both wind and buoyancy pressures. Assuming that the only pressure acting is that due to wind, the internal pressure for a building without internal partitions can be obtained by applying the principle of conservation of mass flow through the flow openings. Taking a four-sided building as an example with four equal area openings (one on each wall), one inlet and three extracts, and assuming an external pressure coefficient, C_{pn}, where $n = 1$–4 is the wall number and an internal pressure coefficient of C_{pi}, then the flow entering or leaving through each opening, Q_n, is:

$$Q_n \propto |(C_{pn} - C_{pi})|^{1/2} \tag{7.6}$$

If the air enters through opening 1 (positive C_p) and leaves through the three remaining openings (negative C_p), then the value of C_{pi} can be calculated from:

$$|(C_{p1} - C_{pi})|^{1/2} = \sum_{n=1}^{n=3} |(C_{pi} - C_{pn})|^{1/2} \tag{7.7}$$

As an example, if $C_{p1} = 0.8$, $C_{p2} = C_{p3} = -0.4$, and $C_{p4} = -0.3$, then the solution of equation (7.7) gives $C_{pi} = -0.24$.

If internal partitions were present and there were openings in the internal partitions as well as on the external surfaces of the building, then the C_p in each internal zone has to be calculated along the same lines above. If the areas of openings are not equal then the area of each opening must be included in the flow equation below (c.f. equation 3.1):

$$Q = A_{eff}\sqrt{(2\Delta p/\rho_0)} \tag{7.8}$$

where A_{eff} is the effective area of the opening (i.e. $A_{eff} = C_d A$ where C_d is the discharge coefficient for the opening), and

$$\Delta p = \tfrac{1}{2}\rho_0 \overline{V}^2 |(C_{pn} - C_{pi})| \tag{7.9}$$

In using the above equations it is imperative that the correct values of wind speed is substituted, which is normally that corresponding to building height. This usually requires the adjustment of meteorological station data to the height of the building as well as any wind shielding adjustments that must be made for the site using equation (7.3).

Example 7.1 Calculate the ventilation rate through two fully open, equal area windows on a building, one is situated on the windward side and the other on the leeward side. The effective open area of each window is $0.5\,\text{m}^2$ and the discharge coefficient is assumed to be 0.8. The building has a square plan of 10 m side and a height of 12 m. It is located in an urban industrial area and the wind speed measured at a nearby meteorological station at a height of 10 m in open country is $5\,\text{m s}^{-1}$.

This example can be solved by either calculating the effective flow area of the two opposite openings and then substituting the required values in the wind flow equation for this case in Table 3.5 (top case) or calculating the internal pressure using equation (7.7) and then using equations (7.8) and (7.9).

(i) Using the wind flow equation in Table 3.5

The wind speed corresponding to the height of the building and the shielding is calculated using equation (7.3) in which case $c = 0.35$ and $a = 0.25$, thus:

$$\frac{\overline{V}}{\overline{V}_r} = cH^a$$

$$\overline{V} = 5 \times 0.35 \times 12^{0.25} = 3.26 \, \text{m s}^{-1}$$

$1/A_w^2 = 1/0.5^2 + 1/0.5^2 = 8$, hence $A_w = (1/8)^{0.5} = 0.353 \, \text{m}^2$.

From Table 7.2 $C_{p1} = 0.2$ (windward wall) and $C_{p1} = -0.25$ (leeward wall)

$$Q_w = C_d A_w \overline{V}(C_{p1} - C_{p2})^{0.5} = 0.8 \times 0.353 \times 3.26 \times (0.2 + 0.25) = 0.618 \, \text{m}^3 \, \text{s}^{-1}$$

This is the air flow rate into the building through the two windows.

(ii) Using the internal pressure and one opening

Using equation (7.7):

$$|C_{p1} - C_{pi}|^{0.5} = |C_{pi} - C_{p2}|^{0.5}$$

i.e. $|0.2 - C_{pi}|^{0.5} = |C_{pi} + 0.25|^{0.5}$ from which we obtain $C_{pi} = -0.025$. From equation (7.9): $\Delta p = 1/2\rho_o \overline{V}^2 |(C_{pn} - C_{pi})| = \frac{1}{2} \times 1.2 \times 3.26^2 \times |0.2 + 0.025| = 1.435 \, \text{Pa}$. Hence, using equation (7.8): $Q = A_{\text{eff}}\sqrt{(2\Delta p/\rho_o)} = 0.5 \times 0.8[2 \times 1.435/1.2]^{0.5} = 0.618 \, \text{m}^3 \, \text{s}^{-1}$. It is not a coincidence that the two approaches give exactly the same answer since both methods are based on the same basic equations and continuity.

7.4 Buoyancy-induced ventilation

The buoyancy or stack pressure at an opening is due to variation in air density as a result of difference in temperature across the opening, and for openings at different heights, the difference in pressure between them is due to the vertical gradient in density and this is calculated using equation (3.13):

$$p_s = -\rho_o gh(1 - T_o/T_i) \tag{3.13}$$

where T_o = reference or outdoor air temperature (K); T_i = internal air temperature (K); ρ_o = air density at reference temperature, T_o (kg m^{-3}); g = acceleration due to gravity (m s^{-2}); h = difference in height between two openings (m). The basic equation (3.13) may be applied to different practical cases involving a known 'linear' temperature variation at each opening or zone but if there is a non-linear variation of temperature with height then the integral form of the equation must be used. Here we shall consider some situations of practical relevance and show how equation (3.13) or its integral form is used.

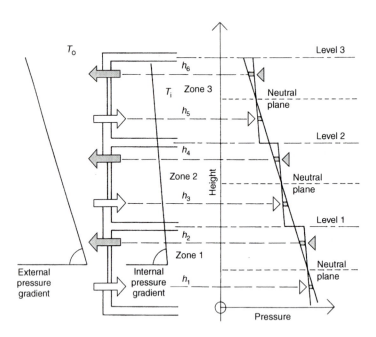

Figure 7.3 Stack pressure in a building with unconnected vertical zones.

Multi-zone building with no permeability between vertical zones

Such a building is illustrated in Figure 7.3, which also shows how the stack pressure would vary in each zone with height when the zones are sealed from each other except to the outside. As shown, each zone has its own neutral plane. In this case the zones are independent of each other with regard to buoyancy-driven flow and equation (3.13) may be applied to each zone separately. The difference in the height between any two openings in one zone is used in the equation. In this respect, it may be convenient to use the height of the lowest opening (or the ground) as datum in calculating the height between two openings in one zone. The difference in pressure between openings 1 and 2 is:

$$p_s = -\rho_0 g(h_2 - h_1)(1 - T_0/T_1) \tag{7.10}$$

where T_1 and T_0 is the temperature of zone 1 and the external temperature respectively (K). This pressure, p_s, is then substituted for Δp in equation (7.8) to calculate the air flow rate. The two openings in this case are in series and the effective area, A_{eff}, is calculated using the first equation in Table 3.5.

Multi-zone building with interconnection between vertical zones

In this case there will be air flow between the building and outside as well as between internal zones in the building if the zone temperatures are not the same. Such a situation is illustrated in Figure 7.4 where air enters the lower zone (zone 1) from outside and leaves through an opening on the upper zone (zone 2) with the flow passing through the opening between the two zones. In general, the temperature of zones 1 and 2

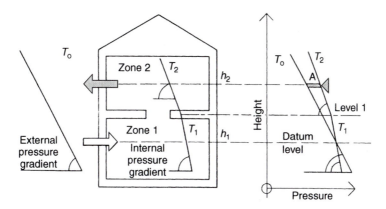

Figure 7.4 Stack pressure in a building with interconnected vertical zones.

are not equal (here $T_2 > T_1$) and this results in different pressure gradients for each zone as shown in the figure. The stack or buoyancy pressure for all the openings is calculated relative to that at the lowest opening. The pressure difference between the two external openings at h_1 and h_2 may be calculated using:

$$p_s = -\rho_o g[(z_1 - h_1)(1 - T_o/T_1) + (h_2 - z_1)(1 - T_o/T_2)] \tag{7.11}$$

where z_1 is the height of the first floor and T_1 and T_2 are the temperatures of zones 1 and 2 (K). To calculate the flow through the three openings (two external and one between the two zones) equation (7.8) is again used and an effective area for the three openings is calculated using the first equation in Table 3.5 similar to the previous case.

Multi-zone building with interconnection between horizontal zones

If internal horizontal zones in a building are at different temperatures and are connected by openings the analysis for buoyancy pressure is the same as in the previous case. The stack pressure is again calculated at each height relative to that at the lowest opening and the effective area for two or more flow openings are calculated as before. Such a situation may occur in a multi-storey building in which a common stairwell connects to all the floors and there is a difference in temperature between the stairwell and the other horizontal zones, as illustrated in Figure 7.5. To simplify the analysis it may be assumed that each zone, including a high zone such as the stairwell, is at uniform 'mean' temperature. The stack pressure difference between zones 1 and 4 is given by:

$$p_s = -\rho_o g(h_5 - h_1)(1 - T_4/T_1) \tag{7.12}$$

where T_1 and T_4 are the temperature of zones 1 and 4 (K) and h_1 and h_5 are the heights of the openings in each zone, respectively. The air flow rate is calculated as before.

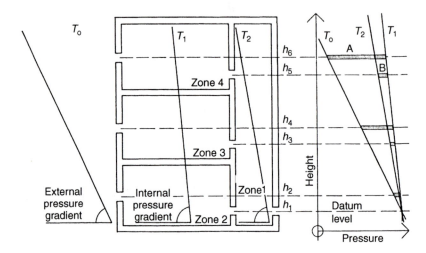

Figure 7.5 Stack pressure in a building with interconnected horizontal zones.

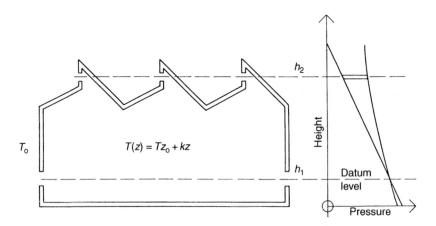

Figure 7.6 Stack pressure in a single-zone building with internal temperature gradient.

Single-zone building with temperature stratification

In a large open enclosure with high roof, such as a large factory or a warehouse, warm air will rise towards the roof resulting in large temperature stratification. As a result, there will be an internal pressure gradient, which affects the stack pressure between roof openings and lower openings, see Figure 7.6. In this case the attack pressure should be written in terms of the vertical air density gradient (namely equation (3.12)) as follows:

$$p_s = -g \int_0^h \rho(z)\, \mathrm{d}z \qquad (7.13)$$

where $\rho(z)$ is the air density at height z and h is the vertical distance between two openings. Since the density is a function of absolute temperature, equation (3.11) in this case becomes:

$$p_s = p_0 - \rho_0 g\, T_0 \int_{z_0}^{h} \frac{dz}{T(z)} \tag{7.14}$$

where T_0 and ρ_0 are the reference temperature and air density, $T(z)$ is the temperature variation with height z, z_0 is a datum height and h is the height of the opening. The stack pressure can be calculated from equation (7.14) providing that the temperature profile function $T(z)$ is known. A linear temperature profile is applicable for most practical situations, i.e.

$$T(z) = T_{z_0} + kz \tag{7.15}$$

where T_{z_0} is the temperature at datum height z_0 (K) and k is the temperature gradient (K m^{-1}). Substituting equation (7.15) into (7.14) gives:

$$p_s = p_0 - \rho_0 g\, T_0 \int_{z_0}^{h} \frac{dz}{T_{z_0} + kz} \tag{7.16}$$

Integrating this equation gives:

$$p_s = p_0 - \rho_0 g\, T_0 \left[\frac{1}{k} \ln(T_{z_0} + kz) \right]_{z=0}^{z=h} \tag{7.17}$$

For $k = 0$, i.e. uniform enclosure temperature, equation (7.16) becomes:

$$p_s = p_0 - \rho_0 g\, T_0 [z]_{z=0}^{z=h} \tag{7.18}$$

The air flow through openings is calculated using these pressure equations as before.

7.5   Combined wind and buoyancy ventilation

When wind and stack pressure act simultaneously on a building, their combination will determine the air flow through building openings. If both pressures have the same sign then the two pressures will increase the air flow but if they have opposite signs the air flow will reduce and in certain circumstances the two pressures can cancel each other to produce no flow through the openings. The combination of these two pressure components has been dealt with in Section 3.2 where it is shown that the quadrature method of combining the flow due to these two pressure components is the simplest and most satisfactory for use in natural ventilation calculations. This is represented by equation (3.17) as given below:

$$Q_t = [Q_w^2 + Q_s^2]^{1/2} \tag{3.17}$$

where Q_w is the flow due to wind pressure and Q_s is that due to the stack pressure as calculated using equation (7.8). If a fan is used for supply or extract then equation (3.14) can be used taking the index $n = 1/2$.

More recent work regarding the interaction of wind and stack pressure is reported by Li *et al.* [9, 10] where the analysis showed some interesting features of the combined flow produced by the action of wind and buoyancy. It was found analytical, and with the support of model experiments using salt water solution in a water channel to represent buoyancy force, that when the wind and buoyancy pressure assist each other the flow is always upwards and the solution is straightforward. However, for opposing pressures, the flow can be either upwards or downwards depending on the combination of wind and buoyancy pressures. In the latter case, the resulting flow rate which was a function of the buoyancy pressure presented some complex features. The analysis, which was also supported by experimental observation in the water channel showed the existence of hysteresis or multiple solutions as the effect of buoyancy was progressively decreased. Similar effects were found for a single-zone enclosure with two openings, a two-dimensional heated channel representing a solar chimney and a two-zone building [10].

7.6 Characteristics of natural ventilation openings

In recent years there have been considerable developments in air inlet components for natural ventilation application, particularly in the areas of air flow and noise control. In the following, some of the devices that are used in natural ventilation are described.

7.6.1 Small openings

These are natural ventilation devices that are designed to provide background ventilation for winter and summer. Considerable progress has been made in recent years in developing devices that are capable of providing flow control. Some types which are in use or have been developed for this purpose are discussed below.

Trickle ventilators

These are air inlet devices that are used to provide a minimum fresh air rate as background ventilation (e.g. about $5\,\mathrm{l\,s^{-1}}$ per occupant). Although some are provided with dampers for opening or closing, in normal use these vents are kept constantly open to provide minimum fresh air for maintaining acceptable quality of air throughout the occupancy periods. The UK Building Regulations Part F [11] for instance recommends a trickle ventilator opening area of $400\,\mathrm{mm^2\,m^{-2}}$ of floor area for spaces in non-domestic buildings of floor area $>10\,\mathrm{m^2}$, with a minimum area for a single room of $4000\,\mathrm{mm^2}$ in domestic and non-domestic buildings. To minimize cold draughts in winter, these should be located at high level, either at the top of window frame or as part of the glazed unit at high level, see Figure 7.7. These devices are not capable of dealing with high pollution loads that may occur during peak occupancy and activity periods and they should be used in conjunction with other types of ventilation openings such as louvers or windows.

One of the problems with natural ventilation openings is the ingress of traffic noise. Some trickle ventilators are provided with noise attenuators but with a penalty for an increased pressure drop and a reduction in air flow through them.

Flow control ventilators

As it is well known, the driving forces for natural ventilation (wind and buoyancy pressures) are continuously varying, which is one of the main drawbacks of natural

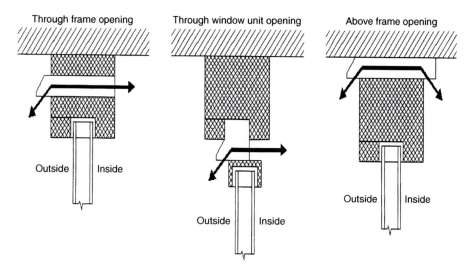

Figure 7.7 Typical trickle ventilator positions.

ventilation. Attempts have been made to develop ventilators that are capable of regulating the air flow through them for the low pressures normally associated with natural ventilation (typically 1–10 Pa). These are usually called *pressure-controlled ventilators*. Figure 7.8 shows such a ventilator, which uses a sensitive damper balanced on a fulcrum to close the opening area as the wind pressure increases. This device can maintain a regular flow rate quite accurately up to a pressure of about 20 Pa. It can also be integrated with a mechanical air extract system but its current cost is at least twice that for a more conventional trickle ventilator.

Other types of flow controllers are based on internal humidity or outdoor temperature, which are respectively called *humidity-controlled ventilators* and *temperature-controlled ventilators* respectively. The former type is designed to respond to changes in the relative humidity (r.h.) in the space with the flow opening increasing as the humidity increases. In this type, a moisture-sensitive tape that changes in length according to the r.h. regulates the flow opening. These are mostly used in dwelling rooms with large moisture generation, such as kitchens or bathrooms but not in commercial buildings.

Temperature-controlled ventilators are more widely used in some Nordic countries where in winter the dominant driving force for ventilation is stack pressure due to the low outside temperature. These ventilators restrict the flow area as the outside temperature reduces by means of a bimetallic temperature sensor. Although the response time of these devices is large they are adequate for general diurnal and seasonal variations in outside temperature.

7.6.2 *Large openings*

Windows

There is a wide range of window types and sizes and their ventilation characteristic varies accordingly. However, knowledge of the performance of a particular window

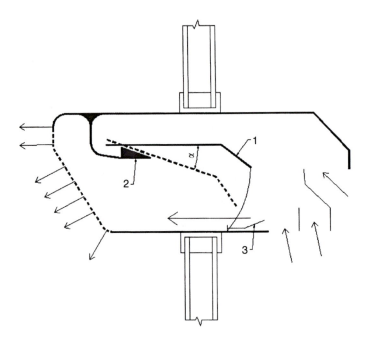

Figure 7.8 Cross-section of a pressure-controlled ventilator. (1) Sensitive damper on a (2) fulcrum, (3) inlet grille.

for natural ventilation is rather limited and is often based on theoretical assumptions of the driving forces and the effective open area and, in practice therefore, it is only possible to make a rough estimate of the air flow rate through a window opening. Some window types are regarded better than others, but this is mainly based on qualitative measures and, on the whole, the difference between and the limitations in the application of window types cannot easily be quantified.

The flow characteristic through an open window depends on the relative values of the wind and buoyancy forces acting upon it. If the wind pressure is dominant and continuous then the flow will be unidirectional, but if the wind speed and turbulence fluctuate greatly or the dominant force is buoyancy, then a bi-direction flow through an open window will occur. In addition, the type of window will affect not only the open area for air flow but also the effect of the wind and buoyancy pressures acting on the window opening itself.

The main types of widely used windows are shown in Figure 7.9 and their characteristics are given below.

Horizontal-pivot window This type has a large flow area which approximates the full area of the window and therefore can provide large air flow rates. When used for single-sided ventilation, air enters at the bottom half and leaves at the top if the external temperature is lower than the room temperature. However, when it is used in cross-flow ventilation air enters the top and bottom halves simultaneously, as illustrated in Figure 7.10. Normally this type of window directs most of the air entering

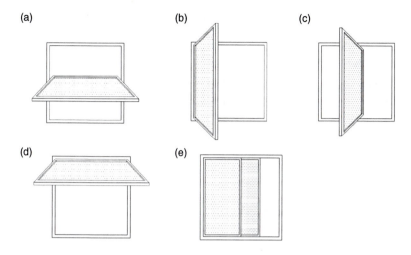

(a)   (b)   (c)

(d)   (e)

Figure 7.9 Types of windows. (a) Horizontal-pivot window; (b) side-hung window; (c) vertical-pivot window; (d) top- or bottom-hung windows and (e) horizontal- and vertical-sliding windows.

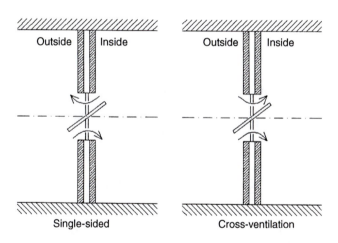

Single-sided   Cross-ventilation

Figure 7.10 Air flow through a horizontal centre pivoted window for single-sided and cross ventilation.

the building upwards, which is more favourable for comfort than say the side-hung windows.

Side-hung window   This type of window is very common but has a smaller effective opening, hence lower air flow than the previous type. In addition, this type of opening could cause draught in the cold season as the air enters the building sideways and at low level. It is also susceptible to be blown wide open by gust and to driving rain than other types. Figure 7.11 shows the effect of pressure difference across a side-hung window on the discharge coefficient (C_d) for different opening areas of the window [12]. The

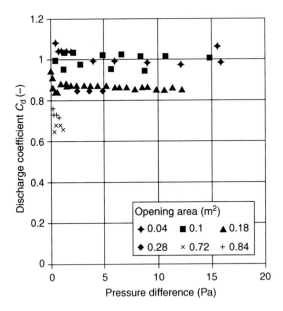

Figure 7.11 Discharge coefficient for side-hung window [12].

opening areas shown in the figure represent the area of the narrowest flow passage. However, the estimation of this area is not easy particularly for small openings of this type of window. Nevertheless, the results show that C_d is not constant at small pressure difference but reaches a constant value at higher pressure difference. In addition, C_d is largest for the smallest opening areas and only approaches the value 0.6, which is often used for sharp-edge openings, for large open areas of the window.

Vertical-pivot window   This is similar to the side-hung window and is even more effective than the latter for 'scoping' the wind and the rain when they are at an angle to the building façade.

Top- or bottom-hung windows   These types have even smaller flow areas than the last two since most of the opening area is at the bottom or top of the window, for top- and bottom-hung windows, respectively. However, they are less prone to produce draught in cold weather, particularly the bottom-hung type, as the flow is restricted to a small area at the top of the window in the bottom-hung type or at the bottom of the window for the top-hung window. Figure 7.12 shows the effect of pressure difference across a bottom-hung window on the discharge coefficient (C_d) for different opening areas of the window [12]. As for the side-hung window, this type also has a lower value of C_d for larger opening areas but is always >0.6. Tests on this type of window [12] have shown that this is the best type to use for single-sided ventilation in cold weather as the cold air at high level is outside the occupied zone of the room. However, this type is less effective in the summer when outdoor air is required to reach the occupied zone, and for best performance all year round, the bottom-hung window should be combined with another type, such as the side-hung window.

Horizontal- and vertical-sliding windows   The vertical-sliding (sash) and horizontal-sliding windows have similar ventilation performance to the horizontal- and

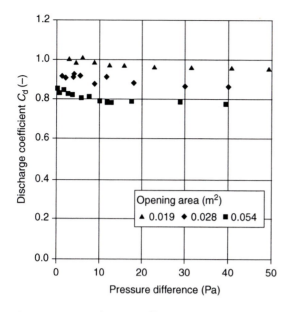

Figure 7.12 Discharge coefficient for bottom-hung window [12].

vertical-pivot windows, respectively. The sash window in particular is convenient for single-sided and cross-ventilation in the summer and winter as the top and lower openings are easily varied.

Louvers

Louvers are adjustable ventilation openings placed on an opening in the building façade at different locations and heights to provide mainly background ventilation. The louver blades are either glass or aluminium with some types incorporate acoustic attenuators. Although the blade openings can be adjusted it is often difficult to achieve an airtight seal when ventilation is not required.

7.6.3   *Ventilation ducts and stacks*

Stack ventilation is used when the types of openings described in Section 7.6.2 do not provide sufficient fresh air supply to the building, either because the building is very deep in plan and/or a high ventilation rate is required. The air flow in the stack is due to: (i) the stack pressure, which is proportional to the vertical distance between the inlet and outlet to the stack, and the temperature difference between the air inside and outside the building as given by equation (7.18); (ii) the wind pressure at the discharge end of the stack, which is proportional to the pressure coefficient and the wind speed, as given by equation (7.4). The performance of the stack ventilation system is most reliable in cold weather and high wind speeds but in milder weather additional openings such as windows are normally needed to supplement the ventilation requirement in the form of single-sided ventilation.

The minimum height of a stack above roof level to avoid back flow (back draught) into the building may be estimated using the empirical relationship given in AIVC TN 44 [13]:

$$h = d[0.5 + 0.16(\theta - 23)] \tag{7.19}$$

where h is the minimum height (m) above the highest intersection point of the stack with the roof, d is the horizontal distance (m) from the centre of the stack outlet and the highest point of the roof and θ is the roof pitch angle (°) that the stack passes through. For a roof pitch $\theta \leq 23°$, h must be at least 0.5 m above the highest intersection point of the duct with the roof, irrespective of its location on the roof.

It is essential that adequate sizing of the ventilation stack is made as unlike a mechanical ventilation duct, large pressure drop in it cannot be tolerated. The pressure losses in a stack is partly due to fluid friction, which is calculated in a similar way to duct friction losses, and partly due to the dynamic pressure loss at the discharge end, which is $\frac{1}{2}\rho v^2$, where v is the efflux velocity. Any other losses such as those due bends, connections, grilles, cowls, etc. must also be added to these two components, see equation (7.29).

7.7   Natural ventilation strategies

A survey of ventilation systems used in a sample of modern buildings in 14 developed countries carried out by the Air Infiltration and Ventilation Centre showed that all commercial buildings included in the survey used air-conditioning systems whereas natural ventilation was still the most common method of ventilating domestic buildings [14]. This would suggest that the potential of natural ventilation for commercial buildings has been overlooked even though it may have been plausible particularly in buildings situated in temperate climates. There may have been a number of reasons for ignoring natural ventilation, such as not giving a 'hi-tech' image to the tenant or owner, undesirable location, traffic noise, etc. but the lack of control of natural ventilation systems may have been a major factor too in most of the buildings surveyed. It is important, therefore, that the most appropriate natural ventilation strategy is selected for the building if natural ventilation is to become a common technique for ventilating commercial as well as domestic buildings. The natural ventilation strategy that is most suited to a particular building can only be arrived at by careful consideration of a number of factors, such as:

- Depth of space with respect to ventilation openings
- Ceiling height
- Thermal mass exposed to the air
- Location of building with respect to environmental pollution sources, e.g. traffic noise, air pollution, etc.
- Heat gain
- Climate.

The most widely used natural ventilation methods are discussed below.

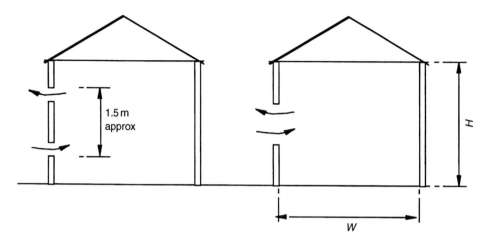

Figure 7.13 Single-sided ventilation ($W_{max} \approx 2.5H$).

7.7.1 Single-sided ventilation

This is usually the simplest form of naturally ventilating a building whereby a simple opening(s) in the form of a window or a ventilation device such as a trickle ventilator on a wall is used to allow outdoor air to enter the building and room air to leave either from the same opening(s) or from another opening(s) situated on the same wall, see Figure 7.13. With single ventilation opening the main driving force is wind, particularly in the case of small openings. Where more than one opening on the same façade is located at different heights, the stack effect can enhance the ventilation rate in addition to the wind.

Although single-side ventilation is a very common and inexpensive strategy the air flow is often uncontrollable, except for an open or a closed position of the ventilator, and is only effective over a distance of about $2.5H$ from the opening itself, where H is the ceiling height. Furthermore, some single-sided openings, e.g. windows, are only suitable in moderate climates and are not always suitable for winter ventilation unless the incoming air is heated, see later sections.

The flow due to buoyancy through a large opening is determined by the pressure difference due to temperature difference across the opening. As shown in [15], the buoyancy flow equations through the opening can be derived as given below:

$$\Delta p(z) = \Delta \rho \, g \, z \tag{7.20}$$

where $\Delta \rho$ is the difference in density across the opening, z is the height and g is the acceleration of gravity. Also, $v(z) = \sqrt{2\Delta p(z)/\rho}$. Hence, $v(z) \propto z^{1/2}$ i.e.:

$$\frac{v(z)}{v_{max}} = \left(\frac{z}{H}\right)^{1/2}$$

The mean velocity ($\bar{v}$) through an opening of height (H), see Figure 7.14, is:

$$\bar{v} = \frac{v_{max}}{H^{1/2}} \int z^{1/2} \, dz = \frac{v_{max}}{H^{1/2}} \frac{2}{3} H^{3/2} = \frac{2}{3} H v_{max} \tag{7.21}$$

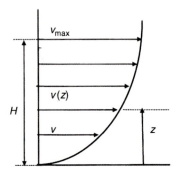

Figure 7.14 Velocity profile across a large opening.

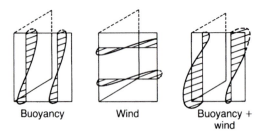

Buoyancy   Wind   Buoyancy +
                                      wind

Figure 7.15 Flow patterns through an open window.

The volume flow rate through the opening (Q) is:

$$Q = C_\mathrm{d} w \bar{v} = \frac{2}{3} C_\mathrm{d} w H v_{max} = \frac{2}{3} C_\mathrm{d} A v_{max} \qquad (7.22)$$

where w is the width of the opening.

However, in a buoyancy-driven flow, equal masses of air enter and leave through the same opening. If H is the total height of the opening, then the influx or efflux flow is:

$$Q = \frac{C_\mathrm{d} A}{3} v_{max} \qquad (7.23)$$

From equation (7.20) it follows that:

$$Q = \frac{C_\mathrm{d} A}{3} \sqrt{\frac{gH\Delta T}{\overline{T}}} \qquad (7.24)$$

where ΔT is the temperature difference across the opening and $\overline{T}$ is the mean temperature (K).

The last equation can be used to estimate the air flow rate through a single-sided opening due to buoyancy only. In the case of a large opening such as a window or a door, the air enters through one part and leaves through another as illustrated in Figure 7.15. The effect of buoyancy and wind on the flow through various types of windows has been investigated by de Gids and Phaff [16]. From air change measurements on site, they derived the following expression for the effective velocity through an open window:

$$V_{eff} = \sqrt{(C_1 \overline{V}^2 + C_2 H \Delta T + C_3)} \qquad (7.25)$$

where V_{eff} = effective velocity through window opening (m s^{-1}); C_1 = dimensionless coefficient depending on the window opening; C_2 = buoyancy constant; C_3 = wind turbulence constant; $\overline{V}$ = mean wind speed measured at a weather station (m s^{-1}); H = window height (m); ΔT = mean temperature difference between inside and outside (K). Using the effective velocity as given by equation (7.25), the flow rate through the window is given by:

$$Q = \tfrac{1}{2} A V_{eff} \qquad (7.26)$$

where A = effective open area of the window (m^2).

From measurements at various wind speeds and temperature differences, de Gids and Phaff obtained the following values for the constants: $C_1 = 0.001$, $C_2 = 0.0035$ and $C_3 = 0.01$. A comparison between the flow rate given by equations (7.25) and (7.26) and that calculated using the expression in Table 3.6 for a buoyancy-driven flow showed a close agreement between the two methods. For single-sided ventilation BRE Digest 399 [17] recommends a window area of about 1/20 floor area and maximum room depth of 2.5 times the ceiling height.

7.7.2 Cross-ventilation

Two-sided or cross-ventilation occurs when air enters the room or building from one or more openings on one side and room air leaves through one or more openings on another side of the room or building, see Figure 7.16. The flow of air in this case is mainly due to wind pressure, and buoyancy pressure becomes important only if there is a significant difference in height between the inflow and outflow openings. The types

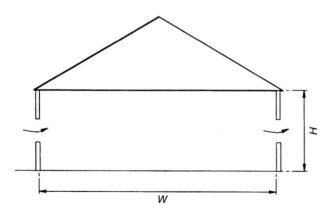

Figure 7.16 Cross-ventilation ($W_{max} \approx 5H$).

of openings that are used for cross-ventilation can be small openings such as trickle ventilators and grilles, or large openings such as windows and doors. Because the air 'sweeps' the room from one side to the opposite side, it has a deep penetration. This strategy is therefore more suitable for ventilating deep-plan rooms. The positioning of openings should be such that some are placed on the windward façade of the building and others placed on the leeward façade, so that a good wind pressure difference is maintained across the inflow and outflow openings. Internal partitions and other obstructions can affect or disturb the air flow pattern in the room and the air penetration depth.

The air flow rate due to cross-ventilation may be estimated using equation (7.8) in which case the effective area A_{eff} is calculated from equations in Table 3.5. The pressure difference across opposite openings Δp is calculated for either the wind effect only or the combined effect of wind and buoyancy, if the latter is significant. The discharge coefficient C_d depends on the type of opening and for windows, Figures 7.11 and 7.12 can be used. However, if no value for C_d is available for the opening, the value of 0.6 for a sharp-edge orifice would give a reasonable estimate. The equations given in Table 3.5 for cross-ventilation can be used to calculate the air flow rate.

Cross-ventilation is suitable for spaces of depth $>2.5H$, where H is the ceiling height and up to $5H$. It is usually more effective than the single-sided ventilation because the wind pressure can be more favourable for providing larger air flow rates hence more suitable for larger heat gains. As the single-sided ventilation however, this ventilation strategy also lacks adequate air flow control unless automated openings are used.

Because of the distortion of the wind flow by the building the reference speed, $\overline{V}$, obtained from (7.3) may be over-estimated. Wind-tunnel measurements for wind profiles representing open country, suburban and urban terrains carried out by Chand *et al.* [18] produced the following correlation for cross-ventilation through two opposite window openings:

$$\frac{V_i}{\overline{V}} = F(1 - 0.82a) \tag{7.27}$$

where V_i is the corrected wind speed at the inlet opening, and a is the terrain exponent. For rectangular openings, the correction factor, F, is given by:

$$F = 1.1 \left[1 + \left(\frac{A_i}{A_e} \right)^2 \right]^{-0.5} \tag{7.28}$$

where A_i and A_e are the inlet and exit areas, respectively.

7.7.3   Stack ventilation

Buildings which require ventilation rates greater than those achievable using either single-sided or cross-ventilation may be ventilated using stacks. In this case, buoyancy is the main driving force and, therefore, the height of the stack becomes significant. The stack pressure is determined by the difference between the internal and external temperature and the height of stack, as given by equation (3.13).

Depending on the position of air inlet and outlet in the building, the wind pressure could assist the stack pressure, reduce its influence or indeed reverse the effect, i.e.

by forcing the air through the outlet. Therefore, when stacks are incorporated in the building, careful design considerations are needed to avoid these adverse effects occurring. This usually requires either a wind tunnel investigation of a scaled model of the building and the stack, or computational fluid dynamics (CFD) analysis of the flow around and within the building, see Chapter 8. The effect of buoyancy cannot be modelled in a wind tunnel but it can be taken into account in a CFD simulation of the air flow.

In buildings which have atria, the stack is most conveniently incorporated with the atrium for two main reasons. First, the solar gain in the atrium causes an elevation of the air temperature and hence there will be more effective stack flow. Second, the atrium will act as a buffer zone between the building and the external environment which can reduce heat losses from the building in winter.

Simple stack

In a simple stack or chimney the driving force is wind as well as buoyancy. To effectively utilize the wind pressure, a correct location of the stack outlet on the building at high level is essential. This requires knowledge of the distribution of wind pressure coefficients around the building and in the case of a roof stack the use of equation (7.19) as a guide.

The wind and buoyancy pressures acting on a stack will balance the pressure losses in the stack to produce a given air flow rate through it. In sizing stacks, therefore, the friction losses, fitting losses and entry and exit losses must be equated to the total pressure difference due to wind and buoyancy acting between the inlet and outlet of the stack. The pressure losses in the stack are given by [19]:

$$\Delta p = \left[4f \frac{z}{D_h} + K_i \left(\frac{A}{A_i} \right) + K_d \left(\frac{A}{A_i} \right) + K_e \left(\frac{A}{A_e} \right) \right] \frac{1}{2} \rho V_m^2 \tag{7.29}$$

where the K's are the pressure loss coefficients; A is the cross-sectional area of the ventilation channel; A_i, A_d, A_e are the areas of inlet, damper (if present) and exit, respectively; z is the height between two openings; ρ is the density; V_m is the mean air speed in the stack; D_h is the hydraulic diameter of the stack and f is the friction factor for the stack wall. For non-circular section stack, the hydraulic diameter is given by:

$$D_h = \frac{2wh}{w+h} \tag{7.30}$$

where w is the stack width and h is the depth. For a narrow channel ($w < 10h$):

$$D_h = 2h \tag{7.31}$$

A system which incorporates ventilation stacks provided with controllable air inlets can, if properly designed, be used for providing large ventilation rates in certain zones of a building, e.g. high moisture or contaminant generation zone in the building, as shown in Figure 7.17. In this case, the inlet to the stack is situated in bathrooms and kitchens with the outlet opening terminating above the roof where negative pressure is generated by the movement of wind over the building. Ventilation is primarily provided by the stack effect and enhanced by the negative static pressure at roof level.

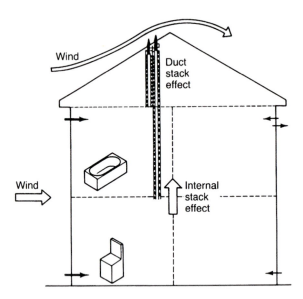

Figure 7.17 Stack ventilation in a house.

As was shown in Section 7.4, the air flow rate due to the stack effect is proportional to the vertical distance between the inlet and outlet to the stack and the temperature difference between the air inside and outside the building. The performance of the stack ventilation system is therefore most reliable in cold climates because it is essentially a temperature-driven system. In milder climates open windows are normally needed to supplement the ventilation requirement in the form of either a single-sided ventilation opening or cross-ventilation arrangement.

A simple and effective passive ventilation system for a house is described by Gaze [20]. It consists of extract ducts rising vertically from the kitchen and bathroom and terminating at a tile ventilator near the ridge with window slot or trickle ventilators providing controllable air inlets. The system operates by the combined effect of stack and wind actions as illustrated in Figure 7.17. The inlet to each duct is fitted with a grille on the ceiling with a large free area (e.g. 90%) to minimize pressure loss. In this case, decoupling of the outlet end of the ducts and the roof tile ventilator should avoid back-draughts and over-extraction as illustrated in Figure 7.18. To reduce the risk of condensation and loss of stack effect the portions of ducts that pass through the loft and other unheated spaces should be insulated.

Sizing of extract ducts can be made using equation (7.29). As a guide recommended UPVC duct sizes for kitchen and bathroom are as follows [20]:

House volume (m³)	<100	100–200	>200
Duct diameter (mm)	110	160	160

This design can be self-throttling under extreme external weather conditions and typical air velocity in the extract ducts was found to be in the range 0.3–$1.3\,\mathrm{m\,s^{-1}}$, depending on the internal–external temperature difference and whether the window

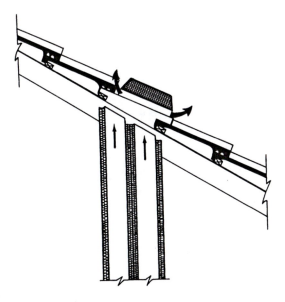

Figure 7.18 Decoupling of ducts and tile ventilator in a passive ventilation system [20].

and door vents were open or shut. Some control over the air flow rate can be achieved by opening or shutting these vents. The opening of windows or the provision of extract fans in the ducts can boost the ventilation rate if necessary.

The above system was tested in a demonstration house as well as in occupied houses [20]. Measurements of temperature, humidity and air flow rate in the ducts were carried out in addition to air leakage and airtightness tests. The occupied houses had low adventitious air leakage and would have been under-ventilated under normal conditions without the stack ventilation. With the ventilation system operating and with the vents open, a total air change rate in the range 0.5–1.0 ach was measured, in which case the system contributed between about 0.3 and 0.6 ach of the total.

Large enclosures

Large enclosures such as atria experience large temperature stratification and this phenomenon can be utilized to ventilate the enclosure naturally. As a special case, glazed atria are becoming increasingly popular structures for creating a microclimate approaching that found indoors, particularly in high northern latitudes. Because of the large glass area in such structures the thermal environment inside is greatly influenced by external weather conditions. A prominent feature of atria is the reliance on natural lighting for much of the day, solar gain through the glazing and radiant heat transfer with surrounding buildings. At high northern latitudes in winter a certain amount of auxiliary heating will be required to maintain acceptable indoor temperatures (e.g. 15 °C) for users of the atrium and also to reduce the risk of water vapour condensation on the internal glass surfaces. In the summer, solar gain through the glazing normally elevates the internal air temperature and causes a large vertical temperature gradient

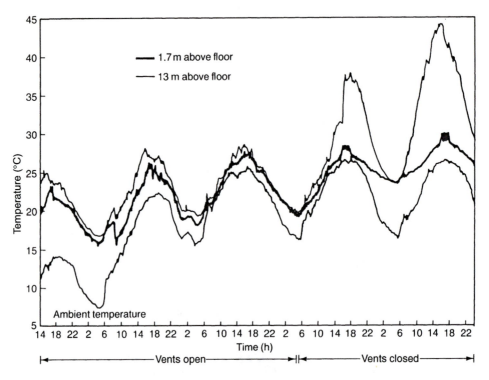

Figure 7.19 Measured air temperature at two heights in a glazed atrium with and without natural ventilation [21].

and sometimes overheating problems at upper levels such as shops and offices. There-fore, mechanical ventilation created by extract fans or buoyancy-driven ventilation through roof hatches is needed to remove the excess heat in the warm seasons.

A field study was carried out in Norway [21] on a glazed atrium of dimensions 46 m × 10 m × 17 m with an effective air volume of 7400 m^3 to measure the internal air temperature in winter and summer and assess the effect of natural ventilation in reducing the internal temperature in the summer season. Using the tracer gas decay method, the air flow rate into the atrium was measured with all the ventilation openings closed to give an air change rate of 0.75 ach in the lower part of the atrium and about 0.55 ach in the upper part. By opening the roof ventilation hatches the air change rate in the summer had increased to between 3 and 4 ach for the whole atrium. In this case the cooler air entering the lower regions of the atrium and the warmer air escaping from the roof vents form an upward displacement system due to the natural driving forces of stack and wind effects. Figure 7.19 shows the effect of opening the ventilation hatches on the internal air temperature at two heights inside the atrium in the summer. The first three days represent open hatches during which the maximum temperature difference at 13 m and 1.7 m above the floor was only 2 K. In the last two days during which the hatches were closed a temperature difference as high as 16 K (representing a maximum temperature at roof level of 46 °C) was recorded.

In calculating the size of roof ventilation openings area the effect of buoyancy (calculated at roof height) and wind is combined (see Sections 7.3–7.5) to produce the desired air change rate for the atrium. The value of air change rate will depend on the solar and internal heat gains for the building, the outside air temperature and local wind speed. However, for a conservative estimation of the air change rate the buoyancy pressure alone is used in sizing the roof openings.

Windcatchers

Windscoops have been employed in buildings in the Middle East for more than three thousands years. They were traditionally constructed from wood-reinforced masonry with openings at height above the building level ranging from 2 m to 20 m. In the modern design of windcatchers, the two ventilation principles of wind-scoop and passive stack are combined in one design around a stack that is divided into two halves or four quadrants/segments with the division running the full length of the stack. Due to manufacturing requirements, the area of each segment is in some cases the same (e.g. a square-section windcatcher) but in other cases (e.g. circular-section windcatcher) not all segments have the same section.

Windcatcher systems are increasingly being installed in buildings to enhance natural ventilation where conventional systems do not provide sufficient air flow. In most of these modern installations the windcatchers terminate at ceiling level with four quadrants acting as air supply/extract ducts, see Figure 7.20.

Circular and a square section windcatchers are available either for ventilation purposes only or are integrated with natural lighting devices. In this case the term *'suncatcher'* is sometimes used to distinguish this variety from standard windcatchers, Figure 7.20(a). In this case, the central part of the device is constructed from polished metal tube with a transparent dome cover to provide natural light to the room. At ceiling level a light prism cover is used to diffuse natural light before entering the building.

Wind tunnel tests [22, 23] have produced some interesting characteristics of windcatchers, most significantly these are:

(a)  (b)

Figure 7.20 Suncatcher and windcatcher. (a) External view of slots and of sun pipes for a suncatcher and (b) The four air flow segments of a windcatcher.

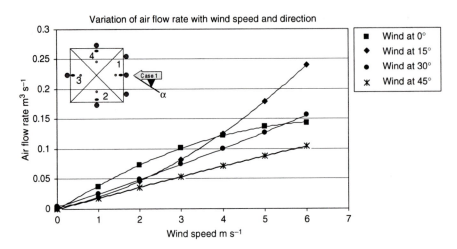

Figure 7.21 Measured air flow through a 0.5 m square section and 1.5 m long windcatcher.

- The windcatcher performance depends, to a certain extent, on the wind direction and, to a greater extent, on the speed of the wind in relation to the windcatcher segments, particularly for a square section windcatcher.
- For larger flow rates, the larger windcatcher segments should be positioned in a way that the prevailing external wind will be blowing through them to allow more wind to enter the building through these segments.
- Windcatchers are capable of providing larger air flows than conventional stacks due to the harnessing of both with and stack pressures more effectively.

Typical flow characteristics for a 0.5 m square section and 1.5 m long windcatcher obtained from wind-tunnel tests are shown in Figure 7.21.

Solar-induced ventilation

Natural ventilation systems are usually designed to utilize both buoyancy and wind pressure under the expected environmental conditions. In situations where the wind assists the buoyancy pressure, the air flow rate that can be supplied to a building is the highest possible for a given ventilation strategy and environmental conditions. However, in cases where the wind effect is not well captured or where the buoyancy pressure is not sufficient to provide the required ventilation rates then solar-induced ventilation may be a viable alternative. This strategy relies upon the heating of part of the building fabric by solar irradiation resulting into a greater temperature difference, hence larger air flow rates, than in conventional buoyancy-driven strategies in which the air flow is due to temperature difference between inside and outside.

There are usually three devices which can be used for this purpose:

- Trombe wall
- Solar chimney
- Solar roof.

The first type incorporates glazed elements in the wall to absorb solar irradiation into the wall structure, whereas the solar chimney and solar roof usually rely on the wall of the chimney and roof tiles to absorb and store solar energy respectively. These devices are governed by the same physical principles and are based on the same fluid flow and heat transfer equations as other natural ventilation systems, although they have certain unique characteristics. These devices have common characteristics with each other, which are described here first before considering each device separately.

Solar-induced ventilation is buoyancy-driven by the use of a solar air collector and, therefore, all the equations derived earlier for buoyancy pressure (equation 7.10) and flow rate through large openings (equation 7.8) also apply here. However, the external temperature in equation (7.10), T_o, is replaced by the exit temperature of the collector, T_e. In addition, there will be pressure losses through the collector as well as pressure losses at the inlet and outlet openings as for the case of a stack, see equation (7.29).

The exit temperature, T_e, of the collector is given by [24]:

$$T_e = A/B + [T_i - A/B] \exp\{-BwH/(\rho_e C_p Q)\} \tag{7.32}$$

where $A = h_1 T_{w1} + h_2 T_{w2}$; $B = h_1 + h_2$; h_1 and h_2 are the surface heat transfer coefficients for the internal surfaces of the channel and T_{w1} and T_{w2} are the temperatures of the corresponding internal surfaces of the channel. T_i = inlet air temperature of collector (°C); Q = volume air flow rate (m^3 s^{-1}); ρ_e = air density at exit (kg m^{-3}); C_p = specific heat of air (J kg^{-1} K^{-1}).

The heat transfer coefficients h_1 and h_2 are usually obtained using:

$$Nu = 0.1 \, Ra^{1/3} \tag{7.33}$$

where $Nu = hH/k$ = Nusselt's number; $Ra = Pr \, Gr$ = Rayleigh number; $Pr = \mu C_p/k$ = Prandtl's number; $Gr = g\beta H^3 (T_w - T_i)/\nu$ = Grashof's number; μ = dynamic viscosity of air (Pa); k = thermal conductivity of air (W m^{-1} K^{-1}); ν = kinematic viscosity of air (m^2 s^{-1}); β = cubic expansion coefficient of air $\approx 1/T_i$ (K^{-1}).

Equation (7.33) applies to vertical and moderately inclined surfaces (< 30° from the vertical) for Ra range $10^{13} > Ra > 10^9$.

Equations (7.29), (7.32), (7.10) and (7.8) are all interconnected and to estimate the air flow rate produced by a collector, Q, it is necessary to solve these four equations by iteration using a computer. In equation (7.8), Δp is the pressure difference due to the stack pressure given by equation (7.10) and the pressure losses in the collector given by equation (7.29). Further details are given in references [19, 24].

Trombe wall ventilator   A Trombe wall collector consists of a wall of moderate thickness (thermal mass) with lower and upper openings covered externally by a pane of glass. A gap of 50–100 mm between the glass and the wall allows the heated air to rise. Trombe wall collectors have traditionally been used for space heating by allowing air from the room to enter at the bottom of the wall which is heated by the collector and then returned back to the room at high level, see Figure 7.22.

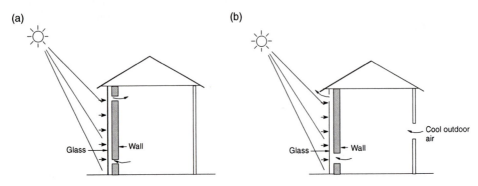

Figure 7.22 Trombe wall ventilator. (a) Collector for winter heating and (b) collector for summer ventilation.

The arrangement shown in Figure 7.22(a) is for the winter situation where the Trombe wall is used to heat room air. However, by putting a high level external opening on the glazing and closing the top opening to the room, this device can be used for cooling the room by drawing outdoor air from another opening into the room and the warm room air is extracted out through the Trombe wall, Figure 7.22(b). To be effective, the wall needs to be placed in a south or south-west facing position in the northern hemisphere.

To calculate the air flow rate through the collector the method described in the previous section is applied. However, the wall and glazing temperature or the corresponding heat fluxes will be required and these can be estimated from knowledge of the solar gain, thermal mass of wall, emissivity of glass and wall, etc. An analytical method has been developed for calculating the air flow rate for winter and summer situations [25]. Additional information on the design of Trombe walls may be found in [26, 27].

Solar chimney   A solar chimney attached to south/south-west facing wall is heated by solar irradiation and the heat stored in its fabric can be utilized for ventilation, Figure 7.23. The heated external surface of the chimney generates a natural convection current by drawing air from the building and extracting it at the top. Outdoor air enters the building to replace the warm, stagnant air inside.

The method described earlier for calculating the air flow rate also applies here but usually only the external surface of the chimney is heated. In this case, equation (7.32) may be simplified to:

$$T_e = T_w + (T_i - T_w)\, \exp\{-hwH/(\rho_e C_p Q)\} \tag{7.34}$$

where T_w = the inside wall (hot) temperature of chimney (°C).

In designing a solar chimney particular attention should be given to the depth of the chimney section. As the gap increases, the air flow rate increases but when the gap exceeds a certain value the flow rate starts to decrease slightly. In an experimental

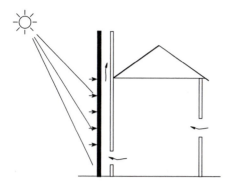

Figure 7.23 Solar chimney.

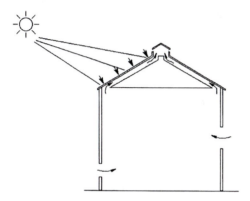

Figure 7.24 Solar roof ventilator.

facility in which two surfaces of the chimney were heated the optimum gap was found to be 200 mm [28].

Solar roof ventilator   In climates where the solar altitude is large, a Trombe wall or a solar chimney may not be very effective collectors of solar energy and, therefore, the ventilation rate that can be achieved with these devices may be limited. In this situation, a sloping roof collector can be more effective in collecting solar energy but because of the sloping surface the height of the collector will be small. A solar roof ventilator is shown in Figure 7.24.

The advantage of a roof collector is that a large surface area is available for collecting the solar energy and hence higher air exit temperatures can be achieved than those for a Trombe wall or a solar chimney. As a result, a roof ventilator could achieve ventilation rates close to a solar chimney or even higher depending on its design and the climate. The estimation of the flow rate is carried out using the method given earlier. Here the height, H, is taken as the vertical distance between the inlet and outlet to the roof and not the length of the roof.

7.8 Combined natural and mechanical (hybrid) ventilation

Natural ventilation alone may not be an adequate strategy for a variety of reasons. There are, however, two basic problems inherent in purely natural ventilation systems:

1 Lack of air flow control.
2 No temperature control.

The first problem can contribute to energy losses from the building in cold weather and overheating in hot weather, whereas the second problem can cause draught in cold weather and discomfort due to higher indoor temperatures in hot weather. Therefore, hybrid ventilation systems (combined natural and mechanical) can help alleviate some of the problems associated with natural ventilation. There are many ways of combining natural and mechanical systems to overcome these difficulties. The basic philosophy is to combine the properties of the mechanical and natural systems to achieve the best performance from both systems and to overcome the problems associated with a natural system. The combination of the two systems varies with demand for outdoor air and variation in outdoor temperature. A good hybrid system should be designed to utilize the natural conditions to the full and incorporate the mechanical component efficiently. This undoubtedly requires an intelligent control strategy for integrating the two systems to minimize energy consumption and maximize indoor air quality and comfort. It is therefore inevitable that a hybrid system should depend upon the building design, external conditions, the desired internal environment, internal thermal load and pollution load. Very often the design of hybrid systems require the use of advanced design tools involving analysis and simulation of the energy flow in the building, the air flow into the building and the air movement inside. In most situations, dynamic building thermal models, zonal air flow models and CFD simulation models are required for the design of hybrid ventilation systems.

There are many ways of combining the driving forces from natural ventilation systems and mechanical systems and the choice depends on the design parameters of the building. Some of these approaches are:

• Mechanical air extraction with natural supply inlets
• Mechanical air supply with natural extraction
• Mechanical cooling or heating combined with natural ventilation.

These strategies can be applied to a whole building or certain zones. In the latter case this is called '*mixed mode ventilation*', which can be used in those parts of a building with particular environmental requirements, different thermal or pollution loads, or used in certain seasons during a year.

7.8.1 Hybrid ventilation in domestic and small buildings

In non-mechanically ventilated buildings such as domestic and small buildings, it is not uncommon to use a mechanical ventilation system to complement natural ventilation in zones where natural ventilation is not capable of providing adequate air flow rates such as kitchens, bathrooms, etc. Such a system can be either an extract only or a balanced system with a heat recovery unit. In an extract system, fans are used to extract air from regions of high odour and moisture concentration, such as kitchens and bathrooms [29], and special ventilation openings are provided for the outside air to replace the stale air inside. In a balanced system, stale air is extracted from kitchens, bathrooms and sometimes bedrooms [30, 31] and passed through a heat recovery device through which outside air is also passed before entering the building at selected areas, such as living rooms and bedrooms. Some 60–70% energy recovery from exhaust air can be achieved using appropriate heat exchangers such as a cross-flow plate heat exchanger or a thermal wheel [32].

Natural or hybrid ventilation is essential in impervious buildings and may be necessary in buildings of moderate airtightness, in which cases adventitious (background) air leakage is insufficient to maintain acceptable standards of indoor air quality. Adventitious or background leakage occurs through cracks, building joints and other openings in the building structure and detailed analysis of their distribution around the structure is seldom possible in practice. Instead, an effective or equivalent leakage area for the whole or a zone of the structure can be determined using the pressurization/ depressurization techniques described in Chapter 3. In some Nordic countries, these tests are mandatory for new buildings and the Swedish Building Regulations [33] stipulate a maximum adventitious leakage rate at a pressure difference $\pm 50\,Pa$ of $0.8\,l\,s^{-1}\,m^{-2}$ of external surfaces for dwellings and $1.6\,l\,s^{-1}\,m^{-2}$ for other spaces. Under normal environmental conditions, the air leakage rate to the buildings under these regulations is of the order of 0.1–0.2 ach, which is well below the minimum required by ventilation standards [34]. That is why the Swedish standard specifies the use of either mechanical, natural or hybrid ventilation in dwellings to achieve a specific air change rate. In the UK however, where houses are generally less airtight, there are no airtightness regulations for dwellings but only for buildings of floor area in excess of $1000\,m^2$. However, the Building Regulations [11] specify the need for purpose-provided ventilators with opening area of $400\,mm^2\,m^{-2}$ of floor area with a minimum area for a single room of $4000\,mm^2$.

Figure 7.25(a) and (b) shows a comparison between the air leakage performances of two classes of buildings representing a very tight (e.g. Swedish) and a moderately tight (e.g. British) construction under different temperature differences and wind speeds. Whereas the moderately tight building can provide acceptable air leakage (≈ 0.5 ach) at a large temperature difference between the air inside and outside or when the wind speed is in excess of $4\,m\,s^{-1}$, the very tight building cannot, under normal circumstances, provide an acceptable air leakage rate. The latter class of buildings therefore requires the provision of a natural, mechanical or a hybrid ventilation system to produce an air change rate of at least 0.5 ach in addition to the adventitious leakage to provide a margin of safety [30, 34].

A natural or hybrid ventilation system should be designed with three factors being considered:

1   Provision of sufficient air leakage rate at different environmental conditions.
2   Ability to control or throttle back the leakage rate under severe weather conditions, e.g. during high winds.
3   Prevention of back-draught or over-extraction through the extract openings, such as stacks, during high winds.

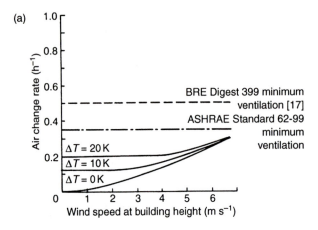

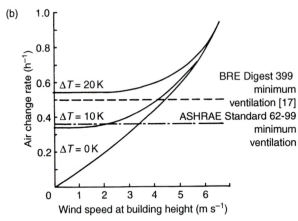

Figure 7.25 Air leakage performance of a tight and a moderately tight building [34].
(a) Tight building (1.5 ach at 50 Pa) and (b) moderately tight building (10 ach at 50 Pa).

In a study involving the control of ventilation rate in an occupied French house [35], humidity and carbon dioxide (CO_2) sensors were used to control a two-speed fan when mechanical ventilation was used and an exhaust duct damper when natural ventilation was used. The aim of the study was to assess the air change rate necessary to achieve an indoor CO_2 concentration limit of 800 p.p.m. and a relative humidity of 60% at 18–20 °C during occupancy, and to evaluate the energy saved in heating the house compared with a constant ventilation rate. This study has shown that CO_2 concentration was directly related to the occupancy behaviour but the correlation with water vapour concentration was not obvious. As a result, CO_2 concentration was regarded as the most suitable parameter for modulating the ventilation rate to the house. A daily average air change rate of 0.5–0.6 ach was sufficient to achieve the CO_2 and H_2O limits specified above, not including the adventitious air flow of about 0.2 ach for the house. It was found that an increase in ventilation rate to 0.8 ach was necessary during meal preparation.

Readers interested in domestic natural ventilation should refer to the excellent report by Axley [36] which covers the theory of domestic ventilation and its applications.

7.8.2 *Hybrid ventilation in commercial and large buildings*

With the popularity of natural ventilation in commercial buildings increasing, designers have quickly realized that natural ventilation alone cannot always provide the desired indoor environment and have, in many situations, resorted to hybrid systems. A survey of hybrid ventilation systems in 22 buildings located in 15 countries, who have participated in the IEA Annex 35 HybVent (Hybrid Ventilation in New and Retrofitted Office Buildings), has been carried and reported in [37]. The survey includes buildings, which vary from low-rise to some of the tallest buildings in Europe. This report represents state-of-the-art assessment of the application of hybrid ventilation systems.

7.9 Ventilation for fire control

Fire statistics in the UK show that most deaths from fire are caused by inhaling smoke and toxic fumes rather than due to the victims being burnt. If smoke can be contained at the start of the fire there will be a greater opportunity to evacuate the building and allow fire-fighting measures to control its spread. In addition to passive measures like fire doors and active measures like water sprinklers, etc., the most effective way of containing smoke is by removing it by a smoke ventilation system. A correctly designed system will remove noxious fumes and smoke, allowing the occupants to escape while providing a clearer view for fire personnel to perform their tasks more effectively. The removal of smoke also takes excessive heat away from the fire zone thus reducing the spread of fire to other zones of the building, as well as lowering the indoor air and radiant temperatures. A smoke extraction system is most needed in public buildings such as shopping malls, theatres, leisure centres and airport terminals.

A smoke ventilation system should be able to contain the fire in the zone where it starts and also remove the smoke at a sufficient rate so that a smoke-free zone will be present over a height of at least 2.5 m above the floor. This can be achieved by confining the smoke to ceiling reservoirs formed by screens, downstands or other features, provided that the gases are removed from the reservoir by ventilation openings and that an equal amount of replacement air is allowed to enter below the main smoke layer. Ceiling reservoirs are used for smoke extraction because of the tendency for the hot gases produced by fire to rise as plumes, entraining surrounding air in the form of an inverted cone of increasing base with height.

The first step in smoke ventilation is to predict the type of fire that is likely to occur in the building and this is primarily dictated by the materials used in the construction and furnishing of the building and the fuel being burnt. The heat and smoke produced will depend on the material used in fuelling the fire. An inflammable liquid quickly produces a vast amount of heat, rubber produces a lot of smoke and modern plastics produce both toxic smoke and heat. The rate of smoke extraction is therefore dependent upon the rate of smoke generated by the fire which is in turn determined by the size and type of fire.

The spread of smoke also depends on the design of building, height of ceiling, distribution of sprinklers and whether smoke curtains are present. The compartmentation of smoke reservoirs to be used for smoke extraction will normally be based on these parameters. Smoke removal can be performed by natural ventilation openings or mechanical roof extracts or a combination of the two. The natural ventilators are normally roof openings fitted with flaps, a weather louver or a cowl and are operated by smoke- or temperature-sensing devices. In sizing the smoke ventilators the natural buoyancy of hot smoke is used as the driving force. This requires the estimation of an average temperature of the hot gases produced by the fire, allowing for the mixing that occurs between the smoke rising from the fire and the colder room air entrained. Once the temperature of a smoke layer reaches about 600 °C, the downward radiation from the layer is usually sufficient to cause ignition of the remaining combustible materials in the room. Where there is sufficient fuel and oxygen in the room the smoke layer will rapidly rise to flame temperature which is about 1000 °C.

Careful positioning of the extract openings will be necessary to avoid down-draughts from surrounding buildings as this can reduce the smoke extraction rate and in extreme situations result in blowback. A mechanical smoke extractor is usually a roof extract fan fitted with an electric motor capable of withstanding very high temperatures over a running period of a few hours. The choice of a natural, mechanical or combined smoke extraction system is dependent on the type of building, its uses and the smoke extraction rate required for it.

An effective smoke ventilation system is a combination of ceiling reservoirs for the containment of smoke, a means of extraction of the hot gases and smoke, and a provision of fresh air inflow at a level below the smoke plumes. The smoke reservoirs are formed with screens, upstands or other suitable building features. These must be of sufficient depth to prevent lateral spread of the smoke and to contain it within the reservoir for extraction outside the building. However, they must be sufficiently high to ensure safe movement for the people beneath them, typically between 2.5 m and 3 m. Typical smoke reservoirs are shown in Figure 7.26. The screens used in forming these reservoirs need not be permanent structural features but can be devices actuated by automatic smoke detectors such as those shown in Figure 7.26(a). According to [38], to prevent excessive heat transfer to the walls and ceilings, which could cause a downward mixing of smoke as a result of cooling, the size of each ceiling reservoir should be limited to 1000 m^2 for natural extraction and 1300 m^2 for mechanical extraction with a length not exceeding 60 m to allow for easy escape. The smoke extract vents must be located in the ceiling reservoir, the number of which is dependent on the smoke extraction rate and the depth of reservoir as discussed later.

7.9.1   Fire growth

A fire is a time dependent phenomenon, which grows to a peak often very rapidly and then gradually extinguishes. A number of time-dependent fire growth models have been proposed, which express the heat release from the fire as a function of time [39]. However, since the primary objective of predicting the heat release from fire is the estimation of smoke production and extraction rates, there is a common approach of using a steady-state 'design' fire for this purpose. This is the largest size the fire is likely to reach during its development (including the influence of active fire prevention or fire extinguishing media which may be in use). In other words, any system that can deal with the steady-state design fire will also be able to deal with the fire as it grows. Table 7.4 summarizes several of the more commonly adopted steady-state design fire sizes in use [39]. The heat release, Q_f, is the total heat generated by combustion in

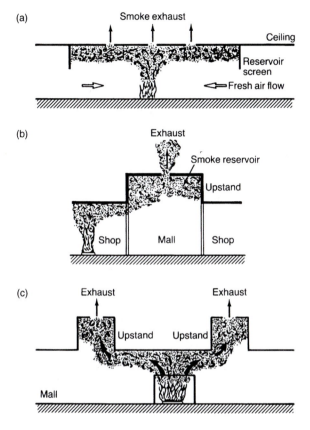

Figure 7.26 Types of smoke reservoirs. (a) Screens; (b) single upstand and (c) two upstands.

kW and the convective heat flux, Q_c, is the net heat remaining in the gases as some heat will be radiated to the surrounding surfaces.

7.9.2 Smoke plume

The mass flow rate in the smoke plume is determined by the rate of air entrainment by the rising plume which in turn is dependent mainly on the height of plume to the base of the smoke layer, see Figure 7.27. This is given approximately by the following expression [39]:

$$M = CPy^{1.5} \tag{7.35}$$

where M = mass flow rate in the plume entering the smoke layer $(kg\,s^{-1})$; P = perimeter of fire (m); y = height of smoke layer from the base of fire (m); $= H - d$, where H is the ceiling height and d is the depth of the smoke layer; C = entrainment coefficient that is dependent on room ceiling height.

The value of C in equation (7.35) is on average 0.21 for a low ceiling (e.g. a large open plan office) and a high ceiling (e.g. an atrium) providing that $y < 3\sqrt{A_f}$, where A_f is the fire area. For a small room or a shop C has a value of 0.34. The equation is

Table 7.4 Steady-state design fires [39]

Occupancy type	Fire area, A_f (m^2)	Fire Perimeter, P (m)	Total heat release, Q_f (kW m^{-2})	Convective heat flux, Q_c (kW m^{-2})
Retail areas				
Standard response sprinklers	10	12	625	500
Quick response sprinklers	5	9	625	500
No sprinklers	Entire room	Width of opening	1200	—
Open-plan offices				
Standard response sprinklers	16	14	255	170[a] or 60[b]
No sprinklers: fuel-bed controlled	47	24	255	170[a] or 130[b]
No sprinklers: full involvement of compartment	Entire room	Width of opening	255	—
Hotel bedrooms				
Quick response sprinklers	2	6	250	200[a] or 150[b]
No sprinklers	Entire room (20 m^2)	Width of opening	100	50[b]
Car parks				
A burning car	10	12	400	300[a]

Notes
a Close to the fire plume.
b At the window.

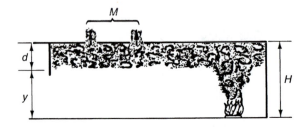

Figure 7.27 Smoke plume height and layer depth.

valid for fires ranging from 8 kW to 18 MW or a heat flux in the range 200 kW m^{-2} to 1.8 MW m^{-2}. Taking a minimum allowable height for the smoke layer of 2.5 m, the following flow rates entering the smoke reservoir can be expected:

A cellular office, shop, etc. with standard response sprinklers, $M = 16.13$ kg s^{-1}

A large open plan office, an atrium, etc. with standard response sprinklers, $M = 11.62$ kg s^{-1}.

Flow rates for other occupancy given in Table 7.4 can also be obtained using equation (7.35).

7.9.3 Smoke extraction

For a natural smoke extraction system the flow is buoyancy driven and therefore the depth of the smoke layer provides the whole of the stack effect and a compromise must be made between the depth of the smoke reservoir and the total area of ventilation openings in the reservoir for extracting the smoke generated by the fire. A relationship between the mass flow rate, the ventilation opening area, the air inlet area and the depth of the smoke layer for a mall fire is given as [39]:

$$M = C_d A_v \rho_a \sqrt{\left(\frac{2gd\Delta T_c T_a}{T_c[T_c + (A_v/A_i)^2 T_a]}\right)} \qquad (7.36)$$

where C_d = discharge coefficient of ventilation opening (usually between 0.5 and 0.7); A_v = total area of ventilation openings in the smoke reservoir (m^2); A_i = total area of all air inlets (m^2); d = depth of smoke layer below the ventilation openings (m); ρ_a = density of ambient air $(kg\,m^{-3})$; T_a = ambient air temperature (K); T_c = temperature of smoke layer in the reservoir (K); $\Delta T_c = T_c - T_a$ (K); g = acceleration due to gravity $(m\,s^{-2})$.

The smoke layer temperature, T_c, for a smoke reservoir can be determined from:

$$\Delta T_c = Q_c/(C_p M) \qquad (7.37)$$

where Q_c is the convective heat flux (kW); M, the mass flow rate of smoke $(kg\,s^{-1})$; C_p, the specific heat capacity of the gases ≈ 1.0 $(kJ\,kg^{-1}\,K^{-1})$.

If a mechanical smoke extract system with exhaust fans and associated ductwork is used to extract the smoke entering the smoke reservoirs then the previous equations for mass flow rate and gases temperature (equations (7.35) and (7.37)) can be used with equation (7.38) below to estimate the fan flow rate:

$$V = MT_c/(\rho_a T_a) \qquad (7.38)$$

where V is the volumetric flow rate of smoke extraction $(m^3\,s^{-1})$.

Here we have presented the basic principles that can be used to design smoke reservoirs and smoke extract openings by natural or mechanical means. In practice, attention should be given to the variation in construction, geometry and usage of different zones in a building. For a more detailed assessment of these parameters and their effect on the design of smoke ventilation systems readers should refer to [39].

References

1. B.S. 6399-2 (1997) Loading for buildings – Part 2: Code of practice for wind loads, British Standards Institution, London.
2. B.S. 5925 (1991) Code of practice for ventilation principles and design for natural ventilation, British Standards Institution, London.
3. CIBSE Guide A (2000) Ch. 4: Air infiltration and natural ventilation, Chartered Institution of Building Services Engineers, London.

4. *ASHRAE Handbook of Fundamental* (2001) Ch. 27: Climatic design information, American Society of Heating, Refrigeration and Air-Conditioning Engineers, Atlanta, GA.
5. ESDU Data Item no. 71016 (1971) Fluid forces, pressures and moments on rectangular blocks, Engineering Sciences Data Unit International, London.
6. Liddament, M. W. (1986) Air infiltration calculation techniques – an application guide, Air Infiltration and Ventilation Centre, International Network for Information on Ventilation, Brussels, Belgium (www.aivc.org).
7. Bala'zs, K. (1989) A wind pressure database from Hungary for ventilation and infiltration calculations, *Air Infiltration Rev.*, 10(4), 1–4.
8. Knoll, B., Phaff, J. C. and de Gids, W. F. (1996) Pressure coefficient simulation program, *Air Infiltration Rev.*, 17(3), 1–5.
9. Li, Y. and Delsante, A. (2001) Natural ventilation induced by combined wind and thermal forces, *Build. Env.*, 36, 59–71.
10. Li, Y. *et al.* (2001) Some examples of solution multiplicity in natural ventilation, *Build. Env.*, 36, 851–8.
11. The Building Regulations, Approved Document F (1995) *Ventilation*, HMSO, London.
12. Heiselberg, P., Svidt, K. and Nielsen, P. V. (2001) Characteristics of air flow from open windows, *Build. Env.*, 36, 859–69.
13. Orme, M., Liddament, M. W. and Wilson, A. (1994) An analysis and data summary of the AIVC's numerical database, TN 44, Air Infiltration and Ventilation Centre, International Network for Information on Ventilation, Brussels, Belgium (www.aivc.org).
14. Limb, M. J. (1994) Current Ventilation and Air Conditioning Systems and Strategies. TN 42, Air Infiltration and Ventilation Centre, International Network for Information on Ventilation, Brussels, Belgium (www.aivc.org).
15. Awbi, H. B. (1996) Air movement in naturally ventilated buildings, *Renewable Energy*, 8, 241–7.
16. de Gids, W. and Phaff, H. (1982) Ventilation rates and energy consumption due to open windows – a brief overview of research in the Netherlands. *Air Infiltration Rev.* 4(1), 4–5.
17. BRE Digest 399 (1994). Natural ventilation in non-domestic buildings, Building Research Establishment, Garston, UK.
18. Chand, I. *et al.* (1992) Studies on the effect of mean wind speed profile on rate of air flow through cross-ventilated enclosures. *Arch. Sci. Rev.*, 35, 83–8.
19. Awbi, H. B. (1994) Design considerations for naturally ventilated buildings, *Renewable Energy*, 5, 1081–90.
20. Gaze, A. L. (1986) Passive ventilation: a method of controllable natural ventilation of housing. *TRADA Res. Rep.* 12/86, Timber Research and Development Association, UK.
21. Jacobsen, T. (1988) Thermal climate and air exchange rate in a glasscovered atrium without mechanical ventilation. *Proc. Healthy Buildings, 88, Stockholm*, Vol. 2, pp. 567–76.
22. Awbi, H. B. and Elmualim, A. A. (2002) Full scale model windcatcher performance evaluation using a wind tunnel, *Proc. 7th World Renewable Energy Congress*, Cologne, Germany, 29 June–5 July, 2002, Pergamon, Oxford, ISBN 0-08-044079-7.
23. Elmualim, A. A. *et al.* (2001) Evaluation of the performance of a windcatcher system using wind tunnel testing. *Proc. 22nd AIVC Conference*, pp. 29.1–29.12, Bath, UK.
24. Awbi, H. B. and Gan, G. (1992). Simulation of solar-induced ventilation. *Proc. 2nd World Renewable Energy Congress*, 4, 2016–30.
25. Rodrigues, A. M., da Piedade, A. C. and Awbi, H. B. (2000) The use of solar air collectors for room ventilation: a study using two numerical approaches, Air Distribution in Rooms: Ventilation for Health and Sustainable Environment, *Proc. ROOMVENT 2000*, H. B. Awbi (ed.), Vol. 2, pp. 1013–18, Elsevier, Oxford.
26. Trombe, F. *et al.* (1977) Concrete walls to collect and hold heat, *Solar Age*, 2, 13–35.
27. Akbarzadeh, A. *et al.* (1982) Thermocirculation characteristics of Trombe wall passive test cell, *Solar Energy*, 28, 461–8.

28. Bouchair, A. *et al.* (1988) Moving air, using stored solar energy. *Proc. 13th National Passive Solar Conference*, Cambridge, Massachusetts.

29. Cornish, J. P., Sanders, C. H. and Garratt, J. (1985) The effectiveness of remedies to surface condensation and mould, *Proc. Condensation and Energy Problems: A Research for an International Strategy*, Leuven, September, pp. 23–5.

30. Awbi, H. B. and Allwinkle, S. J. (1986) Domestic ventilation ventilation with heat recovery to improve indoor air quality. *Energy Build.* 9, 305–12.

31. Awbi, H. B. and Allwinkle, S. J. (1987) Control of condensation in dwellings with mechanical ventilation. *Proc. 3rd mt. Congr. on Building Energy Management ICBEM '87, Lausanne, Switzerland*, Vol. III, pp. 43–51.

32. Liddament, M. W. (1996) A guide to energy efficient ventilation, *Air Infiltration and Ventilation Centre*, International Network for Information on Ventilation, Brussels, Belgium (www.aivc.org).

33. Building Regulations of Swedish Board of Housing (2001) published by *Swedish Board of Housing, Building and Planning*, Karlskrona, Sweden (www.boverket.se).

34. Liddament, M. W. (1987) A comparison of ventilation practices. *Proc. CIBSE Tech. Conf.*, pp. 41–53.

35. Barthez, M. and Soupault, O. (1983) Control of ventilation rate in building using H_2O and CO_2 content. *Proc. Int. Seminar on Energy Saving in Buildings, The Hague, 1983*, Reidel, Dordrecht, pp. 490–4.

36. Axley, J. W. (1998) Residential passive ventilation systems: evaluation and design, AIVC TN 54, Air Infiltration and Ventilation Centre, International Network for Information on Ventilation, Brussels, Belgium (www.aivc.org).

37. Delsante, A. and Vik, T. A. (2001) Hybrid ventilation: state-of-the-art review, IEA Annex 35 HybVent.

38. Morgan, H. P. (1979) Smoke control models in enclosed shopping complexes of one or more storeys: a design summary, Building Research Establishment Report, UK.

39. Morgan, H. P. *et al.* (1999) Design methodologies for smoke and heat exhaust ventilation, Building Research Establishment Report, UK.

8 Computational fluid dynamics in room air flow analysis

8.1 Introduction

In Chapter 6 the factors that influence the air movement in a room were discussed and the methods used to predict the air flow patterns and the air velocity and temperature distributions were described. Design procedures based on data obtained from physical model tests were also presented for various methods of supplying air to the space being ventilated. Although these design procedures are used, in practice they are limited to only a few methods of supplying the air to a generally empty room. For different air supply situations or where obstructions are present, mock-up tests are required to assess the air movement pattern.

During the last two decades there has been great interest in developing computational fluid dynamics (CFD) computer programs for predicting the air flow in ventilated rooms [1–3]. The majority of these CFD programs are based upon the solution of Navier–Stokes equations, the energy equation, the mass and concentration equations as well as the transport equations for turbulent velocity and its scale. The numerical solution of these equations in two- and three-dimensions has been applied to flow problems ranging from the diffusion of jets to the prediction of smoke and fire spread in buildings.

In practice, there are two main numerical techniques that are in use for solving the Navier–Stokes equations: the finite volume method (FVM) and the finite element method (FEM). However, most current CFD codes for room air flow analysis use the former. Consequently, this chapter describes this method of solving the flow equations in some detail. Another important issue in using CFD is the type of turbulence model used. As it will be shown later in this chapter, there are a range of turbulence models to choose from with varying complexity and accuracy. Here again, most of the CFD codes in use for room flow analysis tend to use the standard k–ϵ model or its many variants and this model will therefore be considered in more detail than the other models that are available. In addition to presenting the fundamental flow equations and the modelling of turbulence, in this chapter examples are presented showing the range of application of CFD models for solving ventilation problems of practical interest.

8.2 Transport equations

The equations that describe the flow of a fluid, heat and concentration within an enclosure are all based on the conservation of mass, momentum, thermal energy and concentration of species within the enclosure. Because a microscopic analysis of the distribution of these physical quantities within the enclosure will be necessary for detailed analysis, the equations take the form of partial differential equations (PDEs). Each PDE describes the conservation of one dependent variable within the field and this implies that there must be a balance among the various factors that influence this variable. In ventilated enclosures these dependent variables are normally mass, velocity components (i.e. momentum per unit mass), enthalpy or temperature, particle or concentration mass fraction and some turbulence variables in the case of turbulent flows. Other equations may also be solved for quantifying additional variables such as the mean age of air or thermal radiation exchange between surfaces.

The PDEs that will be presented in this chapter will be in the form of Cartesian coordinate system (i.e. x, y, z) and based on a unit volume of the field. In deriving these equations the conservation law was applied to a control volume of dimensions dx, dy and dz. Such equations are often known as transport equations. The equations may be changed to other coordinate systems if required. For example, when the cylindrical polar coordinate system x, r and θ is used, then in the following equations y is replaced by $r \sin \theta$ and z by $r \cos \theta$. Interested readers may refer to other texts, e.g. [4], where the conservation equations are given in cylindrical polar coordinates. The transport equations are normally presented in time domain to permit transient calculations, which are necessary when simulating fire or smoke spread. The reader who is interested in developing a deeper understanding of CFD principles and other methods of solving the transport equation not covered in this chapter should refer to dedicated texts on the subject, e.g. [5–7].

8.2.1   Conservation of mass

Taking U, V and W to be the velocity components in the x, y and z directions respectively, ρ the fluid density and t the time, then the rate of increase in the density ρ within the control volume $dx\,dy\,dz$ equals the net rate of influx of mass to the control volume, namely:

$$\frac{\partial \rho}{\partial t} + \frac{\partial}{\partial x}(\rho U) + \frac{\partial}{\partial y}(\rho V) + \frac{\partial}{\partial z}(\rho W) = 0 \qquad (8.1)$$

If a turbulent flow is considered (as is the case in most room flow problems) and the velocities in equation (8.1) are replaced by the sum of a time-mean component and a fluctuating component, i.e.:

$$U = u + u' \quad V = v + v' \quad W = w + w'$$

the mass conservation equation becomes:

$$\frac{\partial \rho}{\partial t} + \frac{\partial}{\partial x}(\rho u) + \frac{\partial}{\partial y}(\rho v) + \frac{\partial}{\partial z}(\rho w) = 0 \qquad (8.2)$$

In arriving at this equation it has been assumed that the fluctuations in u', v' and w' occur over a much shorter time interval than ∂t so that $u \approx U, v \approx V$ and $w \approx W$ during this time interval. The density in equation (8.2) or the transport equations given in the succeeding sections may be represented by an equation of state, i.e. $\rho = f(p, T)$.

8.2.2   Conservation of momentum (Navier–Stokes equations)

By applying the law of conservation of momentum (i.e. the net force on the control volume in any direction equals the efflux of momentum minus the influx of momentum in the same direction) in the x, y and z directions the following equations are obtained (Batchelor [8]):

x direction (U-momentum):

$$
\frac{\partial}{\partial t}(\rho U) + \frac{\partial}{\partial x}(\rho UU) + \frac{\partial}{\partial y}(\rho UV) + \frac{\partial}{\partial z}(\rho UW)
$$

$$
= -\frac{\partial P}{\partial x} + \frac{\partial}{\partial x}\left(\mu \frac{\partial U}{\partial x}\right) + \frac{\partial}{\partial y}\left(\mu \frac{\partial U}{\partial y}\right) + \frac{\partial}{\partial z}\left(\mu \frac{\partial U}{\partial z}\right)
$$

$$
+ \frac{1}{3}\frac{\partial}{\partial x}\left[\mu\left(\frac{\partial U}{\partial x} + \frac{\partial V}{\partial y} + \frac{\partial W}{\partial z}\right)\right] + \rho g_x \tag{8.3}
$$

y direction (V momentum):

$$
\frac{\partial}{\partial t}(\rho V) + \frac{\partial}{\partial x}(\rho UV) + \frac{\partial}{\partial y}(\rho VV) + \frac{\partial}{\partial z}(\rho VW)
$$

$$
= -\frac{\partial P}{\partial y} + \frac{\partial}{\partial x}\left(\mu \frac{\partial V}{\partial x}\right) + \frac{\partial}{\partial y}\left(\mu \frac{\partial V}{\partial y}\right) + \frac{\partial}{\partial z}\left(\mu \frac{\partial W}{\partial z}\right)
$$

$$
+ \frac{1}{3}\frac{\partial}{\partial y}\left[\mu\left(\frac{\partial U}{\partial x} + \frac{\partial V}{\partial y} + \frac{\partial W}{\partial z}\right)\right] + \rho g_y \tag{8.4}
$$

z direction (W momentum):

$$
\frac{\partial}{\partial t}(\rho W) + \frac{\partial}{\partial x}(\rho UW) + \frac{\partial}{\partial y}(\rho VW) + \frac{\partial}{\partial z}(\rho WW)
$$

$$
= -\frac{\partial P}{\partial z} + \frac{\partial}{\partial x}\left(\mu \frac{\partial W}{\partial x}\right) + \frac{\partial}{\partial y}\left(\mu \frac{\partial W}{\partial y}\right) + \frac{\partial}{\partial z}\left(\mu \frac{\partial W}{\partial z}\right)
$$

$$
+ \frac{1}{3}\frac{\partial}{\partial z}\left[\mu\left(\frac{\partial U}{\partial x} + \frac{\partial V}{\partial y} + \frac{\partial W}{\partial z}\right)\right] + \rho g_z \tag{8.5}
$$

In equations (8.3)–(8.5), P is the static pressure, μ is the dynamic viscosity of the fluid and ρg_x, ρg_y and ρg_z are the body forces in the x, y and z directions, respectively. Replacing U, V and W by the time-mean and fluctuating components and assuming a similar expression for the pressure P (i.e. $P = p + p'$) then equations (8.3)–(8.5) become:

$$
\frac{\partial}{\partial t}(\rho u) + \frac{\partial}{\partial x}(\rho uu) + \frac{\partial}{\partial y}(\rho uv) + \frac{\partial}{\partial z}(\rho uw)
$$

$$
= -\frac{\partial p}{\partial x} + \frac{\partial}{\partial x}\left(\mu \frac{\partial u}{\partial x}\right) + \frac{\partial}{\partial y}\left(\mu \frac{\partial u}{\partial y}\right) + \frac{\partial}{\partial z}\left(\mu \frac{\partial u}{\partial z}\right) + \frac{1}{3}\frac{\partial}{\partial x}\left[\mu\left(\frac{\partial u}{\partial x} + \frac{\partial v}{\partial y} + \frac{\partial w}{\partial z}\right)\right]
$$

$$
+ \frac{\partial}{\partial x}(-\rho\overline{u'u'}) + \frac{\partial}{\partial y}(-\rho\overline{u'v'}) + \frac{\partial}{\partial z}(-\rho\overline{u'w'}) + \rho g_x \tag{8.6}
$$

$$\frac{\partial}{\partial t}(\rho v) + \frac{\partial}{\partial x}(\rho u v) + \frac{\partial}{\partial y}(\rho v v) + \frac{\partial}{\partial z}(\rho v w)$$

$$= -\frac{\partial p}{\partial y} + \frac{\partial}{\partial x}\left(\mu \frac{\partial v}{\partial x}\right) + \frac{\partial}{\partial y}\left(\mu \frac{\partial v}{\partial y}\right) + \frac{\partial}{\partial z}\left(\mu \frac{\partial v}{\partial z}\right) + \frac{1}{3}\frac{\partial}{\partial y}\left[\mu\left(\frac{\partial u}{\partial x} + \frac{\partial v}{\partial y} + \frac{\partial w}{\partial z}\right)\right]$$

$$+ \frac{\partial}{\partial x}(-\rho\overline{u'v'}) + \frac{\partial}{\partial y}(-\rho\overline{v'v'}) + \frac{\partial}{\partial z}(-\rho\overline{v'w'}) + \rho g_y \tag{8.7}$$

$$\frac{\partial}{\partial t}(\rho w) + \frac{\partial}{\partial x}(\rho u w) + \frac{\partial}{\partial y}(\rho v w) + \frac{\partial}{\partial z}(\rho w w)$$

$$= -\frac{\partial p}{\partial z} + \frac{\partial}{\partial x}\left(\mu \frac{\partial w}{\partial x}\right) + \frac{\partial}{\partial y}\left(\mu \frac{\partial w}{\partial y}\right) + \frac{\partial}{\partial z}\left(\mu \frac{\partial w}{\partial z}\right)$$

$$+ \frac{1}{3}\frac{\partial}{\partial z}\left[\mu\left(\frac{\partial u}{\partial x} + \frac{\partial v}{\partial y} + \frac{\partial w}{\partial z}\right)\right] + \frac{\partial}{\partial x}(-\rho\overline{u'w'})$$

$$+ \frac{\partial}{\partial y}(-\rho\overline{v'w'}) + \frac{\partial}{\partial z}(-\rho\overline{w'w'}) + \rho g_z \tag{8.8}$$

In deriving equations (8.6)–(8.8) the assumptions made in deriving the mass conservation equation (8.2) were also applied and these were extended to include the pressure fluctuation with time. The terms $-\rho\overline{u'u'}, -\rho\overline{v'v'}, -\rho\overline{w'w'}, -\rho\overline{u'v'}, -\rho\overline{u'w'}$, and $-\rho\overline{v'w'}$, are the turbulent (Reynolds) stresses, τ_t. As will be discussed later, a suitable turbulence model is required to represent these terms (closure equations) before the momentum equations can be solved. The terms can be represented by expressions of the time-mean variables already in the equations or by means of other variables not given in equations (8.6)–(8.8). In the latter case, additional PDEs of the new variables must also be solved as well as the other transport equations given earlier.

8.2.3   *Conservation of thermal energy*

The conservation of thermal energy in the control volume dx dy dz stipulates that the net increase in internal energy in the control volume equals the net flow of energy by convection plus the net inflow by thermal and mass diffusion, i.e. the energy equation per unit volume is:

$$\frac{\partial}{\partial t}(\rho \tilde{T}) + \frac{\partial}{\partial x}(\rho U \tilde{T}) + \frac{\partial}{\partial y}(\rho V \tilde{T}) + \frac{\partial}{\partial z}(\rho W \tilde{T})$$

$$= \frac{\partial}{\partial x}\left(\Gamma \frac{\partial \tilde{T}}{\partial x}\right) + \frac{\partial}{\partial y}\left(\Gamma \frac{\partial \tilde{T}}{\partial y}\right) + \frac{\partial}{\partial z}\left(\Gamma \frac{\partial \tilde{T}}{\partial z}\right) \tag{8.9}$$

In this equation Γ is the diffusion coefficient (diffusivity), which is given by:

$$\Gamma = \mu/\sigma$$

where $\sigma = \mu C_p/\lambda$ is the Prandtl or Schmidt number for the fluid.

Again writing $\tilde{T} = T + T'$, where T is the time-mean temperature and T' is the fluctuation from the mean, and making the same assumptions used for deriving the momentum equations, equation (8.9) becomes:

$$\frac{\partial}{\partial t}(\rho T) + \frac{\partial}{\partial x}(\rho u T) + \frac{\partial}{\partial y}(\rho v T) + \frac{\partial}{\partial z}(\rho w T)$$

$$= \frac{\partial}{\partial x}\left(\Gamma \frac{\partial T}{\partial x}\right) + \frac{\partial}{\partial y}\left(\Gamma \frac{\partial T}{\partial y}\right) + \frac{\partial}{\partial z}\left(\Gamma \frac{\partial T}{\partial z}\right)$$

$$+ \frac{\partial}{\partial x}(-\rho \overline{u'T'}) + \frac{\partial}{\partial y}(-\rho \overline{v'T'}) + \frac{\partial}{\partial z}(-\rho \overline{w'T'}) + S_T \qquad (8.10)$$

The terms $-\rho \overline{u'T'}, -\rho \overline{v'T'}$ and $-\rho \overline{w'T'}$ are the turbulent heat fluxes, J_T, and S_T is a source term allowing for the rate of thermal energy production.

8.2.4   The concentration of species equation

The equation for concentration of species is exactly the same as (8.10) in which case S_C represents the rate of production of concentration. The concentration equation is therefore:

$$\frac{\partial}{\partial t}(\rho c) + \frac{\partial}{\partial x}(\rho u c) + \frac{\partial}{\partial y}(\rho v c) + \frac{\partial}{\partial z}(\rho w c)$$

$$= \frac{\partial}{\partial x}\left(\Gamma \frac{\partial c}{\partial x}\right) + \frac{\partial}{\partial y}\left(\Gamma \frac{\partial c}{\partial y}\right) + \frac{\partial}{\partial z}\left(\Gamma \frac{\partial c}{\partial z}\right)$$

$$+ \frac{\partial}{\partial x}(-\rho \overline{u'c'}) + \frac{\partial}{\partial y}(-\rho \overline{v'c'}) + \frac{\partial}{\partial z}(-\rho \overline{w'c'}) + S_C \qquad (8.11)$$

In this equation c is the time-mean concentration and c' is the deviation from the mean. The terms $-\rho \overline{u'c'}, -\rho \overline{v'T'}$ and $-\rho \overline{w'c'}$ are the turbulent diffusion fluxes, J_C.

8.2.5   The mean age of air equation

In Section 2.4 the concepts of age of air and mean age of air are introduced. The local mean age of air is regarded as the average time for air to travel from a supply outlet to any point in a ventilated room. CFD is a useful technique for mean age of air calculations and three methods have been used for its calculation, namely a transient method in which a time-dependent concentration equation is solved, a steady-state method in which a steady-state equation of the mean air age is solved and a particle marker method in which particles are tracked explicitly, e.g. see Li *et al.* [9]. It has been shown that transient and steady-state methods produce identical results and both are reliable for predicting the local mean age. However, the steady-state method requires less computing time and effort than the transient method and this is presented here.

For calculating the local mean age of air an additional transport equation must be solved in addition to the air flow transport equations. The transport equation for local mean age of air, $\bar{\tau}$, in steady-state is given by [10]:

$$u\frac{\partial\bar{\tau}}{\partial x} + v\frac{\partial\bar{\tau}}{\partial y} + w\frac{\partial\bar{\tau}}{\partial z} = \frac{\partial}{\partial x}\left(\Gamma\frac{\partial\bar{\tau}}{\partial x}\right) + \frac{\partial}{\partial y}\left(\Gamma\frac{\partial\bar{\tau}}{\partial y}\right) + \frac{\partial}{\partial z}\left(\Gamma\frac{\partial\bar{\tau}}{\partial z}\right)$$
$$+ \frac{\partial}{\partial x}(-\rho\overline{u'\tau}) + \frac{\partial}{\partial y}(-\rho\overline{v'\tau}) + \frac{\partial}{\partial z}(-\rho\overline{w'\tau}) + 1 \qquad (8.12)$$

In equation (8.12), $\bar{\tau}$ is the local mean age and τ is the deviation from the mean due to turbulence. The terms $-\rho\overline{u'\tau}$, $-\rho\overline{v'\tau}$ and $-\rho\overline{w'\tau}$ are the turbulence diffusion fluxes, J_τ. The age of air is considered a passive quantity and hence does not affect the air flow pattern. The local mean age is then obtained from the values of fluid velocity and viscosity in the control volume.

8.3 Turbulence models

A turbulence model in CFD calculations is a computational procedure to close the system of mean flow equations (8.2) and (8.6)–(8.12) to enable numerical calculation of these equations. For most practical purposes it is not necessary to resolve the details of the turbulent fluctuations and only the effect on the mean flow is all what is normally required. The task of a turbulence model therefore is to express the Reynolds stresses in equations (8.6)–(8.8), the turbulent heat fluxes in equation (8.10) and the turbulent diffusion fluxes in equations (8.11) and (8.12) by a set of auxiliary equations (differential and/or algebraic) containing time-mean quantities of the flow. These auxiliary equations permit the closure of the transport equations and the calculation of the time-mean dependent variables in the transport equations and as a result they are usually called 'closure equations'.

8.3.1 Turbulence viscosity and diffusion coefficients

The majority of the classical turbulence models used in solving practical fluid flow problems are based on the eddy (turbulent) viscosity and the eddy (turbulent) diffusivity concepts. These concepts are based on the presumption that there exists an analogy between the effects of viscous stresses and Reynolds stresses on the mean flow. Both stresses appear on the right side of the momentum equation. Furthermore, Newton's law of viscosity states that the viscous stresses, τ_{ij}, are proportional to the rate of strain (deformation) of a fluid element, i.e. for incompressible flow:

$$\tau_{ij} = \mu\left(\frac{\partial u_i}{\partial x_j} + \frac{\partial u_j}{\partial x_i}\right) \qquad (8.13)$$

where μ is the viscosity, u_i and u_j are the velocity components (u, v, w) in x_i and x_j directions respectively, in which case $i \neq j$. For example the shear stress in the x–y plane is:

$$\tau_{xy} = \mu\left(\frac{\partial u}{\partial y} + \frac{\partial v}{\partial x}\right)$$

It has been established that turbulent stresses increase as the mean rate of strain increases. In 1877, J. Boussinesq proposed that the Reynolds stresses could be linked to the mean rate of strain. This led to the concept of turbulent viscosity, which was further extended by O. Reynolds for turbulent diffusivity. If an isotropic turbulence is assumed, i.e. at any point in the flow the mean-square values of the three fluctuating velocity components are equal, $\overline{u'^2} = \overline{v'^2} = \overline{w'^2}$, the Reynolds stresses and the turbulent fluxes can be represented by the following equations:

$$-\rho\overline{u'u'} = 2\mu_t\frac{\partial u}{\partial x} - \frac{2}{3}\rho k$$

$$-\rho\overline{v'v'} = 2\mu_t\frac{\partial v}{\partial y} - \frac{2}{3}\rho k$$

$$-\rho\overline{w'w'} = 2\mu_t\frac{\partial w}{\partial z} - \frac{2}{3}\rho k$$

$$-\rho\overline{u'v'} = \mu_t\left(\frac{\partial u}{\partial y} + \frac{\partial v}{\partial x}\right) \tag{8.14}$$

$$-\rho\overline{v'w'} = \mu_t\left(\frac{\partial v}{\partial z} + \frac{\partial w}{\partial y}\right)$$

$$-\rho\overline{u'w'} = \mu_t\left(\frac{\partial u}{\partial z} + \frac{\partial w}{\partial x}\right)$$

$$-\rho\overline{u'\phi'} = \Gamma_t\partial\phi/\partial x$$

$$-\rho\overline{v'\phi'} = \Gamma_t\partial\phi/\partial y \tag{8.15}$$

$$-\rho\overline{w'\phi'} = \Gamma_t\partial\phi/\partial z$$

In equations (8.14) μ_t is the turbulent or eddy viscosity and k is the kinetic energy of the turbulent velocities, i.e.:

$$k = \tfrac{1}{2}[\overline{(u')^2}] + \overline{(v')^2} + \overline{(w')^2}] \tag{8.16}$$

and in equations (8.15) ϕ represents a scalar quantity such as temperature, concentration, etc., and Γ_t, is the turbulent diffusion coefficient or turbulent diffusivity that is given as:

$$\Gamma_t = \mu_t/\sigma_t$$

where σ_t is the turbulent Prandtl or Schmidt number.

In the transport equations for turbulent flows the turbulent stress or turbulent diffusion terms are added to the laminar stress or laminar diffusion terms by using the concept of effective viscosity or effective diffusion coefficients. The effective viscosity coefficient, μ_e, is:

$$\mu_e = \mu + \mu_t \tag{8.17}$$

The effective diffusion coefficient for a scalar, Γ_e, is:

$$\Gamma_e = \Gamma + \Gamma_t = \frac{\mu}{\sigma} + \frac{\mu_t}{\sigma_t} \tag{8.18}$$

The effective diffusion coefficient can also be written as:

$$\Gamma_e = \frac{\mu_e}{\sigma_e} = \frac{\mu + \mu_t}{\sigma_e} \tag{8.19}$$

where σ_e is the effective Prandtl or Schmidt number that is given by:

$$\sigma_e = (\mu + \mu_t)^{-1} \left(\frac{\mu}{\sigma} + \frac{\mu_t}{\sigma_t} \right)^{-1} \tag{8.20}$$

Using the effective viscosity in the transport equations for u, v and w, equations (8.6)–(8.8) can be written, with some simplifying, as:

$$\frac{\partial}{\partial t}(\rho u) + \frac{\partial}{\partial x}(\rho u u) + \frac{\partial}{\partial y}(\rho u v) + \frac{\partial}{\partial z}(\rho u w)$$
$$= -\frac{\partial p}{\partial x} + \frac{\partial}{\partial x}\left(\mu_e \frac{\partial u}{\partial x} \right) + \frac{\partial}{\partial y}\left(\mu_e \frac{\partial u}{\partial y} \right) + \frac{\partial}{\partial z}\left(\mu_e \frac{\partial u}{\partial z} \right)$$
$$+ \frac{\partial}{\partial x}\left(\mu_e \frac{\partial u}{\partial x} \right) + \frac{\partial}{\partial y}\left(\mu_e \frac{\partial v}{\partial x} \right) + \frac{\partial}{\partial z}\left(\mu_e \frac{\partial w}{\partial x} \right) \tag{8.21}$$

$$\frac{\partial}{\partial t}(\rho v) + \frac{\partial}{\partial x}(\rho u v) + \frac{\partial}{\partial y}(\rho v v) + \frac{\partial}{\partial z}(\rho v w)$$
$$= -\frac{\partial p}{\partial y} + \frac{\partial}{\partial x}\left(\mu_e \frac{\partial v}{\partial x} \right) + \frac{\partial}{\partial y}\left(\mu_e \frac{\partial u}{\partial y} \right) + \frac{\partial}{\partial z}\left(\mu_e \frac{\partial w}{\partial z} \right)$$
$$+ \frac{\partial}{\partial x}\left(\mu_e \frac{\partial u}{\partial y} \right) + \frac{\partial}{\partial y}\left(\mu_e \frac{\partial v}{\partial y} \right) + \frac{\partial}{\partial z}\left(\mu_e \frac{\partial w}{\partial y} \right) - g(\rho - \rho_0) \tag{8.22}$$

$$\frac{\partial}{\partial t}(\rho w) + \frac{\partial}{\partial x}(\rho u w) + \frac{\partial}{\partial y}(\rho v w) + \frac{\partial}{\partial z}(\rho w w)$$
$$= -\frac{\partial p}{\partial z} + \frac{\partial}{\partial x}\left(\mu_e \frac{\partial w}{\partial x} \right) + \frac{\partial}{\partial y}\left(\mu_e \frac{\partial w}{\partial y} \right) + \frac{\partial}{\partial z}\left(\mu_e \frac{\partial w}{\partial z} \right)$$
$$+ \frac{\partial}{\partial x}\left(\mu_e \frac{\partial u}{\partial z} \right) + \frac{\partial}{\partial y}\left(\mu_e \frac{\partial v}{\partial z} \right) \frac{\partial}{\partial z}\left(\mu_e \frac{\partial w}{\partial z} \right) \tag{8.23}$$

In deriving equations (8.21)–(8.23) the terms:

$$\frac{1}{3}\frac{\partial}{\partial x}[\mu(\ )], \qquad \frac{1}{3}\frac{\partial}{\partial y}[\mu(\ )], \qquad \frac{1}{3}\frac{\partial}{\partial z}[\mu(\ )]$$

in equations (8.6)–(8.8) have been neglected because, for a turbulent flow, $\mu \ll \mu_t$. The terms $\frac{2}{3}\rho k$, etc. and the body forces in the x and z directions (i.e. ρg_x and ρg_z) were also neglected. The body force in the y direction has been written as a buoyancy force where ρ_o is the density at a reference temperature in the flow, T_o. Similarly, the transport equations for temperature (8.10), concentration (8.11) and mean age of air (8.12) can be written as:

$$\frac{\partial}{\partial t}(\rho T) + \frac{\partial}{\partial x}(\rho u T) + \frac{\partial}{\partial y}(\rho v T) + \frac{\partial}{\partial z}(\rho w T)$$

$$= \frac{\partial}{\partial x}\left(\Gamma_e \frac{\partial T}{\partial x}\right) + \frac{\partial}{\partial y}\left(\Gamma_e \frac{\partial T}{\partial y}\right) + \frac{\partial}{\partial z}\left(\Gamma_e \frac{\partial T}{\partial z}\right) + S_T \qquad (8.24)$$

$$\frac{\partial}{\partial t}(\rho c) + \frac{\partial}{\partial x}(\rho u c) + \frac{\partial}{\partial y}(\rho v c) + \frac{\partial}{\partial z}(\rho w c)$$

$$= \frac{\partial}{\partial x}\left(\Gamma_e \frac{\partial c}{\partial x}\right) + \frac{\partial}{\partial y}\left(\Gamma_e \frac{\partial c}{\partial y}\right) + \frac{\partial}{\partial z}\left(\Gamma_e \frac{\partial c}{\partial z}\right) + S_C \qquad (8.25)$$

$$u\frac{\partial \bar{\tau}}{\partial x} + v\frac{\partial \bar{\tau}}{\partial y} + w\frac{\partial \bar{\tau}}{\partial z} = \frac{\partial}{\partial x}\left(\Gamma_e \frac{\partial \bar{\tau}}{\partial x}\right) + \frac{\partial}{\partial y}\left(\Gamma_e \frac{\partial \bar{\tau}}{\partial y}\right) + \frac{\partial}{\partial z}\left(\Gamma_e \frac{\partial \bar{\tau}}{\partial z}\right) + 1 \qquad (8.26)$$

8.3.2 *The turbulence viscosity models*

To solve the transport equations (8.21)–(8.26) mathematical expressions for the effective viscosity μ_e and the effective diffusion coefficient Γ_e will be required. Current turbulent viscosity models assume that μ_t is isotropic, i.e. that the ratio of Reynolds stress and mean rate of strain is the same in three directions. This assumption is approximately valid in some flow situations but fails in others. Since in a turbulent flow these stresses can also be transported by the fluid, in some cases a solution of the Reynolds stress equations will be necessary for accurate prediction of the flow. In this section a brief description is given of the most well-known turbulent viscosity models but for a thorough treatment of this subject readers should consult a specialist textbook such as Abbott and Basco [5], Launder and Spalding [11] and Rodi [12].

The Mixing length (zero-equation) model

In 1925 L. Prandtl suggested the first turbulence model, which came to be known as the 'mixing length hypothesis'. In analogy to the kinetic theory of gases, Prandtl assumed that the turbulent viscosity can be represented by the following expression:

$$\mu_t = c_1 \rho l v_t \qquad (8.27)$$

where l is a length scale, v_t is a lateral turbulence velocity and c_1 is a dimensionless constant. To evaluate μ_t at each point in the flow the quantities c_1, l and v_t must be ascribed values. Most of the kinetic energy of turbulence is contained in the largest eddies that exert the greatest influence on the mean flow. The mean flow is therefore influenced by the characteristics of the largest eddies and the connection between them can be found in isotropic turbulent flow, where the Reynolds stress in all directions is $\tau = -\overline{\rho u' u'}$ and the mean velocity gradient is $\partial u/\partial y$. Considering the flow in shear layers, which show the characteristics just described, Prandtl postulated that the lateral fluctuating velocity is proportional to a length scale and the time-mean velocity gradient. Thus:

$$v_t = c_2 l |\partial u/\partial y| \qquad (8.28)$$

Table 8.1 Mixing length values [8, 11]

Type of flow	L_m/δ
Plane mixing layer	0.07
Plane wall jet	0.09
Plane jet	0.09
Circular jet	0.075
Radial jet	0.125
Plane wake	0.16

Hence combining equations (8.27) and (8.28) gives:

$$\mu_t = \rho L_m^2 |\partial u/\partial y| \tag{8.29}$$

where L_m is the 'mixing length'. Turbulence is dependent on the type of flow and it is therefore necessary to account for different flows by varying the value of the mixing length, L_m. It is generally assumed that L_m is proportional to the distance from the nearest wall or the width of a shear layer, a wake, a jet, etc. Table 8.1 gives typical values of L_m in terms of a characteristic length of the flow, δ. For a plane mixing layer and a wall jet δ is taken as the local thickness of the layer or jet whereas for a plane jet, a free circular jet or a radial jet in a stagnant fluid δ is half the local thickness of the jet. The thickness of the layer or jet is taken to be the point where the velocity is within 1% of the free stream velocity.

The buoyancy forces caused by vertical changes in fluid density due to temperature or concentration gradient (stratification) can affect the diffusion of turbulence, i.e. the length scale. In this case, empirical correction formulae involving the Richardson number, Ri (ratio of gravity force to inertia force) can be applied to determine L_m. For stable stratification ($Ri > 0$), Rodi [12] gives:

$$L_m/L_{m0} = 1 - \beta_1 Ri \tag{8.30}$$

and for unstable stratification ($Ri < 0$):

$$L_m/L_{m0} = (1 - \beta_2 Ri)^{-0.25} \tag{8.31}$$

where

$$Ri = -(g/\rho)\frac{(\partial\rho/\partial y)}{(\partial u/\partial y)^2} \tag{8.32}$$

L_{m0} is the mixing length when $Ri = 0$, $\beta_1 \approx 7$ and $\beta_2 \approx 14$.

The mixing length model can also be used to predict turbulence transport of scalar quantities as given by equations (8.15) but with the assumption that $u' = v' = w'$, i.e. only one equation is used to present the variations in the x, y and z directions. Values of the turbulent Prandtl or Schmidt number, σ_t, are 0.9 for near wall flows (boundary layer or wall jets), 0.5 for plane jets and mixing layers and 0.7 for axisymmetric jets [12].

Although the mixing length hypothesis has successfully been applied to solving numerous turbulent flow problems, it has little application in complex flows because of the difficulty in specifying L_m and the isotropic assumption of turbulence. Furthermore, the method has been found to be unsuitable when processes of turbulence convection or diffusion are significant as in recirculating flows in rooms.

Since in applying the mixing length model no additional transport equations need to be solved it is sometimes referred to as a *zero-equation turbulence model*. Other eddy-viscosity based models that require the solution of one or more additional transport equations will now be presented.

The Kolmogorov–Prandtl (one-equation) model

A major defect of the mixing length hypothesis is in prescribing the turbulence velocity v_t proportional to the time-mean velocity gradient $\partial u/\partial y$ (equation (8.28)). In this case, the turbulent viscosity μ_t vanishes when the velocity gradient is zero, which is clearly unjustifiable. A classical example is the turbulent flow at the centre of a pipe or duct where $\partial u/\partial y = 0$ but μ_t is very large.

A turbulence model, suggested by A. N. Kolmogorov in 1942 and independently by Prandtl in 1945, represents the turbulence velocity by $\sqrt{k}$ where k is the kinetic energy of turbulence defined by equation (8.16). It follows that the turbulent viscosity is:

$$\mu_t = C_D \rho L \sqrt{k} \tag{8.33}$$

where C_D is an empirical constant (≈ 1.0) and L is a length scale.

Equation (8.33) is known as the Kolmogorov–Prandtl equation. The distribution of k in the field is obtained by solving a transport equation with k as the dependent variable derived from the general Navier–Stokes equation. For high Reynolds number flow the transport equation for k is:

$$\frac{\partial}{\partial t}(\rho k) + \frac{\partial}{\partial x}(\rho u k) + \frac{\partial}{\partial y}(\rho v k) + \frac{\partial}{\partial z}(\rho w k)$$

$$= \frac{\partial}{\partial x}\left(\Gamma_k \frac{\partial k}{\partial x}\right) + \frac{\partial}{\partial y}\left(\Gamma_k \frac{\partial k}{\partial y}\right) + \frac{\partial}{\partial z}\left(\Gamma_k \frac{\partial k}{\partial z}\right)$$

$$+ \mu_t \left\{ 2\left[\left(\frac{\partial u}{\partial x}\right)^2 + \left(\frac{\partial v}{\partial y}\right)^2 + \left(\frac{\partial w}{\partial z}\right)^2\right] + \left(\frac{\partial u}{\partial y} + \frac{\partial v}{\partial x}\right)^2 \right.$$

$$\left. + \left(\frac{\partial u}{\partial z} + \frac{\partial w}{\partial x}\right)^2 + \left(\frac{\partial w}{\partial y} + \frac{\partial v}{\partial z}\right)^2 \right\} - C_\mu \rho \frac{k^{1.5}}{L} + \beta g \frac{\mu_t}{\sigma_t} \frac{\partial T}{\partial y} \tag{8.34}$$

In this equation $\Gamma_k = \mu_e/\sigma_k$ where $\sigma_k \approx 1$, σ_t is the turbulent Prandtl or Schmidt number (0.5–0.9) and C_μ, is a constant ≈ 0.09. The last term represents the effect of buoyancy, with β as the coefficient of volumetric expansion.

Although the one-equation model is an improvement on the mixing length hypothesis (zero-equation) it still relies on a value for the length scale L, which is not always definable particularly in flows involving separation and recirculation.

The k–ε (two-equation) model

Although the previous one-equation model is an improvement on the mixing length hypothesis, one has to estimate the length scale for the determination of the turbulent viscosity. In simple well-defined flows this may not create a problem but in a number of other more complex flows the estimation of L is no better than a guess. This is true for flows governed by elliptic equations, such as recirculating flows, where convective transport is important and the determination of L is difficult even by measurement. For such flows, a two-equation turbulence model that describes both the turbulence velocity and length scale by transport equations will be more appropriate. There are a number of two-equation models that can be applied to solve recirculating flows, most of which do not solve for the length scale as the dependent variable but for another variable, z, related to it by an expression of the form:

$$z = k^m L^n$$

where m and n are indices, the values of which vary from one model to another. A more detailed description of two-equation models (which also use the turbulent viscosity concept) can be found in [11] and [12], but the model that uses the turbulence energy, k, for the turbulence velocity and its dissipation rate, ϵ, as an implicit scale is described here. This is usually known as the 'k–ϵ turbulence model' and currently it is the most widely used model because of its applicability to wide-ranging flow situations and its lower computational demand than more complex models that are available.

In the k–ϵ model, z in the preceding equation represents the kinetic energy dissipation rate, ϵ, i.e.:

$$\epsilon = C_\mu k^{1.5}/L \tag{8.35}$$

It follows from equation (8.33) that after substituting for $C_D = 1$:

$$\mu_t = C_\mu \rho k^2/\epsilon \tag{8.36}$$

The transport equation for ϵ is as follows:

$$\frac{\partial}{\partial t}(\rho\epsilon) + \frac{\partial}{\partial x}(\rho u\epsilon) + \frac{\partial}{\partial y}(\rho v\epsilon) + \frac{\partial}{\partial z}(\rho w\epsilon)$$

$$= \frac{\partial}{\partial x}\left(\Gamma_\epsilon \frac{\partial\epsilon}{\partial x}\right) + \frac{\partial}{\partial y}\left(\Gamma_\epsilon \frac{\partial\epsilon}{\partial y}\right) + \frac{\partial}{\partial z}\left(\Gamma_\epsilon \frac{\partial\epsilon}{\partial z}\right)$$

$$+ C_1 \frac{\epsilon}{k}\mu_t \left\{2\left[\left(\frac{\partial u}{\partial x}\right)^2 + \left(\frac{\partial v}{\partial y}\right)^2 + \left(\frac{\partial w}{\partial z}\right)^2\right] + \left(\frac{\partial u}{\partial y} + \frac{\partial v}{\partial x}\right)^2 \right.$$

$$\left. + \left(\frac{\partial u}{\partial z} + \frac{\partial w}{\partial x}\right)^2 + \left(\frac{\partial w}{\partial y} + \frac{\partial v}{\partial z}\right)^2\right\} - C_2\rho\frac{\epsilon^2}{k} + C_3\beta g\frac{\epsilon}{k}\Gamma_t\frac{\partial T}{\partial y} \tag{8.37}$$

Here, $\Gamma_\epsilon = \mu_e/\sigma_\epsilon$ where σ_ϵ is a constant equal to 1.22, $C_1 = 1.44, C_2 = 1.92$ and $C_3 = 1.0$. The transport equation for the second turbulence scale k is the same as equation (8.34) with ϵ substituted for the last but one term on the r.h.s.

The k–ϵ model has been applied to numerous room air movement problems with good predictive accuracy to the flow. Some typical cases are presented in Section 8.5. Although other more advanced models are available, as highlighted in the following sections, the majority are not thoroughly proven for predicting flows with large strain or deformation, such as a recirculating flow found in room ventilation, and, in any case, most of these models require considerably more computation than the k–ϵ model.

8.3.3 *The renormalization group (RNG) model*

One of the difficulties with the k–ϵ model is the inadequate representation of the effect of small-scale turbulence, which is not usually very significant in high Reynolds number (Re) flows but its effect begins to be prominent in low Re flows or in some local domains of high Re flows, such as close to a surface where the large eddies are reduced by the damping effect of the surface. A number of turbulence models have been developed to take account of low-scale turbulence particularly when heat convection from surfaces is being calculated [13, 14]. A large number of these are based on the k–ϵ model and involve a modification of some terms in the equations for k and ϵ (equations 8.34 and 8.37). A model that is gaining popularity in room air movement simulation is the renormalized group (RNG) model due to Yakhot *et al.* [15, 16]. In this model, they represent the effects of small-scale turbulence by means of a random forcing function in the ϵ-equation. The procedure involves the representation of the effects of the small-scale turbulent motion by a large-scale turbulence and a modified viscosity. This is carried out by adding an additional source term to the ϵ-equation (8.37), which is: $-\rho R$ where R is the rate of strain written as:

$$R = \frac{C_\mu \eta^3 (1 - \eta/\eta_0)}{1 + \beta \eta^3} \frac{\epsilon^2}{k} \tag{8.38}$$

where η_0 is a constant $= 4.38$, $\beta = 0.012$ for near wall flow but may have a different value for other flow situations and:

$$\eta = Sk/\epsilon \tag{8.39}$$

In equation (8.39) $S = v(2S_{ij}S_{ij})$, where, using tensor notation:

$$S_{ij} = \tfrac{1}{2}(\partial U_i/\partial x_j + \partial U_j/\partial x_i) \tag{8.40}$$

Equation (8.40) represents the first four terms on the right-hand side of equation (8.37) in PDE notation.

In addition to the extra term $(-\rho R)$ in the ϵ-equation, the turbulent Prandtl or Schmidt numbers in equations (8.34 and 8.37) are modified and computed as follows:

$$\left| \frac{\alpha - 1.393}{\alpha_0 - 1.393} \right|^{0.632} \left| \frac{\alpha + 1.393}{\alpha_0 + 1.393} \right|^{0.368} = \frac{\mu}{\mu_e} \tag{8.41}$$

where $\alpha = 1/\sigma_k = 1/\sigma_\epsilon = 1/\sigma_t$, $\alpha_0 = 1/\sigma$ (σ being the laminar Prandtl number for the fluid). Furthermore, the constants in equations (8.34 and 8.37) of the k–ϵ model are assigned slightly different values, as given in the following, with the one between brackets referring to the standard k–ϵ model value:

$$C_\mu = 0.085(0.09)$$
$$C_1 = 1.42(1.44)$$
$$C_2 = 1.68(1.92)$$

The RNG model has been evaluated for some complex flows with improved results to those obtained using the standard k–ϵ model, e.g. for backward-facing step [16], room air flow [17], and cavity flow [18]. However, the model is still relatively new to be extensively validated for other types of flow. It should be noted here that the computation effort associated with the RNG model is only marginally more intensive than the standard k–ϵ model calculations.

8.3.4 The low Reynolds number turbulence models

As will be shown later, most CFD codes use 'logarithmic wall functions' to describe the momentum and heat transfer near a solid boundary. The majority of these wall functions have been derived from flows in pipes and boundary layers and their use in other types of flow may not produce accurate results, particularly where the Reynolds number is low and/or heat transfer takes place from a surface to the fluid. In addition, close to a surface the turbulence scale is reduced and a standard turbulence model, such as the k–ϵ model, may not represent the effect of the small-scale turbulence well. Consequently, a number of modifications to the k–ϵ model have been devised to allow for such effects. These are normally called 'low Reynolds number turbulence models'. The majority of these models are based on providing algebraic expressions to represent the damping effect of the wall on the turbulence fluctuations [19–21]. These are used in the equations for k and ϵ and furthermore the expression for the turbulent viscosity in the standard k–ϵ equation (8.36) is modified to:

$$\mu_t = f_\mu C_\mu \rho k^2 / \epsilon \tag{8.42}$$

Also, the constants C_1 and C_2 in equation (8.37) are multiplied by the functions f_1 and f_2, respectively. The functions f_μ, f_1 and f_2 are each represented by an expression depending on the low Reynolds number model used. Some of the models available also add a source term in the ϵ-equation (8.37). The functions that are used by eight low Re models can be found in a paper by Patel *et al.* [22]. The paper also gives a comparison between the results from these models with experimental results for standard test cases. Because no wall functions are used with low Re models, a very fine computational grid is needed, particularly near solid boundaries, so that the Navier–Stokes equations can be solved to the boundary. In three-dimensional problems the solution becomes very demanding in computational terms and most of the low Re model applications are limited to two-dimensional problems.

8.3.5 The Reynolds stress (RSM) and algebraic stress (ASM) models

All the turbulence models described in the previous sections employ the turbulent viscosity in the time-mean momentum equations, in which case the turbulent viscosity is related to the time-mean velocity gradient. However, in some flows, particularly those involving boundary layer separation or reattachment, the shear stress may vanish where the mean velocity gradient is non-zero and vice versa. These situations cannot be predicted by the Boussinesq turbulent viscosity concept. In reality, the turbulent stresses are not directly related to the mean velocity gradient and consequently some turbulence models relate these shear stresses to the Reynolds (turbulent) stresses – hence the name 'Reynolds stress model' (RSM). This is the most complex classical turbulence model, which is also known as the 'second-order' or 'second-moment' closure model. Here the exact Reynolds stress equations (8.14), which account for directional effects of the Reynolds stress field, are solved rather than a single stress equation in the case of isotropic turbulence assumption, such as in the k–ϵ model.

The RSM model rely on the derivation of a set of transport equations for the Reynolds stresses, which normally contain unknown variables that must be defined before a solution can be achieved. These additional variables may be represented either by other, but determinable, quantities or by making them dependent variables of other sets of transport equations. The latter option clearly increases the number of equations to be solved and therefore at some appropriate stage, the set of transport equations are 'closed' and the unknown terms appearing in the new transport equations are represented by calculable quantities. For this reason this multi-equation model is sometimes known as 'closure model'. A full description of the different Reynolds stress models available and the relevant transport equations is beyond the scope of this book and interested readers should consult a specialist text on turbulence such as [6, 7, 11, 12, 23].

A Reynolds stress model that describes the six Reynolds stresses in equation (8.14), the three turbulent fluxes in equation (8.15) and the scalar fluctuation $-\rho\overline{\phi'\phi'}$, requires the solution of ten transport equations for the turbulence quantities alone. This is clearly not a trivial task even with modern computers. Attempts have therefore been made to simplify the transport equations by approximating the convection and diffusion terms (the terms nearest to the equal sign on both sides of the equations) into algebraic expressions. This will reduce the computational time required to obtain a solution of the transport equations. Such models are usually known as 'algebraic stress models' (ASM) and their treatment can be found in Rodi [12].

The RSM and ASM models are beginning to find some application in solving engineering fluid dynamic problems although they are still not as commonly used as the k–ϵ model. This is mainly because of their complexity and the vast computing effort needed and also because they do not always produce better predictions than the k–ϵ turbulence model. Although a number of commercial CFD codes have an option for using RSM and/or ASM models, they are still mainly used by researchers and CFD modellers.

8.3.6   *The large-eddy simulation (LES) model*

In Reynolds-averaged Navier–Stokes computations described so far, the turbulence fluctuations are ruled out by a time-averaging process, leaving only the mean flow to be simulated. The statistical effect of turbulence on the mean flow is then fully modelled using the eddy viscosity or Reynolds stress models described earlier. Large-eddy simulation (LES) on the other hand, is a modelling technique that accounts for turbulent flows where the large, energy-carrying turbulent eddies are computed directly and the effect of the small-scale eddies below the computational grid size, which contribute to the turbulence energy dissipation, are modelled through a subgrid-scale stress term. The large turbulent scales in LES are often distinguished from the small scales by using a spatial filtering process, which determines to what degree the large-scale eddy motions in a turbulent flow are explicitly resolved. The non-linear interaction between the large-scale and small-scale turbulence motion is approximated through a subgrid-scale turbulent viscosity model. The width of the spatial filter, which is usually related to the numerical mesh size, determines the largest size of the subgrid scales. Generally, the modelling for the subgrid-scale turbulent stresses can be made more universal than that obtained from other turbulence models using transport equations, provided that the Reynolds number of the flow is sufficiently large and the computational grid lies in the inertial subrange where energy cascade takes place (the transfer of momentum from larger eddies to smaller eddies) and the dissipation rate, ϵ, has a constant value. Because the LES method is based on time-domain solution of the Navier–Stokes equations, it has the ability to freeze the flow at any moment in time and if mean flow quantities are required the calculations must be conducted over a sufficiently large time scale. However, this advantage is at the expense of much more intensive computational requirements than those required for the solution of the time-average equations. It is not intended here to delve into the theory of LES but interested readers should consult specialist CFD publications on the subject, e.g. [5, 24].

The application of LES has so far been limited to mainly flow problems in channels [24] and over a cube [25] and more recently in room air flow [26]. The method is still mainly used by CFD modellers and its general use in engineering applications, and particularly in room air flow simulation, is rather limited. There are different LES techniques that are being developed for resolving the subgrid stresses [27] and until more robust methods become available so that the LES method can be applied universally to solve turbulent flow in rooms, it will remain as a tool for researchers.

8.3.7 *The direct numerical simulation (DNS) method*

In all the turbulence models described earlier, the Navier–Stokes equations are solved with different degrees of approximation and number of empirical constants. In the direct numerical simulation (DNS) method the equations are solved directly using a numerical method without the reliance on a turbulence model. In this way, all the essential turbulence fluctuations are resolved from the largest to the smallest eddies and consequently the technique requires a sufficiently fine computational grid. It is not uncommon for a DNS solution to require 10^6 computational cells just to resolve the flow in a large single eddy. Because the size of the eddies is Reynolds number dependent and the larger the eddies are the larger the computational grid must become, current computer capacities only permit the application of the DNS method to low Reynolds number turbulent flows and often in two dimensions. Examples of the application of DNS solutions for a heated cavity and a backward-facing step in two dimensions are given in [28] and [29] respectively.

8.3.8 *Comparison between turbulence models*

Many attempts have been made to shed some understanding on the performance of the turbulence models discussed in the previous sections and their variants in predicting flows that are encountered in ventilation studies. Due to the often complex nature of room flows, there is unfortunately not a single model that produces accurate predictions in all situations except perhaps the DNS model, which is beyond the reach of most room flow modellers. Murakami [30] studied the performance of the standard k–ϵ, the ASM and the LES models for predicting the air flow around a model of a building (bluff body) with boundary layer separation at the top, and mixed convection in an enclosure ventilated by an air jet. Chen [31] compared five eddy-viscosity turbulence models, viz. the standard k–ϵ model, a low Reynolds number k–ϵ model, a two-layer model (one equation for k near the wall region and a standard two-equation k–ϵ away from the wall region) a two-scale k–ϵ model (divides the turbulence energy spectrum

Table 8.2 Comparison between the performances of different turbulence models

Turbulence model	Standard k–ε	Low Re k–ε	RNG	ASM	LES
Internal flows with weak streamline curvature (e.g. duct flow)	1	1	1	1	1
Flows with strong streamline curvature (e.g. external flow, separation, etc.)	3, 4	3, 4		1	1
Jets:					
Normal	1	1	1	1	1
Swirl	3, 4	3, 4		1	1
Impinging	2, 3	2, 3	1, 2	1	1
Cavity flow (non-isothermal):					
Weak stratification	1	1	1	1	1
Strong stratification (e.g. negatively buoyant)	3, 4	3, 4	3, 4	1, 2	1
Forced convection	2, 3	1, 2	2, 3	2, 3	1, 2
Mixed convection	2, 3	1, 2	1	2, 3	2, 3
Low Reynolds number flow	3, 4	2	2	3, 4	1
Unsteady flow (e.g. vortex shedding)	4	2		2	1

Note
Where no score is given there is little or no information available.

into a production region and a transfer region), and the RNG model, in four different flow situations involving natural convection, forced convection, mixed convection and impinging jet. Chen and Chao [32] used four turbulence models, namely a standard k–ϵ model, a two-scale k–ϵ model, an RNG model (i.e. three eddy-viscosity models) and an RSM for predicting the buoyant plume in a displacement flow. Leschziner [33] studied the effect of turbulence level and streamline curvature (due to buoyancy and/or flow separation) on the accuracy of the k–ϵ model and second-moment closure (ASM and RSM) models that include adaptations to the low Reynolds number wall region. The results from these and other studies is summarized in Table 8.2 to provide the reader with some guidance on the suitability of different turbulence models for various flows experienced in room ventilation studies. The table shows four categories: 1 (excellent), 2 (good), 3 (fair) and 4 (poor/unacceptable), but where information is not available no score is given.

8.4   Solution of transport equations

8.4.1   *The general transport equation*

The transport equations for momentum (8.21–8.23), temperature (8.24), concentration (8.25), mean age (8.26) and the turbulence scales k (8.34) and ϵ (8.37) all have the general form:

$$
\underbrace{\frac{\partial}{\partial t}(\rho\phi)}_{\text{Transient}} + \underbrace{\frac{\partial}{\partial x}(\rho u\phi) + \frac{\partial}{\partial y}(\rho v\phi) + \frac{\partial}{\partial z}(\rho w\phi)}_{\text{Convection}}
$$

$$
= \underbrace{\frac{\partial}{\partial x}\left(\Gamma_\phi \frac{\partial\phi}{\partial x}\right) + \frac{\partial}{\partial y}\left(\Gamma_\phi \frac{\partial\phi}{\partial y}\right) + \frac{\partial}{\partial z}\left(\Gamma_\phi \frac{\partial\phi}{\partial z}\right)}_{\text{Diffusion}} + \underbrace{S_\phi}_{\text{Source}} \qquad (8.43)
$$

where ϕ is the dependent variable and S_ϕ is the source term, which has different expressions for different transport equations. The convection and diffusion terms for all the transport equations are identical, with Γ_ϕ representing the diffusion coefficient for scalar variables and the effective viscosity μ_e for vector variables, i.e. the velocities. This characteristic of the transport equations is extremely useful when the equations are discretized (reduced to algebraic equations) and solved numerically since only a solution of the general equation (8.43) is required. In fact this equation also represents the continuity equation when $\phi = 1$ and $S_\phi = 0$. Table 8.3 gives the expressions for the source terms S_ϕ for each dependent variable that is likely to be needed in solving ventilation problems.

The discretized form of equation (8.43) can be solved by one of the well-established numerical procedures such as the FVM or the FEM. In computational fluid dynamics the FVM is more popular than the FEM because it is generally more robust and economical in computational time. The majority of commercially available CFD codes use the FVM solution techniques and therefore this is the method that will be described in the next section. The readers who want to familiarize themselves with the application of FEM in solving CFD problems should consult specialist texts, e.g. [5, 34].

8.4.2 The finite volume method

Before equation (8.43) can be solved numerically it is necessary to use a computational grid with the dependent variables, ϕ's, to be evaluated given discrete values at the grid points (points of intersection of grid lines). This means that the distribution of ϕ in the flow domain is discretized and such a numerical procedure is known as a 'discretization'. To distribute ϕ at the grid points it is necessary to use an algebraic equation that expresses the same physical quantities as in the transport equation describing the variation of ϕ in the flow domain or more practically in a subdomain, i.e. element of control volume. The equivalent algebraic equation is the discretization equation connecting the values of ϕ for a group of grid points; usually the grid points of the control volume. Such an equation can be solved relatively easily using standard solution techniques. As the number of grid points in the flow domain becomes larger and larger the solution of the discretization equation gets closer and closer to the exact solution of the partial differential equation. However, the computational task becomes greater as the number of grid points increases.

The one-dimensional discretization equation

To illustrate the discretization procedure, a one-dimensional transport equation containing convection and diffusion terms will be chosen. Although such an equation has a limited practical application, nevertheless the same procedure also applies to two- and three-dimensional transport equations.

The equation to be discretized is:

$$\frac{\partial}{\partial x}(\rho u \phi) = \frac{\partial}{\partial x}\left(\Gamma \frac{\partial \phi}{\partial x}\right) \tag{8.44}$$

To derive the equivalent discretization equation we shall employ the grid-point convention shown in Figure 8.1 in which x increases from W (west) to E (east). The dashed lines represent the faces of the control volume (control surfaces), which are

Table 8.3 Source terms in the transport equations

Equation	ϕ	Γ_ϕ	S_ϕ
Continuity	1	0	0
u-momentum	u	μ_e	$-\dfrac{\partial p}{\partial x} + \dfrac{\partial}{\partial x}\left(\mu_e \dfrac{\partial u}{\partial x}\right) + \dfrac{\partial}{\partial y}\left(\mu_e \dfrac{\partial v}{\partial x}\right) + \dfrac{\partial}{\partial z}\left(\mu_e \dfrac{\partial w}{\partial x}\right)$
v-momentum	v	μ_e	$-\dfrac{\partial p}{\partial y} + \dfrac{\partial}{\partial x}\left(\mu_e \dfrac{\partial u}{\partial y}\right) + \dfrac{\partial}{\partial y}\left(\mu_e \dfrac{\partial v}{\partial x}\right) + \dfrac{\partial}{\partial z}\left(\mu_e \dfrac{\partial w}{\partial y}\right) - g(\rho - \rho_o)$
w-momentum	w	μ_e	$-\dfrac{\partial p}{\partial z} + \dfrac{\partial}{\partial x}\left(\mu_e \dfrac{\partial u}{\partial z}\right) + \dfrac{\partial}{\partial y}\left(\mu_e \dfrac{\partial v}{\partial z}\right) + \dfrac{\partial}{\partial z}\left(\mu_e \dfrac{\partial w}{\partial z}\right)$
Temperature	T	Γ_e	q/C_p
Concentration	c	Γ_e	ρc
Local mean age	$\bar{\tau}$	Γ_e	1
Kinetic energy	k	Γ_k	$G_S - \rho\epsilon + G_B$
Dissipation rate	ϵ	Γ_ϵ	$\dfrac{\epsilon}{k}(C_1 G_S + C_3 G_B) - C_2 \rho \dfrac{\epsilon^2}{k}$

Notes

$$\frac{q}{C_p} = \frac{\text{Energy production rate } (\text{W m}^{-3})}{\text{specific heat } (\text{J kg}^{-1}\text{K}^{-1})}$$

$\rho c = $ concentration production rate $(\text{kg s}^{-1}\text{m}^{-3})$

$$G_S = \mu_t \left\{ 2\left[\left(\frac{\partial u}{\partial x}\right)^2 + \left(\frac{\partial v}{\partial y}\right)^2 + \left(\frac{\partial w}{\partial z}\right)^2 \right] + \left(\frac{\partial u}{\partial y} + \frac{\partial v}{\partial x}\right)^2 + \left(\frac{\partial u}{\partial z} + \frac{\partial w}{\partial x}\right)^2 + \left(\frac{\partial v}{\partial z} + \frac{\partial w}{\partial y}\right)^2 \right\}$$

$\qquad = $ kinetic energy generation by shear

$$G_B = \beta g \frac{\mu_t}{\sigma_t}\frac{\partial T}{\partial y} = \text{kinetic energy generation by buoyancy}$$

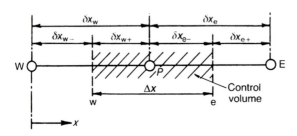

Figure 8.1 Grid points for a one-dimensional field.

denoted by e and w. Assuming a unit thickness in the y and z directions the volume of the control volume is Δx.

Integrating equation (8.44) over the control volume and assuming ϕ is a scalar we obtain:

$$(\rho u \phi)_e - (\rho u \phi)_w = (\Gamma \partial \phi / \partial x)_e - (\Gamma \partial \phi / \partial x)_w \qquad (8.45)$$

The discretization of this equation requires the application of a finite difference scheme. There are a number of differencing schemes that may be used but here only three schemes will be described: the central difference, the upwind difference and the hybrid schemes. Other schemes may be found in [35–37].

The central difference scheme Using the central difference and assuming uniform grid spacing, the values of a scalar variable ϕ at the control surfaces are given by:

$$\phi_e = 0.5(\phi_E + \phi_P)$$
$$\phi_w = 0.5(\phi_P + \phi_W) \tag{8.46}$$

Substituting for ϕ_e and ϕ_w in equation (8.45) gives:

$$\frac{1}{2}(\rho u)_e(\phi_E + \phi_P) - \frac{1}{2}(\rho u)_w(\phi_P + \phi_W) = \frac{\Gamma_e}{\delta x_e}(\phi_E - \phi_P) - \frac{\Gamma_w}{\delta x_w}(\phi_P + \phi_W) \tag{8.47}$$

Following Patankar's recommendations [35] the diffusion coefficients at the control surfaces are represented by:

$$\Gamma_e = \left(\frac{1-f_e}{\Gamma_P} + \frac{f_e}{\Gamma_E}\right)^{-1} \qquad \Gamma_w = \left(\frac{1-f_w}{\Gamma_P} + \frac{f_w}{\Gamma_W}\right)^{-1} \tag{8.48}$$

where the interpolation factors are given by (see Figure 8.1):

$$f_e = \delta x_{e+}/\delta x_e \qquad f_w = \delta x_{w+}/\delta x_w$$

If a uniform grid is used, as assumed here, then: $f_e = f_w = 0.5$. Hence,

$$\Gamma_e = \left(\frac{0.5}{\Gamma_P} + \frac{0.5}{\Gamma_E}\right)^{-1} \qquad \Gamma_w = \left(\frac{0.5}{\Gamma_P} + \frac{0.5}{\Gamma_W}\right)^{-1}$$

The integration of the one-dimensional steady-flow continuity equation gives:

$$\int \frac{\partial}{\partial x}(\rho u)\,dx = 0$$

i.e. ρu is a constant.

We can then define the following convection and diffusion terms:

$$F = \rho u, \quad D = \Gamma/\delta x \tag{8.49}$$

Both F and D have the same units although, whilst D is always positive, F may be either positive or negative depending on the direction of u. With these definitions the discretization equation (8.47) can be written as:

$$a_P\phi_P = a_E\phi_E + a_W\phi_W \tag{8.50}$$

where

$$a_E = D_e - 0.5F_e$$
$$a_W = D_w + 0.5F_w$$
$$a_P = D_e + 0.5F_e + D_w - 0.5F_w = a_E + a_W + F_e - F_w \tag{8.51}$$

But by continuity $F_e = F_w$ and therefore: $a_P = a_E + a_W$.

Using equations (8.50) and (8.51) it is possible to determine the value of the variable at the node point, ϕ_P from its values at the neighbouring points. This scheme is suitable for solving diffusion dominated problems.

The upwind difference scheme   It is well recognized that a central differencing scheme is unable to identify flow direction and when the convective term is significant (such as in room flow), the scheme does not provide a converged solution. This occurs when the control volume Reynolds number or Peclet number is high because of the possibility that a_E or a_W may become negative with the consequence that an unrealistic solution will be obtained [35]. The Peclet number, Pe, is the ratio of convection flux to diffusion flux and is defined as:

$$Pe = F/D = \rho u \Delta x / \Gamma \tag{8.52}$$

To overcome this difficulty an upwind differencing scheme (sometimes called the donor-cell method) is used. In this scheme the value of ϕ at the control surfaces is taken to be the same value as at the upstream node point, i.e.:

$$\phi_e = \phi_P \quad \text{for } F_e > 0 \qquad \phi_e = \phi_E \quad \text{for } F_e < 0 \tag{8.53}$$

and similarly for ϕ_w.

Using the notation $[\![A, B]\!]$ to denote the greater of A and B (equivalent to the FORTRAN language statement AMAX1 (A, B)), equation (8.53) may be written as:

$$F_e \phi_e = \phi_P [\![F_e, 0]\!] - \phi_E [\![-F_e, 0]\!] \tag{8.54}$$

Using the upwind scheme, the coefficients of the discretization equation (8.50) become:

$$a_E = D_e + [\![-F_e, 0]\!]$$
$$a_W = D_w + [\![F_w, 0]\!] \tag{8.55}$$
$$a_P = a_E + a_W + F_e - F_w$$

It is clear from the above equations that the coefficients a_E, a_W and a_P are always positive and a meaningful solution can be obtained in this case. The accuracy of the upwind scheme is examined in detail by Raithby [38].

The hybrid scheme   The hybrid scheme combines the central and upwind difference schemes and can be applied more generally to solve equations that contain both convection and diffusion terms. It is identical to the central difference when the control volume Peclet number is $-2 \le Pe \le 2$ and to the upwind scheme when $|Pe| > 2$. The coefficients of the discretization equations are as follows:

$$a_E = [\![0, D_e - 0.5F_e]\!] + [\![0, -F_e]\!]$$
$$a_W = [\![0, D_w - 0.5F_w]\!] + [\![0, F_w]\!] \tag{8.56}$$
$$a_P = a_E + a_W + F_e - F_w$$

It should be noted that the expressions in the three difference schemes presented here have been derived for a scalar variable in which case, in the finite volume solution, a grid node is the centre of the control volume. Because a staggered grid is usually used for solving vector quantities, the centre of the control volume for vector quantities (i.e. velocities) is at the midpoint between two adjacent grid nodes (see Section 8.4.3).

In this case, the discretization equation and its coefficients will be for the midpoint between the grid points (see Figure 8.1), i.e. capital letter subscripts are used as follows:

$$a_E = [\![0, D_E - 0.5F_E]\!] + [\![0, -F_E]\!]$$
$$a_W = [\![0, D_P - 0.5F_P]\!] + [\![0, F_P]\!] \tag{8.57}$$
$$a_P = a_E + a_W + F_E - F_P$$

The QUICK scheme The accuracy of the schemes described so far is only first order in terms of Taylor series truncation error and this makes them prone to numerical errors. These errors can be reduced by using a higher-order discretization scheme that involves more points. One such scheme is the 'quadratic upstream interpolation for convective kinetics' (QUICK) that was developed by Leonard [36]. The scheme uses three-point upstream-weighted quadratic interpolation for cell face values. The cell face value of ϕ is obtained from a quadratic function through two nodes on either side of the cell face and a node upstream, see Figure 8.2. The following interpolation formula is used for positive ($F_e > 0, F_w > 0$) and negative ($F_e < 0, F_w < 0$) flow directions for one-dimensional convection-diffusion problems:

$$a_P\phi_P = a_W\phi_W + a_E\phi_E + a_{WW}\phi_{WW} + a_{EE}\phi_{EE} \quad \text{for central coefficients} \tag{8.58}$$

$$a_P = a_W + a_E + a_{WW} + a_{EE} + (F_e - F_w) \quad \text{for neighbouring coefficients} \tag{8.59}$$

where

$$a_W = D_w + \tfrac{6}{8}\alpha_w F_w + \tfrac{1}{8}\alpha_e F_e + \tfrac{3}{8}(1 - \alpha_w)F_w \tag{8.60}$$

$$a_{WW} = -\tfrac{1}{8}\alpha_w F_w \tag{8.61}$$

$$a_E = D_e - \tfrac{3}{8}\alpha_e F_e - \tfrac{6}{8}(1 - \alpha_e)F_e - \tfrac{1}{8}(1 - \alpha_w)F_w \tag{8.62}$$

$$a_{EE} = \tfrac{1}{8}(1 - \alpha_e)F_e \tag{8.63}$$

$$\alpha_w = 1 \text{ for } F_w > 0 \quad \text{and} \quad \alpha_e = 1 \text{ for } F_e > 0$$
$$\alpha_w = 0 \text{ for } F_w < 0 \quad \text{and} \quad \alpha_e = 0 \text{ for } F_e < 0$$

Due to the appearance of negative coefficients (equations 8.60–8.63) the QUICK scheme is prone to instability. A number of procedures have been developed to overcome this problem, all involving placing negative coefficients in the source term to

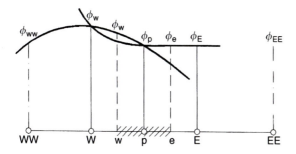

Figure 8.2 Grid points for quadratic interpolation using the QUICK scheme.

retain positive main coefficients. This approach was generalized by Hayase *et al.* [39], which is summarized below.

$$\phi_w = \phi_W + \tfrac{1}{8}(3\phi_P - 2\phi_W - \phi_{WW}) \quad \text{for } F_w > 0 \tag{8.64}$$

$$\phi_e = \phi_P + \tfrac{1}{8}(3\phi_E - 2\phi_P - \phi_W) \quad \text{for } F_e > 0 \tag{8.65}$$

$$\phi_w = \phi_P + \tfrac{1}{8}(3\phi_W - 2\phi_P - \phi_E) \quad \text{for } F_w < 0 \tag{8.66}$$

$$\phi_e = \phi_E + \tfrac{1}{8}(3\phi_P - 2\phi_E - \phi_{EE}) \quad \text{for } F_e < 0 \tag{8.67}$$

The resulting discretized equation takes the form:

$$a_P\phi_P = a_W\phi_W + a_E\phi_E + \bar{S} \tag{8.68}$$

The central coefficient is:

$$a_P = a_W + a_E + (F_e - F_w) \tag{8.69}$$

where

$$a_W = D_w + \alpha_w F_w \tag{8.70}$$

$$a_E = D_e - (1 - \alpha_e)F_e\bar{S} \tag{8.71}$$

$$\bar{S} = \tfrac{1}{8}(3\phi_P - 2\phi_W - \phi_{WW})\alpha_w F_w + \tfrac{1}{8}(\phi_W + 2\phi_P - 3\phi_E)\alpha_e F_e$$
$$+ \tfrac{1}{8}(3\phi_W - 2\phi_P - \phi_E)(1 - \alpha_w)F_w + \tfrac{1}{8}(2\phi_E + \phi_{EE} - 3\phi_P)(1 - \alpha_e)F_e \tag{8.72}$$

$$\alpha_w = 1 \text{ for } F_w > 0 \quad \text{and} \quad \alpha_e = 1 \text{ for } F_e > 0$$
$$\alpha_w = 0 \text{ for } F_w < 0 \quad \text{and} \quad \alpha_e = 0 \text{ for } F_e < 0$$

In the modified scheme, the coefficients are always positive and this should overcome the instability problem of the original QUICK scheme. The QUICK scheme has greater accuracy than the central difference or hybrid schemes with a smaller false diffusion. It makes it more accurate for use in coarse grid computation. It can, however, produce minor 'overshoots' and 'undershoots' during iterations and this ought to be considered when interpreting the results.

The three-dimensional discretization equation

The general transport equation (8.43) can be written as:

$$\frac{\partial}{\partial t}(\rho\phi) + \frac{\partial J_x}{\partial x} + \frac{\partial J_y}{\partial y} + \frac{\partial J_z}{\partial z} = S_\phi \tag{8.73}$$

where

$$J_x = \rho u\phi - \Gamma_\phi \partial\phi/\partial x$$
$$J_y = \rho v\phi - \Gamma_\phi \partial\phi/\partial y \tag{8.74}$$
$$J_z = \rho w\phi - \Gamma_\phi \partial\phi/\partial z$$

Expressing the source or generation term S_ϕ as a linear expression:

$$S_\phi \Delta V = b\phi_P + c \tag{8.75}$$

and, using the control volume shown in Figure 8.3 in the x–y plane, the integration of equation (8.73) over the control volume yields:

$$(\rho_P \phi_P - \rho_P^\circ \phi_P^\circ)(\Delta V / \Delta t) + J_e - J_w + J_n - J_s + J_r - J_l = (b\phi_P + c)\Delta V \tag{8.76}$$

where n (north) and s (south) are the neighbouring points of P in the y direction, l (left) and r (right) are the neighbouring points of P in the z direction, $\Delta V = \Delta x \Delta y \Delta z$ is the volume of the control volume, the symbols with a superscript ($^\circ$) represent the node values at time t (i.e. old values), and the J's are the integrated total fluxes (i.e. convection plus diffusion) at the corresponding control surface. All the symbols without superscripts in this equation are regarded as new values, i.e. at time $t + \Delta t$.

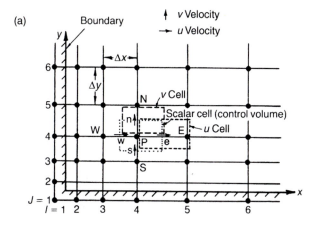

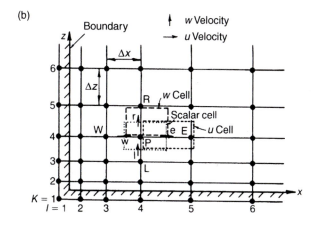

Figure 8.3 Grid and control volume for a three-dimensional field.

Similarly the integration of the continuity equation (8.2) over the control volume yields:

$$(\rho_P - \rho_P^o)(\Delta V/\Delta t) + F_e - F_w + F_n - F_s + F_r - F_l = 0 \tag{8.77}$$

where the F's are the mass flow rates over the control surfaces.

The general three-dimensional discretization equation can be obtained from equations (8.76) and (8.77) and is given by:

$$\left(\sum_i a_i + a_P^o - b\right)\phi_P = \sum_i a_i\phi_i + c \tag{8.78}$$

where

$$\sum_i a_i = a_P = a_E + a_W + a_N + a_S + a_R + a_L$$

$$\sum_i a_i\phi_i = a_P\phi_P = a_E\phi_E + a_W\phi_W + a_N\phi_N + a_S\phi_S + a_R\phi_R + a_L\phi_L$$

$$a_P^o = \rho_P \Delta V/\Delta t$$

$$b = S_P \Delta V$$

$$c = S_U \Delta V + (\rho_P^o \Delta V/\Delta t)\phi_P^o$$

in which case S_P and S_U are coefficients in the source terms b and c respectively, which can be obtained from Table 8.3 and equation (8.75). All the positive terms in the source expressions are taken as S_U and all the negative terms are divided by ϕ and taken to be S_P. In solving the momentum equations, the pressure terms are not included in S_U but added to the right-hand side of the discretization equation (see Section 8.4.3).

Applying the hybrid scheme the coefficients a_i can be written as follows [40]:

$$a_E = [\![0, D_e - 0.5|F_e|]\!] + [\![0, -F_e]\!]$$
$$a_W = [\![0, D_w - 0.5|F_w|]\!] + [\![0, F_w]\!]$$
$$a_N = [\![0, D_n - 0.5|F_n|]\!] + [\![0, -F_n]\!]$$
$$a_S = [\![0, D_s - 0.5|F_s|]\!] + [\![0, F_s]\!] \tag{8.79}$$
$$a_L = [\![0, D_l - 0.5|F_l|]\!] + [\![0, -F_l]\!]$$
$$a_R = [\![0, D_r - 0.5|F_r|]\!] + [\![0, F_r]\!]$$

The convection and diffusion fluxes are as follows:

$$F_e = (\rho u)_e \Delta A_e \qquad D_e = \Gamma_e \Delta A_e/\delta x_e$$
$$F_w = (\rho u)_w \Delta A_w \qquad D_w = \Gamma_w \Delta A_w/\delta x_w$$
$$F_n = (\rho v)_n \Delta A_n \qquad D_n = \Gamma_n \Delta A_n/\delta y_n$$
$$F_s = (\rho v)_s \Delta A_s \qquad D_s = \Gamma_s \Delta A_s/\delta y_s \tag{8.80}$$
$$F_l = (\rho w)_l \Delta A_l \qquad D_l = \Gamma_l \Delta A_l/\delta z_l$$
$$F_r = (\rho w)_r \Delta A_r \qquad D_r = \Gamma_r \Delta A_r/\delta z_r$$

where $\Delta A_e = \Delta A_w = \Delta y \Delta z$; $\Delta A_n = \Delta A_s = \Delta x \Delta z$; and $\Delta A_r = \Delta A_l = \Delta x \Delta y$.

The coefficients *a* in equation (8.79) are for the scalar variables $T, c, \bar{\tau}, k$ and ϵ. By referring to Figure 8.3(a) and (b) similar expressions for the velocity coefficients may be obtained. It should be noted, however, that the centre of the control volumes for the *u, v* and *w* velocity components lie between two grid points along the *x, y* and *z* grid lines respectively. The discretization equations for the velocity components are presented in Section 8.4.3.

8.4.3   Solution of the discretization equations

Because of the non-linearity of the transport equations, an iterative method of solving the discretization equations to achieve a converged solution is the most plausible approach. An iterative solution starts from guessed values of the dependent variables for the whole field. In deriving the transport equations and their discretized forms there is no equation for pressure except that the pressure gradient is added to the source terms. However, to achieve a convergent solution it is obvious that for the velocity components *u, v, w* (obtained from a solution of the momentum equation) to satisfy continuity the correct pressure field must be used in the momentum equations. This link between velocity and pressure can be employed in the iterative solution without the necessity for solving a discretization equation for the pressure.

Pressure-linked discretization methods

One of the most widely used methods that links the velocity to the pressure in order to satisfy continuity is the SIMPLE (semi-implicit method for pressure-linked equations) procedure of Patankar and Spalding [41]. This method uses a staggered grid so that correction to the velocity components can be made using the pressure values at the neighbouring grid nodes such that continuity is observed. To achieve this, the positions of the velocity components must be displaced from the grid nodes (pressure positions) so that the pressure forces act at the surfaces of the velocity control volumes, as shown in Figure 8.3.

Excluding the pressure term from the coefficient of the source term S_U, the discretization equations (obtained by integration) for the three velocity components u_e, v_n, w_r at the velocity points e, n and r (see Figure 8.3) are:

$$a_e u_e = \sum_i a_i u_i + c + (p_P - p_E)\Delta A_e$$

$$a_n v_n = \sum_i a_i v_i + c + (p_P - p_N)\Delta A_n \qquad (8.81)$$

$$a_r w_r = \sum_i a_i w_i + c + (p_P - P_R)\Delta A_r$$

where $\sum_i a_i u_i, \sum_i a_i v_i, \sum_i a_i w_i$ represent summation of the values at the neighbouring locations of the centres of the velocity components, which are two, four and six for one-, two- and three-dimensional flow fields respectively, i.e. *i* represents e, w, n, s, l and r for a three-dimensional field. The velocities in equation (8.81) will only satisfy continuity if the pressures at the grid points are correct. At the start of calculation, only guessed values of pressure (and velocity) are available and continuity will not be satisfied. Therefore, some means of correcting the pressures is needed to achieve a solution. This is the essence of the SIMPLE procedure.

In the SIMPLE procedure the pressure and velocity components are written as:

$$
\begin{aligned}
p &= p^* + p' \\
u &= u^* + u' \\
v &= v^* + v' \\
w &= w^* + w'
\end{aligned}
\tag{8.82}
$$

where the asterisk represents a guessed value and the prime is the correction necessary to satisfy continuity. The discretization equations for the guessed velocities are then:

$$
\begin{aligned}
a_e u_e^* &= \sum_i a_i u_i^* + c + (p_P^* - p_E^*)\Delta A_e \\
a_n v_n^* &= \sum_i a_i v_i^* + c + (p_P^* - p_N^*)\Delta A_n \\
a_r w_r^* &= \sum_i a_i w_i^* + c + (p_P^* - p_R^*)\Delta A_r
\end{aligned}
\tag{8.83}
$$

Subtracting the first of equations (8.83) from the first of equations (8.81) we obtain:

$$
a_e u_e' = \sum_i a_i u_i' + (p_P' - p_E')\Delta A_e
\tag{8.84}
$$

For computational convenience the term $\sum a_i u_i'$ may be dropped from the last equation and implicitly included in the pressure correction p' (hence the semi-implicit method) to give:

$$
u_e' = (\Delta A_e / a_e)(p_P' - p_E')
$$

which is the velocity correction formula and can be written for all the three components as:

$$
\begin{aligned}
u_e &= u_e^* + (\Delta A_e / a_e)(p_P' - p_E') \\
v_n &= v_n^* + (\Delta A_n / a_n)(p_P' - p_N') \\
w_r &= w_r^* + (\Delta A_r / a_r)(p_P' - p_R')
\end{aligned}
\tag{8.85}
$$

If we can now substitute the velocity components given by equation (8.85) into the discretized continuity equation (8.77) and rearrange the terms, the following discretization equation for pressure correction is obtained:

$$
\sum_i a_i p_P' = \sum_i a_i p_i' + c
\tag{8.86}
$$

where

$$\sum_i a_i = a_E + a_W + a_N + a_S + a_R + a_L$$

$$\sum_i a_i p_i' = a_E p_E' + a_W p_W' + a_N p_N' + a_S p_S' + a_R p_R' + a_L p_L'$$

$$a_E = (\rho_e/a_e)(\Delta A_e)^2$$

$$a_W = (\rho_w/a_w)(\Delta A_w)^2$$

$$a_N = (\rho_n/a_n)(\Delta A_n)^2$$

$$a_S = (\rho_s/a_s)(\Delta A_s)^2$$

$$a_R = (\rho_r/a_r)(\Delta A_r)^2$$

$$a_L = (\rho_l/a_l)(\Delta A_l)^2$$

$$c = (\rho_P^0 - \rho_P)(\Delta V/\Delta t) + F_w^* - F_e^* + F_s^* - F_n^* + F_l^* - F_r^*$$

$$F_w^* = (\rho u^*)_w \Delta A_w \quad F_e^* = (\rho u^*)_e \Delta A_e$$

$$F_s^* = (\rho u^*)_s \Delta A_s \quad F_n^* = (\rho u^*)_n \Delta A_n$$

$$F_l^* = (\rho u^*)_l \Delta A_l \quad F_r^* = (\rho u^*)_r \Delta A_r$$

In equation (8.86) the term c is the negative discretized continuity equation (8.77) in terms of the starred (guessed) velocities. When c is zero the guessed velocities will satisfy the continuity equation and no pressure correction is needed.

The procedure for applying the SIMPLE method can be summarized as follows:

1   the pressure field p^* is guessed;
2   the velocities u^*, v^* and w^* are obtained from equation (8.83);
3   the pressure correction p' is obtained from equation (8.86);
4   the pressure field p is obtained by adding p' to p^*;
5   the velocity components u, v and w are obtained from equations (8.81);
6   the discretization equations for the scalar variables ϕ (e.g. T, k, ϵ) are solved using equation (8.78) if they influence the flow field through fluid properties, source terms, etc., otherwise the variables (e.g. $\bar{\tau}$ and probably c) are solved after a converged solution of the flow field is obtained;
7   p^* is replaced by p in step 1 and the same procedure is repeated until a converged solution is obtained.

To speed up convergence and improve the stability of the numerical solution various attempts have been made to modify the SIMPLE algorithm. Because the term $\sum_i a_i u_i'$ (representing the velocity corrections at the neighbouring points to P) has been ignored when equation (8.84) is used for correcting the velocity, an exaggerated pressure correction results, which must be compensated for by underrelaxation (see section 8.4.4). Patankar [35] has revised SIMPLE so that the term $\sum_i a_i u_i'$ is included in the velocity correction equation and the pressure field is obtained from an initially guessed velocity field; velocity corrections are obtained from the pressures. The remaining steps in the SIMPLE procedure are unchanged. This is called the SIMPLE Revised or SIMPLER algorithm. The revised scheme normally gives faster convergence but at the same time it involves more computational effort per iteration because the pressure equations are solved twice.

Van Doormaal and Raithby [42] devised the SIMPLEC (C denotes consistent approximation) method, which is a modification of the original SIMPLE algorithm but without the requirement for underrelaxing the pressure equations. In this method the velocity correction term $\sum_i a_i u_i'$ is transferred to the left-hand side of equation (8.84) and the velocity correction equation becomes:

$$u_e' = \frac{\Delta A_e}{a_e - \sum\limits_i a_i}(p_P' - p_E')$$

All the other equations in SIMPLEC are identical to those given earlier in SIMPLE with a_e replaced by $a_e - \sum_i a_i$. It is claimed [42] that some computing economy can be made if SIMPLEC is used instead of SIMPLE but this method has not been as popular as the original algorithm.

Spalding [43] refined the SIMPLE procedure by including all the diffusion terms in the coefficients of the discretization equation for momentum and all the convection terms in the linearized source term of the equation. This so-called SIMPLEST algorithm is particularly suited to flows with large convection terms and very small diffusion terms in which case the Jacobi point-by-point procedure is used instead of the line-by-line procedure in the iteration process. In such cases faster convergence is claimed [44]. A further extension of the SIMPLE algorithm is the PISO (*pressure implicit with splitting operators*), which involves one predictor step and two corrector steps. Details of this procedure can be found in [6], which also gives a comparison between all the algorithms described earlier.

The iterative solution

The general discretization equation (8.78) may be written as a recurrence formula that is suitable for an iterative solution as follows:

$$\phi_j = A_j\phi_{j+1} + B_j\phi_{j-1} + C_j \tag{8.87}$$

where $\phi_p = \phi_j$, $\phi_N = \phi_{j+1}$ and $\phi_s = \phi_{j-1}$, and the coefficients A_j, B_j and C_j take the forms:

$$A_j = \frac{a_n}{\sum\limits_i a_i + a_P^o - b}$$

$$B_j = \frac{a_s}{\sum\limits_i a_i + a_P^o - b}$$

$$C_j = \frac{a_E\phi_E + a_W\phi_W + a_L\phi_L + a_R\phi_R + c}{\sum\limits_i a_i + a_P^o - b}$$

The recurrence equation (8.87) can be solved conveniently using the tri-diagonal matrix algorithm (TDMA) for all the dependent variables (ϕ). A line-by-line (LBL) iterative procedure is used, starting in each plane from top to bottom of each grid line in succession until the whole field is swept plane by plane from east to west. Initial values of ϕ in the field are guessed and these are improved by solving the equations along each grid line. When the equations are solved along a grid line it is assumed that the values of ϕ at the neighbouring lines are known and this procedure is repeated for each line until the whole flow field is swept. Further details of this solution procedure may be found in [45].

Because of the inherent non-linearity of the discretization equation it is sometimes necessary to slow down the change in ϕ from one iteration to the next, i.e. underrelaxing, to avoid divergence to the solution. Rearranging the discretization equation (8.78) we obtain:

$$\phi_P = \frac{\sum_i a_i\phi_i + c}{\sum_i a_i + a_P^o - b} \qquad (8.88)$$

Adding and subtracting the value of ϕ_P from the previous iteration, ϕ_P^o, we obtain:

$$\phi_P = \phi_P^o + \left[\frac{\sum_i a_i\phi_i + c}{\sum_i a_i + a_P^o - b} - \phi_P^o \right]$$

The terms in the square brackets represent the change in ϕ_P produced by the current iteration and this change may be reduced by a relaxation factor, $\alpha(< 1)$, so that:

$$\phi_P^r = \phi_P^o + \left[\frac{\sum_i a_i\phi_i + c}{\sum_i a_i + a_P^o - b} - \phi_P \right]$$

where ϕ_P^r is the new underrelaxed value of ϕ_P. If ϕ_P is substituted from equation (8.88) into this equation, we obtain:

$$\phi_P^r = \alpha\phi_P + (1 - \alpha)\phi_P^o \qquad (8.89)$$

In equation (8.89), ϕ_P^o is the value of ϕ_P from the previous iteration, ϕ_P is the value obtained from equation (8.88) and ϕ_P^r is the new value. The value of the underrelaxation factor α can be in the range $0 < \alpha \le 1$ and the optimum value for each dependent variable depends on the problem being solved and can only be found by trial and error. Some typical values for a three-dimensional, non-isothermal room flow problem will be $\alpha = 0.5$ for u, v, w, k and ϵ, and $\alpha = 1.0$ for p, T, $\bar{\tau}$ and c.

In a non-isothermal flow with strong influence of buoyancy, numerical stability may not be achieved by reducing the underrelaxation factor or increasing the number of iterations. The use of an 'inertia' type of relaxation procedure such as that suggested by Ideriah [46] may enhance stability and assist convergence. In this method, a variable or inertia relaxation factor α_I is used in solving the discretization equation for the vertical component of velocity, v. By writing the velocity discretization equation (8.81) in the form of the general discretization equation (8.78), we get:

$$\phi_P = \frac{\sum_i a_i\phi_i + c}{\sum_i a_i - b}$$

where the linearized source term c includes the pressure and buoyancy terms of the momentum equation for v (equation (8.22)). Introducing the inertia relaxation factor the above equation may be written as:

$$\phi_P = \frac{\sum_i a_i\phi_i + c + \alpha_I\phi_P^o}{\sum_i a_i - b + \alpha_I} \qquad (8.90)$$

where ϕ_p^o represents the value of ϕ_p from the previous iteration. α_I is defined by the following expression:

$$\alpha_I = \frac{C_I \rho g \beta (T_p - T_o)}{\sqrt{(g\beta L T_o)}} \tag{8.91}$$

where $\rho g \beta (T_p - T_o)$ is the buoyancy term in the momentum equation for v, T_p is the temperature of the control volume at P, T_o is a reference temperature (e.g. inlet or outlet temperature), L is a characteristic length (e.g. length of flow field) and C_I is a relaxation constant taken as 0.2 [46]. For stability, only the absolute value of α_I is used in equation (8.90). If α_I is very large then according to the latter equation only a small change in ϕ_p will occur between two successive iterations. On the other hand, if $\alpha_I = 0$ (i.e. for isothermal flow) then no under-relaxation is provided and in this case equation (8.89) should be used instead.

8.4.4   Boundary conditions

The accuracy of the solution of the discretization equations presented in the previous section will depend inter alia on the accuracy of specifying the physical quantities at the boundary of the flow domain and on the methods of linking these quantities to the bulk of the flow. For example, close to a solid boundary (e.g. a room surface) the local Reynolds number ($Re = uy/v$ where u is the velocity component parallel to the boundary at a distance y from it) is extremely small and turbulent fluctuations are damped out by the proximity of the surface so that laminar shear becomes a locally dominant force in the diffusion process because of the steep velocity gradient. In addition to the solid boundaries, the flow conditions at other boundaries, e.g. inlet, outlet, free boundary, etc., must be specified. The main types of boundaries usually encountered in solving room flow problems are described in the following.

Wall boundary

Because of the damping effect of the wall the transport equations for the turbulence quantities (i.e. k and ϵ) in their standard forms do not apply close to the wall. One way of dealing with this problem is to add extra source terms to the standard transport equations for k and ϵ (assuming the k–ϵ model is used) and use an extremely fine grid close to the surface so that the first few points are within the laminar sublayer. This is referred to as 'low Reynolds number turbulence modelling'. Various models for treating the flow near a boundary are available as discussed in Section 8.3.4. Although many of the low Reynolds number models can adequately describe the damping effect of the wall, a vast number of grid points will be needed (i.e. large computing resources) and this is often impractical for the complex three-dimensional geometries of room enclosures.

The alternative is to extend the Couette flow analysis and apply algebraic relations, the so-called logarithmic laws or wall functions, close to the surface. This approach does not require an ultra-fine grid near the surface and is considered later for impermeable wall with negligible streamwise pressure gradient. At a point close to the wall (P in Figure 8.4) the momentum equation is reduced to a one-dimensional form with gradients in the direction normal to the surface. Constant shear will be assumed throughout the wall region, i.e. $\tau_p \approx \tau_w$. Within the laminar sublayer region, i.e. $y_p^+ < 8$ in Figure 8.5 (but sometimes quoted to be ≤ 11.63), where $y_p^+ = u_\tau y_p / v$ and $u_\tau = \sqrt{(\tau_w / \rho)}$, viscous

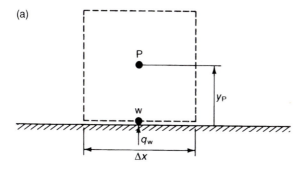

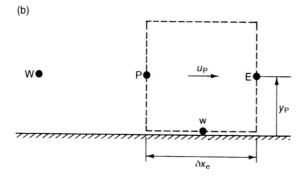

Figure 8.4 Near-wall control volumes. (a) Scalar control volume and (b) velocity (vector) control volume.

diffusion predominates and the wall shear stress, τ_w, is described by the usual Couette flow expression, i.e.:

$$u_P^+ = y_P^+, \qquad T_P^+ = \sigma y_P^+ \tag{8.92}$$

where σ is the Prandtl number, y_P^+ is a dimensionless velocity defined as:

$$u_P^+ = u_P/u_\tau \tag{8.93}$$

and T_P^+ is a non-dimensional temperature defined by:

$$T_P^+ = \rho u_\tau C_p(T_w - T_p)/q_w \tag{8.94}$$

where q_w is the wall heat flux (W m^{-2}).

Transition of the boundary layer takes place over the region $8 < y^+ < 40$ (buffer zone), which leads to the inner turbulent region (log-law region) represented by $40 < y^+ < 130$ and then the outer region, which is given by $y^+ > 130$, see Figure 8.5. The flow in a plane wall jet is similar to a boundary layer flow up to $y^+ \approx 130$ but is different for greater values. At a grid point where $y^+ > 30$ the laminar shear stress, τ_l, is small in comparison with the turbulent stress, τ_t, and the former may be neglected. When the generation and dissipation of kinetic energy of turbulence are in balance

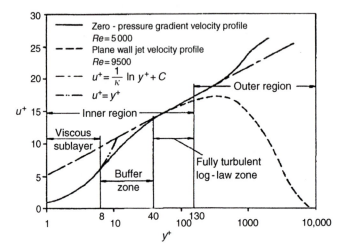

Figure 8.5 Non-dimensional velocity profile in turbulent boundary layer and plane wall jet.

then $\tau_p = C_\mu^{1/2} k$ and the following boundary layer expressions for momentum and heat fluxes may be used:

$$u^+ = 1/\kappa \, \ln y^+ + C \qquad (8.95)$$

where κ is the Karman constant ($= 0.4187$) and C is a constant that ranges in value between 4.9 and 5.6.

The heat transfer in the turbulent boundary layer is mainly by convection and the heat diffusion may be ignored. The Reynolds analogy between heat and momentum transfer leads to the following expression for T^+:

$$T^+ == 1/\kappa_h \, \ln y^+ + C_h \qquad (8.96)$$

where κ_h is the Karman constant for heat transfer (≈ 0.44) and C_h is a function of laminar and turbulent Prandtl numbers, σ and σ_t. Usually C_h is expressed as a function of σ and σ_t, viz:

$$C_h = f(\sigma/\sigma_t)$$

A widely used expression for C_h is that due to Jayatilaka [46], i.e.,

$$f(\sigma/\sigma_t) = 9.24[(\sigma/\sigma_t)^{0.75}) - 1]\{1 + 0.28 \exp[-0.007(\sigma/\sigma_t)]\} \qquad (8.97)$$

For air σ_t is in the range 0.7–0.9. Equations (8.95) and (8.96) are usually called the 'logarithmic wall functions' for turbulent transport of momentum and heat respectively. These expressions are valid for $30 < y^+ < 130$. It is known that the calculation of heat transfer from a surface is particularly sensitive to the distance of point P from the surface, y_P, where the wall function is applied, i.e. the value of y^+ at the point. For natural convection from a heated or cooled surface, the optimum value of y_P is about 5 mm [14]. Other expressions for the non-dimensional velocity u^+ and temperature

T^+ may be used instead of those given as equations (8.95) and (8.96). A comparison of various velocity expressions is given by Hammond [47], of temperature expressions by Nizou [48] and wall functions of velocity and temperature for natural convection along a vertical plate are given in [49].

At the wall boundary itself (point w in Figure 8.4) the no-slip condition applies and if the wall is stationary then $u_w = 0$ and the shear stress is calculated from the expressions for u_p^+ (equations (8.92) or (8.95)). The shear stress is added to the linearized source term, i.e. to b and c of the momentum equation (equation (8.75)). If P is in the laminar sublayer, the appropriate sink term coefficient is given by:

$$S_P = -\mu A_w / y_P \tag{8.98}$$

where A_w is the wall surface area of the control volume. If P is within the turbulent region, then:

$$S_P = -\rho C_\mu^{1/4} k_P^{1/2} A_w / u_P^+ \tag{8.99}$$

Similarly the heat flux and generation of species (when the concentration equation is solved) from the wall is calculated from the expressions for T_p^+ (equations (8.92) or (8.96)) and added to b and c of the energy equation. If P is within the laminar sublayer region:

$$S_P = -\mu C_p A_w / (\sigma_l y_P)$$
$$S_U = \mu C_p T_w A_w / (\sigma_l y_P) \tag{8.100}$$

If P is within the turbulent layer:

$$S_P = -\rho C_\mu^{1/4} k_P^{1/2} C_p A_w / T_P^+$$
$$S_U = \rho C_\mu^{1/4} k_P^{1/2} C_p T_w A_w / T_P^+ \tag{8.101}$$

If an adiabatic wall is specified then $S_P = S_U = 0$.

For turbulent flow the kinetic energy at the wall, k_w, is zero but the value at point P, k_p, is calculated from the discretization equation for k with the generation terms (equation (8.34)) represented by:

$$\int_V \mu_t \left(\frac{\partial u}{\partial y} + \frac{\partial v}{\partial x} \right)^2 dV \approx \frac{\tau_P u_P \delta V}{y_P}$$

and so on for the other generation terms. The coefficients of the sink and source terms are given by:

$$S_P = -\rho C_\mu^{3/4} k_P^{1/2} u_P^+ \Delta V / y_P$$
$$S_U = \tau_w u_P^+ \Delta V / y_P \tag{8.102}$$

Unlike k, which falls to zero at a solid boundary, ϵ reaches a maximum value near the boundary, i.e. $(\partial \epsilon / \partial y)_w = 0$. Hence, the dissipation at the first computational point P, ϵ_p, cannot be calculated from the discretization equation as in the case of k_p. Instead, ϵ_p should be calculated using an appropriate expression that depends on whether P is located within the viscous sublayer ($y^+ < 8$) or within the turbulent layer ($y^+ > 30$). For the former, ϵ_p is given by [50]:

$$\epsilon_p = 2\nu_1 k / y_P^2$$

If P is located outside the laminar sublayer, then the expressions for S_P and S_U are multiplied by a large arbitrary value of the order 10^{10}–10^{30} to account for the large dissipation of turbulence energy at the surface, i.e.:

$$S_P = -10^{30}$$

$$S_U = C_\mu^{3/4} k_P^{3/2} / (\kappa y_P) \cdot 10^{30} \tag{8.103}$$

Free boundary

A free boundary in a flow domain is used in situations where the boundary pressures are known such as for a free stream at the boundary or a stagnant surrounding fluid. Applications include an air jet diffusing in stagnant surroundings, external flows around objects, buoyancy-driven flows, etc. Normally the velocity, temperature, pressure (p = constant), turbulence quantities, etc. are specified. The pressure correction at the boundary nodes is set to zero by specifying $S_U = 0$ for the source term and a large negative number, e.g. $S_P = -10^{30}$, for the sink term (see equation 8.78). If the flow in a free or wall jet is being solved the free boundary must be specified a long distance from the flow where no flow due to jet entrainment crosses the free boundary, or alternatively an entrainment velocity (normal to jet axis) may be specified at the free boundary as a fraction of the maximum jet velocity, i.e.;

$$V_e = C_e U_m$$

where V_e is the entrainment velocity, U_m is the maximum jet velocity across the plane and C_e is an entrainment coefficient for the jet, which may be obtained from Rajaratnam [51]. The temperature at the free boundary is the bulk temperature of the surrounding fluid and the pressure is the static pressure of the fluid at the boundary.

Conditions at supply outlets

The physical parameters at the supply outlet are problem specific and are usually known or may be calculated from other quantities. It is necessary to specify the velocity components, fluid temperature, concentration level and the turbulence quantities at the outlet. Using subscript 'o' to indicate supply values the following quantities are required: U_o, V_o and W_o as the velocity components, T_o as the outlet temperature and k_o and ϵ_o as the kinetic energy and dissipation rate. These are usually taken as being uniformly distributed across the supply unless assumed otherwise. The turbulence kinetic energy may be calculated from the known velocity components and the turbulence intensity of the supply stream. Thus:

$$k = \tfrac{1}{2}\left[(\overline{u'})^2 + (\overline{v'})^2 + (\overline{w'})^2\right] = \tfrac{1}{2}\left[I_u^2 U_o^2 + I_u^2 V_o^2 + I_u^2 W_o^2\right]$$

where the turbulence intensities are defined by:

$$I_u = \overline{u'}/U_o \qquad I_v = \overline{v'}/V_o \qquad I_w = \overline{w'}/W_o$$

If the turbulence at the outlet is assumed to be isotropic (a reasonable assumption for most practical cases), then:

$$\overline{u'} = \overline{v'} = \overline{w'}$$

and

$$k_o = 1.5I_u^2 U_o^2 \tag{8.104}$$

The dissipation rate at the supply may be calculated using either of the following expressions:

$$\epsilon_o = k_o^{1.5}/\lambda H \tag{8.105}$$

$$\epsilon_o = C_\mu k_o^{1.5}/l_o \tag{8.106}$$

In equation (8.105), H is the height of the enclosure and λ is a constant (≈ 0.005). However, l_o in equation (8.106) is the mixing length at the supply (taken as a ratio of the height of the supply opening) and C_μ is a constant ($= 0.09$). The difference between these two expressions is that, in the first, the dissipation of turbulence energy of the supply is related to the size of the flow field, whereas in the second it is related to the size of the supply opening itself. There does not seem to be any clear guidance on which of these two formulae should be used but in this author's experience there is little difference between the two on the final solution. This is confirmed by Saïd *et al.* [52] who reviewed the expressions used for describing the turbulence intensity, k_o, and its dissipation rate, ϵ_o, at the inlet, and studied their effect on the air flow pattern, velocity and turbulence intensity in a room served by a supply diffuser. The study showed that the computed results converged to values close to those measured at a faster rate when lower values of k_o and ϵ_o are used, but the final solution was not influenced by the values of I_u in the range 4–37% used in the study. Similarly, the computed results did not seem to be affected by value of ϵ_o used.

The pressure at the supply outlet only needs to be specified and because the pressure is linked to the velocity it is not required at other boundaries in the field except in the case of a free boundary (see previous subsection). Usually, the supply static pressure can be set as zero unless another value is required to be specified.

Exit conditions

The transverse velocity components at the exit are normally set to zero and the longitudinal exit velocity, U_e is calculated from mass balance, i.e.:

$$U_e = U_o \left(\frac{A_o \rho_o}{A_e \rho_e} \right) \tag{8.107}$$

where subscripts e refer to values at the exit opening(s) and o refer to supply opening(s). Similarly, the exit temperature T_e or concentration c_e are calculated from an energy or concentration balance for the whole flow field respectively, taking into consideration heat transfer or concentration generation flux at all boundaries. Exit values for k and ϵ are not required because the Reynolds number at the exit is usually large and an upwind difference scheme is used. The gradients (normal to flow direction) of the dependent variables may also be set to zero at the exit plane.

Obstacle boundaries

Within the domain of an obstacle a false source term is added to the discretization equation of each dependent variable, ϕ, so that the coefficients of the source and sink terms are given an arbitrary large value, i.e. $S_P = -10^{30}$ and $S_U = 10^{30}$. At the boundaries of the obstacle the same wall function treatment as a solid boundary may be used for velocity, temperature, concentration, $\bar{\tau}$, k and ϵ. However, this treatment may not be valid where no established boundary layer exists on the surface as in the case of flow separation over an obstacle.

8.4.5 *Accuracy of CFD results*

The accuracy of an iterative solution of the transport equations using computational fluid dynamics (CFD) depends on a number of factors and in particular the following:

- The discretization scheme
- The computational grid
- The near-wall boundary conditions
- The convergence criteria.

We shall consider some of the issues that impact on the accuracy of the final solution in the FVM. Additional information is found in Roache [53, 54].

The discretization scheme

Both the central difference (CDS), upwind difference (UDS) and the hybrid (HS) schemes are based on first-order terms, i.e. second-order truncation of the Taylor series term, which is equivalent to the diffusion terms in the transport equations. This truncation, therefore, often leads to 'numerical' or 'artificial' diffusion, corresponding to the viscosity in the Navier–Stokes equations. The resulting solution inherently produces the sum of the true viscosity and the numerical viscosity. The UDS also produces erroneous results when the flow is not aligned with the grid lines, in which case the numerical diffusion due to the crossflow will be significant in comparison with the effective diffusion. This leads to 'numerical viscosity' that will diminish the influence of the turbulent viscosity calculated by the turbulence model. Higher-order schemes, such as QUICK and its derivatives, often produce non-physical 'overshoots' in the solution domain, which can contribute to errors that are comparable to those from a first-order scheme. Errors due to both numerical diffusion and overshoots are reduced by selecting a finer grid.

The computational grid

The effort used in discretizing the computational domain to construct a suitable grid is often the most demanding task of CFD computation. Although each problem has its unique computational grid requirement, there are however, general computational errors that will be present even with a reasonably well-constructed grid. Errors associated with grid resolution arise from interpolation between neighbouring grid points and these errors may be minimized by increasing the number of grid points. An ideal

CFD solution should be independent of the computational grid, i.e. the same results will be obtained if the grid used to obtain them is further refined. However, because of the higher computational demand associated with fine grids, in practice a solution that is completely grid-independent is strictly unattainable and the concept of 'grid convergence' is applied. Obtaining grid convergence implies that the solution obtained asymptotically approaches the exact solution as given by the transport equations. However, often an exact solution is not known and other concepts need to be used. Roache [53] has introduced the concept of 'grid convergence index' (GCI) that provides an estimate of the error between the solution and the 'unknown' exact solution. This is defined as:

$$GCI = F|(f_2 - f_1)/(1 - r^p)| \qquad (8.108)$$

where $r = h_2/h_1$, h_1 and h_2 being the cell size for fine and coarse grids respectively (i.e. $r > 1$, typically $= 2$); f_1 and f_2 are the solutions obtained (e.g. value of velocity, temperature, concentration, etc.) by the fine and coarse grids respectively, at a certain point in the computational domain; p is the order of accuracy of the differencing scheme; and F is a factor ($= 3$).

An estimation of p may be made by considering a third coarser grid that produces a solution f_3 as follows:

$$p = \frac{\ln((f_3 - f_2)/(f_2 - f_1))}{\ln(r)} \qquad (8.109)$$

The aim is to achieve the smallest GCI value that is practical.

A less-formal approach towards a grid-independent solution than the one suggested by Roache can also be pursued to achieve the desired result. The user of CFD code would typically perform a simulation with a coarse grid first to get an impression of the overall features of the solution. Subsequently, the grid is refined in stages until no 'significant' difference in the results is found between successive grid refinements. The final results are then taken to be 'grid independent'. In the case of a non-uniform grid, experience shows that the expansion ratio (i.e. dimension ratio of two adjacent cells) should be limited to 1.5.

The near-wall boundary conditions

If a low Reynolds number turbulence model is used then it is essential that a number of computational cells are placed within the laminar sublayer ($y^+ < 8$). Normally, y_p^+ is chosen to be < 1.0. This inevitably requires an extremely fine grid, which is uneconomical for three-dimensional flow problems. If the wall functions discussed in Section 8.4.4 are to be used near solid boundaries, then it is important that the position of y_p is such that $30 < y_p^+ < 130$ to ensure that the wall functions for turbulent flow are invoked in the solution for improved accuracy, particularly the turbulent energy dissipation, ϵ, as was discussed.

The convergence criteria

It is clear that in an iterative solution, the variables change from one iteration to another and a residual (error) is defined as indicator of this change. A convergence criterion may be based on an acceptable value of the largest residual in the discretization equation or an acceptable difference in the value of ϕ between two successive iterations. Most CFD codes have a default convergence criterion, which assumes that a converged solution is achieved when reached. The residual approach is to be recommended because every iterative solution produces residual sources. The residual sources for the discretization equation (8.78) can be defined by:

$$R_\phi = \left(\sum_i a_i + a_P^0 - b \right) \phi_P - \sum_i a_i \phi_i - c \tag{8.110}$$

To achieve a converged solution of ϕ we require that the sum of the residues for all computational cells be less than a given value, i.e.:

$$\sum |R_\phi| < \lambda F_\phi \tag{8.111}$$

where F_ϕ is a known flux of ϕ in the field (e.g. at the inlet) and λ is an arbitrary factor of the order of 0.001. This normalized residual-source test ensures that the finite difference equations have been solved.

8.5   CFD applications to room air movement

In the last three decades there has been extensive activity in the use of commercial CFD software and in developing special programs for room air movement applications. The applications of CFD range from the prediction of air jet diffusion, calculation of air velocity and temperature distribution in rooms, spread of contamination in enclosures, natural ventilation assessment, to predicting fire and smoke spread inside buildings. In most cases the predicted results have been promising when compared with available experimental data. Although considerable progress has been made in the numerical modelling of building ventilation, more research and development work is still needed, particularly in establishing more robust computational schemes, better irregular griding techniques, improved turbulence modelling, and more universally applicable wall functions in order to establish CFD as a reliable design and research tool. Developments in some of these areas are reported by e.g. Patankar [55].

In this section some typical applications of CFD to ventilation studies are given and, where possible, a comparison is made between prediction and measurement. It should be noted that the investigations reported here are mostly based on the finite volume solution procedure described in the previous sections. The cases presented here by no means represent a complete listing of the numerous studies that have been carried out in the last three decades but they intend to give an appreciation of where CFD can be used and what degree of confidence one can expect in the results.

8.5.1   *Air jets*

Launder and Spalding [11] used a finite difference marching integration procedure, described in Patankar and Spalding [56], and the mixing length turbulence model to predict the velocity and temperature profiles and the decay of the maximum axial velocity for a free circular jet and a plane wall jet. They found that in order to correlate their predictions with experimental data it was necessary to adjust the mixing length ratio L_m/δ, the Karman constant κ and the turbulent Prandtl number σ_t. The parabolic form of the two-dimensional momentum equations (i.e. $\partial p/\partial y = 0$ and $\partial p/\partial x = f(x)$) were used in their numerical solution.

Hjertager and Magnussen [57] solved the transport equation for momentum and the two turbulence parameters (k and ϵ) using an upwind difference scheme and the SIMPLE algorithm to calculate the three-dimensional isothermal flows of a turbulent free jet and a turbulent wall jet in enclosures. The first case was of a jet supplied through a square slot at the centre of a wall in a square-section enclosure of length to width (height) ratio 3. The ratio of width of the supply opening to the width of the enclosure was 0.1. The overall dimensions of the enclosure were 0.1 m × 0.1 m × 0.3 m and only a quarter of the flow field was modelled using a computational grid of 9 × 9 × 9 with 4 × 4 grid points used at the supply opening. The turbulence kinetic energy at the supply (k_o) was obtained from a measurement of the turbulence velocity and equation (8.104) and the dissipation rate ϵ_o was calculated from equation (8.106) using a length scale of 0.1 × width of the opening. For the wall jet case, a high-level sidewall slot of height $h = 0.056\,H$ (where H is the enclosure height, 89.3 mm) and width $b = 0.5H$ was located in the middle of a square-section enclosure of length $L = 3H$. An effective supply velocity of 18.5 m s^{-1} (i.e. $Re = U_o d/\nu \approx 1.2 \times 10^4$) was used for the free jet and 15 m s^{-1} (i.e. $Re = U_o h/\nu \approx 5 \times 10^3$) for the wall jet. Figure 8.6 shows a comparison between the predicted and experimental results for the free jet case of the decay with axial distance of the maximum axial velocity of the jet and the root mean square (r.m.s.) of the axial fluctuating velocity u'_m. The predicted velocity profiles across the enclosure at distances of $x/H = 1$ and 2 are shown in Figure 8.7 for the wall jet case. In this case the experimental results of Nielsen *et al.* [58] were used for comparison. Both these figures show good agreement between the computed and measured data.

An extensive investigation into the flow of a wall jet and the interference of obstacles on the jet has been carried out by Awbi and Setrak [59–62] both numerically and experimentally. The numerical solution was carried out using a finite volume computer program developed by the present author, which is an earlier version of the VORTEX© (Ventilation Of Rooms with Turbulence and Energy eXchange) [63]. The program solves the three-dimensional momentum equations, the energy equation,

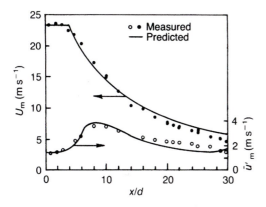

Figure 8.6 Predicted and measured decay of the axial velocity and turbulence velocity for a free jet in an enclosure [57].

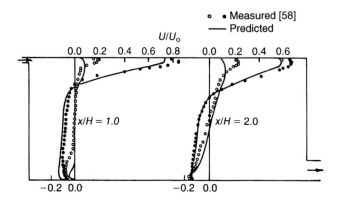

Figure 8.7 Predicted and measured velocity profiles across a ceiling jet in an enclosure [57].

the concentration equation the mean age equation and the two equations for k and ϵ, including the buoyancy terms in these equations. Because the emphasis in this work was on the prediction of velocity and temperature profiles of a wall jet and how these are influenced by the presence of an obstacle on the wall, it was necessary to use a fine computational grid close to the wall where large diffusion occurred within the boundary layer region. However, gradients within the shear layer region were less steep and a coarser grid was adequate there. This griding was accomplished using an exponentially expanding spacing both in the axial direction (x direction) and normal to the wall (y direction) but a uniform grid was used in the lateral direction (z direction). For a plane wall jet (two-dimensional flow), a grid of 41×27 was used and for a three-dimensional wall jet a grid of $20 \times 20 \times 11$ was employed.

The experimental measurements were carried out in an open-top air jet test rig 5 m long $\times$ 1.03 m wide $\times$ 0.65 m high side walls. The air was supplied through a converging rectangular nozzle of variable geometry giving a maximum slot height of 35 mm for the whole width of the nozzle or less as required. The measurements were carried out using Dantec hot-wire anemometers and low-velocity hot-film anemometers. The instruments were precisely traversed across the jet using a stepper motor. Full details of the test rig are given by Setrakian [61].

Predicted and measured normalized velocity and temperature profiles (see Chapter 4) in the fully developed region (i.e. downstream of the core) of a plane wall jet are shown in Figure 8.8(a) and (b). The experimental data of Rajaratnam [51] and Albright and Scott [64] are also given in the figure for comparison. As shown the predicted results are very close to the measured values. Figure 8.9(a) and (b) shows normalized velocity profiles for a three-dimensional wall jet produced by a rectangular nozzle of height $b = 19$ mm and width $b = 62.7$ mm at a distance $x/\sqrt{A}$ of 14 (where x is the axial distance from the nozzle and A is the area of the supply opening), which is in the fully developed region of the jet. Figure 8.9(a) is a profile in an x–y plane (normal to the wall) through the centre of the nozzle and Figure 8.9(b) is an x–z plane (parallel to the wall) where the jet velocity is a maximum. Here again, the predictions are close to the measurements carried out in the air jet test rig. The decay of the maximum velocity in the axial direction of the same jet is shown compared with the experimental data of Rajaratnam and Pani [65] in Figure 8.10.

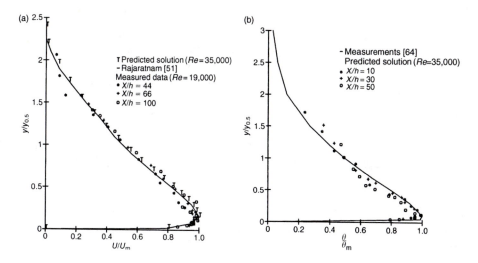

Figure 8.8 Normalized velocity (a) and temperature (b) profiles for a plane wall jet.

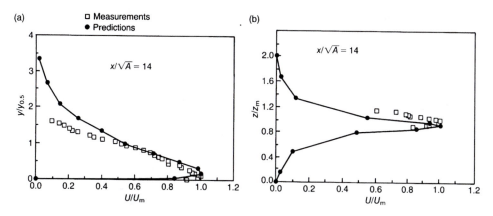

Figure 8.9 Non-dimensional velocity profiles for a three-dimensional wall jet [61]. (a) Plane vertical to wall and (b) plane parallel to wall.

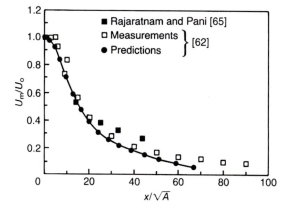

Figure 8.10 Decay of maximum velocity of a three-dimensional wall jet.

As explained in Chapter 4, the presence of an obstacle in the path of a plane wall jet can either cause a local separation of the jet followed by reattachment to the wall downstream of the obstacle or the jet may completely leave the wall. These types of flow were investigated numerically and experimentally by Awbi and Setrak [60, 62] and by Setrakian [61]. Figures 8.11 and 8.12 show comparisons between calculated and measured non-dimensional resultant velocity profiles ($U_r = \sqrt{(u^2 + v^2)}$) at various positions upstream, at, and downstream of, an obstacle for the two cases of reattaching and separating plane wall jet. In both figures the distance of the obstacle from the supply nozzle is the same, but, in Figure 8.12 the height of the obstacle, d, is larger, i.e. $d/h = 2.68$ in Figure 8.12 and 1.3 in Figure 8.11. The agreement between prediction and measurement is good for most profiles. There are some discrepancies for the profiles in Figure 8.11(a) (i.e. upstream of the obstacle) and 8.11(c) at the top of the obstacle. In the first case the discrepancy was attributed to the failure of the hot-wire probe to differentiate between positive and negative velocities and as a result velocities outside the shear layer (a region of recirculation) were taken to be positive but the computed results showed negative velocities in that region. The discrepancy in Figure 8.11(c) is caused by the underprediction of the curvature of the streamlines at the top of the obstacle by the CFD solution. This was caused by imposing boundary

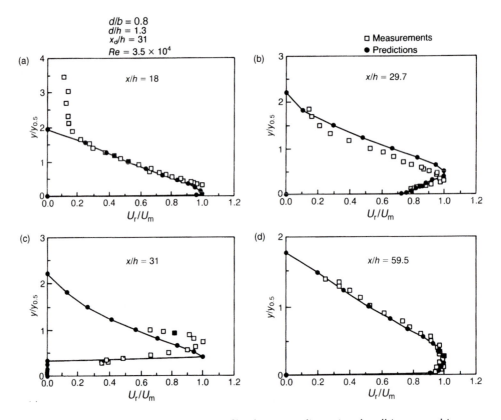

Figure 8.11 Non-dimensional velocity profiles for a two-dimensional wall jet reattaching over an obstacle [61].

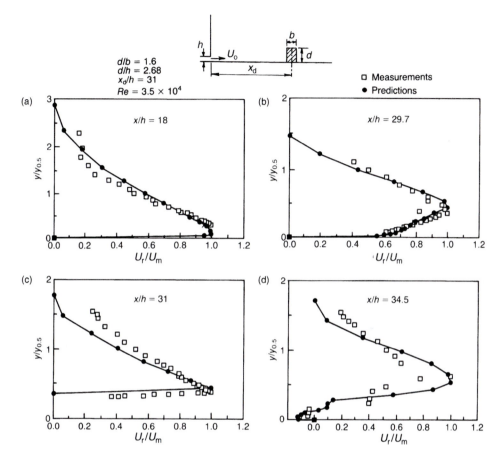

Figure 8.12 Non-dimensional velocity profiles for a two-dimensional wall jet separating over an obstacle [62].

layer wall functions at the top surface (and other surfaces) of the obstacle that formed an unrealistic boundary layer there instead of the recirculation region, which normally exists between the separating shear layer and the top of an obstacle. Clearly, better wall functions were needed in this case to improve the CFD prediction.

Using a three-dimensional isothermal solution the flow of a plane wall jet over a finite length obstacle was investigated by Setrakian [61]. Figure 8.13 shows normalized velocity profiles in a vertical x–y plane near the edge of the obstacle and at different axial positions upstream and downstream of the obstacle. The computed velocities are close to those measured in an air jet rig. The length of the obstacle was $0.3 \times$ the width of the channel and it was not possible in this case to achieve a complete separation of the jet from the surface. The flow of a three-dimensional wall jet over an obstacle spanning the width of the jet channel was also investigated. Typical computed and measured velocity profiles in an x–y plane through the centre of the nozzle are shown in Figure 8.14.

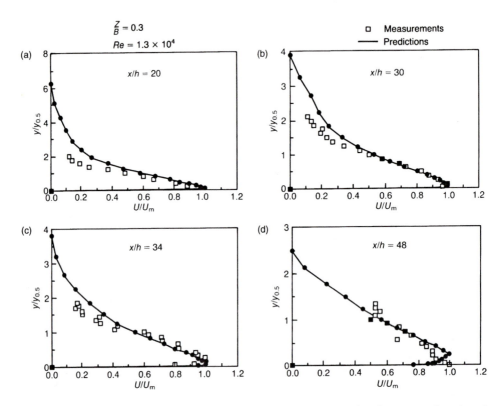

Figure 8.13 Non-dimensional velocity profiles for a two-dimensional wall jet over a finite length obstacle: *xy*-plane near the obstacle edge [61].

The effect of surface roughness on the diffusion of a two-dimensional wall jet was investigated numerically [60] by distributing rectangular-section roughness blocks along the surface. The height ratio of the roughness block *d/h* (which was varied in the range $0.19 < d/h < 1.3$) was found to influence the decay of the maximum jet velocity as shown in Figure 8.15. However, the effect of the pitch of the blocks was not significant for the two pitch ratios, *p/h*, of 5.1 and 6.6 that were used in the investigation.

In the testing of air terminal devices the wall facing the terminal causes interference to the jet well upstream of the wall. To evaluate the interference distance experimentally is not an easy task because of the complex three-dimensional flow, which exists close to the far wall. Awbi and Setrak [59] obtained a numerical solution of the influence of the far wall on the diffusion of a two-dimensional ceiling jet. The effect of the wall on the decay of maximum jet velocity is shown in Figure 8.16 for rooms of different lengths. A correlation was obtained between the length of the jet that is free from wall interference, x_f, and the room length, *L*, both non-dimensionalized with respect to the height of the supply slot, *h*. This is given by equation (4.66) as follows:

$$x_f/h = 0.52(L/h)^{1.09}$$

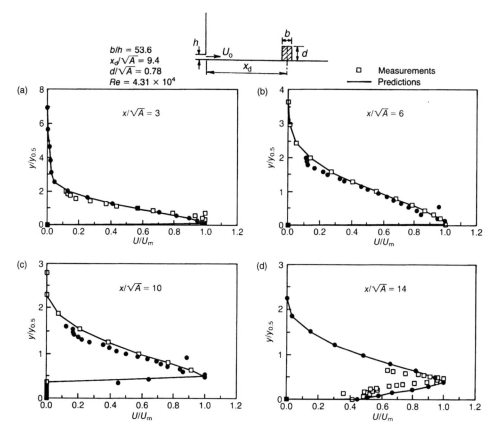

Figure 8.14 Non-dimensional velocity profiles for a three-dimensional wall jet over a long obstacle: central *xy*-plane. (a) Core region; (b) upstream of obstacle; (c) above obstacle; and (d) downstream of obstacle [62].

Mohammed [66] studied the diffusion of a two-dimensional non-buoyant turbulent jet discharging parallel to and offset from a solid boundary using a two-dimensional finite volume program that solves, in addition to the transport equations for the two velocity components and temperature, the transport equations for k and ϵ. The flow in an offset jet that attaches to the surface as a result of the Coanda effect may be divided into three regions: (i) the reverse flow region in the corner, (ii) the reattachment region, and (iii) the wall jet region. Among other things, Mohammed calculated the decay of the maximum axial velocity of the jet in the three regions. Figure 8.17 shows a comparison between prediction and measurement for a jet of an offset ratio (distance from surface to centre of slot divided by slot height), $H/h = 5.7$. The initial decay in velocity up to the turning point corresponds to the recirculation region that ends at the turning point. A slight increase in the axial velocity occurs in the reattachment zone followed by a gradual decrease in the wall jet region similar to that for a plane wall jet. As shown in Figure 8.17 the axial velocity is not very well predicted near the reattachment point. This was attributed to the underprediction of the reattachment distance by the $k-\epsilon$ model, which is now recognized as one of the limitations of the latter.

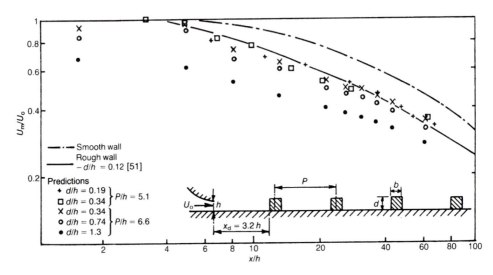

Figure 8.15 Effect of wall roughness on the decay of maximum velocity of a two-dimensional jet [61].

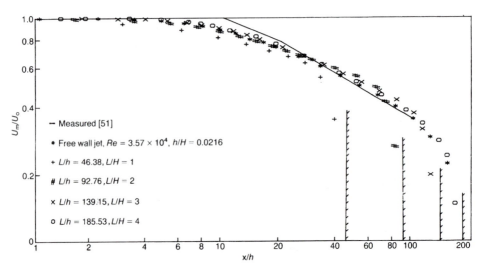

Figure 8.16 Effect of the far wall on the decay of a ceiling jet for rooms of different length to height ratios [59].

8.5.2   *Air flow in rooms*

One of the earliest attempts to simulate the air flow in rooms numerically was by Nielsen [67]. He used the stream function, ψ, and vorticity, ω, as the dependent variables in the two-dimensional Navier–Stokes equations and applied the finite difference solution procedure developed by Gosman *et al.* [68] to solve these two equations and other transport equations for temperature and mass concentration as well as k and ϵ.

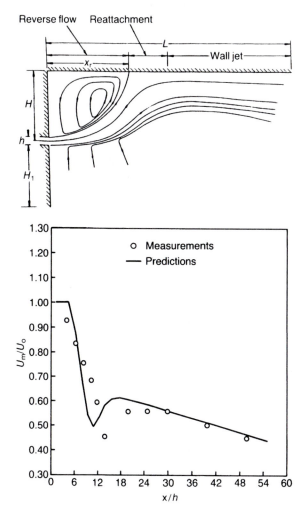

Figure 8.17 Predicted and measured decay of the maximum axial velocity for a two-dimensional offset jet of $H/h = 5.7$ [66].

The two velocity components were calculated from the field values of the stream function and vorticity. This approach produced realistic predictions of room flows that were approximately two-dimensional. Nielsen [69] also applied this method to predict moisture distribution in cold stores with the buoyancy terms ignored in the transport equation for ψ.

Timmons *et al.* [70] developed an inviscid flow model for two-dimensional slot-ventilated enclosures. The model solved the vorticity equation (Poisson's equation) for the stream function, which is given by:

$$\frac{\partial^2 \psi}{\partial x^2} + \frac{\partial^2 \psi}{\partial y^2} = f(\psi)$$

The function $f(\psi)$ was derived from semi-empirical expressions based on jet theory relating the stream function of a jet to the entrainment, expansion and vorticity of the jet. The vorticity equation was then solved iteratively using a uniform grid of 21×21 and a successive overrelaxation scheme to obtain the values of ψ at each node. The two velocity components u and v were derived from the stream function field by using a forward difference solution of $\partial\psi/\partial y$ and $-\partial\psi/\partial x$. Timmons *et al.* applied this method to examine the effect of enclosure length (in the jet direction) on the recirculating flow and to determine jet attachment lengths to the surface of the enclosure. The prediction of recirculating flow patterns in the enclosures was generally in agreement with those obtained by flow visualization tests. However, the velocity values close to the enclosure surfaces were overpredicted because of the breakdown of the inviscid flow assumption (perfect slip) in the wall region.

Since the work of Nielsen and Timmons *et al.*, few new developments were made in using the stream function solution of the flow equations mainly because of two major drawbacks associated with this numerical approach. First, the flow in a ventilated enclosure is governed by elliptic equations and is turbulent and any solution based on parabolic equations, such as those solved using the stream function method, will not produce accurate predictions, particularly if inviscid flow is assumed as was the case with Timmons *et al.*'s solution. Second, this approach is only suitable for two-dimensional flows and in most enclosures the flow is invariably three-dimensional even when linear slot diffusers span the width of the room.

One of the earliest reported numerical studies of room air movement applying the SIMPLE algorithm developed by Patankar and Spalding [41] was that by Hjertager and Magnussen [71]. They solved the three-dimensional transport equations for the three velocity components u, v and w, the energy equation and the turbulence energy, k, and its dissipation rate, ϵ, using the finite volume procedure. The buoyancy effect was included in the equation for the vertical components of the velocity but not for k and ϵ. They applied this solution to predict the air velocity and temperature distribution in a room 5.6 m long $\times$ 2.9 m wide $\times$ 2.4 m high, supplied by an air jet from a 243 mm $\times$ 35 mm rectangular nozzle on a 2.9 m $\times$ 2.4 m wall at ceiling level and the extraction was from two nearby ceiling locations, see Figure 8.18. Both isothermal and non-isothermal (cooling) flows were predicted and in the latter case the room heat load was in the form of partly heated floor and partly heated far wall. A nine-point non-uniform grid was used in each direction with a larger concentration of points near the room surfaces. Figure 8.18 shows a comparison between predicted and measured velocities in a plane of symmetry through the supply nozzle for a supply jet temperature 11 K below room temperature and a supply velocity of 2.42 m s^{-1} giving a Reynolds number and an Archimedes number based on the hydraulic diameter of the nozzle of 9800 and 0.0038, respectively. It is shown that the point of jet separation from the ceiling has been well predicted, but the predicted velocity of the jet as it enters the lower region of the room was higher than the measured value. However, the predicted velocity fields for the isothermal case were closer to measurements than the non-isothermal case. No plausible explanation was given by Hjertager and Magnussen for the higher predicted velocities in the non-isothermal case but this may have been attributed to the authors ignoring the buoyancy terms for k and ϵ.

Nielsen *et al.* [58] solved the transport equations for velocities, k and ϵ, in two-dimensions using the TEACH computer code [45], which employs the SIM-PLE algorithm and a hybrid computational scheme. The purpose of this work was

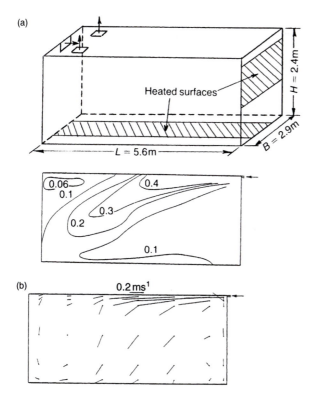

Figure 8.18 Comparison of measured velocity contours and predicted velocity vectors in a vertical plane through the centre of a room supplied by a cool jet at high level [71]. (a) Measurement and (b) prediction.

two-fold. The first was to assess the accuracy of the numerical procedure in predicting two-dimensional isothermal flows in ventilated enclosures and the second was to establish the extent to which a two-dimensional solution may be used to represent three-dimensional ventilation problems. The numerical predictions were compared with measurements made in a Perspex model enclosure of 89.3 mm square section (i.e. equal height, H, and width B) and a length, L, three times this value. The air was supplied from a high-level slot adjacent to the ceiling with a fixed height, h, of 5 mm and two widths: one equal to the width of the enclosure and the other half of that. The air extraction was from a slot at floor level on the far wall spanning the width of the room. Measurements of air velocity were made with a laser Doppler anemometer. A range of Reynolds numbers between 5000 and 10,000 was used. To avoid using a very fine computational grid at the supply opening, Nielsen *et al.* specified the supply conditions using empirical wall jet data that enabled them to employ a coarse grid at the supply. Figure 8.19 shows a comparison between calculated and measured velocity profiles in an *x*–*y* plane through the centre of the supply aperture at a distance of 2/3*L*. As shown the velocities have been generally well predicted apart from some discrepancy in the reverse flow region, which Nielsen *et al.* attributed to three-dimensional flow effects in the physical model. In Figure 8.20 the effect of the discharge angle of the jet, α, on

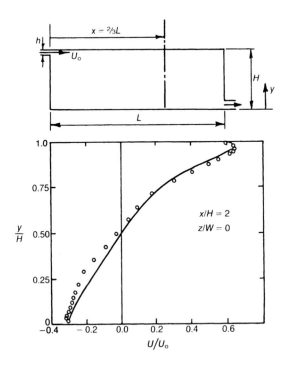

Figure 8.19 Comparison between predicted and measured velocity profiles in a two-dimensional flow [58]. o, measured; −, predicted.

the maximum reverse velocity, U_{rm}, is shown using the same numerical solution. It is obvious that U_{rm}/U_o decreases as the discharge angle increases and negative angles produce larger velocities in the reverse flow. The effect of the ratio of the slot height to enclosure height, h/H, on the maximum reverse velocity for a horizontal air jet supply was also predicted by Neilsen *et al.* and the results are shown in Figure 8.21 for three length to height ratios, L/H, of the enclosure. It is clear that the larger h/H and the smaller L/H are the greater is the reverse velocity.

Davidson and Nielsen [72] used the FVM with an LES turbulence model to predict the mean and r.m.s. velocity profiles across the room previously studied by Nielsen *et al.* [58] and shown in Figure 8.19. They used two different subgrid turbulence models to model the turbulent scales that are smaller than the computational cells. Using three different grid sizes ($72 \times 52 \times 25$, $72 \times 42 \times 52$ and $102 \times 52 \times 52$), it was found that the results were grid dependent, particularly the r.m.s. velocity, $\sqrt{u'^2}$. The results for the largest grid are shown in Figure 8.22 compared with the measurements of Nielsen *et al.* [58]. The mean velocity in this figure ($x/H = 2$) may be compared with that in Figure 8.19 as they represent exactly the same location in the room.

Nielsen *et al.* [73] extended the two-dimensional isothermal flow solution to a non-isothermal flow in which case an additional transport equation for temperature was solved and buoyancy terms were added to the transport equations for v-velocity, k and ϵ given as equations (8.22), (8.34) and (8.37). The room heat load was uniformly distributed over the floor and a cool jet was supplied to offset this load. The predicted

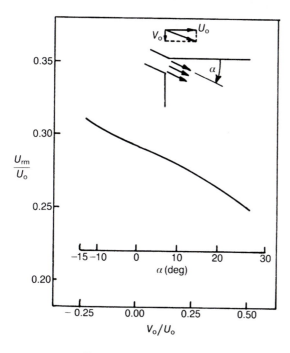

Figure 8.20 Effect of discharge angle on the predicted maximum velocity in the reverse flow of a two-dimensional enclosure [58].

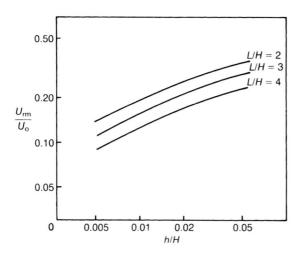

Figure 8.21 Influence of slot height and room length on the predicted maximum velocity in the reverse flow of a two-dimensional enclosure [58].

results were compared with experimental data, where available, for a room of $L/H=3$. The Reynolds number was about 7000. The effect of Archimedes number, based on the slot height, on the predicted maximum reverse flow velocity is shown in Figure 8.23 for two h/H ratios. As would be expected, the higher Ar is the greater is the reverse flow

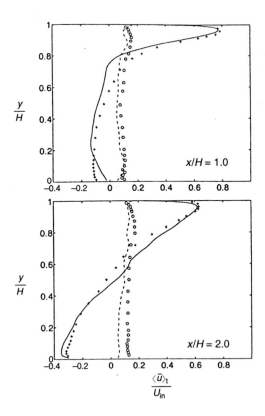

Figure 8.22 Time average and r.m.s. velocity profiles in the room shown in Figure 8.19 [72]. o, measured; ——, predicted mean velocity; +, measured; - - -, predicted r.m.s. velocity.

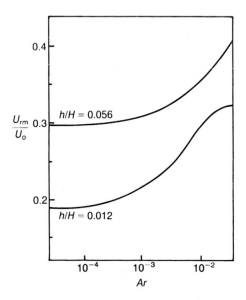

Figure 8.23 Effect of Archimedes number on the maximum reverse flow velocity in a two-dimensional flow (predicted) [73].

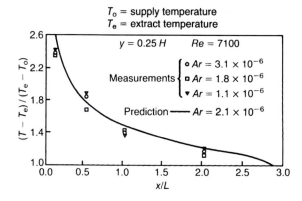

Figure 8.24 Predicted and measured non-dimensional temperature profiles in a horizontal plane [73].

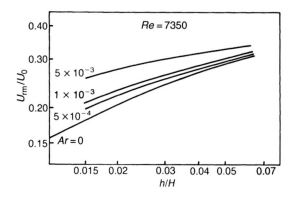

Figure 8.25 Effect of slot height and Archimedes number on the maximum reverse velocity in a two-dimensional enclosure (predicted) [73].

velocity because of greater convection. The non-dimensional horizontal temperature profiles for the room are shown in Figure 8.24 for a plane at a height of $0.25H$ from the floor. The predicted temperatures are shown to correlate well with measurements. The effect of h/H and Ar on the predicted maximum velocity in the reverse flow is shown in Figure 8.25 normalized with reference to U_0.

An interesting phenomenon sometimes experienced in VAV systems is the hysteresis effect during a gradual increase or decrease in the value of Ar when jet separation from the ceiling occurs (i.e. dumping). Nielsen *et al.* [73] predicted this numerically by using as initial field values the results from a converged solution at different Ar (i.e. higher or lower Ar than the one to be solved). Figure 8.26(a) shows this effect on the distance of jet penetration on to the floor, x_r, and although there is a considerable difference between the calculated and measured values the main pattern was correctly reproduced numerically. The effect of Reynolds number on the penetration length is shown in Figure 8.26(b) [74].

Gosman *et al.* [75] extended their two-dimensional computer code to solve three-dimensional flows in enclosures. They used data from their Perspex model, with a

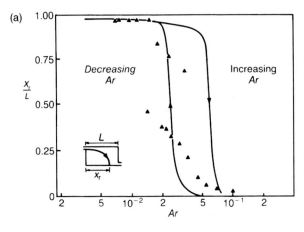

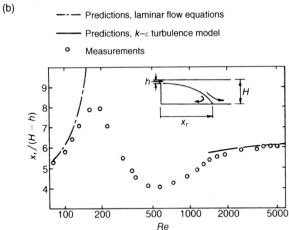

Figure 8.26 (a) Dependence of a jet penetration length on the Archimedes number in two-dimensional enclosures [73]. Δ, measured; ———, predicted. (b) Effect of Reynolds number on the jet penetration length for two-dimensional enclosures [74].

high-level square opening air supply of side $0.1\,H$ and other published data to validate their CFD predictions. A computational grid of $13 \times 15 \times 8$ was used and isothermal flow only was considered. Applying similar jet boundary conditions as in their two-dimensional work, they achieved good correlations of velocity profiles and maximum jet velocity decay with measurements. Figure 8.27 shows a comparison between the predicted maximum reverse velocity U_{rm} obtained in their enclosure for two- and three-dimensional jets. These results show that, for the same ratio of supply area to room area, a/A, a two-dimensional jet produces a larger U_{rm} because of lower entrainment. The velocity ratio, U_{rm}/U_o in both cases is approximately proportional to $\sqrt{(a/A)}$.

Sakamoto and Matsuo [76] used the marker and cell (MAC) computational method to predict three-dimensional isothermal flow in a 2 m cubic room ventilated from a square supply opening in the centre of the ceiling and an identical floor-level side wall

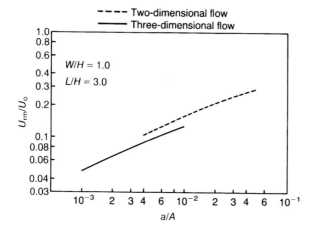

Figure 8.27 Effect of supply area ratio on the maximum reverse velocity for two- and three-dimensional flow [75].

extract. A uniform cubic grid of 18 nodes in each direction was used together with two turbulence models: the k–ϵ model and the LES due to Deardorff [24]. The MAC method is a convenient way of solving time-dependent problems, which are needed when the LES turbulence model is used even when time-mean quantities only are required. In addition to solving directly for the velocity components and pressures, the MAC method also uses marker particles in the computational cells that are convected in the field by the motion of the fluid. The marker particles themselves do not participate in the calculation but only serve to indicate fluid configuration. This is particularly useful in solving problems with a free surface, which can be used to define the free boundary as that where the particle concentration is zero. The particles also serve as a flow visualization coordinate where the positions of fluid elements can be traced. The MAC method can be used in both free boundary and confined flows and further details of the method are found in [77, 78].

Figure 8.28 represents a typical comparison between the k–ϵ model, the LES model and the experimental measurements by Sakamoto and Matsuo [76] for their cubic room. The figure shows the contours of the resultant velocity normalized with the inlet velocity, i.e. V_r/V_o, for the three cases in a vertical x–y plane through the supply and extract openings. The Reynolds number is estimated to be 1.36×10^5. Although the measured profiles show a more rapid diffusion of the air jet into the room than the prediction profiles indicate, the agreement between the k–ϵ solution and the LES solution is reasonable. As a result of this work, Sakamoto and Matsuo recommended the use of the k–ϵ model for room flow predictions because it is simpler to use and requires less computing effort than the LES model.

Murakami *et al.* [79] investigated the three-dimensional air flow and contamination dispersal in six types of ceiling-supply clean rooms both numerically and experimentally under isothermal conditions. They used the MAC method for defining the dependent variables, a central difference computational scheme for the three velocity components and the QUICK scheme (see section 8.4.2) for the scalar variables (k, ϵ and c) and for the velocity near the extract openings. The QUICK scheme was used

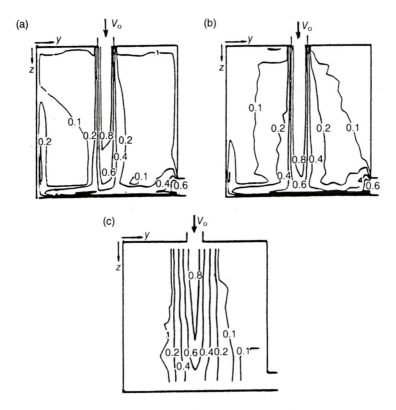

Figure 8.28 Predicted and measured resultant velocity contours in a vertical plane in the middle of a cubic room – the velocities normalized with the supply velocity V_o [76]. (a) k–ϵ model; (b) LES; and (c) measurements.

near the extracts because this improved numerical stability where convection is large. Each transport equation was written in a time-dependent form and numerical time integration, using the Adams–Bashforth scheme, was carried out over sufficiently long time periods until the solution became that for a steady state. The physical models of the six clean rooms were a sixth scale of normal size clean rooms. The ceiling height in all the models was the same and equal to 0.45 m, which corresponds to a fullscale height of 2.7 m. All the models except No. 3 were of a square plan with floorlevel extract ports, one at each corner. The plan area and number of ceiling supply ports (all square) were different for each model. The specification of each model is given in Table 8.4 where all dimensions are normalized with respect to the width of outlet opening. Velocity measurements in the physical models were carried out using two tandem-wire hot-wire anemometers that are capable of measuring the magnitude and direction of velocity.

A comparison between the predicted and measured velocity vectors in different planes across a type 1 clean room (one middle supply opening) is shown in Figure 8.29(a) and (b). The results show a good agreement between prediction and measurement. Some interesting flow phenomena are found in these plots. The central jet reaches the floor with a centreline velocity equal to the supply velocity because the core has not been consumed as the jet reaches the floor ($y/d = 4.5$). After impinging on the floor the jet deflects upwards at the side walls and is then entrained by the central jet creating recirculation zones between the central jet and the side walls.

The flow in a type 2 clean room with four ceiling openings is shown in Figure 8.30(a) and (b). Here again the numerical solution produced a good prediction of the flow

Table 8.4 Specifications of clean-room models [79]

Room type	Floor area (m × m)	Ceiling height (m)	No. of supply openings	No. of extracts	Area of supply opening (m × m)	Supply velocity (m s^{-1})
1	0.5 × 0.5	0.45	1	4	0.1 × 0.1	6
2	0.8 × 0.8	0.45	4	4	0.1 × 0.1	6
3	1.1 × 1.1	0.45	6	4	0.1 × 0.1	6
4	1.1 × 1.1	0.45	9	4	0.1 × 0.1	6
5	1.1 × 1.1	0.45	4	4	0.1 × 0.1	6
6	1.1 × 1.1	0.45	1	4	0.1 × 0.1	6

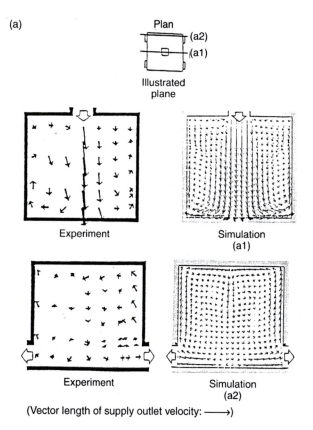

Figure 8.29 Comparison between predictions and measurements in a clean room of type 1 in Table 8.4. (a) Vertical sections showing velocity vectors; (b) horizontal sections showing velocity vectors [79].

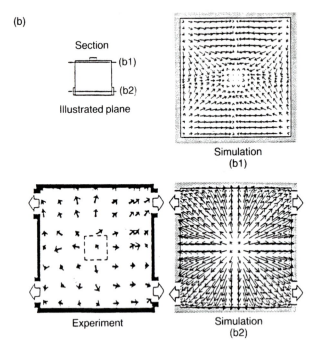

Figure 8.29 Continued.

pattern, which is similar to four type 1 flows superimposed together. Figure 8.30(c) shows a comparison between the predicted and measured normalized concentrations in the room due to a pollution source below one of the inlet openings [80]. Here again the predictions using the k–ϵ model and measurements are good.

A comparison between the velocity vectors in a type 3 clean room with six ceiling openings is given in Figure 8.31. The correlation between prediction and measurement is good except at the centre of the room (between two jets) where section (a) shows that the deflected jet from the floor appears to reach the ceiling according to measurements but not according to predictions. Murakami *et al.* attributed the sensitivity of the flow in this region to the air flow rate supplied to the room, which may not have been equal in both the experimental and the numerical models.

Finally a comparison for a type 4 clean room with nine ceiling openings is given in Figure 8.32. As in the previous rooms a good prediction of the flow was achieved using the CFD solution. The flows for the other two clean-room types, 5 and 6, are very similar to those of types 2 and 1 respectively, except that the rooms in this case have a larger plan area.

Reinartz and Renz [81] investigated the flow in a rectangular room 4.7 m × 3 m × 2 m high ventilated by a plate diffuser positioned in the centre of the ceiling for heating and cooling modes. They solved the transport equations for the velocity components, temperature, k and ϵ in the axial and the radial directions of a vertical plane of symmetry through the diffuser, i.e. a two-dimensional solution. They included the effect of buoyancy on the vertical component of velocity but ignored it in the equations for k and ϵ because it was claimed to be insignificant. Reinartz and Renz probably made

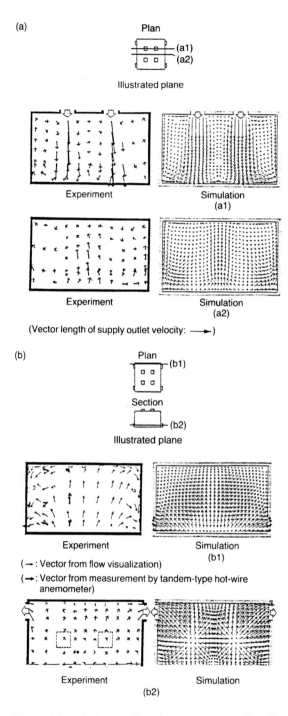

Figure 8.30a, b Comparison between predicted and measured velocity vectors in a clean room of type 2 in Table 8.4 [79, 80]. (a) Vertical sections; (b) horizontal sections.

(c)

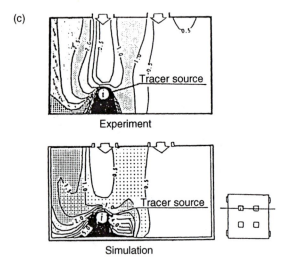

Experiment

Simulation

Figure 8.30c Vertical section showing normalized contaminant concentration.

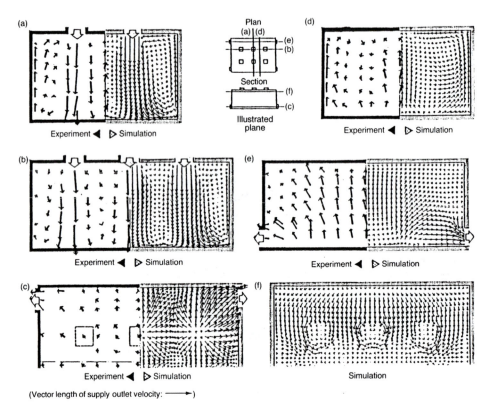

Figure 8.31 Comparison between predicted and measured velocity vectors in a clean room of type 3 in Table 8.4 [79]. (a) Centre of room; (b) including supply outlets; (c) near floor; (d) centre of supply outlets; (e) including exhaust inlets and (f) near ceiling.

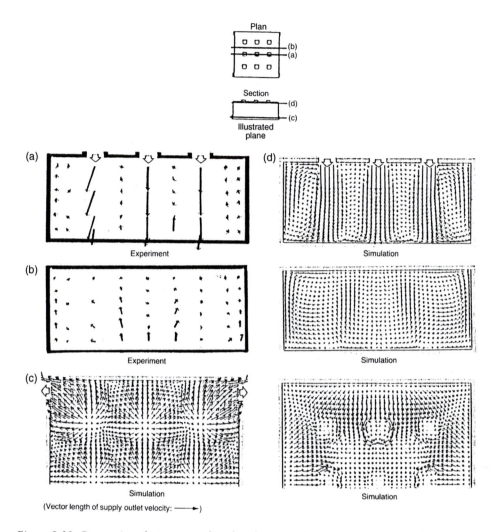

Figure 8.32 Comparison between predicted and measured velocity vectors in a clean room of type 4 in Table 8.4 [79]. (a) Including supply outlets; (b) centre of supply outlets; (c) near floor and (d) near ceiling.

the first attempt to obtain a numerical solution of the flow within the diffuser as well as in the room. A grid of 50×50 was used for this purpose. The inlet conditions to the duct supplying the diffuser were used as boundary conditions for the flow field. This is in contrast to Nielsen and co-workers [73–75] who used measured velocity values in the jet region as boundary values for computing the flow field. The diffusion of the radial jet on the ceiling was well predicted by the CFD solution when compared with experimental data obtained for the room. It was found that when the height of the circular diffuser exceeded 15 mm ($h/r_0 = 0.2$, where r_0 is the radius of the plate) a recirculation zone appeared between the jet and the ceiling in the outlet region of the diffuser. The recirculation flow between the middle of the room and a side wall was also well predicted on comparison with flow visualization tests. The predicted and

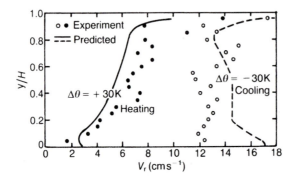

Figure 8.33 Vertical velocity distribution in a room with a circular ceiling diffuser [81].

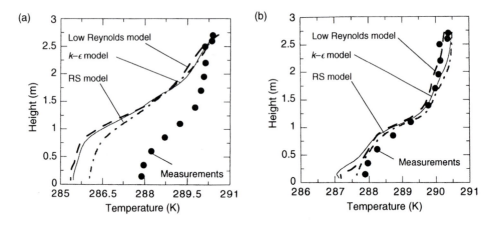

Figure 8.34 Vertical temperature profile in the centre of a room with displacement ventilation. (a) No wall-to-wall radiation allowance and (b) wall-to-wall radiation is considered [82].

measured vertical distributions of the resultant velocity in the room are compared in Figure 8.33 for heating and cooling. These velocities represent the average velocity in a horizontal plane. The larger room velocities in the cooling mode than in the heating mode have been reasonably well predicted by the numerical solution.

More recently Müller and Renz [82] applied three turbulence models to study the air flow and temperature distribution in a ventilated room (6 m × 4 m × 3 m) with two heat sources due to an occupant and a personal computer. Both mixing ventilation (high-level slot diffuser) and displacement ventilation were used in the study, which also involved measurements of air temperature and air movement in the room using particle streak tracking (PST). Poor agreement between measurements and computation were obtained with the three models when the effect of radiation heat exchanged was not properly taken into account (equal radiation distribution with no wall-to-wall radiation assumed), see Figure 8.34 (a) for temperature profile in the centre of room. However, the correlation greatly improved when the radiant heat source was uniformly distributed over the floor and the wall-to-wall exchange is taken into consideration, as shown in Figure 8.34(b). In the latter case, good agreement between the three turbulence models was also achieved.

Awbi [83] used an earlier version of the VORTEX© code [63] to predict the air movement in a 4.2 m square plan room and height 2.8 m ventilated by a continuous slot diffuser across the width of the ceiling and at a distance 1.2 m from a wall discharging towards the far wall. A two-dimensional non-isothermal solution was used and this produced realistic predictions of the vertical velocity and temperature profiles when compared with measurements in the room. The effect of Archimedes number (based on the supply jet) on the mean velocity in the occupied zone for the cooling mode is shown in Figure 8.35 and the effect of a uniformly distributed room load over the floor on the mean room velocity is shown in Figure 8.36. The load was numerically represented by a uniform heat flux in the source term of the temperature equation (8.24) and experimentally by an electrically heated carpet covering the floor. Both figures show a good prediction with a two-dimensional solution of the three-dimensional room supplied by a continuous slot diffuser. However, the experimental measurements showed some variation in the velocity in the lateral direction (diffuser length) particularly near the floor and close to the side walls. Figure 8.36 shows that for the isothermal cases (zero load) the room velocity is influenced by the supply velocity of the jet as one would expect. However, for large room loads the effect of jet velocity on the average room velocity is very small, which indicates that at high loads buoyancy is the major influence on the room air movement.

Ideriah [46] used a similar numerical procedure to the one previously described to compute the turbulent flow and convective heat transfer in a ventilated square

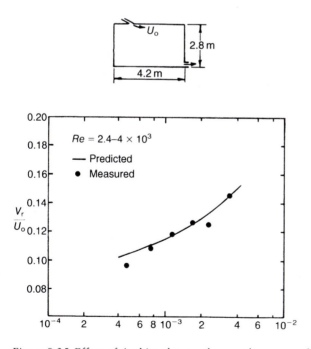

Figure 8.35 Effect of Archimedes number on the mean velocity in the occupied zone of a room ventilated by a slot diffuser.

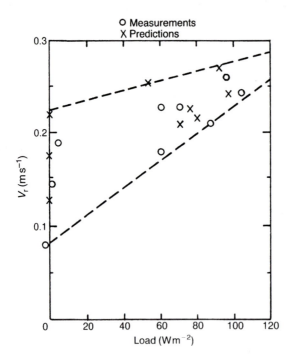

Figure 8.36 Effect of load on mean velocity in the occupied zone of a room ventilated by a slot diffuser for a range of jet velocities.

cavity with the main driving shear provided by the ventilation jet at the bottom and buoyancy from a heated top surface of the cavity. The k–ϵ turbulence model was used and allowance was made for the effect of buoyancy on the vertical component of velocity, k and ϵ. Figure 8.37(a) and (b) shows velocity vector plots and temperature contours for different Reynolds numbers, $Re = U_oL/\nu$, and Archimedes numbers, $Ar = g\beta L\Delta T/U_o^2$, where L is the height of cavity, U_o is the supply velocity to the cavity and ΔT is the difference in temperature between the heated wall and the air supply. It should be noted that the velocity vectors marked ⟶ have not been drawn to scale and therefore they represent higher velocities than indicated. From Figure 8.37(a) it can be seen that as Ar increases (constant Re) the centre of the main vortex in the cavity shifts downwards and in the downstream direction and the strength of the vortex decreases due to the stratification caused by buoyancy. The increase in the extent of the stably stratified region at the upper part of the cavity due to increase in buoyancy is also illustrated by the non-dimensional temperature contours $[T^*(T-T_o)/\Delta T]$ shown in Figure 8.37(b).

Figure 8.38 shows the profiles of the mean horizontal and vertical velocity components normalized with the inlet velocity U_o. As shown, increasing Ar reduces the maximum velocity in the boundary layer regions on the four walls particularly at the ceiling and the two side walls. The predicted profiles compare favourably with measurements.

Jacobsen and Nielsen [84] used three types of turbulence models: a standard k–ϵ model (model 1), a low Reynolds number model (model 2), and a modified low

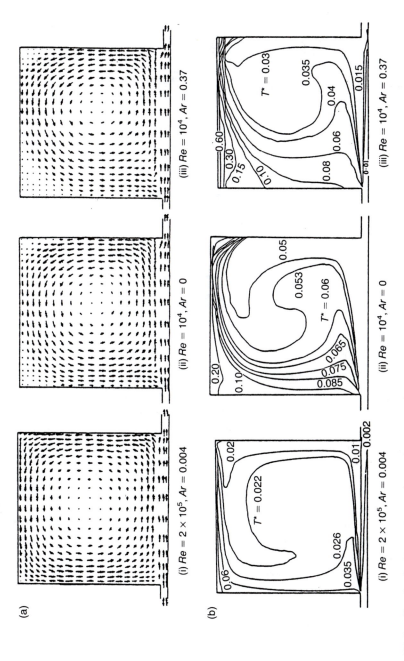

(a)

(i) $Re = 2 \times 10^5$, $Ar = 0.004$ (ii) $Re = 10^4$, $Ar = 0$ (iii) $Re = 10^4$, $Ar = 0.37$

(b)

(i) $Re = 2 \times 10^5$, $Ar = 0.004$ (ii) $Re = 10^4$, $Ar = 0$ (iii) $Re = 10^4$, $Ar = 0.37$

Figure 8.37 Predicted velocity vectors and non-dimensional temperature contours for a square ventilated cavity. Reproduced from [46] by permission of the Council of the Institution of Mechanical Engineers, London.

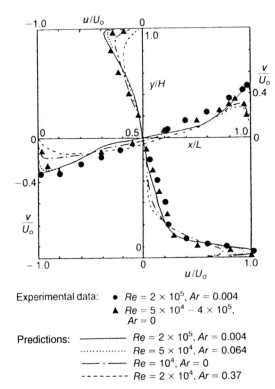

Experimental data: • $Re = 2 \times 10^5$, $Ar = 0.004$
▲ $Re = 5 \times 10^4 - 4 \times 10^5$, $Ar = 0$

Predictions: ——— $Re = 2 \times 10^5$, $Ar = 0.004$
·············· $Re = 5 \times 10^4$, $Ar = 0.064$
— · —— $Re = 10^4$, $Ar = 0$
- - - - - - - $Re = 2 \times 10^4$, $Ar = 0.37$

Figure 8.38 Profiles of the mean horizontal and vertical velocity components in a square ventilated cavity. Reproduced from [46] by permission of the Council of the Institution of Mechanical Engineers, London.

Reynolds number model that includes a buoyancy damping function to improve stability (model 3), to study the flow in a room of 8 m × 6 m × 4 m ceiling height ventilated with displacement ventilation and containing heat sources. Because the flow was mainly buoyancy-driven, the predictions of temperature and velocity gradients were the criteria used for comparison between the three turbulence models and measurements in the room. Figure 8.39 shows the velocity and temperature profiles at two different locations in the room (centre line of diffuser for velocity but off-centre for temperature) produced by the three models and compared with measurements. Generally, the results show that the measured velocity is higher than the computed values. This was partly attributed to the boundary conditions used to describe the thermal condition of the room surfaces in the CFD computation and also the boundary conditions at the air outlet diffuser.

8.5.3 *Thermal comfort prediction*

As discussed in Chapter 1, thermal comfort may be evaluated in terms of the predicted mean vote (*PMV*) and the predicted percentage of dissatisfied (*PPD*). These comfort indices take into account the combined effect of environmental conditions (air velocity, air temperature, mean radiant temperature and partial vapour pressure)

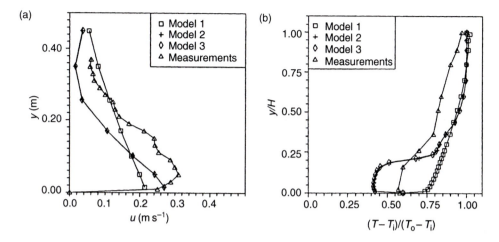

Figure 8.39 Predicted and measured velocity and temperature profiles in a room with displacement ventilation [84]. (a) Velocity profile and (b) temperature profile. H is the ceiling height and y is the distance from the floor.

and occupant's conditions (clothing and activity). Normally, the environmental conditions need to be measured or calculated and the occupant's conditions may be specified, depending on the clothing worn and activity. All CFD codes will predict the distribution of air velocity, air temperature and possibly the vapour pressure. Some CFD codes also include an algorithm for calculating thermal radiation (either surface and/or gas) from which the local mean radiant temperature may be obtained. Hence, it is possible to predict the *PMV* and *PPD* distribution in a space using CFD.

In normal indoor environment the contribution of gas (air) radiation is insignificant in comparison with that between room surfaces themselves or between hot/cold objects and room surfaces. It is therefore only necessary to estimate the thermal radiation between internal surfaces in order to calculate local mean radiation temperature to use with the other variables for obtaining the local *PMV* and *PPD*. One such radiation model is based on the calculation of radiosity at room surfaces. Radiosity is the rate at which total radiant energy leaves a surface per unit area, $W\,m^{-2}$. In this case, each room surface is divided into cells (as one uses in CFD calculations) and the determination of radiant heat exchange between any two surface cells is done through the use of a geometric shape factor. For a cell, i, on surface 1 exchanging radiation with another cell, j, on surface 2, the radiosity leaving cell i to cell j is:

$$J_{1i} = \epsilon_{i1}\sigma T_{i1}^4 + (1 - \epsilon_{i1})F_{i1-j2}J_{j2} \tag{8.112}$$

where J_{i1} is the radiosity of cell i on surface 1; J_{j2} is the radiosity of cell j on surface 2; ϵ_{i1} is the emissivity of cell i on surface 1; F_{i1-j2} is the geometric shape factor between cells i and j and σ is the Stephan–Boltzmann constant $= 5.67 \times 10^{-8}\,W\,m^{-2}\,K^{-4}$.

The temperature of cell i is then given by:

$$T_{i1} = \left[\frac{1}{\epsilon_{i1}\sigma}\left(J_{i1} - (1 - \epsilon_{i1})F_{i1-j2}J_{j2}\right)\right]^{1/4} \tag{8.113}$$

If a heat flux, q_{i1} (W m^{-2}), is present on cell i, then the radiosity is obtained from:

$$J_{i1} = q_{i1} + F_{i1-j2}J_{j2} \tag{8.114}$$

And the cell temperature is then given as:

$$T_{i1} = \left[\frac{1}{\sigma} \left(J_{i1} + \frac{1 - \epsilon_{i1}}{\epsilon_{i1}} q_{i1} \right) \right]^{1/4} \tag{8.115}$$

Equations (8.112)–(8.115) are iteratively solved for each iteration of the air flow equations to produce the temperature of the room wall cells. The plane radiant temperature, T_{pr}, for a surface of the computational grid cell within the room is:

$$T_{pr} = \left[\frac{1}{\sigma} \sum_{i=1}^{n} \sum_{k=1}^{m} F_{p-ik}J_{ik} \right] \tag{8.116}$$

where F_{p-ik} is the radiation shape factor between grid cell face, p, and room surface cell ik; n is the number of room surfaces; and m is the number of cells on a room surface.

The mean radiant temperature, T_{mr}, for a grid cell, assuming a rectangular parallelepiped, is taken as the mean of the six plane radiant temperatures, T_{pr}, for each face, weighted by the corresponding surface area.

Similarly, calculation may be carried out to allow for thermal radiation exchange between heat sources or sinks within the room and the room surfaces and between the sources or sinks themselves. However, such calculations become much more complicated due to the shielding effect when more than one source is present.

The *PMV* and *PPD* can be then calculated at each computational grid from knowledge of all the environmental variables, see Section 1.41. Figure 8.40 shows *PMV* and *PPD* contours in a centre plane of a room of dimensions 4.9 m × 3.7 m × 2.75 m obtained using the VORTEX© code [85]. Due to the vertical temperature stratification, the lower part of the room represent an area of relative discomfort, i.e. *PPD* > 10%.

CFD can also be used to predict draught risk (see section 1.5.4), which requires knowledge of the local turbulence intensity, *TI*, air speed and air temperature. This is known as the percentage of dissatisfied, *PD*. The turbulence intensity can easily be obtained from the kinetic energy of turbulence, k, using:

$$TI = \frac{\sqrt{2k}}{V} \times 100 \tag{8.117}$$

where V is the local speed.

Iwamoto [86] used CFD to model the flow and temperature distribution around a human occupant. The complex computational grid of a human body was represented by tetrahedral cells. The k–ϵ turbulence model and the SIMPLER algorithms were used in the calculations. The seated human body (1 met and 1 clo) was in a room 4.36 m × 2.20 m × 2.26 m ceiling height with a heated floor (30 °C). Figure 8.41(a–c) shows the body grid and the flow around the body in two vertical planes.

Murakami *et al.* [87] carried out a numerical simulation of the combined air flow, thermal radiation and moisture diffusion (based on a human thermo-physiological model) of a naked human. A low Reynolds number k–ϵ turbulence model was used

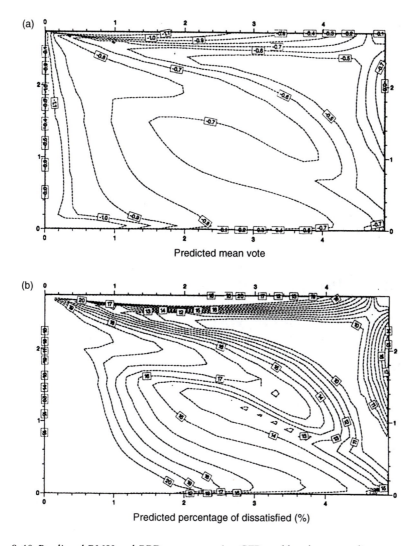

Figure 8.40 Predicted *PMV* and *PPD* contours using CFD and local mean radiant temperature.

with body-fitted (generalized curvilinear coordinate) system to represent the body shape. The metabolic heat production and thermo-regulatory control processes used were based on the two-node thermal model of Gagge (see Section 1.3.1). The body surface temperature and the convective, radiative and evaporative heat transfer rates were well predicted by the model when compared to known expressions for human metabolism.

8.5.4 *Prediction of contamination dispersal*

The contamination produced in a ventilated room will quickly spread over the entire occupied zone, particularly in a mixing ventilation system with a large rate of entrainment and a circulatory motion created by the jet. In some cases it may

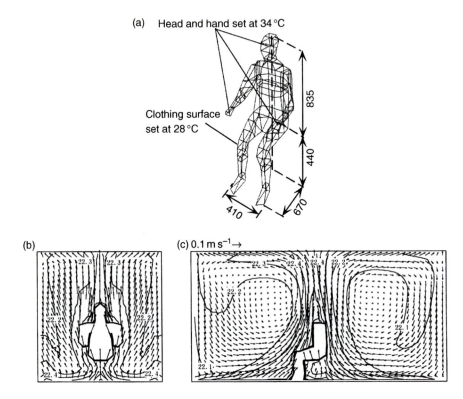

Figure 8.41 Flow around a human body [86]. (a) Body grid; (b) air flow and temperature distribution in *x–z* plane and (c) air flow and temperature distribution in *y–z* plane.

be necessary to extract the contaminant at source using extract hoods before it is given time to disperse by the air motion in the room. Extract hoods are discussed in Section 5.3.4. In all cases involving the presence of a pollutant source in a room, it will be necessary to determine the contaminant concentration for different parts of the room to ensure that nowhere within the occupied zone is the concentration higher than the limit for that pollutant specified by standards and regulations operating in a particular country. Apart from measuring concentrations throughout the room (a tedious exercise), one can apply CFD to predict concentration distribution. Normally, the transport equation for concentration (equation (8.25)) is solved either in time-average form or time-dependent form after a converged solution has been achieved for the other transport equations, i.e. velocity, temperature and turbulence variables. In practice, for most gases and vapours and because of the low concentration levels that normally exist in a room environment (i.e. a few hundred parts per million), the difference in density between the contaminant and air may conveniently be ignored without causing significant errors. The distribution of concentration in a room, c, is normally expressed as normalized concentration with respect to that at the exhaust, c_e, i.e. c/c_e. The definitions of ventilation effectiveness are given in Section 2.4.3.

Nielsen [88] studied the concentration distribution in a two-dimensional enclosure of length to height ratio $L/H = 3$ using the same numerical procedure that he used

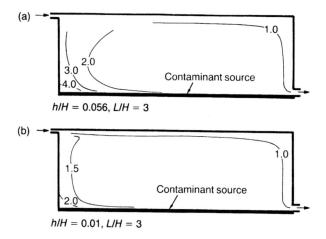

Figure 8.42 Predicted distribution of normalized concentration (c/c_e) in a two-dimensional enclosure with two different heights of the supply slot [88].

to predict the velocity and temperature distribution discussed in Section 8.5.2. An additional transport equation for concentration was solved. The effect of the height of the air supply slot on the normalized concentration distribution, c/c_e, in the enclosure when a uniform contaminant source is evenly distributed over the floor is shown in Figure 8.42. This figure shows that a decrease in the height of the supply slot produces a decrease in concentration in the enclosure, i.e. an increase in ventilation effectiveness, caused by an increase in the air circulation in the room. Clearly, the use of narrower slots will be more desirable so far as ventilation effectiveness is concerned but may at the same time create draughts, i.e. deterioration in thermal comfort. Figure 8.43 shows the effect of location on the floor of a point source on the normalized concentration distribution. In Figure 8.43(a) the source is placed in a region of large recirculation velocity $U_{rm}/U_o = 0.3$, in Figure 8.43(b) it is placed in a region where $U_{rm}/U_o = 0.26$ and in Figure 8.41(c) the source is in a low recirculation velocity region $U_{rm}/U_o = 0.1$. It is clear from these predictions that higher room concentrations will occur when the contamination source is placed in a relatively stagnant region in the room. This finding, which is also supported by experiments, emphasizes the importance of placing contamination sources in regions of high velocities in the room.

Murakami *et al.* [89] solved the three-dimensional transport equations for the velocity components, k, ϵ, time-average concentrations, c, and the r.m.s. concentration, $\sqrt{(c'^2)}$, for cubic and rectangular rooms using a similar computational scheme to that used for their clean-room study, described in Section 8.5.2, with an additional equation for $\sqrt{(c'^2)}$. A comparison between predicted and measured concentration distribution in the rectangular room, normalized with the extract concentration c_e, is shown in Figures 8.44 and 8.45 for a room air change rate, N, of 8, 24, 48 and 96 per hour for the experimental cases but unspecified air change rate for the numerical case. The contaminant source was placed in the middle of the floor and the results shown are for a vertical plane through the middle of the room. The air was supplied and extracted through square openings at the top and bottom of the middle of a small wall respectively. In Figure 8.44, the predicted concentrations are generally greater than

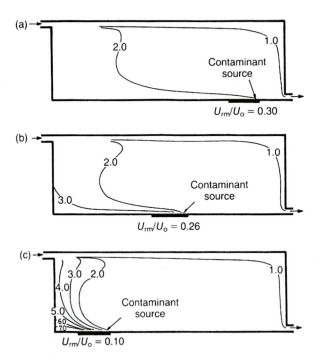

Figure 8.43 Predicted distribution of normalized concentration (c/c_e) in a two-dimensional enclosure for different positions of contaminant source [88].

the measured concentrations and in both cases the largest spread is in the direction of the reverse flow and towards the extract port. The larger the air change rate the larger is the area of low concentration close to the ceiling and down the far wall. The numerically predicted concentrations correspond better to the experimental ones at large air change rates ($N = 48$ and 96). The low concentration in the middle of the room at $N = 8$ was due to the separation of the jet from the ceiling into the occupied zone. Similar patterns of the distribution of $\sqrt{(c'^2)}/c_e$ to those for c/c_e can be seen from Figure 8.45. Murakami *et al.* also attempted to study the effect of buoyancy on the contaminant spread in the rectangular room but the numerical results showed a much larger buoyancy effect than observed by experiments.

Kurabuchi and Kusuda [90] applied a numerical solution similar to that used by Murakami *et al.* [79] to predict the concentration distribution in a parametric study of a chemistry laboratory, which was naturally ventilated by opening two windows and a door opposite and fitted with a mechanically ventilated fume cupboard (draught chamber) as shown in Figure 8.46. The experimental study was carried out on a 1 : 6 scale model of the laboratory. The calculations were carried out in three dimensions using a rectangular grid of $30 \times 20 \times 15$.

Figure 8.47 shows a comparison between measured and predicted normalized concentration distribution when the extract fan in the fume cupboard was not operating and Figure 8.48 is a comparison when the fan was operating. The location of the pollution source in both figures is in front of the fume cupboard and the concentration

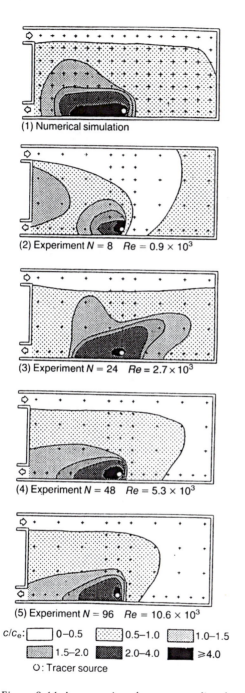

(1) Numerical simulation

(2) Experiment $N = 8$ $Re = 0.9 \times 10^3$

(3) Experiment $N = 24$ $Re = 2.7 \times 10^3$

(4) Experiment $N = 48$ $Re = 5.3 \times 10^3$

(5) Experiment $N = 96$ $Re = 10.6 \times 10^3$

c/c_e: ☐ 0–0.5 ▦ 0.5–1.0 ▨ 1.0–1.5
▧ 1.5–2.0 ■ 2.0–4.0 ■ ≥4.0
○ : Tracer source

Figure 8.44 A comparison between predicted and measured normalized mean concentration in the centre of a three-dimensional room [89].

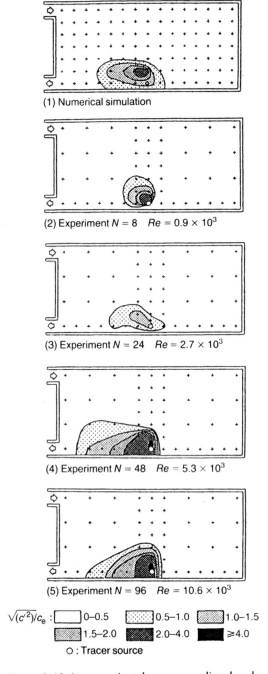

(1) Numerical simulation

(2) Experiment $N = 8$ $Re = 0.9 \times 10^3$

(3) Experiment $N = 24$ $Re = 2.7 \times 10^3$

(4) Experiment $N = 48$ $Re = 5.3 \times 10^3$

(5) Experiment $N = 96$ $Re = 10.6 \times 10^3$

$\sqrt{(c'^2)}/c_e$: ☐ 0–0.5 ▦ 0.5–1.0 ▦ 1.0–1.5
▦ 1.5–2.0 ▦ 2.0–4.0 ■ ≥4.0

O : Tracer source

Figure 8.45 A comparison between predicted and measured normalized r.m.s. concentration in the centre of the three-dimensional room [89].

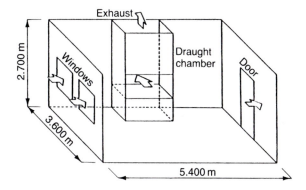

Figure 8.46 An isometric view of a laboratory mock-up.

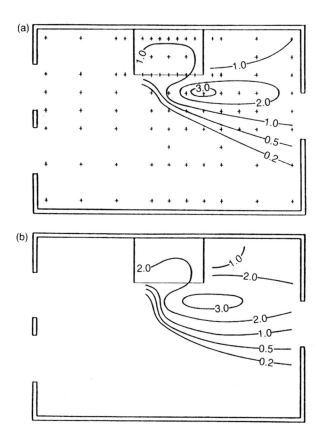

Figure 8.47 Comparison between measured and predicted normalized concentration in a laboratory mock-up at a horizontal plane 1.26 m above floor – extract fan in fume cupboard unoperational [90]. (a) Model test and (b) prediction.

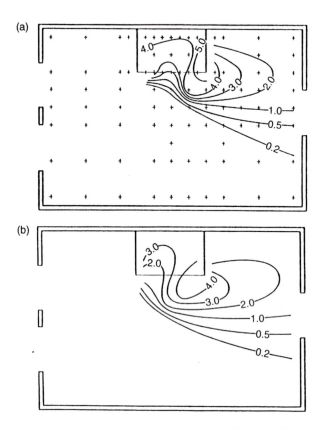

Figure 8.48 Comparison between measured and predicted normalized concentration in a laboratory mock-up at a horizontal plane 1.26 m above floor – extract fan in fume cupboard operational [90]. (a) Model test and (b) prediction.

profiles are for a horizontal plane 1.26 m above the floor. Both figures show a good correlation between predicted and measured concentrations with no allowance given to the buoyancy effect in the numerical solution (the tracer gas used experimentally was C_2H_4). When the extract fan was unoperational the peak concentration was just downstream of the source, but when the fan was operational the peak concentration moved towards the edge of the fume cupboard.

Lee and Awbi [91] studied the effects of internal room partitions on the ventilation performance in terms of room air change efficiency and ventilation effectiveness in the presence of a contaminant source (CO_2) at two different locations in the room. A model test room $1.6(L) \times 0.8(W) \times 0.7 \, m(H)$ was used and the physical test conditions were simulated numerically using the VORTEX© CFD code [63]. The air supply opening was located on the ceiling at $0.25L$ (L being the room length) and the exhaust opening was also on the ceiling at $0.75L$; other test configurations used are described in Section 6.2.2. Figure 8.49 shows the velocity vector plots and concentration contours when a pollution source is located below the exhaust opening and for different locations of the room partition, i.e. $0.4L$, $0.5L$ and $0.6L$, without a gap below the partition. As shown in the figure, the concentration in the supply side of the room

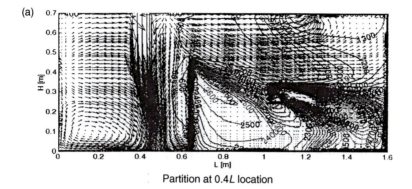

Partition at 0.4*L* location

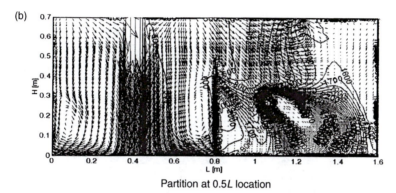

Partition at 0.5*L* location

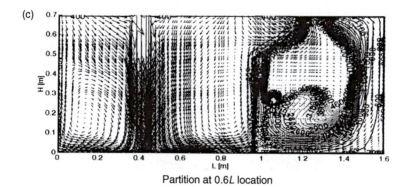

Partition at 0.6*L* location

Figure 8.49 Velocity vectors and concentration contours for different partition locations with isothermal contaminant source in the exhaust zone, $V_{\text{supply}} = 1.54\,\text{m s}^{-1}, 0.7H$ partition height and no gap.

(left side) decreases as the partition moves towards the pollution source because the pollutant becomes trapped on the other side of the partition.

The effect of gap underneath the partition on the air movement and concentration in the room is shown in Figure 8.50 for the case of a partition located at 0.4*L*. It can be seen that the concentration in the room improves considerably when the partition

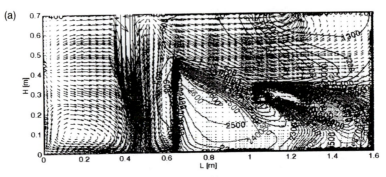

Partition with no gap

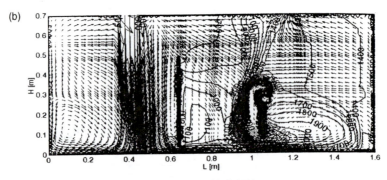

Partition with 0.05*H* gap

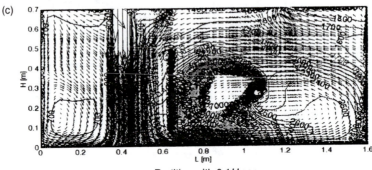

Partition with 0.1*H* gap

Figure 8.50 Velocity vectors and concentration contours for different partition locations with isothermal contaminant source in the exhaust zone: 0.4*L* partition location and 0.7*H* height.

is raised above the floor to create a gap underneath of 0.05*H* (*H* being the room height) but as the gap increases to 0.1*H* the room concentration increases as a result of the backflow over the partition transporting pollution into the supply side, see Figure 4.50(c). The results in Figures 8.49 and 8.50 are also confirmed by the values of ventilation effectiveness and the air change efficiency [91].

8.5.5 *Prediction of aerosol movement*

As mentioned in Section 2.2.8, there are many types of airborne particulates indoors with different composition and size. Their effect on health not only depend on their composition (chemical, biological, radioactivity, etc.) but also on their size. Ultrafine particles can settle in the respiratory system much easier than medium or large particles. The movement of particles in room air is dependent not only on their size but also the momentum of the flow, turbulence, buoyancy, electrostatic charges, etc. CFD has been used to study the movement of particles in which case the dynamics of the particles movement have to be modelled in addition to the fluid movement. Generally, there are two approaches. One deals with the fluid phase as a continuum and the particle phase as individual particles, the so-called Lagrangian approach, in which case particle trajectories are predicted from the forces acting on the particles. The second approach, the so-called Eulerian approach, treats both the solid phase (particles) and the fluid as continuums, in which case the transport equations for the fluid and particles are solved.

In studying the movement of indoor airborne particles it is generally assumed that the particles do not always follow the air stream. This approach is usually valid for at least medium and large particles found indoors. A simplification that is often used for modelling airborne particles movement is to assume a small settling velocity of the particles, so that the effect of particles on the air turbulence can be neglected. This approach allows turbulence to be treated as a one-way coupling between the particles and the fluid and therefore allow any size particle to be modelled. This approach is also known as the 'drift-flux' model [92], which leads to transport equations similar to the Navier–Stokes equations with extra source/sink terms for the body forces, resulting from the difference between particle and fluid densities and any thermally related body forces, see Sections 8.2.2 and 8.4.2.

To demonstrate the movement of particles in a mechanically ventilated room, Holmberg and Li [92] applied the drift-flux model to predict the distribution of mono-disperse particles (i.e. particles of a given size) in the size range 0.5–5 µm. The air to the room, with particle concentration of 10 p.p.m., was supplied in the centre of a wall close to the ceiling ($A_{\text{inlet}} = 1.5 \times 0.5\,\text{m}$) and air extract was from an opening ($A_{\text{out}} = 1.5 \times 1.0\,\text{m}$) in the centre of the opposite wall close to the floor. The simulations were carried out using the k–ϵ model under isothermal conditions where a well-mixed flow was present throughout the room. The results for a vertical plane through the centre of the room are shown in Figure 8.51, where it can be seen that the smaller particles distribute uniformly in the room whereas the larger particles tend to settle in the lower (low velocity) regions of the room.

In another study, Holmberg and Li [93] investigated the distribution of particles in a room ($2.4 \times 1.2 \times 2.4\,\text{m}$ height) with displacement ventilation that included two standing heated mannequins (each $0.18 \times 0.36 \times 1.68\,\text{m}$, $1.67\,\text{m}^2$ body surface area and giving 50 W). The low-level supply terminal was 0.24 m wide $\times$ 0.72 m high supplying air at 17 °C with an air change rate of $7\,\text{h}^{-1}$, and two air extracts of $0.18 \times 0.24\,\text{m}$ at a height of 2.1 m on two opposite walls. Particles mixed with air were supplied into the room from an opening $0.12 \times 0.12\,\text{m}$ in the centre of the floor at the rate of $4.7\,\text{g}\,\text{h}^{-1}$. Figure 8.52(a) shows the flow pattern in a central plane in the room passing through the inlet, the two outlets and the particle supply opening. Figure 8.52(b) shows the predicted distribution of particles of 0.3 µm diameter in a vertical plane through

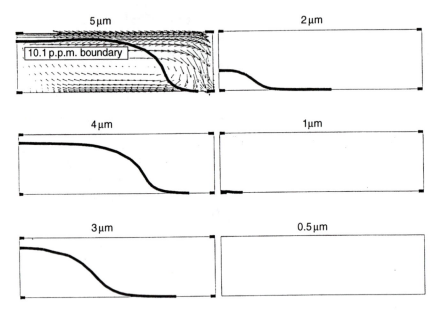

Figure 8.51 Concentration of particles of size range 0.5–5 μm in a mechanically ventilated room. The black lines are the boundaries for particle concentration >10.1 p.p.m., i.e. in the zone below the line there is a large particle concentration of the size shown above each plot [92].

the centre of the room. The particle concentration distribution is normalized with the concentration for the total particle and air supplies. The entrainment of the particles by the heated mannequins is evident due to the large concentration values around them. The concentration at the two extract openings is 1.0 suggesting that no particles have settled on the room surfaces. Figure 8.52(c) for 20 μm particles diameter, show lower concentrations around the mannequin (particularly the one on the right) than those for Figure 8.52(b) and also the concentration at the right-hand extract is lower than the one on the left. These differences are attributed to the larger particles size and the particular flow pattern in the room.

8.5.6 Predicting emission from materials

Certain materials, in particular building materials, such as paints, carpets, floor coverings, etc., are potential indoor air pollutants, see Section 2.2. Volatile organic compounds (VOCs) are emitted from such materials at rates that are determined by the concentration at the surface of the material (p.p.m. or mg m^{-3}), the concentration in the air, air speed, temperature, humidity, etc. This is given by:

$$G(t) = G_o e^{-Kt} \qquad (8.118)$$

where $G(t)$ and G_o are the emission rates (mg m^{-2} s^{-1}) at time $= t$ and 0, respectively, and K is emission decay constant. This equation can be used to predict emission rates from materials at a given time, $G(t)$, if G_o and K are known. Normally, these are

(a)

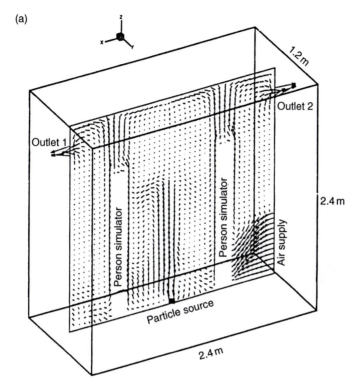

(b) Particle density = 1000 kg . m^{-3}, diameter = 0.3 m^{-6}

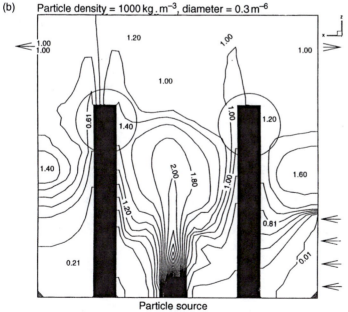

Particle source

Figure 8.52a, b Flow pattern and particles concentration distribution in a room with two heated mannequins and displacement ventilation [93]. (a) Air flow pattern showing air supply, particle supply (floor) and two extracts and (b) distribution of 0.3 μm particles diameter.

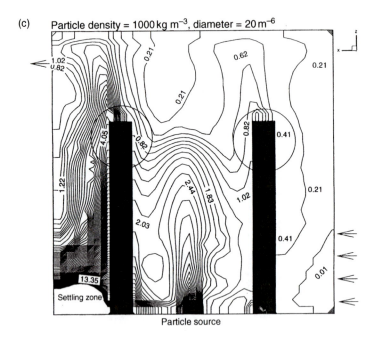

Figure 8.52c Distribution of 20 μm particles diameter [93].

determined from experiments in small-scale test chambers under controlled environmental condition and uniform concentration. However, the results from a chamber test may not be applicable to real buildings because of scale effects and the existence of non-uniform concentration due to incomplete mixing with air in real buildings. Therefore, CFD has been used to model the diffusion within the material and into the room air. In general, emission processes involve both diffusion through the solid material and evaporation from the surface to air. Usually, the latter occurs when a liquid film, e.g. paint, is freshly applied to the surface but after an initial period of intensive evaporation the surface emission is mainly by diffusion through the material surface.

The general emission modelling approach is to apply Fick's law for the diffusing within the material and from the surface of the material to the air, i.e. the vapour pressure or diffusion boundary layer (surface boundary conditions), see Figure 8.53. Most building materials are permeable such that the internal diffusion of VOCs in the material is represented by a one-dimensional diffusion process for the solid-phase concentration and if the solid-phase concentration is expressed in terms of the equivalent air-phase concentration [94], then Fick's law gives:

$$\partial c_m / \partial t = \partial / \partial y (D_m \partial c_m / \partial y) + S_c \qquad (8.119)$$

where c_m is the concentration (mg m^{-3}) in the material, y is the distance from the surface (m), t is time (s), D_m is the diffusion coefficient of the material (m^2 s^{-1}) and S_c is the rate of generation of VOC within the material (mg m^{-3} s^{-1}). If material of homogeneous diffusivity with no internal source, $S_c = 0$, is assumed, then equation (8.119)

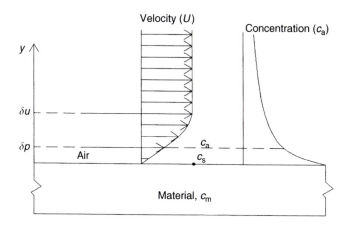

Figure 8.53 Diffusion through material and air vapour layer.

is written as:

$$\partial c_m / \partial t = D_m (\partial^2 c_m / \partial y^2) \tag{8.120}$$

Similarly, the emission of VOC from the surface is by molecular diffusion through the vapour boundary layer at the surface–air interface and Fick's law applied to the vapour boundary layer gives:

$$\partial c_a / \partial t = D_a (\partial^2 c_a / \partial y^2) \tag{8.121}$$

where c_a is the concentration in the air (boundary layer), D_a is the diffusion coefficient of the air ($m^2 \, s^{-1}$) and y is the distance from the surface. The emission rate, $G(t)$, at the surface of the material is equal to the transportation by diffusion through the material, hence by integration with y equations (8.120) and (8.121) give:

$$G(t) = -D_m (\partial c_m / \partial y) = -D_a (\partial c_a / \partial y) \tag{8.122}$$

where $G(t)$ is the emission from the surface ($mg \, m^{-2} \, s^{-1}$). In terms of the concentration at the surface, c_s, and the background (outside the vapour layer) concentration, c_∞, the equation gives:

$$G(t) = k_c (c_s - c_\infty) \tag{8.123}$$

where k_c is the mass transfer coefficient ($m \, s^{-1}$) $= D_a / \delta_D$ and δ_D is the vapour boundary layer thickness (m). δ_D depends on the nature of air flow over the surface, such as velocity and turbulence intensity as well as the concentration gradient between the surface and background.

A number of models have been developed based on the forementioned concepts and used in a CFD code to solve the concentration gradient simultaneously within the solid phase and in the air [95–100]. The solution of the equations within the solid requires a very fine resolution (fine grid) to eliminate false numerical diffusion because of the high concentration gradient near the surface. The solution of the equations in the vapour boundary layer also requires the use of a low Reynolds number turbulence model (a fine grid) because of the damping effect of the surface. In addition to emission, building materials also act as adsorbers and desorbers of VOCs. The modelling of emission from certain materials in a room, e.g. floor covering, needs to take account of the adsorption and desorption (usually referred to as 'sorption') processes of room surfaces [98–100]. These surfaces are treated in the same way as an emitting material, the difference being that these can desorb (emit) and adsorb VOCs at the same time.

Yang et al. [95] used CFD to predict the emission from wood stain and compared the predictions with measurements in a 1.0 m × 0.8 m × 0.5 m ventilated test chamber (0.5 ach) under controlled conditions and with results from an empirical first-order decay model. A zero-equation turbulence model was used in the CFD simulation and boundary conditions for the surface emission were determined experimentally. The CFD simulations were performed for the first 5 h from the stain application, as most of the emission was expected to occur during the first few hours. Figure 8.54 shows the predicted and measured TVOC (total VOC) concentration at the chamber exhaust and, as shown, the CFD model predicted the decay well whereas the empirical model produced a large discrepancy.

Topp et al. [97] predicted the emission from the ceiling and floor of a 'two-dimensional' room 9 m long × 3 m high, ventilated by a high-level slot on one wall (see also Section 8.5.4). The inlet slot height was 0.168 m and the exhaust slot on the opposite wall at floor level was 0.48 m high. The inlet velocity was 0.45 m s^{-1} giving an air change rate of 10 ach. As the interest was focused on the near-wall region, a low Reynolds number k–ϵ model was used in the CFD simulations. The locations of the emitting surfaces are shown in Figure 8.55(a) and the contours of concentration

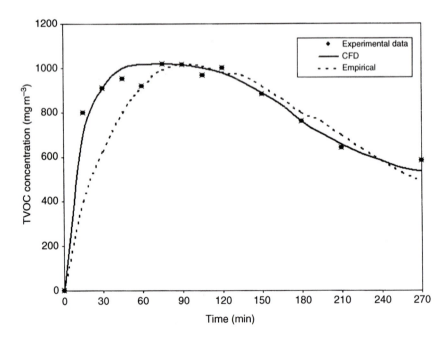

Figure 8.54 TVOC emission from wood stain [95].

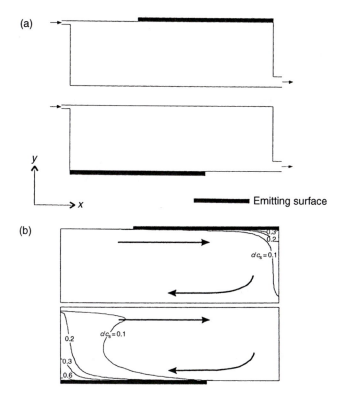

Figure 8.55 Emission from ceiling and floor of a ventilated room; c_s is the emitting surface concentration [97]. (a) Room and emitting surfaces; (b) concentration distribution.

normalized with the emission at the surface, c/c_s, are shown in Figure 8.55(b). With the emission source on the ceiling the zone of high concentration is limited to the far corner of the room as the contaminant is removed by the exhaust below, but when the source is spread over part of the floor, higher room concentrations were obtained. The results presented in Figures 8.54 and 8.55 do not account for the sorption of room or chamber surfaces but it is possible to allow for this effect as has been demonstrated in [98–100].

8.5.7 Mean age prediction

Davidson and Olsson [101, 102] investigated numerically the local purging flow rate and the local age of concentration in a two-dimensional ventilated room (isothermal flow) and the local age in a three-dimensional room (non-isothermal flow). The local mean age (i.e. mean age at a point) is 'the average time that has elapsed since the concentration molecules passing the point entered the room'. This definition applied in the case of step-up (concentration is carried to the room by the supply air) or step-down (room initially filled with concentration and supplied with contaminant-free air) concentration dispersion with time. The step-down approach was used by Davidson and Olsson. Local purging, $\dot{m}_p$, is defined either as the net rate at which a dynamically passive contaminant (the flow field is not influenced by the contaminant) is purged out of the system (room) from a point, or the rate at which contaminant-free air is supplied to a point within the system (room). They used the FVM to solve the transport equations for the velocity components, temperature, concentration and the turbulence variables, k and ϵ, in the three-dimensional case, and a length scale, obtained using a one-equation turbulence model developed by them, for the two-dimensional case.

Figure 8.56 shows the two rooms investigated. The three-dimensional room was supplied with warm air at high level with cold windows (winter situation) and with cold air at low level with warm windows (summer situation). The predicted normal-ized local mean age distribution $\bar{\tau}_p/\tau_n$ (where τ_n is the nominal time constant of the room equal to $\rho \dot{V}/\dot{m}_a$, ρ is air density, $\dot{V}$ is room volume and $\dot{m}_a$ is the mass of air supply to the room) for a vertical central plane in the heated room (high-level supply) is shown in Figure 8.57. Experimentally determined local ages at five points are also indicated on the contour plot. The predicted age contours show a steady decrease in the vertical direction whereas measurements show maximum age at about mid-room height. The predicted distribution of local mean age for the cooled room (low-level supply) is shown compared with measurements in Figure 8.58 for the same vertical plane as in Figure 8.57. The predicted local mean age near the floor is considerably higher than that indicated by measurement. Davidson and Olsson attributed this dif-ference to a number of possible causes ranging from false numerical diffusion in the computational scheme, inadequacy of turbulence model, to insufficient experimental measurements. The latter cause may have made some contribution to the discrepancy since experimental measurements were only carried out once.

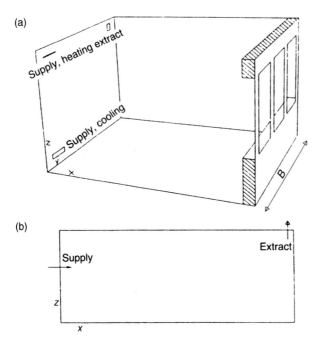

Figure 8.56 Configuration of the two rooms investigated [101]. (a) Three-dimensional room and (b) two-dimensional room.

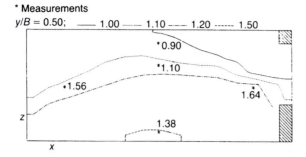

Figure 8.57 Predicted and measured normalized local age ($\bar{\tau}_p/\tau_n$) in a vertical plane through the middle of a three-dimensional heated room [101].

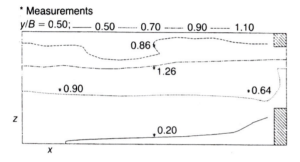

Figure 8.58 Predicted and measured normalized local age ($\bar{\tau}_p/\tau_n$) in a vertical plane through the middle of a three-dimensional cooled room [101].

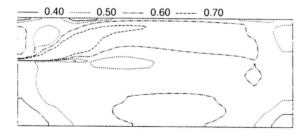

Figure 8.59 Predicted normalized local purging contours ($\dot{m}_p/\dot{m}_a$) for a two-dimensional isothermal room [101].

The predicted local purging contours normalized with mass of air flow rate to the room, $\dot{m}_p/\dot{m}_a$, for the two-dimensional room are shown in Figure 8.59, which indicates that the flow in the lower half of the room is more stagnant than in the upper half. No comparison with measurements was given since there were no adequate means of measuring local purging at the time Davidson and Olsson carried out their investigation.

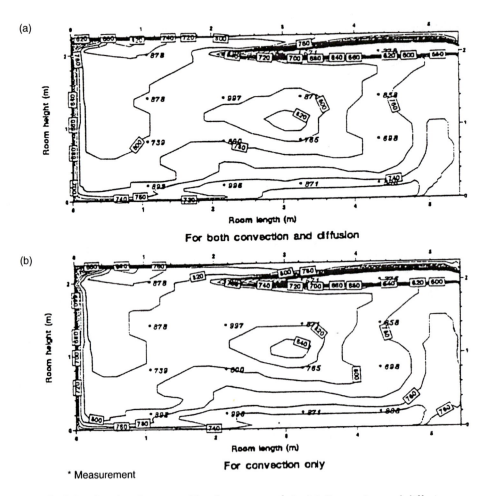

Figure 8.60 Predicted and measured local mean age of air. (a) Convection and diffusion terms
        solved; (b) diffusion term ignored.

Gan and Awbi [10] used the VORTEX© code to predict the local mean age of air
distribution in a room of 5.4 m × 2.7 m × 2.25 m high, which was used by Matsumoto
for experimental measurements of local mean age by tracer gas injection. They carried
two types of simulation: one by solving the full age of air equation (8.12) and another
ignoring the diffusion terms on the right-hand side of the equation. The air was supplied from 0.1 m diameter opening, the centre of which was 0.25 m below the ceiling in
the centre of one wall, and an extract opening facing the supply opening on the opposite wall having the same diameter. Figure 8.60 shows the computed mean age of air
contours in seconds for a vertical plane in the centre of the supply opening compared
with measurements. As can be seen, the pattern of local mean age distribution with
and without the diffusion term is very similar although there is some difference in the
values between the two simulations. These results would suggest that the transport of
age element is convection dominant. The agreement with measurements is reasonably
close.

8.5.8 *Natural ventilation predictions*

Computational fluid dynamics has been extensively used for predicting naturally driven flows through buildings due to the effect of wind and/or buoyancy because of the difficulties involved in obtaining reliable data experimentally. For buoyancy-driven ventilation, the modelling can be either based on the internal flow in the building by specifying certain boundary conditions at the inlet and/or outlet openings or a combined simulation of the internal and external flow. For wind-driven ventilation however, the modelling normally involves both the internal and the external flows simultaneously. Many investigations have been carried out to predict the internal or the internal/external flow in and around buildings. Kurabuchi *et al.* [103] used four turbulence models to compare the flow and pressure distribution around a building with two opposite openings (cross-ventilation) with wind-tunnel measurements. The model building is shown in Figure 8.61(a) and the velocity distribution from the for turbulence models are compared with wind-tunnel measurements in Figure 8.61(b). The first model is the standard $k–\epsilon$, the second and third are $k–\epsilon$ models with modified kinetic energy generation term, P_k (the terms between { } in equation 8.34) and the fourth is an LES model. The results show that the LES model produced closer agreement with wind-tunnel data than the other three models, however, the results from the standard $k–\epsilon$ model are not too different from the results obtained from the two modified $k–\epsilon$ models. A comparison between the pressure distributions around the building obtained using the $k–\epsilon$ and the LES models and the wind-tunnel measurements is shown in Figure 8.62 (a) and (b). Here too, the LES model produced closer agreement with measurements but the $k–\epsilon$ model results are not too different from the wind-tunnel results.

The flow through a single-sided opening has been studied using CFD for buoyancy-driven flow through the opening [104–106]. Schaelin *et al.* [104] solved the flow through a large opening by coupling the internal and the external flows to allow bidirectional flow through the opening. They compared the two and three-dimensional simulations with results from analytical equations for the buoyancy-driven flow only and buoyancy with wind through a large opening, with good agreement. Gan [105] studied the buoyancy-driven flow using the RNG turbulence model (see Section 8.3.3) for the internal flow only. The purpose was to investigate the penetration depth into the room of the flow through the opening. The simulations were carried out for a room 3 m wide, 3 m high and 15 m deep with an open window 1.5 m high and 3 m wide on the 3 m × 3 m wall. Figure 8.63 shows the velocity vectors, air temperature and normalized (with the exit values) mean age of air contours in the room for an outside air temperature of 20 °C and 15 W m^{-2} convective thermal load uniformly distributed over the floor. The results show that the air penetrated into the room through the lower part to almost 11 m from the opening.

Allocca *et al.* [106] simulated the flow in a three-storey building with buoyancy alone and with the combined effect of buoyancy and wind. The internal flow was coupled with the external flow. Each room (4.7 m × 2.9 m × 2.8 m high) had an upper and a lower opening of 0.4 m^2 each on one wall. The room contained furniture and had a total internal heat load of 700 W (350 W assumed convective and the rest radiation to the room surfaces). The external temperature used for the simulations was 25.5 °C. Comparisons were made between the CFD results and those calculated using semi-analytical models giving agreement within 10% for the buoyancy-driven flows

(a)

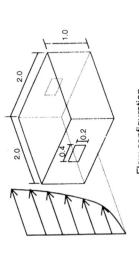

Flow configuration

(b)

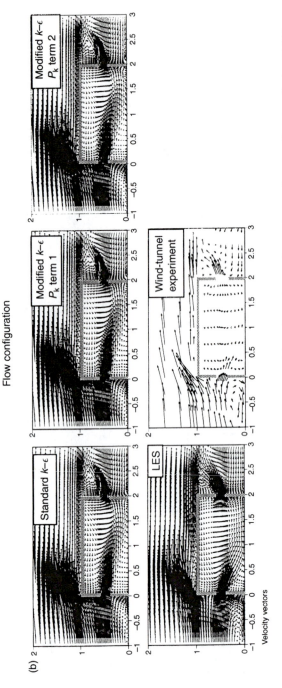

Figure 8.61 Distribution of velocity vectors using four turbulence models and wind-tunnel experiments for crossflow ventilation [103].

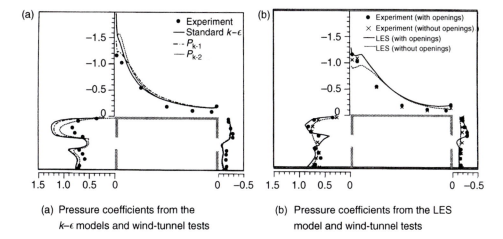

(a) Pressure coefficients from the
k–ε models and wind-tunnel tests

(b) Pressure coefficients from the LES
model and wind-tunnel tests

Figure 8.62 Distribution of pressure coefficients around the building [103].

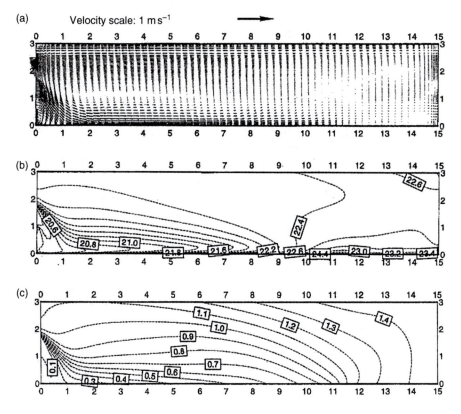

Figure 8.63 CFD prediction of the flow through single-sided ventilation due to buoyancy only [108]. (a) Air velocity; (b) air temperature and (c) normalized mean age of air. Vertical axes are room height (m) and horizontal axes are room length (m).

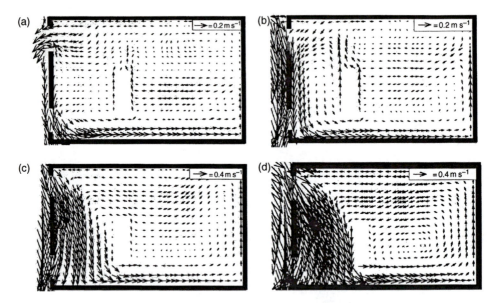

Figure 8.64 Flow through two single-sided openings due to buoyancy and wind. Reference wind speed: (a) $2\,\mathrm{m\,s^{-1}}$; (b) $4\,\mathrm{m\,s^{-1}}$; (c) $6\,\mathrm{m\,s^{-1}}$ and (d) $8\,\mathrm{m\,s^{-1}}$ [106].

and within 25% for the combined wind- and buoyancy-driven flows. An interesting outcome of this investigation is the interaction between the wind and buoyancy forces that can cause the flow through the two single-sided openings to change direction. Figures 8.64(a)–(d) clearly show this effect. For wind speeds of $2\,\mathrm{m\,s^{-1}}$, buoyancy is dominant and the flow enters at the bottom opening and leaves at the top, see Figure 8.64(a). At wind speed of $4\,\mathrm{m\,s^{-1}}$ the effect of buoyancy diminishes and almost equals the wind force, resulting in minimum ventilation rate through the two openings, see Figure 8.64(b). As the wind speed increases to $6\,\mathrm{m\,s^{-1}}$ and above, the flow changes direction by entering the room at the top opening and leaving at the bottom, see Figures 8.64(c) and (d).

In Section 7.7.3 the buoyancy-driven ventilation due to heated surfaces, i.e. solar-induced ventilation, is discussed and simple models to study the various systems are presented. There have been various attempts at modelling the flow in such systems using CFD [107–110]. The CFD results have been compared with some experimental data as well as results with data using simplified integral models [109, 110]. The CFD predictions are able to account for temperature stratification in the solar collector whereas the integral models assume uniform temperature on the collector surface as well as constant heat transfer rate from the surface. Figure 8.65(b) shows results for the air flow rate with wall heat flux produced by the collector shown in Figure 8.65(a) obtained from CFD simulation using the VORTEX© code and compared with those from an integral model [110] for summer 'S' and winter 'W' scenarios. As illustrated in Figure 8.65(a), in the winter scenario, outside air is warmed by the solar collector before it is supplied to the room, whereas in the summer scenario the collector is used to draw air from outside directly into the room and out through the collector to increase the buoyancy force. The gap between the collector's surfaces used in this

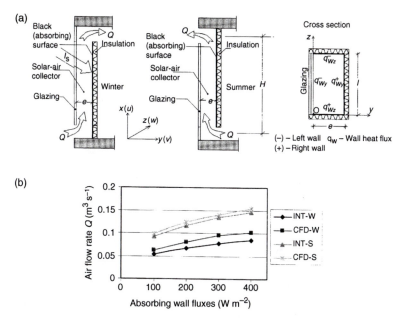

Figure 8.65 Solar-induced ventilation [110]. (a) Solar air collector and (b) air flow rate for winter and summer scenarios, W – winter; S – summer.

study is 0.2 m, the collector width is 0.7 m and its height is 2.9 m. The openings at the top and bottom are 0.3 m high and 0.7 m wide (= width of collector). The difference between the results from the integral model and the CFD predictions is due to the simplifications assumed in the integral model, such as assuming uniform heat transfer between air and collector surfaces for the whole height of the collector, which is not the case in the CFD calculations. The higher flow rates in the summer cases for the same heat flux as in the winter cases, is due to the lower resistance of the air flow path in the summer case. In the winter case the air is extracted from the room through a stack (higher flow resistance) but in summer case the air enters the room through a large open window.

8.5.9 Prediction of fire and smoke spread

One of the most difficult tasks that building designers have to face is to predict the manner in which fire and smoke would spread within a building. In particular, designers need to estimate the time interval between a fire starting in any location within the building and the concentration of smoke and toxic fumes reaching threshold levels for the safety of the occupants. This time interval is also critical for providing adequate fire escape routes to complete evacuation of the building. The extent and manner in which the smoke would spread are needed to specify fire safety equipment such as the capacity and location of smoke extractors. In predicting these parameters designers rely on codes of practice, e.g. [111], which are based on practical experience of fire spread and its control. However, because of the wide differences in building construction, materials inside and their use, these codes can never provide a fully comprehensive guidance.

(a)

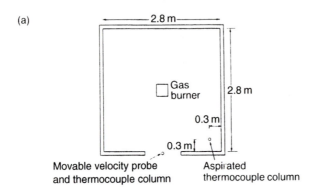

Movable velocity probe
and thermocouple column

Aspirated
thermocouple column

(b)

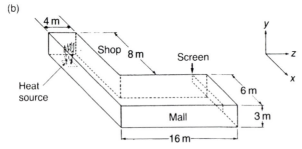

Figure 8.66 Enclosures used for predicting smoke spread [112]. (a) Plan of room and (b) shopping mall.

A design procedure for extracting smoke based on mock-up tests is described in Section 7.9. The other alternative is to apply a numerical solution of the flow and energy equations to the propagation of fire and smoke. However, because of the physical complexity of these problems the results from a numerical solution should only be considered as complementary to other design procedures.

Markatos and Cox [112] extended the PHOENICS CFD code to model the steady-state and transient spread of smoke in a room and an L-shaped shopping mall. This version of PHOENICS, which was called JASMINE, solved the transport equations for the three velocity components, temperature, k and ϵ, using the control volume method and the SIMPLEST algorithm [43]. The use of this algorithm, which adds the coefficients of the convective terms in the momentum equations to the linearized source term, was found to eliminate the need for severe underrelaxation of the pressure correction because of the large convection in this type of problem, thus accelerating convergence.

Markatos and Cox studied the smoke spread in two enclosures: a room 2.8 m square by 2.18 m high with a door way opening 1.83 m high by either 0.74 m or 0.99 m wide, and a 16 m long, 6 m wide and 3 m high L-shaped shopping mall, see Figure 8.66. A steady-state solution was used for the room and both steady-state and transient solutions were used for the mall. A gas burner of output ranging from 31.6 to 158 kW was situated at the centre of the room covering a floor area of $0.09 \, \text{m}^2$ as shown in Figure 8.66(a). A comparison between the predicted and measured profiles of the horizontal velocity component, U_e, and temperature, T_e, at the doorway for a fire

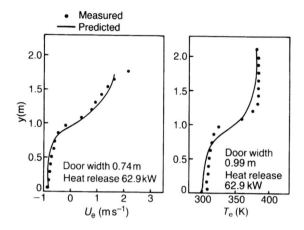

Figure 8.67 Predicted and measured velocity and temperature profiles at the door of room of Figure 8.66(a) [112].

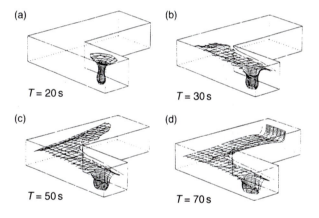

Figure 8.68 Smoke spread in the shop and mall of Figure 8.64(b) [112]. Crown Copyright reproduced by permission of the Controller of HMSO.

source of 62.9 kW (1 MW m^{-2}) is shown in Figure 8.67. The agreement between measurement and prediction is generally satisfactory with small differences in the velocity profiles near the top of the doorway opening and in the temperature profiles at the centre of the doorway in the region where the cold air entering the lower region meets the hot smoke leaving through the upper region of the doorway.

The computations for the shopping mall were carried out for either a fixed heat release rate of 3.2 MW distributed over $1.6 \text{ m}^2 (2 \text{ MW m}^{-2})$ or a linearly growing fire reaching the same rate at 3 min from ignition. The transient spread of smoke in the mall is illustrated in Figure 8.68, which shows the development of the 100 K temperature increase surface contour for four time steps. After 20 s the 100 K contour was inside the shop and had just reached the ceiling. After a further 10 s this had reached the mall zone, and by 70 s from the start of the fire the mall had been engulfed by smoke at a temperature of at least 100 K above the initial temperature.

The results presented in Figures 8.67 and 8.68 show the correct trends of smoke spread in two enclosures and are therefore encouraging. Similar predictions were made by Rosemann and Moser [113] for a fire in a room with an open door in which case comparison of the CFD results and measurements were found to be reasonable. The use of numerical solutions in the prediction of fire and smoke spread can therefore provide an insight into the most appropriate measures that are likely to help speed up the evacuation of the building as well as control the spread of fire.

References

1. ASHRAE (1989) *Building Systems: Room Air and Air Contaminant Distribution* (ed. L. L. Christianson), American Society of Heating, Refrigeration and Air-Conditioning Engineers, Atlanta, GA.
2. Proceedings of Seminar on Computational Fluid Dynamics – Tool or Toy? (1991) Institution of Mechanical Engineers, London.
3. Heiselberg, P. K., Murakami, S. and Roulet, C.-A., eds (1996) Ventilation of large spaces in buildings – Part 3: Analysis and prediction techniques, *Final Report of IEA-ECB&CS Annex 26*.
4. Gupta, A. K. and Lilley, D. G. (1985) *Flowfield Modelling and Diagnostics*, Ababcus, Tunbridge Wells, Kent.
5. Abbott, M. B. and Basco, D. R. (1989) *Computational Fluid Dynamics – An Introduction for Engineers*, Longman Scientific & Technical, England.
6. Versteeg, H. K. and Malalasekera, W. (1995) *An Introduction to Computational Fluid Dynamics – The Finite Volume Method*, Longman Scientific & Technical, England.
7. Ferziger, J. H. and Perić, M. (1996) *Computational Methods for Fluid Dynamics*, Springer-Verlag, Berlin.
8. Batchelor, G. K. (1970) *An Introduction to Fluid Dynamics*, Cambridge University Press, London.
9. Li, Y., Fuchs, L. and Holmberg, S. (1992) Methods for predicting air change efficiency, *Proc. 3rd International Conference on Air Distribution in Rooms* (Roomvent '92), Aalborg, Denmark, pp. 255–71.
10. Gan, G. and Awbi, H. B. (1994) Numerical prediction of the age of air in ventilated rooms, *Proc. 4th International Conference on Air Distribution in Rooms* (Roomvent '94), Krakow, Poland, pp. 16–27.
11. Launder, B. E. and Spalding, D. B. (1972) *Mathematical Models of Turbulence*, Academic Press, London.
12. Rodi, W. (1984) *Turbulence Models and Their Application in Hydraulics – A State of the Art Review*, International Association for Hydraulic Research, Delft, The Netherlands.
13. Henkes, R. A. W. M. and Hoogendoorn, C. J. (1989) Comparison of turbulence models for the natural convection boundary layer along a heated vertical plate. *Int. J. Heat Mass Transfer*, 32, 157–69.
14. Awbi, H. B. (1998) Calculation of heat transfer coefficients of room surfaces for natural convection. *Energy and Buildings*, 28, 219–27.
15. Yakhot, V. and Orszag, A. (1986) A renormalized group analysis of turbulence. *J. Sci. Comput.*, 1, 3–51.
16. Yakhot, V., Orszag, A., Thangam, S., Gatski, T. B. and Speziale, C. G. (1992) Development of turbulence models for shear flows by double expansion technique. *Phys. Fluids, Part A*, 4, 1510–20.
17. Chen, Q. (1995) Comparison of different k–ϵ models for indoor air flow computations. *Numer. Heat Transfer, Part B*, 28, 353–69.

18. Gan, G. (1998) Prediction of turbulent buoyant flow using RNG k–ϵ model. *Numer. Heat Transfer, Part A*, **33**, 169–89.
19. Lam, C. K. G. and Bermhorst, K. (1981) A modified form of the k–ϵ model for predicting wall turbulence. *Trans. ASME, I. Fluids Eng.*, **103**, 456–60.
20. Nagano, Y. and Hishida, M. (1987) Improved form of the k–ϵ model for wall turbulent shear flows. *Trans. ASME, J. Fluids Eng.*, **109**, 156–60.
21. Jones, W. P. and Launder, B. E. (1972) The prediction of laminarization with a two-equation model of turbulence. *Int. J. Heat Mass Transfer*, **15**, 301–14.
22. Patel, V. C., Rodi, W. and Scheuerer, G. (1985) Turbulence models for near-wall and low Reynolds number flows: a review. *AIAA J.*, **23**, 1308–19.
23. Hinze, J. O. (1975) *Turbulence*, McGraw-Hill, New York.
24. Deardorff, J. W. (1970) A numerical study of three-dimensional channel flow at large Reynolds numbers. *J. Fluid Mech.*, **41**, 453–80.
25. Murakami, S., Mochida, A. and Hibi, K. (1987) Three-dimensional numerical simulation of air flow around a cubic model by means of large eddy simulation. *J. Wind Eng. Ind. Aerodyn.*, **25**, 291–305.
26. Peng, S-H. and Davidson, L. (2000) The potential of large eddy simulation techniques for modeling indoor air flows (ed. H. B. Awbi), *Proc. Roomvent 2000*, Vol. 1, pp. 295–300, Elsevier, Oxford.
27. Müller, D. and Davidson, L. (2000) Comparison of different subgrid turbulence models and boundary conditions for large-eddy simulations (ed. H. B. Awbi), *Proc. Roomvent 2000*, Vol. 1, pp. 301–6, Elsevier, Oxford.
28. Xin, S. and le Quéré, P. (1995) Direct numerical simulations of two-dimensional chaotic natural convection in a differentially heated cavity of aspect ration 4. *J. Fluid Mech.*, **304**, 87–118.
29. Le, H., Moin, P. and Kim, J. (1997) Direct numerical simulation of turbulent flow over a backward-facing step. *J. Fluid Mech.*, **330**, 349–74.
30. Murakami, S. (1992) Prediction, analysis and design for indoor climate in large enclosures, *Proc. 3rd International Conference on Air Distribution in Rooms (Roomvent '92)*, Vol. 1, pp. 1–30, Aalborg, Denmark.
31. Chen, Q. (1995) Comparison of different k–ϵ models for indoor air flow computations, *Numer. Heat Trans.*, Part B, **28**, 353–69.
32. Chen, Q. and Chao, N-T. (1997) Comparing turbulence models for buoyant plume and displacement ventilation simulation. *Indoor Built Environ.*, **6**, 140–9.
33. Leschziner, M. A. (1992) Turbulence modeling challenges posed by complex flows, *Proc. 3rd International Conference on Air Distribution in Rooms (Roomvent '92)*, Vol. 1, pp. 31–58, Aalborg, Denmark.
34. Baker, A. J. (1985) *Finite Element Computational Fluid Mechanics*, McGraw-Hill, New York.
35. Pantankar, S. V. (1980) *Numerical Heat Transfer and Fluid Flow*, Hemisphere, New York.
36. Leonard, B. P. (1979) A stable and accurate convective modelling procedure based on quadratic upstream interpolation. *Comput. Methods Appl. Mech. Eng.*, **19**, 59–98.
37. Huang, P. G., Launder, B. E. and Leschziner, M. A. (1985) Discretization of nonlinear convection processes: a broad-range comparison of four schemes. *Comput. Methods Appl. Mech. Eng.*, **48**, 1–24.
38. Raithby, G. D. (1976) A critical evaluation of upstream differencing applied to problems involving fluid flow. *Comput. Methods App. Mech. Eng.*, **9**, 75–103.
39. Hayase, T., Humphrey, J. A. C. and Greif, R. (1992) A consistently formulated QUICK scheme for fast and stable convergence using finite-volume iterative calculations. *J. Comput. Phys.*, **98**, 108–18.
40. Rosten, H. I. and Spalding, D. B. (1987) The PHOENICS equations. Concentrated Heat and Mass Ltd, Repeort No. TR/99, London.

41. Patankar, S. V. and Spalding, D. B. (1972) A calculation procedure for heat, mass and momentum transfer in three-dimensional parabolic flows. *Int. J. Heat Mass Transfer*, **15**, 1787.
42. van Doormaal, J. P. and Raithby, G. D. (1984) Enhancement of the SIMPLE methods for predicting incompressible fluid flows. *Numer. Heat Trans.*, 7, 147–63.
43. Spalding, D. B. (1980) Mathematical modelling of fluid-mechanics, heat-transfer and chemical-reaction processes, CFDU Report No. HTS/80/1, Imperial College, London.
44. Markatos, N. C. and Pericleous, K. A. (1984) Laminar and turbulent natural convection in an enclosed cavity. *Int. J. Heat Mass Transfer*, **27**, 755–72.
45. Gosman, A. D. and Ideriah, F. J. K. (1983) TEACH-2E: A general computer program for two-dimensional, turbulent recirculating flows, Report no. FM-83-2, Mechanical Engineering Department, Imperial College, London.
46. Ideriah, F. J. K. (1980) Prediction of turbulent cavity flow driven by buoyancy and shear. *J. Mech. Eng. Sci.*, **22**, 287–95.
47. Hammond, G. P. (1982) Complete velocity profile and 'optimum' skin friction formulas for the plane wall-jet. *Trans. ASME, J. Fluid Eng.*, **104**, 59–66.
48. Nizou, P. Y. (1981) Heat and momentum transfer in a plane wall jet. *Trans. ASME, J. Heat Transfer*, **103**, 138–40.
49. Yuan, X., Moser, A. and Sutter, P. (1993) Wall functions for numerical simulation of turbulent natural convection along vertical plates. *Int. J. Heat Mass Transfer*, **36**, 4477–85.
50. Chen, Q. (1990) Construction of a low Reynolds number k–ϵ model. *Phoenics J. Comput. Fluid Dynamics*, **3**, 288–329.
51. Rajaratnam, N. (1976) *Turbulent Jets*, Elsevier, Amsterdam.
52. Saïd, M. N. A., Jouini, D. B. and Plett, E. G. (1993) Influence of turbulence parameters at supply inlet on room air diffusion, *Proc. ASME Winter Meeting*, Paper 93-WA/HT-67, p. 14, New Orleans, Louisiana, USA.
53. Roache, P. J. (1994) Perspective: a method for uniform reporting of grid refinement studies. *J. Fluids Engg.*, **116**, 405–13.
54. Roache, P. J. (1997) Quantification of uncertainty in computational fluid dynamics. *Annu. Rev. Fluid Mech.*, **29**, 123–60.
55. Patankar, S. V. (1988) Recent development in computational heat transfer. *Trans. ASME, J. Heat Transfer*, **110**, 1037–45.
56. Patankar, S. V. and Spalding, D. B. (1970) *Heat and Mass Transfer in Boundary Layers*, Intertext Books, London.
57. Hjertager, B. H. and Magnussen, B. J. (1981) Calculation of turbulent three-dimensional jet induced flow in rectangular enclosure. *Comput. Fluids*, **9**, 395–407.
58. Nielsen, P. V., Restivo, A. and Whitelaw, J. H. (1978) The velocity characteristics of ventilated rooms. *Trans. ASME, J. Fluids Eng.*, **100**, 291–8.
59. Awbi, H. B. and Setrak, A. A. (1986) Numerical solution of ventilation air jet, *Proc. 5th CIB Int. Symp. on the Use of Computers for Environmental Engineering Related to Buildings*, Bath, UK, pp. 236–46.
60. Awbi, H. B. and Setrak, A. A. (1987) Air Jet interference due to ceiling-mounted obstacles. *Air Distribution in Ventilated Spaces Symp.*, Roomvent '87, Stockholm, session 1.
61. Setrakian, A. A. (1988) *The Effect of Rectangular Obstacles on the Diffusion of a Wall Jet*, PhD thesis, Napier Polytechnic.
62. Awbi, H. B. and Setrak, A. A. (1989) Numerical solution of wall jets, *Building Systems: Room Air and Air Contaminant Distribution* (ed. L. L. Christianson), American Society of Heating, Refrigeration and Air-Conditioning Engineers, Atlanta, GA.
63. Awbi, H. B. (2001) Vortex© version 3.1 user manual, Reading, UK, www.personal.rdg.ac.uk/~kcsawbi.

64. Albright, L. D. and Scott, N. R. (1974) The low-speed non-isothermal wall jet. *J. Agric. Eng. Res.*, **19**, 25–34.

65. Rajaratnam, N. and Pani, B. (1974) Three- dimensional turbulent wall jets. *ASCE Trans., J. Hydraul. Div.*, **100**, 69–83.

66. Mohammad, W. S. (1986) *Space Air-Conditioning of Mechanically-Ventilated Rooms: Computation of Flow and Heat Transfer*, PhD thesis, Cranfield University, UK.

67. Nielsen, P. V. (1974) *Flow in Air-Conditioned Rooms*, PhD thesis, Technical University of Denmark.

68. Gosman, A. D. *et al.* (1969) *Heat and Mass Transfer in Recirculating Flows*, Academic Press, London.

69. Nielsen, P. V. (1974) Moisture transfer in air-conditioned rooms and cold stores, *Proc. 2nd Int. CIB/RILEM Symp. on Moisture Problems in Buildings, Rotterdam*, Paper 1.2.1.

70. Timmons, M. B. *et al.* (1980) Experimental and numerical study of air movement in slot-ventilated enclosures, *ASHRAE Trans.*, **86** (1), 221–40.

71. Hjertager, B. H. and Magnussen, B. F. (1977) Numerical prediction of three-dimensional turbulent buoyant flow in a ventilated room, *Heat Transfer and Turbulent Buoyant Convection* (eds D. B. Spalding and N. Afgan), Vol. 11, pp. 429–41, Hemisphere, Washington, DC.

72. Davidson, L. and Nielsen, P. V. (1996) Large eddy simulations of the flow in a three-dimensional ventilated room, *Proc. 5th International Conference on Air Distribution in Rooms Roomvent 1996*, Vol. 2, pp. 161–8, Yokohama, Japan.

73. Nielsen, P. V., Restivo, A. and Whitelaw, J. H. (1979) Buoyancy-affected flows in ventilated rooms. *Numer. Heat Transfer*, **2**, 115–27.

74. Nielsen, P. V. (1989) Numerical predictions of air distribution in rooms – Status and potentials, *Building Systems: Room Air and Air Contaminants Distribution* (ed. L. L. Christianson), pp. 31–8, American Society of Heating, Refrigeration and Air-Conditioning Engineers, Atlanta, GA.

75. Gosman, A. D., Nielsen, P. V., Restivo, A. and Whitelaw, J. H. (1980) The flow properties of rooms with small ventilation openings. *Trans. ASME, I. Fluid Eng.*, **102**, 316–23.

76. Sakamoto, Y. and Matsuo, Y. (1980) Numerical predictions of three-dimensional flow in a ventilated room using turbulence models. *Appl. Math. Modelling*, **4**, 67–71.

77. Harlow, F. H. and Welch, J. E. (1965) Numerical calculation of time-dependent viscous incompressible flow of fluid with free surfaces. *Phys. Fluids*, **8**, 2182–9.

78. Hirt, C. W. and Cook, J. L. (1972) Calculating three-dimensional flows around structures and over rough terrain. *J. Comput. Phys.*, **10**, 324–40.

79. Murakami, S., Kato, S. and Suyama, Y. (1987) Three-dimensional numerical simulation of turbulent air flow in a ventilated room by means of a two-equation model. *ASHRAE Trans.*, **93** (2), 621–42.

80. Murakami, S. and Kato, S. (1989) Current status of numerical and experimental methods for analyzing flow field and diffusion field in a room. *Building Systems: Room Air and Air Contaminants Distribution* (ed. L. L. Christianson), pp. 39–56, American Society of Heating, Refrigeration and Air-Conditioning Engineers, Atlanta, GA.

81. Reinartz, A. and Renz, U. (1984) Calculations of the temperature and flow field in a room ventilated by a radial air distributor. *Int. J. Refrig.*, **7**, 308–12.

82. Müller, D. and Renz, U. (1998) Measurements and predictions of room air flow patterns using different turbulence models. *Proc. of 6th International Conference on Air Distribution in Rooms (Roomvent '98)*, Vol. 1, pp. 109–23, Stockholm.

83. Awbi, H. B. (1989) Application of computational fluid dynamics in room ventilation. *Build. Environ.*, **24**, 73–84.

84. Jacobsen, T. V. and Nielsen, P. V. (1993) Numerical modeling of thermal environment in a displacement-ventilated room. *Proc. 6th International Conference on Indoor Air Quality and Climate (Indoor Air '93)*, Vol. 5, pp. 301–6, Helsinki.

85. Awbi, H. B. and Gan, G. (1994) Predicting air flow and thermal comfort in offices, *ASHRAE J.*, February 1994, pp. 17–21.

86. Iwamoto, S. (1998) A study on numerical prediction method of indoor thermal environment including human body, *Proc. of 6th International Conference on Air Distribution in Rooms (Roomvent '98)*, Vol. 2, pp. 167–72, Stockholm.

87. Murakami, S., Kato, S. and Zeng, J. (1998) Combined simulation of air flow, radiation and moisture transport for heat release from human body. *Proc. of 6th International Conference on Air Distribution in Rooms (Roomvent '98)*, Vol. 2, pp. 141–50, Stockholm.

88. Nielsen, P. V. (1981) Contamination distribution in industrial areas with forced ventilation and two-dimensional flow, *Int. Inst. of Refrigeration, Joint Meeting El, Essen*, West Germany, pp. 223–30.

89. Murakami, S., Tanaka, T. and Kato, S. (1983) Numerical simulation of air flow and gas diffusion in room model – correspondence between numerical simulation and model experiment, *Proc. 4th CIB Int. Symp. on the Use of Computers for Environmental Engineering Related to Buildings*, Tokyo, pp. 90–5.

90. Kurabuchi, T. and Kusuda, T. (1987) Numerical prediction for indoor air movement. *ASHRAE J.*, 26–30, December.

91. Lee, H. and Awbi, H. B. (2003) Effect of internal partition on indoor air quality of rooms with mixing ventilation – basic study, submitted for publication in *Building and Environment*.

92. Holmberg, S. and Li, Y. (1998) Modelling of the indoor environment – particle dispersion and deposition. *Indoor Air*, 8, 113–22.

93. Holmberg, S. and Li, Y. (1998) Non-passive particle dispersion in a displacement ventilated room – A numerical study. *Proc. of 6th International Conference on Air Distribution in Rooms (Roomvent '98)*, Vol. 1, pp. 467–73, Stockholm.

94. Sparks, L. E., Tichenor, B. A., Chang, J. and Guo, Z. (1996) Gas-phase mass transfer for predicting volatile organic compound (VOC) emission rates from indoor pollutant sources. *Indoor Air*, 6, 31–40.

95. Yang, X., Chen, Q. and Zhang, J. S. (1997) Study of VOC emissions from building materials using computational fluid dynamics. *Proc. Healthy Buildings/IAQ '97*, (eds J. E. Woods *et al.*), Vol. 3, pp. 587–92.

96. Topp, C., Nielsen, P. V. and Heiselberg, P. (1997) Modelling emission from building materials with computational fluid dynamics, *Proc. 8th International Conference on Indoor Air quality and Climate (Indoor Air '99)*, Vol. 4, pp. 737–42.

97. Topp, C., Nielsen, P. V. and Heiselberg, P. (1999) Evaporation controlled emission in ventilated rooms. *Proc. Healthy Buildings/IAQ '97*, (eds. J. E. Woods *et al.*), Vol. 3, pp. 557–63.

98. Yang, X. and Chen, Q. (1999) A model for numerical simulation of VOC sorption by building materials. *Proc. 8th International Conference on Indoor Air quality and Climate (Indoor Air '99)*, Vol. 4, pp. 797–802.

99. Murakami, S., Kato, S., Ito, K. and Yamamoto, A. (1999) Analysis of chemical pollutants distribution based on coupled simulation of CFD and emission/sorption processes. *Proc. 8th International Conference on Indoor Air quality and Climate (Indoor Air '99)*, Vol. 4, pp. 725–30.

100. Murakami, S., Kato, S., Kondo, Y., Ito, K. and Yamamoto, A. (2000) VOC distribution in a room based on CFD simulation coupled with emission/sorption analysis, (ed. H. B. Awbi), *Proc. of 7th International Conference on Air Distribution in Rooms (Roomvent 2000)*, Vol. 1, pp. 473–8, Elsevier, Oxford.

101. Davidson, L. and Olsson, E. (1987) Calculation of age and local purging flow rate in rooms. *Build. Environ.*, 22, 111–27.

102. Davidson, L. (1989) *Numerical Simulation of Turbulent Flow in Ventilated Rooms*, PhD thesis, Chalmers University of Technology, Götenborg.

103. Kurabuchi, T., Ohba, M., Arashiguchi, A. and Iwabuchi, T. (2000) Numerical study of air flow structure of a cross-ventilated model building, (ed. H. B. Awbi), *Proc. of 7th International Conference on Air Distribution in Rooms (Roomvent 2000)*, Vol. 1, pp. 313–8, Elsevier, Oxford.

104. Schaelin, A. J., Orzag, S. A., van der Mass, A. J. and Moser, A. (1992) Simulation of air flow through large openings in buildings. *ASHRAE Transactions*, 92, 319–28.

105. Gan, G. (1999) Numerical determination of the effective depth for single-sided natural ventilation. *Proc. 8th International Conference on Indoor Air quality and Climate (Indoor Air '99)*, Vol. 4, pp. 354–9.

106. Allocca, C., Chen, Q. and Glicksman, R. (2003) Design analysis of single-sided natural ventilation, to be published in *Energy and Buildings*.

107. Awbi, H. B. and Gan, G. (1992) Simulation of solar-induced ventilation, *Proc. 2nd World Renewable Energy Congress*, (ed. A. A. M. Sayigh), Vol. 4, pp. 2016–30, Elsevier.

108. Gan, G. (1998) A parametric study of Trombe walls for passive cooling of buildings, *Energy and Buildings*, 27, 37–43.

109. Rodrigues, A. M., de Piedade, A. C. and Awbi, H. B. (2000) Evaluation of vertical solar-air collectors for natural ventilation using integral and CFD models, *Proc. 6th World Renewable Energy Congress*, (ed. A. A. M. Sayigh), Vol. 1, pp. 389–94, Elsevier.

110. Rodrigues, A. M., de Piedade, A. C. and Awbi, H. B. (2000) The use of solar air collectors for room ventilation: A study using two numerical approaches, (ed. H. B. Awbi), *Proc. of 7th International Conference on Air Distribution in Rooms (Roomvent 2000)*, Vol. 1, pp. 281–7, Elsevier, Oxford.

111. BS 5588 (1997) Fire precautions in the design and construction of buildings – Part 11: Code of practice for shops, offices, industrial, storage and other similar buildings, British Standards Institution, London.

112. Markatos, M. C. and Cox, G. (1984) Hydrodynamics and heat transfer in enclosures containing a fire source. *Phys. Chem. Hydrodyn.*, 5, 53–66.

113. Rosemann, P. and Moser, A. (1999) Smoke movement in buildings. *Proc. 8th International Conference on Indoor Air quality and Climate (Indoor Air '99)*, Vol. 4, pp. 773–8.

9 Measurement of indoor climate

9.1 Introduction

It is evident from the previous chapters that the way buildings are ventilated determines the levels of thermal and air qualities as well as the energy requirement to achieve them. To quantify these levels requires the measurement of many physical parameters either in physical models of the ventilation system and building module or on site during commissioning and post-commissioning to ensure that the design targets are achieved. The evaluation of the thermal quality, air quality and energy consumption depends on the measurement of parameters such as temperature, vapour pressure, air velocity, air flow rate, thermal comfort, air contamination and the visualization of the air flow in the space being ventilated. The measurement of any physical parameter requires the use of a sensor, i.e. an input transducer and an output device, i.e. an output transducer. A transducer, whether of the input or the output type, is basically a device, which converts energy from one form to another. For example, the junction of a thermocouple represents an input transducer and the temperature indicator is the output transducer.

In this chapter, devices that are often used in indoor climate measurements are described and, where possible, their range and accuracy are given. Some of the devices described are standard instruments used for general scientific and engineering measurements whereas others are specific to the measurement of the physical parameters of the indoor climate. The devices featured here cover measurements within the ventilated space as well as the plant. It is well known that all measurements lead to certain errors associated with uncertainties in instrument calibration, environmental factors, the use of constants and factors in the measurement, uncertainties in the manufacture and use of the instrument, user-related errors in installing the instrument, data recording and analysis, etc. It is not the intention here to deal with such uncertainties and how to obtain their limits and the interested reader should consult specialist texts on error analysis, e.g. Taylor [1].

9.2 Measurement of air temperature

The fundamental meaning of temperature lies within the principle of thermal equilibrium, which is usually described by the 'zeroth law of thermodynamics'. The basis of temperature measurement is based upon the principle of thermal equilibrium where a suitable pre-calibrated device makes contact with the body of which the temperature is required, and when equilibrium is reached the temperature of the body is determined using a standard scale of temperature. A temperature-measuring sensor is called a 'thermometer' although this term is often taken to mean the liquid-in-glass type. In general, a thermometer can be any device that exhibits a reproducible variation of any physical property of materials with temperature, e.g. length, volume, electrical resistance, electromotive force (e.m.f.), etc. The output of a thermometer must be related to an acceptable temperature scale. In ventilation work the Celsius (C) and Fahrenheit (F) scales are most widely used for environmental temperature measurement and the Kelvin scale (K) is sometimes used in calculations involving the gas laws or for defining difference in temperature. The Kelvin scale is related to the Celsius scale by:

$$K = 273.15 + °C$$

A thermometer is calibrated against a scale of temperature, which consists of a number of reproducible temperature reference points, all at standard atmospheric pressure, such as the triple point of water (273.16 K or 0.01 °C); the boiling points of liquid oxygen (-182.97 °C), water (100 °C) and sulphur (444.60 °C); and the melting points of antimony (630.5 °C), silver (960.8 °C) and gold (1063.0 °C). The equilibrium temperatures of these thermodynamic states have all been determined accurately by gas thermometry. A thermometer is normally calibrated to include at least two of these reference temperatures and if necessary interpolation between the reference points is accomplished by the use of resistance thermometers. Those temperature-measuring devices most commonly used in ventilation work are described in the next section.

9.2.1 Temperature-measuring devices

Liquid-in-glass thermometers

A number of commonly used temperature-measuring devices utilize the physical phenomenon of thermal expansion. The expansion of solids is utilized in the bi-metallic thermometer by the differential expansion of bonded strips of two metals; the expansion of a gas is used in the gas thermometer; and the expansion of a liquid is used in the liquid-in-glass thermometer. The latter is the most common type of thermometer utilizing the thermal expansion principle. By using different liquids a range of liquid-in-glass thermometers are available for measuring temperatures in the range of about -70 to 540 °C. Mercury is the most common thermometer liquid used in HVAC environments as its temperature range (-39 to 537 °C) covers most temperatures experienced in the built environment.

Thermometers are calibrated during manufacture for two or more temperatures, which are usually the triple and boiling points of water for climatic temperature measurements. The distance between the calibration points is divided into a linear scale and a probable measuring error is plus or minus one scale division. However, the accuracy obtainable depends on the instrument quality and temperature range. For good quality calibrated thermometers the error can be as low as 0.2 K over the range 0–100 °C. When measuring temperatures in pipes or ducts, accuracy can be improved by fitting the instrument in a thermometer well. A thermometer measuring air or gas temperatures can be affected by radiation from surrounding surfaces and this should be minimized if the temperature of the gas differs substantially from that of the surrounding surfaces. This may be achieved by shielding or aspiration. Shielding is accomplished by placing highly reflective surfaces between the thermometer bulb and the surrounding surfaces but care must be taken so that air movement around the bulb is not appreciably restricted. Aspiration is attained by drawing the surrounding air or gas over the bulb using a fan if necessary. Because of the relatively large thermal mass of the instrument, ample time is required to attain temperature equilibrium with the surrounding fluid before reading the temperature scale. Parallax may be avoided by keeping the eye level with the top of the liquid column.

Pressure thermometers

A pressure thermometer consists of a bulb and a pressure-measuring device (such as a Bourdon gauge, a diaphragm or bellows) interconnected by a capillary tube. The thermometer fluid can be a liquid, a combination of liquid and vapour, or a gas. For a liquid-filled pressure thermometer, a change in the bulb temperature causes essentially a change in the volume of liquid, ignoring compressibility effects. However, for the other two thermometers a temperature change corresponds to a pressure change at an essentially constant volume of the fluid. A pressure thermometer using a Bourdon gauge is shown in Figure 9.1.

Because the pressure gauge is normally located away from the bulb (temperature-measuring point) a long capillary tube is sometimes needed. As a result, measurement errors are introduced by the temperature variation along the capillary and at the pressure gauge and these can be minimized by compensation. A common method of compensation is to link the primary pressure gauge to an auxiliary gauge and capillary. Compensation is normally justified in liquid-filled pressure thermometers but is not usually necessary in gas or vapour-filled thermometers.

Pressure thermometers produce a linear output over a wide temperature range. Different fluids can be used to cover a temperature range suitable for most engineering applications. However, the accuracy of these devices is not as good as other temperature-measuring sensors discussed here, and under the best conditions it is in the region of $\pm0.5\%$ of full scale.

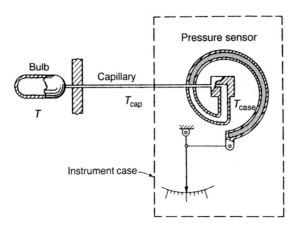

Figure 9.1 Pressure thermometer.

Thermocouples

A thermocouple is a sensor, which converts thermal energy to electrical energy using two junctions of two dissimilar metals. The e.m.f. generated will be proportional to the difference in temperature between the two junctions, the so-called Seebeck effect.

In addition to the Seebeck effect there are two other phenomena associated with thermocouples: the so-called Peltier effect, which causes heat generation or absorption at a junction where an electric current passes; and the Thompson effect, which causes heat generation at a point on a conductor where the flow of current is in the same direction as the heat flow. Because of these phenomena a thermocouple material must be calibrated over the complete range of temperature in which it is to be used. During calibration only the overall voltage is measured and the contributions of the Peltier and Thomson effects are lumped with the Seebeck effect.

A typical thermocouple circuit is shown in Figure 9.2 and some commonly used thermocouple materials and their temperature range are listed in Table 9.1. In Figure 9.2, the two thermocouple metals are A and B but a third metal, C (e.g. copper wire), is introduced to make the electrical connection to a millivoltmeter. The e.m.f. measured by the millivoltmeter is the same as that generated at the junction of A and B (hot junction) provided that the two other junctions of A and C and of B and C are maintained at the same temperature, i.e. a reference temperature. In most applications the reference temperature is taken to be the triple point of water (0.01 °C) and is referred to as the cold-junction temperature. A mixture of ice and water can be used as an accurate cold junction or alternatively a compensation for the variation in the reference temperature can be provided by a thermistor (described in the section on 'Thermistors') albeit less accurately.

When the triple point of water is used as a reference temperature a standard calibration table for the thermocouple type may be used to obtain the hot-junction temperature. In most applications the thermocouple output is amplified and linearized to produce a direct reading of the hot-junction temperature or a constant sensitivity (μV K^{-1}) throughout the temperature-measuring range. This is due to the low sensitivity of most thermocouples (see Table 9.1) and the non-linear output from them. When choosing a thermocouple, both the sensitivity and the temperature range must be taken into consideration as well as other characteristics like durability, stability and resistance to oxidation.

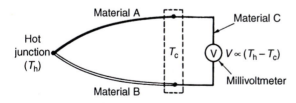

Figure 9.2 The principle of a thermocouple.

Table 9.1 Commonly used thermocouple materials

International type designation	Conductor material	Temperature range (°C)	Approximate sensitivity (μVK^{-1}) at 100 °C (0 °C reference junction)	Accuracy (0 °C reference junction)[a]
K	Nickel chromium Ni–Cr (+) versus Nickel aluminium Ni–Al (−)	0 to +1250	42	±2.2 K or ±0.75%
T	Copper, Cu (+) versus Constantan, Cu–Ni (−)	−185 to +350	46	±1.0 K or ±0.75%
J	Iron, Fe (+) versus Constantan, Cu–Ni (−)	+20 to +750	46	±1.1 K or ±0.4%
E	Nickel chromium Ni–Cr (+) versus Constantan, Cu–Ni (−)	0 to +900	68	±1.7 K or ±0.5%
R	Platinum and 13% rhodium, Pt–13%Rh (+) versus Platinum, Pt (−)	0 to +1450	8	±1.5 K or ±0.25%

Note
a This is the standard accuracy for a thermocouple; the greater of the two tolerances should be used.

Resistance thermometers

The electrical resistance of some materials changes in a reproducible manner with temperature and this characteristic forms the basis of the resistance thermometer. Metals such as platinum, nickel, copper, tungsten and some alloys are often used but because platinum is inert and stable it produces a repeatable change of resistance with temperature and is therefore the most widely used. The sensor is made using a coil of platinum wire on a glass or ceramic former or as a thin film of platinum on a suitable substrate.

The variation of platinum resistance with temperature can be represented by the quadratic equation:

$$R_t/R_0 = 1 + At + Bt^2 \tag{9.1}$$

where R_t = thermometer resistance at temperature, $t(\Omega)$; R_0 = thermometer resistance at $0\,°C(\Omega)$; t = temperature(°C) and A and B are temperature coefficients. For commercially produced platinum resistance thermometers, standard tables of resistance versus temperature are available based on $R_0 = 100\,\Omega$ and $R_{100} = 138.5\,\Omega$.

The resistance of the sensor is measured using either a resistance bridge, such as a Wheatstone bridge, or a potentiometer and a constant current source for energizing the thermometer. A bridge circuit is the most widely used, in which the unknown resistance of the thermometer forms one arm of the bridge and the other three resistances are known or may be determined. The unknown resistance may be determined either by using a balanced bridge or using an unbalanced bridge and measuring the out-of-balance voltage, V_t, that results in a bridge of three fixed resistances as in Figure 9.3(a). This is a simple two-wire connection, which is used where high accuracy is not required

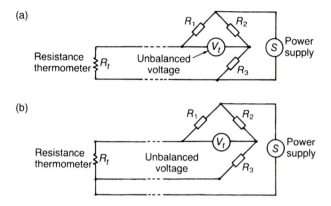

Figure 9.3 Resistance thermometer bridge circuits. (a) Two-wire connection and (b) three-wire connection.

as the resistance of the wires connecting the sensor to the bridge is included with that of the sensor. A better scheme is to use a three-wire connection, Figure 9.3(b), in which case the two leads connecting the sensor are on either side of the bridge and thus effectively cancel, while the third lead function as an extended supply lead. A more accurate connection still is the four-wire connection in which the resistance of the leads has a negligible effect on the measurement. These instruments can give measurement accuracy within ± 0.1 K. The supply voltage to the bridge can be either a.c. or d.c. From the value of the unbalanced voltage, V_t and the known resistances of the bridge, R_1, R_2 and R_3, the resistance and hence the temperature of the sensor can be determined. A resistance thermometer is usually more accurate than a thermocouple and is less influenced by mains current interference.

Thermistors

The generic name thermistor is derived from '*thermally sensitive resistor*' and it is based on the same principles as a resistance thermometer except that the sensor is made of a semi-conducting material. Most common thermistors are made of sintered oxides of manganese, nickel and cobalt which undergo a large resistance change with temperature. The sensor can be made in the form of rods, discs or small beads, but for sensing purposes the small bead shape is used where faster response is required. The sensor is encased by a glass or stainless steel sheath. Unlike resistance thermometers, thermistors exhibit a very non-linear characteristic of resistance with temperature which is expressed by:

$$R_t = R_0 \exp\left[B\left(\frac{1}{T} - \frac{1}{T_0}\right)\right] \tag{9.2}$$

where R_t = resistance at sensor temperature $T(\Omega)$; R_0 = resistance at reference temperature $T_0(\Omega)$; B = temperature coefficient of sensor (K) and T and T_0 are absolute temperatures (K). The temperature coefficient, B, is usually of the order 3000–4000, and it can be seen from equation (9.2) that the resistance of the thermistor decreases with increasing sensor temperature in contrast with the resistance thermometer, which exhibits an increase in resistance with temperature. However, some thermistors which exhibit an increase in resistance are also available.

The bridge circuits for measuring the sensor resistance, and hence temperature, are essentially the same as for resistance thermometers (Figure 9.3), although the greater non-linearity of thermistors makes the measurement of a wide temperature range less desirable. The resistance non-linearity may be reduced by shunting the thermistor with a fixed resistor (i.e. connecting the resistor in parallel with the sensor). Although thermistors are less accurate than a platinum resistance thermometer they are widely used because they can be manufactured in different shapes and sizes and small thermistors have a faster response than a resistance thermometer. Thermistors can also be used for measuring the velocity of air and other fluids when a reference thermistor is added to the circuit of a Wheatstone bridge in addition to the active thermistor. This is discussed in Section 9.6.

9.2.2   *Precautions in air temperature measurement*

In measuring the static temperature of moving fluids using the sensors described earlier errors may arise as a result of: (i) heat conduction from the probe to a surface in contact with it but at a different temperature from the fluid; (ii) velocity pressure of the moving fluid which results in the measurement of the stagnation temperature of the fluid instead of the static temperature; (iii) radiation heat exchange between the probe and surrounding surfaces at different temperatures from the fluid and (iv) dynamic response of the temperature-measuring device. For most environmental and ventilation temperature measurements the values of the first two errors are usually negligible and therefore only the last two sources of error are discussed here.

Radiation error

When measuring the temperature of a moving fluid using a probe, the sensor is influenced by convective heat exchange with the fluid and radiant heat exchange with the surrounding surfaces as well as with the fluid. However, when gases are involved the radiation heat transfer with the fluid may be neglected, unless they contain particulates as in flue gases. If conduction errors are also neglected then the temperature recorded by a probe is determined by balancing the convective and radiative heat exchange. Thus:

$$h_c A_p (T_p - T_f) = \epsilon_p \sigma A_p (T_w^4 - T_p^4) \tag{9.3}$$

where A_p = surface area of probe (m^2); T_f = fluid temperature (K); T_w = temperature of surrounding surfaces (K); T_p = probe temperature (K); ϵ_p = emissivity of probe surface; σ = Stefan–Boltzmann constant = 5.67×10^{-8} W m^{-2} K^{-4} and h_c = convective heat transfer coefficient of probe surface (W m^{-2} K^{-1}).

Equation (9.3) assumes that the probe is completely enclosed by surfaces at constant temperature, T_w, and ignores gas radiation. The error due to surface radiation is:

$$T_p - T_f = (\epsilon_p \sigma / h_c)(T_w^4 - T_p^4) \tag{9.4}$$

It is evident from this equation that the radiation error may be reduced by taking one or more of the following precautions:

1   Reducing the difference between T_W and T_p by isolating the surrounding surfaces, which may not always be plausible unless reflective screens are used between the surfaces and the probe.
2   Increasing the convective heat transfer coefficient, h_c, by probe aspiration or by reducing the physical size of the sensor.
3   Reducing the emissivity, ϵ_p, of the probe surface by using a polished probe surface if it is metallic or coating it with reflective paint if it is made of a non-metallic material.

The most common alternative to these measures is to surround the probe by a cylindrical shield of polished metal. If adequately ventilated, the shield temperature becomes closer to the fluid temperature than are the surrounding surfaces of the enclosure and the probe then exchanges radiation with the shield, thus reducing the radiation error.

Dynamic response

A temperature sensor is a device that converts thermal energy into other forms of energy such as mechanical or electrical. This conversion is essentially instantaneous. However, the dynamic characteristics of the sensor are almost entirely determined by the heat transfer and thermal storage processes that occur in the sensor. It is these thermal processes that cause the sensor temperature to lag behind that of the measured medium.

Neglecting the heat loss from the sensor and equating the heat transfer during time interval $d\theta$ to the thermal storage, the energy balance equation for a sensor becomes:

$$U_p A_P (T_f - T_p)\, d\theta = mC dT_p \tag{9.5}$$

where A_p = surface area of probe (m^2); U_p = overall heat transfer coefficient for the probe (W m^2 K^{-1}); m = mass of probe (kg); C = specific heat of probe (J kg^{-1}K^{-1}); T_f = fluid temperature (°C) and T_p = probe temperature (°C).

From equation (9.5) it can be deduced that the time constant, τ, of the sensing element is given by:

$$\tau = mC/(U_p A_p) \tag{9.6}$$

Clearly, a low time constant can be achieved by decreasing the mass, m, and thermal capacity, C, of the sensor and increasing the overall heat transfer coefficient, U_p and the heat transfer area, A_p. The parameters m, C and A_p are sensor dependent whereas U_p is dependent upon the physical properties and velocity of the fluid in contact with the sensor. Therefore, the time constant of a particular sensor is not a constant value for the sensor but rather dependent on how it is used to measure the temperature of a fluid.

9.3   Measurement of radiant temperature

9.3.1   *Mean radiant temperature devices*

The mean radiant temperature is defined as the temperature of a uniform radiantly black enclosure in which an occupant would exchange the same amount of thermal radiation as in the actual non-uniform space [2]. In practice this is represented by the *'effective'* average temperature of the surfaces of an enclosure *'seen'* at a given point within the enclosure. In this section the methods used for determining this temperature are described. These methods are based on using a sphere as the probe because it is able to integrate the radiant effect of the surrounding surfaces.

Globe thermometer

This is the earliest and still the most commonly used instrument for measuring the mean radiant temperature of an enclosure. It consists of a thin-walled black sphere in the centre of which is placed a temperature sensor such as the bulb of a liquid-in-glass thermometer, a thermocouple junction or a resistance thermometer. To increase the absorptivity of the sphere the outer surface is darkened either by means of electro-chemical coating or, more commonly, by a layer of matt black paint. The instrument was first developed by Vernon in 1932 [2] to study radiation heat losses from humans. When the instrument is suspended within an enclosure and sufficient time elapses to reach equilibrium between convective and radiant heat transfer from the surface of the sphere, the temperature recorded by the temperature sensor is called the 'globe temperature'. Depending on the value of radiant and convective heat transfer, the globe temperature lies between the mean radiant temperature of the enclosure and the temperature of air surrounding the globe. The smaller the diameter of the sphere, the greater is the effect of the air temperature and air velocity, thus reducing the accuracy of measuring the mean radiant temperature. ISO Standard 7726 [3] recommends a sphere of 150 mm diameter but measurements using spheres as small as 38 mm (table tennis ball) have produced results similar to those obtained from a sphere of 150 mm diameter [4].

The principle behind the globe thermometer is the equilibrium of heat transfer by convection and radiation from the surface of the sphere. The mean radiant temperature, T_r, may be expressed in terms of the air temperature, T_a and the globe temperature, T_g, using the expression (9.3). Thus:

$$T_r = \sqrt[4]{T_g^4 + \frac{h_c}{\sigma \epsilon}(T_g - T_a)} \qquad (9.7)$$

This expression is derived from equation (9.3) by writing T_r for T_w, T_g for T_p and T_a for T_f. The convective heat transfer coefficient, h_c, may be determined from the following expressions [3]:

For natural convection:

$$h_c = 1.4\sqrt{(\Delta t/d)} \qquad (9.8)$$

For forced convection:

$$h_c = 6.3 v^{0.6} d^{-0.4} \qquad (9.9)$$

where Δt = difference between air and globe temperatures (K); d = globe diameter (m) and v = air velocity (m s^{-1}).

Equation (9.9) applies for a sphere Reynolds number ($Re = \rho v d/\mu$) of $10^2 < Re < 10^5$ and equation (9.8) for $Re < 10^2$ [4]. The emissivity of a black sphere is $\epsilon \approx 0.95$.

From equations (9.7)–(9.9) it is obvious that the evaluation of the mean radiant temperature is based on the measurement of air temperature and air velocity as well as the globe temperature.

Two-globe radiometer

To eliminate the effect of convective heat transfer on the globe temperature reading, two globes of different emissivity (one black and one polished) heated to the same temperature are used. Since the convective heat loss from the two spheres is equal, the difference in heat input to the two spheres is equal to the difference in radiant heat loss between the two spheres. The heat input to each sphere is given by:

$$Q_b = h_c A(T_g - T_a) + \sigma \epsilon_b A(T_g^4 - T_r^4) \tag{9.10}$$

$$Q_p = h_c A(T_g - T_a) + \sigma \epsilon_p A(T_g^4 - T_r^4) \tag{9.11}$$

where subscripts b and p refer to the black and polished spheres, respectively. By eliminating the convective terms from these equations, the expression for the mean radiant temperature becomes:

$$T_r = \sqrt[4]{T_g^4 + \frac{Q_p - Q_b}{\sigma A(\epsilon_b - \epsilon_p)}} \tag{9.12}$$

Typical values of ϵ_b and ϵ_p may be 0.95 and 0.02, respectively. Non-spherical shapes may also be used, such as an ellipsoid, which are closer in shape to the human body, to produce devices that are more suitable for thermal comfort investigations.

Constant air temperature globe

The convective heat transfer from a globe thermometer may be eliminated by maintaining the globe temperature at the same temperature as the air. This requires heating or cooling of the globe in which case the mean radiant temperature is then given by:

$$T_r = \sqrt[4]{T_g^4 - \frac{Q}{\epsilon \sigma A}} \tag{9.13}$$

where Q = heat supplied or extracted (W); T_g = globe temperature (K); A = surface area of sphere (m^2) and ϵ = emissivity of sphere surface.

Another method used to reduce the convective component of heat transfer is to enclose the sphere by a polyethylene shield. Because polyethylene is transparent to infrared radiation over a wide range of wavelengths, about 90% of room thermal radiation may be transmitted through a thin film of polyethylene. However, the disadvantage of such an instrument is that the radiant temperature recorded is not absolute (but dependent on the dimensions and geometry) of the shield. Consequently calibration of the device is needed. A shielded globe thermometer is described in [5].

9.3.2 Radiant temperature devices

The plane radiant temperature and the radiant temperature asymmetry were defined in Chapter 1. Here, methods are described for measuring these temperatures, which are important parameters for evaluating the thermal comfort in a room as was discussed previously.

Net radiometer

This instrument can be used to measure radiant temperature asymmetry or the plane radiant temperature on one side of an enclosure. It consists of a small black plane element with a heat flow meter (thermopile) for measuring the net heat flow between the two sides of the element. These two sides are separated by an insulation material and each side is enclosed by a polyethylene hemisphere to eliminate convective heat transfer from the element. This device was originally developed by Funk [6] and is sometimes known as the Funk radiometer. The instrument is widely used in indoor and outdoor environment measurements and is available commercially. An adaptor (a metal enclosure that replaces one of the polyethylene hemispheres) is sometimes used for measuring the plane radiant temperature uni-directionally instead of the radiant temperature asymmetry.

The net radiant heat flow across the radiometer, which is measured by a thermopile is given by:

$$Q = A\sigma(T_{pr1}^4 - T_{pr2}^4) = Ah_r(T_{pr1} - T_{pr2}) \tag{9.14}$$

where Q = net radiant heat measured (W); A = surface area of element (m²); T_{pr1} = plane radiant temperature of side 1 (K); T_{pr2} = plane radiant temperature of side 2 (K) and h_r = radiant heat transfer coefficient for the radiometer (W m⁻² K⁻¹).

From equation (9.14) the radiant heat transfer coefficient, h_r, is given by:

$$h_r = 4\sigma T_n^3 \tag{9.15}$$

where $T_n = (T_{pr1} + T_{pr2})/2$ is the mean temperature of the net radiometer (K). Substituting for h_r in equation (9.14) gives the following expression for the radiant temperature asymmetry:

$$\Delta t_{pr} = T_{pr1} - T_{pr2} = Q/(4A\sigma T_n^3) \tag{9.16}$$

In addition to the net heat flow the average temperature of the net radiometer must be measured to determine Δt_{pr}. This is easily accomplished by a thermocouple.

If measurement of the plane radiant temperature is desired then one of the polyethylene hemispheres is replaced by a reference cavity with a metallic shield. The net heat transfer across the radiometer is then:

$$Q = \sigma T_{pr1}^4 - \sigma \epsilon T_n^4 \tag{9.17}$$

where ϵ is the emissivity of the radiometer which is approximately equal to 0.95 for black surfaces. From equation (9.17) it follows that the plane radiant temperature is given by:

$$T_{pr1} = \sqrt[4]{[\epsilon T_n^4 + (Q/\sigma)]} \tag{9.18}$$

This expression can also be used to measure Δt_{pr} by measuring T_{pr} in two opposite directions, i.e. T_{pr1} and T_{pr2}.

Two heated radiometers

Another method of measuring the plane radiant temperature is to use two disc radiometers heated to the same temperature, one with a black surface and the other with a highly polished (reflective) surface, both pointing in the same direction. Similar to the two-globe thermometer described earlier, the polished sensor will lose heat by convection only and the black sensor by both convection and radiation. The expression for the plane radiant temperature is similar to equation (9.12), i.e.:

$$T_{pr} = \sqrt[4]{T_s^4 + \frac{Q_p - Q_b}{\sigma A(\epsilon_b - \epsilon_p)}} \qquad (9.19)$$

where T_{pr} = plane radiant temperature (K); T_s = sensor temperature (K); Q_b = heat supply to black sensor (W); Q_p = heat supply to polished sensor (W); A = surface area of each sensor (m^2); ϵ_b = emissivity of black sensor and ϵ_p = emissivity of polished sensor.

Constant air temperature radiometer

This sensor uses the same principle as the constant air temperature globe in which convective heat transfer is eliminated by heating (or cooling) the radiometer to the same temperature as the air. The plane radiant temperature is then given by:

$$T_{pr} = \sqrt[4]{T_s^4 - Q/(\epsilon \sigma A)} \qquad (9.20)$$

where Q is the heat supplied or extracted from the sensor (W) and ϵ is the emissivity of the sensor.

9.3.3 Precautions in mean radiant temperature measurement

When a globe thermometer is at thermal equilibrium with the surroundings, the temperature recorded by the globe, T_g, can be expressed by:

$$T_g = \frac{h_c}{h_c + h_r} T_a + \frac{h_r}{h_c + h_r} T_r \qquad (9.21)$$

where T_g = globe temperature (K); T_a = air temperature (K); T_r = mean radiant temperature (K); h_c = convective heat transfer coefficient (W m^{-2} K^{-1}); h_r = radiant heat transfer coefficient = $4\epsilon\sigma T_n^3$ (W m^{-2}K^{-1}) and $T_n = \frac{1}{2}(T_g + T_r)$(K).

Defining a radiant response ratio, R, by:

$$R = \frac{h_r}{h_c + h_r} \qquad (9.22)$$

Equation (9.21) can be written as:

$$T_g = (1 - R)T_a + RT_r \qquad (9.23)$$

It is clear from the above expressions and equations (9.8) and (9.9) that the globe temperature is not only determined by the mean radiant temperature, T_r, but also by the globe diameter, d and the air velocity, v. Assuming a mean temperature $T \approx 295\,\text{K}$ and $\epsilon = 0.95$, gives $h_r = 5.53\,\text{W}\,\text{m}^{-2}\,\text{K}^{-1}$. If forced convection is assumed and equation (9.9) is used for h_r, equation (9.22) becomes:

$$R = 1/(1 + 1.14v^{0.6}d^{-0.4}) \tag{9.24}$$

Equation (9.24) shows that the radiant response ratio, R, decreases with increasing v and decreasing d, and these factors must be taken into consideration when a measurement of T_g is made.

Traditionally in room air movement studies a globe thermometer has been used as an indicator of human thermal comfort. The above analysis has shown this is not strictly valid. Humphreys [7] analysed human response data in natural convection environments with an air velocity between 0.1 and $0.15\,\text{m}\,\text{s}^{-1}$ and obtained a value of $R = 0.46$ with a mean standard deviation of 0.02. He then concluded that a globe thermometer of about 40 mm diameter would best describe the human response at low air speeds [4]. He found that a table tennis ball of diameter 38 mm painted matt black and a mercury-in-glass thermometer of the quick response type inserted inside the ball produced a satisfactory globe thermometer for assessing human thermal response. An additional characteristic of this device is its faster response compared with a standard 150 mm diameter globe. The new device reached thermal equilibrium in less than 10 min compared with about 20 min for the standard globe thermometer.

A major disadvantage of the globe thermometer is its slow response, which makes it unattractive when measurements of mean radiant temperature are required at a number of locations in an enclosure. Investigations have been carried out to improve the response of the instrument, i.e. reduce the time required for equilibrium and measurement of the globe temperature [8, 9]. Hellon and Crockford [8] studied the effect of the globe thickness (thermal mass) and found little effect on the response of a 150 mm globe thermometer. However, replacing the standard mercury-in-glass thermometer by a thermocouple reduced the equilibrium time from about 20 to 10 min. The emissivity of the interior surface of the globe had a small influence on the response, when a sphere with blackened interior reached thermal equilibrium 4 min sooner than a polished surface of very low emissivity. Both references [8] and [9] achieved about a 50% reduction in the equilibrium time when a small fan was used to stir the air inside the globe and thus increase the heat exchange between the sphere and the temperature sensor at its centre. Using a thin-walled sphere with black internal and external surfaces fitted with a stirrer fan and a thermocouple, Hellon and Crockford were able to reduce the equilibrium time to 6 min.

9.3.4 Infrared thermography

A thermal imaging or infrared thermographic system provides a two-dimensional colour image of the temperature distribution of a surface using an infra-red camera and associated hardware and software. Each colour scale on the image represents a certain temperature range when the emissivity of the surface to infrared radiation is known. This usually requires a calibration of the camera by measuring the temperature at one point of the surface by another device in order to estimate the emissivity of the surface.

The basis of infrared thermography is the emission of electromagnetic radiation within the infrared range from a surface, which is at a temperature above a certain reference value. Theoretically, an object emits infrared radiation at temperatures above 0 K, but in practice a reference temperature>0 K is usually used. The reference temperature is normally provided by an electronic temperature setting and control device. A thermographic recording camera, which includes a lens and moving mirrors or more commonly rotating prisms, that scan the images, focuses the heat wave on an infrared photoelectric detector. The detector is cooled to the required reference temperature hence producing an electrical signal proportional to the incident radiant flux. Radiation from successive small scanned areas is converted into electrical signals that are amplified and used to modulate an electron beam or digital signal scanned over a visual display unit in synchronization with the movement of the scanning system in the camera. In this way, a television-like image is seen on the display in real time. A thermal image (thermogram) is made by simply photographing the image on the display unit or recording it on a video tape recorder (VTR) or digital video recorder (DVD). In modern thermographic systems, image processing techniques are used in the measurement, calibration and image analysis.

Applications of thermographic scanning techniques include the detection of heat loss from buildings, monitoring industrial process installations, testing heating and cooling systems, and many other applications. In buildings, thermography is used in detecting building defects and areas of air leakages around the building envelop that are characterized by different temperature scales from other areas on the image.

9.4   Measurement of humidity

Any instrument capable of measuring the humidity (relative or specific) of air is called a 'hygrometer'. There are a number of hygrometers available, which are based on different principles of operation. A description of the different types of hygrometer and their characteristics is given by Hurley and Hasegawa [10] and the accuracy of such devices is specified in the ASHRAE Handbook [11]. In this section only a brief description of the most common hygrometers is given.

9.4.1   Psychrometer

A *psychrometer* is an instrument consisting of two temperature sensors: one for measuring the dry-bulb temperature of the air and the other, fitted with a cotton wick that is kept moist, for measuring the wet-bulb temperature. The wick is usually equipped with a water reservoir filled with distilled water and ventilated with air moving at a velocity of between 3.5 and $10\,\mathrm{m\,s^{-1}}$ relative to the instrument. The evaporation cooling by the air stream produces a temperature close to the adiabatic saturation temperature. The difference between the dry- and wet-bulb temperatures is sometimes referred to as the 'wet-bulb depression'.

In a psychrometer, a wide range of temperature sensors are used such as liquid-in-glass thermometers, thermocouples, resistance thermometers and thermistors. The air movement over the wick is produced either by a small fan or by whirling the instrument through the air using a sling, i.e. in a sling psychrometer.

The specific humidity or percentage saturation of the air may be obtained from a psychrometric chart (Figure 2.5) by plotting the dry- and wet-bulb temperatures on the chart. Alternatively, a sliding rule is sometimes provided with psychrometers to facilitate a quick estimation of the relative humidity (r.h.). A calculation of the r.h. of air, ϕ, can be made using the equations below:

$$\phi = p_s/p_{gdb} \tag{9.25}$$

$$p_s = p_{gwb} - 66.7(t_{db} - t_{wb}) \tag{9.26}$$

$$p_g = 100\exp\{18.956 - [4030.18/(t + 235)]\} \tag{9.27}$$

where p_s = ambient water vapour pressure (Pa); p_g = saturation water vapour pressure at temperature $t\,°C$ (Pa); p_{gwb} = saturation water vapour pressure at wet-bulb temperature $t_{wb}\,°C$ (Pa) and p_{gdb} = saturation water vapour pressure at dry-bulb temperature $t_{db}\,°C$ (Pa). Both p_{gwb} and p_{gdb} can be calculated from equation (9.27), p_s is obtained from equation (9.26) and then ϕ is obtained from equation (9.25).

The accuracy of a psychrometer is better than ±3% r.h. providing care is taken in maintaining and using the instrument, ISO 7726 [3]. The instrument is not suitable for measuring r.h. at temperatures below 0 °C because of water freezing and the difference between t_{wb} for water and ice. At r.h. < 20% the accuracy of the instrument deteriorates because of the difficulty in cooling the wet bulb to its full depression.

9.4.2 *Electrical conductivity hygrometer*

Some substances are hygroscopic, i.e. sensitive to changes in relative humidity, and their electrical conductivity varies with the amount of moisture absorbed. Lithium chloride (LiCl) exhibits a reproducible change in impedence with changes in relative humidity and is therefore a commonly used salt in electrical conductivity hygrometers. One such hygrometer was developed by F. W. Dunmore in the USA in 1944 and is often known as the Dunmore hygrometer. The sensor is made up of two winding electrodes on an insulating tubular or flat substrate coated with a film which is impregnated with an aqueous solution of LiCl. The electrical conduction between the two electrodes is formed by the LiCl solution, the resistance of which is affected by the relative humidity of the air, and this forms the sensing element. An a.c. bridge-type resistance-measuring circuit is normally used to measure element resistance. The relation between sensor resistance and relative humidity is non-linear which is often corrected electronically. Because a Dunmore sensor can only be used over a small range of r.h. (10–20%), seven or eight sensors (each covering a certain range) are needed to cover a range of 7–98% r.h. This instrument can measure r.h. to an accuracy of 1.5% and has a time constant of about 3 s. However, as this type of sensor is sensitive to temperature changes, temperature compensation is needed. Furthermore, since the response of this instrument is directly related to the moisture in the salt solution, adequate protection is required from particulate matter and chemical vapours that may be present in the air. As a result, periodic calibration of the sensor is needed. The instrument has the advantage of being inexpensive, reliable and with a quick response, which makes it suitable for HVAC application.

The Dunmore r.h. sensor can be modified to give an output signal related to the dew-point temperature. The two electrodes in this case are connected to a low-voltage (e.g. 24 V) a.c. power supply causing a heating of the LiCl film (Figure 9.4). LiCl solution exhibits a sharp decrease in electrical resistance when its moisture content increases beyond about 11%. The resulting increase in the current causes the solution to heat up and the water absorbed from the air to evaporate, thus increasing the resistance and reducing the heating. A balance is quickly established between the water vapour

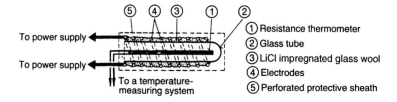

Figure 9.4 A LiCl specific humidity sensor. Reproduced from ISO 7726 [3] by permission of the International Organization for Standardization.

content of the air, the heating power and the temperature of the sensor to maintain the moisture content of the LiCl solution at about 11%. The equilibrium temperature, which is measured by a resistance thermometer is therefore dependent only on the water vapour pressure of the air. Calibration of the sensor produces a relationship between the sensor temperature and the dewpoint temperature of the air. This type of sensor is suitable for a dewpoint temperature measurement in the range −40 to +65 °C with an error of the order of 0.5–1 K. It falls within the medium price range of hygrometers. The response time of this sensor is about 6 min. In order to reduce convective heat loss from the sensor (causing a lowering of equilibrium temperature and shifting the calibration) the air velocity should not exceed a certain value, which is dependent upon the type of screen used to protect the sensor. The manufacturer's recommendations should therefore be adhered to in using the instrument.

Other types of r.h. sensors use a thin film of polymer in which changes in the polymer electrical resistance or capacitance occur as it absorbs or desorbs moisture from the air. The polymer film is usually sandwiched between two electrodes and as the moisture in the polymer changes, the capacitance between the two electrodes also changes. The change in capacitance is then converted into voltage or current when a variable-frequency signal is applied at the electrodes. These instruments have a time constant of about 16 s and an r.h. measuring accuracy in the region of ±2% and fall within the medium to low price range. They can be used within a temperature range of approximately +5 to +55 °C.

9.4.3 Dewpoint hygrometer

The determination of the dewpoint temperature of air can be made by cooling a mirror until condensation of water vapour occurs on its surface; the surface temperature of the mirror can then be converted to the specific or r.h. of air by using tables or charts. The mirror is usually cooled thermoelectrically and the formation of dew on the mirror is detected using a light source and photoelectric cells. The temperature of the mirror is regulated to maintain a constant film thickness or percentage dew coverage of the mirror surface. Under these conditions the mirror surface temperature is measured using a resistance thermometer, which is approximately equivalent to the dewpoint temperature. This type of sensor has a dewpoint measurement accuracy of 1 K or better but this could be influenced by deposits of other vapours and particulate matter on the mirror surface. A regular cleaning of the surface is therefore essential. The device is suitable for measuring the frost-point as well as the dewpoint temperatures over a temperature range of −40 to +65 °C. These types of hygrometer are classified in the high to medium price range.

9.5   Measurement of pressure

Pressure-measuring devices fall into two categories: absolute pressure measurement and differential pressure measurement. Pressure measurements involving air flow are normally of the second category. Some widely used differential pressure-measuring devices are described in this section.

9.5.1   *The manometer*

A manometer is a simple and accurate device for measuring differential pressure that is based on the deflection of a column of liquid, which has a density much greater than the density of the fluid for which the pressure is required. A transparent tube of glass or a plastic material is used to measure the deflection of the manometer liquid. A common type of manometer is the U-tube manometer that consists of a U-shaped transparent tube partially filled with liquid. When a pressure difference is applied at the two legs of the tube a deflection of the two liquid columns occurs such that the difference between the heights of the two columns is proportional to the pressure difference, i.e.:

$$\Delta p = gh(\rho_m - \rho) \tag{9.28}$$

where Δp = pressure difference across the manometer (Pa); h = vertical distance between two manometer liquid columns (m); ρ_m = density of manometer liquid ($kg\,m^{-3}$); ρ = density of fluid ($kg\,m^{-3}$) and g = acceleration due to gravity = 9.806 ($m\,s^{-2}$). When the pressure of gases is measured, ρ in equation (9.28) can usually be neglected without introducing a significant error. If a temperature variation occurs correction must be made to the density of the manometer liquid. Manometers can be used for measuring partial vacuums as well as positive pressure in relation to atmospheric pressure. They are also used for measuring the pressure difference across flow-measuring devices such as pitot-static tubes, venturis, orifice plates and so forth. By using liquids of different densities (e.g. alcohol, water, mercury, etc.) a wide range of pressures can be covered. Due to surface tension, a meniscus forms in the manometer tube, the extent of which is affected by the bore diameter. A bore diameter between 5 and 13 mm is recommended for accurate measurement.

   Pressure differentials in ventilation measurements are often of the order of a few pascals (1 Pa = 0.102 mm water) and to measure such low pressures using a manometer, the tubes must be inclined to increase the deflection of the liquid column. An inclined tube manometer of small bore with one of the legs replaced by a reservoir is used for measuring low pressures. One such device, shown in Figure 9.5, has a manometer angle that can be varied to cover a wide pressure range. Other types of manometers for low pressure measurements are also available and these are described in Ower and Pankhurst [12].

   When pressure measurements are made in a turbulent flow using a manometer, oscillation of the liquid column(s) can occur, which causes difficulty in observing the correct height of the liquid column(s). The fluctuations can be reduced by damping the manometer by using a large reservoir between the manometer and the pressure

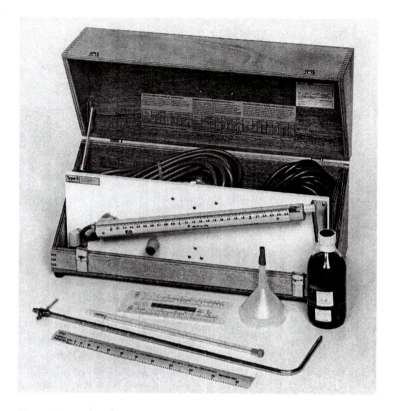

Figure 9.5 Inclined manometer. Courtesy of Airflow Development Ltd.

hole, but this greatly increases the response time of the instrument. For interested readers, the dynamics of a manometer is treated in Doebelin [13]. If the aim is to record the fluctuating pressure signal then a pressure transducer should be used (see Section 9.5.3).

9.5.2 Pressure gauges

Pressure gauges are differential-pressure measuring devices that utilize a form of Bourdon tube, diaphragm or bellows as the pressure-sensing element. The deflection of these elements due to an applied pressure is measured using a suitable mechanical linkage or may be converted to an electrical signal. By suitable calibration, the deflection is directly related to the applied pressure. These devices can be used to measure either gauge pressure (i.e. relative to atmospheric pressure) or pressure differential. Typical pressure gauge elements are shown in Figure 9.6. The most common element is the Bourdon tube, which is a thin-walled tube of oval cross-section curved along its length into a C or spiral shape with one end sealed and the other connected to the pressure tap. As the pressure increases, the tube tends to straighten, and vice versa, and the resulting movement is used as a measure of the change in pressure.

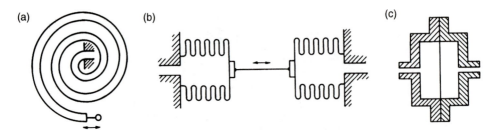

Figure 9.6 Common types of pressure gauges. (a) Spiral Bourdon tube; (b) differential pressure gauge using bellows and (c) differentiated pressure gauge using a diaphragm.

Bourdon-type pressure gauges are suitable for medium to high pressure measurements with an accuracy range of 0.05–5%.

9.5.3   Pressure transducers

The basic sensing element of a pressure transducer is a very thin metal diaphragm (e.g. stainless steel), which deflects or buckles when a pressure differential is applied across it. The amount of deflection is proportional to the pressure difference across the diaphragm. The mechanical deflection of the diaphragm can be converted to an electrical output using various devices such as a capacitor, a strain gauge, a piezoelectric crystal, an induction coil or a magnet. The first three electric converters are the most widely used but the capacitance transducer is most suitable for the low pressure measurements normally experienced in ventilation studies. The main components of this type of pressure sensor are shown in Figure 9.7. Such transducers are available for measuring differential pressures as low as 0.1 μm water ($\approx$0.001 Pa) and as high as 30 bars.

A pressure-measuring system utilizing a transducer requires a power supply, an amplifier and an output (display) device. A capacitance transducer uses a signal conditioner with a potentiometer, which requires a d.c. supply in the region of $\pm$15 V. A strain gauge transducer uses a bridge amplifier and a piezoelectric crystal transducer utilizes an electrostatic charge amplifier. The output from these devices, which represents absolute or differential pressure is usually in the form of a d.c. voltage or current, which may be converted into units of pressure and displayed or inter-faced with a data acquisition system for analysis.

Pressure transducers have a number of advantages over other pressure-measuring devices such as manometers and pressure gauges. Because of their rapid response, they are suitable for measuring pressure fluctuations, which are present in turbulent flows, as well as mean pressures. Because transducers can be mounted directly at the pressure-measuring point without the need for connecting tubes, maximum dynamic performance can be achieved. A connecting tube or a volume chamber causes damping and time lag of the signal at the transducer end and should be discarded if a true dynamic signal is required. Other pressure-measuring devices normally rely on a connecting tube to the measuring point. Furthermore, the electrical output from a

(a)

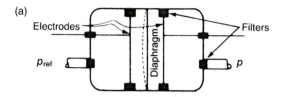

(b)

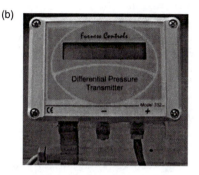

Figure 9.7 Double-side capacitance-type differential pressure transducer. (a) Internal section and (b) transducer with display. The transducer shown in (b) is courtesy of Furness Controls Ltd.

transducer can be displayed at a long distance from the measuring location and is suitable for storage and analysis in a data acquisition system.

Owing to recent advances in electronic microchip technology, very accurate and stable pressure transducers are now commercially available, which are compatible in price to more conventional instruments, such as a manometer. Self-contained battery-operated units with a digital display are also available, which make them particularly useful for measurements on site.

9.6 Measurement of air velocity

9.6.1 Mean air velocity

Measurement of the magnitude and direction of air velocity at a point is often required for determining air flow rates, evaluating air flow patterns or assessing the thermal comfort in a room. In ventilation work, such measurements are often carried out at or close to atmospheric pressure and, at such pressures, air can be treated as an incompressible gas and simplified expressions are then used to determine the air velocity. There are many instruments that can be used for measuring air velocities but most of these are only suitable for a relatively small velocity range. Air velocity measuring devices most commonly used in ventilation measurements are discussed here. For a fuller description of the subject readers should consult a specialist text, e.g. Ower and Pankhurst [12].

Pitot-static tube

If the direction of the velocity vector is known, a pitot-static tube with a differential-pressure measuring device can be used to measure the magnitude of mean air velocity at a point in the flow. The tube measures the difference between the stagnation pressure, which is sensed by the stagnation hole at the nose, and the static pressure, which is sensed by several pressure taps distributed around the head, as shown in Figure 9.8. The stagnation pressure is the sum of the velocity and static pressures and when the static pressure is subtracted from it, the pressure difference as given by the pressure sensor is the velocity pressure $\frac{1}{2}\rho v^2$. However, the static pressure measured by a pitot-static tube is influenced by the position of the static pressure taps between the nose and the stem and therefore the pressure difference recorded by the instrument will be:

$$\Delta p = K \frac{1}{2} \rho v^2 \qquad (9.29)$$

where Δp = measured pressure differential (Pa); ρ = density of moving fluid (kg m^{-3}); v = flow velocity (m s^{-1}) and K = correction factor determined by calibration. The value of K is determined by the position of static taps on the head, the shape of the nose (e.g. hemispherical, ellipsoidal, tapered, etc.) and the Reynolds number $Re = \rho v d / \mu$, where d is the diameter and μ the viscosity of the fluid. The optimum position of the static taps has been found to be $6d$ from the nose and $8d$ from the stem [12]. In this case, the value of $K \approx 1$ for $Re > 670$, but for lower Re, K increases and the error in velocity measurement also increases. A pitot-static tube is suitable for measuring air velocities greater than about 1 m s^{-1} with an accuracy of 1–5% depending on the velocity range, i.e. lowest accuracy at small velocity. A pitot-static tube is sensitive to misalignment with the flow direction, i.e. yaw. The error in measuring the velocity due to yaw, increases as the ratio of inner to outer diameters of the tube decreases. To reduce yaw errors the tube wall should be as thin as possible but without being too vulnerable to damage. Other types of pitot-static tube (e.g. shielded heads) which are less sensitive to yaw are available [12].

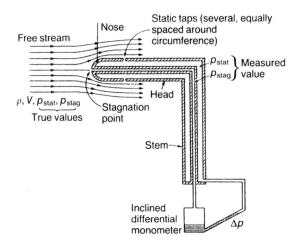

Figure 9.8 Pitot-static tube.

Figure 9.9 Vane anemometers. Courtesy of Airflow Development Ltd.

Vane anemometer

The vane anemometer consists of a number of lightweight inclined vanes attached to a hub to form an impeller similar to a miniature wind turbine. The impeller shaft rotates in two low-friction bearings. The number of vanes varies from four to eight according to the impeller size. Figure 9.9 shows two types of hand-held vane anemometers. The air movement over the vanes causes the impeller to rotate at a speed, which is directly related to the air velocity over the vanes. The impeller rotation is measured by a magnetic pick-up fixed to the impeller housing. The magnetic pick-up usually counts the speed of rotation of the impeller which is converted to air speeds by calibration and this is displayed on the unit.

Vane anemometers are normally suitable for low or medium air speeds but the range of speeds is dependent on the size of impeller. Generally, a large impeller device is suitable for measuring lower speeds (0.1 up to $20\,\mathrm{m\,s^{-1}}$) whereas small impeller instruments can measure velocities up to about $40\,\mathrm{m\,s^{-1}}$. Anemometers are available with impeller diameters between 15 and 150 mm. At very low air speeds the bearing friction introduces errors and the accuracy near the low range of velocity is less than in the middle or higher ranges. A well-maintained and calibrated instrument can have an accuracy of 12% of measuring value, but the accuracy of a less well-maintained device can be up to about ±20%.

The vane anemometer is a useful instrument for measuring the mean air velocity over the surface of a supply or extract grille and, in this case, a large diameter impeller gives a more accurate measure of the mean air velocity. The instrument is also used for measuring velocities in ducts and ventilation shafts where the direction of flow is known. The device is, however, very sensitive to flow direction and this characteristic can, in some cases, be used to determine the direction of flow at a point by realigning the impeller axis until maximum velocity is recorded by the instrument, at which point the flow direction is in the direction of the impeller axis. The theory of a vane anemometer is described in detail by Ower and Pankhurst [12].

Heated-bead (thermistor) anemometer

Thermistors in the form of small beads are sometimes used for measuring mean air velocity over a wide range. This can be accomplished by adding a reference thermistor to the opposite arm of a Wheatstone bridge. When the bridge is activated by a d.c. supply and the active thermistor is exposed to an air current the e.m.f. across the bridge becomes a measure of air velocity. Although a measurable e.m.f. can still be obtained at a low air velocity, the relation between voltage and air velocity is non-linear. Usually, a thermistor bead is protected by a sheath which prevents the use of the device as an omnidirectional anemometer and it is therefore unsuitable for room air velocity measurement unless the velocity direction is known. Omnidirectional anemometers are commercially available and these are described in the next section.

9.6.2 Fluctuating air velocity

It is shown in Section 1.5.4 that air turbulence has a major impact on people's thermal comfort hence it is necessary to evaluate the turbulence characteristics of air velocity in a ventilated room. An important turbulence parameter is the turbulence intensity, which is the ratio of the r.m.s. or standard deviation of the fluctuating air velocity to the mean velocity. To evaluate turbulence intensity as well as mean velocity, rapid response anemometers are necessary. Such instruments are either of the hot-wire or hot-film types, ultrasonic type or of the laser Doppler type. Several hot-wire and hot-film sensors for room air velocity measurements are reviewed in [14]. The basic principles behind these devices are described in the following sections.

Hot-wire anemometer

This is the fundamental instrument for detailed investigation of the microstructure of turbulent fluid flow. The sensor consists of a very fine tungsten or platinum wire, which is heated electrically and when exposed to a fluid flow the wire cools at a rate, which is related to the velocity of the fluid. For a flow normal to the wire axis, L. V. King, see Hinze [15], derived the formula below which is now known as King's law:

$$Nu = 1 + \sqrt{(2\pi Pe)}\tag{9.30}$$

where Nu = Nusselt's number = hd/λ; Pe = Peclet's number = $\rho C_p v d/\lambda$; h = heat transfer coefficient for the wire (W m^{-2} K^{-1}); d = wire diameter (m); v = velocity component normal to wire axis (m s^{-1}); ρ = fluid density (kg m^{-3}) and C_p = specific heat of fluid (J kg^{-1} K^{-1}); λ = thermal conductivity of fluid (W m^{-1} K^{-1}). Equation (9.30) is valid for $Pe > 0.08$ and has been verified experimentally by King and others. In hot-wire applications the equation is normally written in the following form:

$$q = A\sqrt{(v)} + B\tag{9.31}$$

where q = heat loss per unit length of wire (W m^{-1}) and A and B are constants which depend on the wire diameter, wire temperature and fluid properties.

Equation (9.31) applies to forced convective heat transfer from thin heated wires and will not hold for very low velocity where the contribution of the natural convective currents due to buoyancy around the wire may be significant. It has been found experimentally that the equation applies for [15]:

$$Gr < (Re)^3 \qquad\qquad (9.32)$$

where Re = Reynolds number = $\rho v d/v$; Gr = Grashof number = $g\beta d^3 \Delta T/v^2$; v = kinematic viscosity of fluid (m^2 s^{-1}); β = thermal expansion coefficient of fluid (K^{-1}); ΔT = the difference between the wire and fluid temperature = $T_w - T_f$ (K) and g = acceleration due to gravity (m s^{-1}).

Taking a standard hot wire of diameter 5 μm for measurements in air at room temperature, the relationship (9.32) gives:

$$v > 7.93 \times 10^{-3}(\Delta T)^{1/3}$$

i.e. the range of velocity measurements of a hot-wire anemometer circuit based on King's law can be extended to cover low velocities provided that the overheat ratio of the wire ($\Delta T/T_f$) is kept small. Taking $\Delta T = 20$ K, the minimum velocity that a 5 μm wire can measure is about 0.02 m s^{-1}. Some common types of hot-wire probes are shown in Figure 9.10. The wire length is in the region 1–2 mm and a typical diameter

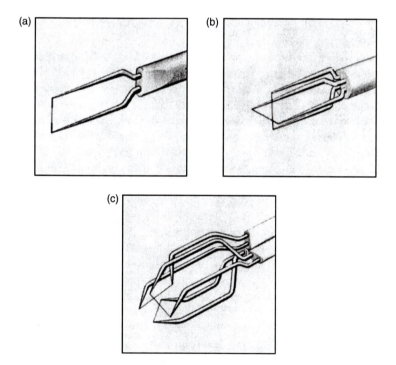

Figure 9.10 Hot wire probes. (a) Single wire for one velocity component; (b) X-probe for two velocity components and (c) triple sensor probe for three velocity components.

is 5 μm giving an aspect ratio (length to diameter) in the region 200–400. A high aspect ratio wire is necessary to reduce the effect of heat conduction to the prongs (wire supports).

There are two basic types of hot-wire anemometer systems: *constant current* and *constant temperature*. Both systems utilize the same physical principles described earlier but in different ways. In both cases, the heat dissipated from the wire to the fluid by convection, q (W m^{-1}), is equal to I^2R where I is the current through the wire (A) and R is the resistance (Ω m^{-1}) which is related to the wire temperature. In the constant-current system, the electrical circuit is designed so that a constant current is maintained through the wire and the wire temperature adjusts itself to the change in convective heat loss until equilibrium is reached. Since the convective heat loss is a function of velocity the equilibrium wire temperature, and hence the resistance, is a measure of velocity. In the constant-temperature anemometer the temperature and resistance of the hot-wire sensor are kept constant by varying the current so that the current through the wire becomes a measure of fluid velocity.

A hot-wire anemometer system consists of the hot-wire sensor, an electronic anemometer module containing a power supply, amplifiers and feedback circuits, filters and suitable output devices such as d.c. and r.m.s. voltmeters. The circuits are designed so that in both the constant-current and the constant-temperature anemometers, the instantaneous voltage drop across the heated wire is proportional to the component of fluid velocity normal to the wire axis. The relationship between voltage and velocity is obtained by calibration in special low-turbulence wind (or water) tunnels at the same fluid temperature as in the actual measurements. If variation in fluid temperature is expected, as is often the case in ventilation work, a special type of hot-wire probe with a temperature compensation sensor is recommended.

A constant-temperature hot-wire anemometer is more widely used than a constant-current anemometer for a number of reasons, which are explained in [13]. The main advantage of the former is the suitability of the system for measuring a wide range of velocity without any circuit adjustments and the fact that the anemometer can operate at low overheat ratios makes it more suitable for measuring low velocities. A simplified constant-temperature anemometer circuit is shown in Figure 9.11. The hot-wire, R_w, is connected to one arm of a Wheatstone bridge, which also has two equal resistances, R, and a variable resistance, R_b. The variable resistance of the bridge, R_b, is manually adjusted so that $R_b > R_w$ and the unbalanced bridge then produces an unbalanced voltage E_e, which is amplified by the d.c. amplifier supplying the bridge excitation current. The current flowing through R_w, I_w increases its temperature and resistance and as it approaches R_b the bridge-unbalanced voltage, E_e, decreases. A high-gain d.c. amplifier is used so that $R_w \approx R_b$ always and hence a constant wire temperature is maintained. The bridge excitation current, I, increases as the fluid velocity over the wire increases and this produces a change in voltage, E_o, across the output resistance, R_o. With calibration, a relationship between fluid velocity, v and output voltage, E_o, can be obtained.

By using a sensor of small thermal mass and frequency compensation with feedback circuits, a flat frequency response can be achieved up to about 50 kHz for measuring high fluid velocities. This characteristic makes a hot-wire anemometer suitable for measuring turbulence quantities as well as mean velocities. Using two- or three-wire probes it is possible to measure two or three components of velocity simultaneously. However, a major disadvantage of the hot-wire anemometer is its sensitivity to flow

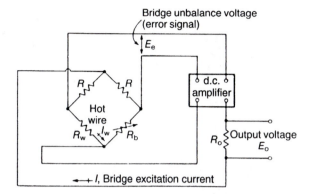

Figure 9.11 A simplified circuit of a constant-temperature hot-wire anemometer.

direction as the cosine law, i.e. it only senses the component of velocity normal to the wire axis. In some instances, such as air velocity measurement in a room, the flow direction is unknown and a hot-wire probe, is, therefore, not suitable. An omnidirectional hot-film anemometer, an ultrasonic anemometer or a laser anemometer will be more suitable for measuring the magnitude of velocity of unknown direction at a point.

Hot-film anemometer

Finkelstein *et al.* [14] described the characteristics of hot-wire/-film sensors required for measurement of room air movement. If the mean velocity and the turbulence intensity are required, such a sensor should be able to measure accurately very low air velocities ($0.1–0.5\,\mathrm{m\,s^{-1}}$), should have a time constant less than 3 s, should not be sensitive to flow direction and should have temperature compensation. A normal hot-wire sensor does not satisfy these requirements because of its sensitivity to flow direction and its inaccuracy at low air velocities. An omnidirectional hot-film sensor with a low thermal mass can be a suitable instrument for measuring room air velocities. One such sensor (Dantec type 56C15) and shown in Figure 9.12, consists of a glass sphere of diameter 3 mm with a thin nickel film deposited around the sphere in a helix pattern and then coated with a $0.5\,\mu\mathrm{m}$ layer of quartz.

An identical sphere is used for air temperature sensing and automatic compensation for variation in air temperature during measurement. The probe is connected to an electronic bridge, which contains overheat resistors and temperature compensation circuitry. The output from the bridge is non-linear but this is an advantage in measuring air velocity at the lower range of the instrument where the gradient dE/dv (E is the voltage output and v is the air velocity) is greatest. A typical calibration curve is shown in Figure 9.13 for a velocity range between 0.05 and $2\,\mathrm{m\,s^{-1}}$. The frequency response of this particular instrument for the -3 dB limit is between 2 and 7 Hz for velocities in the range $0.075–0.9\,\mathrm{m\,s^{-1}}$ and the time constant is 0.03 s. The directional sensitivity is reduced by the choice of a spherical sensor. It is claimed that the effect of roll (rotation about stem axis) to be less than 2% on the velocity and the effect of yaw (rotation of stem relative to flow direction) to be less than $\pm10\%$ on the velocity for yaw angles up to $135°$ measured from a flow direction parallel to the stem.

 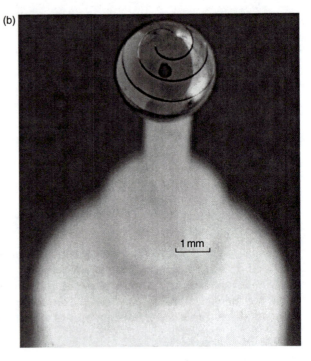

Figure 9.12 Omnidirectional hot-film anemometer probe. (a) Velocity sensor (upper sphere) and temperature sensor/compensator (lower sphere) and (b) close up of sensor showing helical nickel film. Courtesy of Dantec Dynamics A/S.

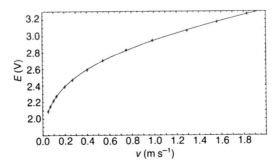

Figure 9.13 Calibration for omnidirection aneometer probe type Dantec 56C15.

Another type of omnidirectional anemometer sensor has been developed by TSI Inc. (type 1620), which uses a copper sphere for velocity measurement and a resistance thermometer for temperature compensation. Because the whole sphere is heated, the response of this sensor is 2 s, which is considerably greater than that of the Dantec sensor. However, this response is only suitable for air movement measurements where the frequency of velocity fluctuation is low, i.e. measurement in large-scale turbulence.

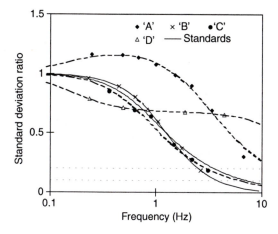

Figure 9.14 Comparison between the dynamic response characteristic of four low velocity thermal anemometers (A, B, C and D) with one of a time constant of 0.5 s for different velocity fluctuation frequencies [16].

Melikov [16] investigated the performance of four types of low velocity thermal anemometers and compared their output with that from a hot-wire anemometer (considered to have a much faster response). Figure 9.14 shows the dynamic response of the instruments tested in terms of the standard deviation ratio as a function of the frequency of the velocity fluctuations. The standard deviation ratio is defined here by 'the standard deviation of the velocity measured with the thermal anemometer divided by the standard deviation of the velocity measured with the hot-wire anemometer as a reference'. The curve representing the 'standard' anemometer is one that has a time constant of 0.5 s, which is recommended by the ASHRAE Standard 55 [17] and ISO Standard 7726 [3] for measuring room air turbulence. As the Figure 9.14 shows, the standard deviation of the four anemometers decreases with the frequency of velocity fluctuations, with all other conditions being identical. The results shown are for a mean air velocity of $0.4\,\mathrm{m\,s^{-1}}$ although the tests showed an insignificant effect of mean velocity on the standard deviation. It is clear that for three of the anemometers tested as well as the 'standard' one, the standard deviation ratio falls rapidly for a fluctuating velocity frequency >1 Hz. Melikov also investigated the flow-directional sensitivity of the four instruments, i.e. roll and yaw effects. He found that the instruments had acceptable roll characteristics, whereas, the yaw characteristics showed rather large deviations. The accuracy of the sensors tested was found to improve as the time constant of the sensor decreased and the overheat ratio increased. However, as the overheat ratio increases the free convection from the heated sensor also increases giving rise to additional errors particularly at low velocities ($<0.15\,\mathrm{m\,s^{-1}}$). A low overheat ratio, however, produces a negative effect on the velocity response of the instrument. The overheat ratio must, therefore, be carefully considered to achieve a good compromise between free convection errors and dynamic response errors.

In an earlier study than Melikov's, Welling *et al.* [18] compared the performance of the Dantec and TSI omnidirectional thermal anemometers and an ultrasonic anemometer (see the following subsection) in laminar and turbulent flows. It was found that for laminar flow, the velocity measured with the three probes were close to each other, however in the case of turbulent flow, the mean velocity measured with the two thermal anemometers were quite close but about twice the value measured with the ultrasonic anemometer. The turbulent intensity measured by the three probes was quite different and in particular the ultrasonic anemometer gave much higher turbulence intensity than the other two anemometers. This was due to the difference in the response time of the sensors, in which case the Dantec probe produced a turbulent flow signal for frequency even higher than 10 Hz, but the signals from the two other probes were not as would be expected for the flow under which they were tested, with the TSI probe producing highly time-averaged signal and the ultrasonic probe a little faster. The study concluded that a hot-sphere anemometer, i.e. of the TSI type, is not suitable for measurement in turbulent flows for the purpose of assessing the draught risk.

Ultrasonic anemometer

An ultrasonic anemometer consists of 2, 4 or 6 prongs that emit a sonic signal in 1, 2 or 3 directions. Figure 9.15 shows a 6-prong ultrasonic anemometer for measuring three orthogonal velocity components. The instrument provides a direct measurement of velocity components and, with adequately high frequency sampling, their fluctuations with time. The time of flight of pulsed sound waves between each two prongs of the instrument (about 100 mm apart) is measured and the velocity in the direction of the signal is determined with a programmable sampling rate that may be varied from about 1 to 100 Hz. To measure the air velocity on each axis, two ultrasonic signals are pulsed in opposite directions. The time of flights of the out signal, t_{out}, and the back signal, t_{back}, are given by:

$$t_{out} = d/(c + u) \tag{9.33a}$$

$$t_{back} = d/(c - u) \tag{9.33b}$$

where u is the nonorthogonal velocity component along the axis of two opposite prongs (m s^{-1}), d is the distance between the tips of the two prongs (m), and c is the speed of sound (m s^{-1}). Using equations (9.33a) and (9.33b), the velocity component along one axis of the probe, u_1, is then:

$$u_1 = 0.5d(1/t_{out} - 1/t_{back}) \tag{9.34}$$

The three nonorthogonal components (in the case of a 3-component instrument) are transformed into orthogonal air velocity components, u_x, u_y and u_z that refer to the three axes of anemometer head using a 3×3 transformation matrix, A, given by:

$$\begin{bmatrix} u_x \\ u_y \\ u_z \end{bmatrix} = A \begin{bmatrix} u_1 \\ u_2 \\ u_3 \end{bmatrix} \tag{9.35}$$

The magnitude of velocity for each sample of measurement i is then calculated from the three orthogonal components using:

$$u_i = \sqrt{u_{ix}^2 + u_{iy}^2 + u_{iz}^2} \tag{9.36}$$

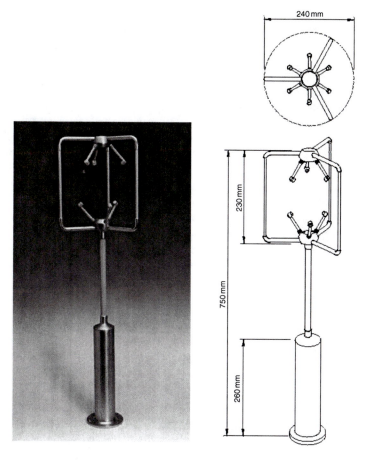

Figure 9.15 Ultrasonic anemometer for measuring three velocity components. Courtesy of Gill Instruments Ltd.

The time-average of velocity, $\bar{u}$, for N samples is then:

$$\bar{u} = \frac{1}{N} \sum_{i=1}^{N} u_i \qquad (9.37)$$

The turbulence intensity, TI, can be obtained from the standard deviations of the three orthogonal velocity components. The standard deviation may be defined by the root mean square of the fluctuations in the velocity components, u'_x, u'_y and u'_z, i.e.:

$$u'_x = \sqrt{\overline{u_x^2}}, \; u'_y = \sqrt{\overline{u_y^2}} \text{ and } u'_z = \sqrt{\overline{u_z^2}}$$

and the turbulence intensity is then given by:

$$TI = \frac{\sqrt{u'^2_x + u'^2_y + u'^2_z}}{\bar{u}} \times 100 \quad (\%) \qquad (9.38)$$

The evaluation of an ultrasonic anemometer for room air velocity measurements has been undertaken by [18, 19] and generally it is concluded that a high sampling frequency is needed for accurate evaluation of mean and fluctuating velocities.

Laser Doppler anemometer

Laser Doppler anemometers (LDAs) are non-contact optical instruments for investigating flow fields in gases and liquids. Velocity measurement is carried out by focusing a laser beam on to the measuring point and observing the frequency shift of the light scattered by moving particles. As a result, very rapid fluctuations in the fluid velocity can be accurately measured. However, it is essential that the measuring point be transparent to the laser beam.

Laser Doppler anemometer devices are particularly useful for measurements in inaccessible regions of the flow or where a normal hot-wire or hot-film probe cannot be used. Unlike other methods of velocity measurements, an LDA system uses no probe at the measuring point and therefore no disturbance to the flow occurs. This is a major advantage of the LDA system over other systems. The disadvantage, however, is that the LDA system is bulky and expensive in comparison with, say, a hot-wire/film anemometer.

A typical LDA system consists of a laser beam(s), optical components, photodetectors and signal processors. The system's optics focus the laser beam on to the measuring point, collect light scattered from moving particles in the flow and focus the scattered light on to a photodetector, which converts it to an electrical signal. The scattered light experiences a frequency change given by the Doppler shift but this is difficult to measure in practice. More commonly, the scattered light is mixed with another light beam split from the same laser source on to a photodetector which measures the difference in frequency between the two light beams. In practice, the splitting of the incident beams and the collection of the scattered beams can be arranged in different modes, most commonly the reference beam mode and the differential or fringe mode. A vector diagram showing beam splitting and scattering is shown in Figure 9.16.

In the reference beam mode, the light scattered from the illuminating beam is combined with a reference beam, which is taken directly from the incident beam by means of a beamsplitter. The light scattered from the particles in the direction of the reference beam and the reference beam itself mix on to the photodetector surface and the difference in frequency is indicated by a change in detector current. In the differential or fringe mode two scattered beams derived from two different incident beams are combined and the photodetector measures the difference between the two

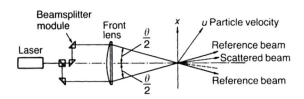

Figure 9.16 Vector representation of laser anemometer light scattering by a moving particle.

Doppler-shifted scattered beams. In both modes the Doppler frequency, f_D is given by:

$$f_D = (2u_x/\lambda) \sin(\theta/2) \qquad (9.39)$$

where u_x = the particle velocity component in the x direction, i.e. normal to laser beam (m s^{-1}); λ = wavelength of laser light (m) and θ = angle between the incident beams (°). Knowing the angle θ and the wavelength of laser light, λ, the flow velocity in a direction normal to the original beam can be calculated once the Doppler frequency is measured by the photodetector. A popular laser light is argon, which can emit light of a single frequency.

It has been shown that the principle of a laser anemometer is based on the shift in the light frequency produced by a moving particle, hence the LDA system can only operate if the flow is seeded with particles of the correct size. These particles must be sufficiently small to track the flow accurately yet large enough to scatter sufficient light for proper operation of the photodetector and the signal processor. For measurement of air velocity, small droplets of water or dust can usually provide the seeding particles.

By using three- or five-beam LDA it is possible to measure two or three orthogonal components of velocity. However, these multi-component systems are extremely bulky and very expensive and are seldom used in ventilation investigations.

9.6.3 Turbulence parameters

A hot-wire/film anemometer of a sufficiently small time constant or a LDA can accurately record the velocity fluctuations of turbulent flow in a room. The recorded signal can be analysed by using electronic hardware or software to find the turbulence characteristics. Those characteristics relevant to ventilation studies are described below.

Time-mean velocity

The instantaneous velocity in a turbulent flow can be represented by:

$$V = \overline{V} + v \qquad (9.40)$$

where V = instantaneous velocity (m s^{-1}); $\overline{V}$ = time-mean velocity (m s^{-1}); v = velocity fluctuation from the mean (m s^{-1}).

The time-mean velocity, $\overline{V}$, represents the average velocity of the flow over a time interval T and can be expressed by:

$$\overline{V} = \frac{1}{T} \int_t^{t+T} V \, dt \qquad (9.41)$$

Hence, to find $\overline{V}$ the velocity signal is integrated over a sufficiently long time interval T.

Turbulence intensity (TI)

Defining the root mean square of the fluctuating velocity, v, by:

$$v = \sqrt{\overline{v^2}} \qquad (9.42)$$

which is also the standard deviation, then the turbulence intensity, *TI*, is given by:

$$TI = v'/\overline{V} \tag{9.43}$$

Turbulence energy spectrum

A turbulent motion is composed of eddies of different sizes and frequencies. Normally it is difficult to distinguish between the various eddy frequencies, but it is possible to define the kinetic energy of turbulence over a distinct range of frequencies and this is called an energy spectrum.

The kinetic energy of turbulence for a velocity component *v* at a fixed point in the flow can be obtained from the expression:

$$\overline{v^2} = \int_0^\infty E(f)\,df \tag{9.44}$$

where *f* is the frequency and $E(f)$ is the energy density spectrum of *v*.

Scales of turbulence

There are two turbulence scales: the time scale and the length scale. The time scale characterizes the average duration of the effect of disturbance at a point. The length scale gives a comparative measure of the average size of eddies in a given direction. These scales can be obtained by integrating the correlation functions between two turbulence signals either with time (autocorrelation) for the time scale or with distance (spatial) for the length scale, over an appropriate time or spatial range. In room air movement the length scale can be an important parameter because it gives an indication of the average size of eddies. It is expressed by:

$$L = \int_0^\infty C_x\,dx \tag{9.45}$$

where L = length scale (m); x = distance between velocity measuring points (m) and C_x = correlation coefficient between the velocity components at points 1 and 2 defined as:

$$C_x = \frac{\overline{v_1 v_2}}{v_1' v_2'} \tag{9.46}$$

where v_1' and v_2' are the r.m.s. values of v_1 and v_2, respectively.

9.7   Measurement of volume flow rate

9.7.1   *Flow rate in ducts*

The most widely used flow-metering principle is based on placing a fixed restriction of known geometry in the pipe or duct carrying the fluid. The flow restriction causes a pressure drop, which is related to the flow rate or generates vortices at a frequency, which is proportional to the flow rate. These are the most common methods of metering the flow rate of air in ducts. There are other techniques of non-obstructive flow metering, such as magnetic flow meters and ultrasonic flow meters, but these are only suitable for measuring the flow of liquids.

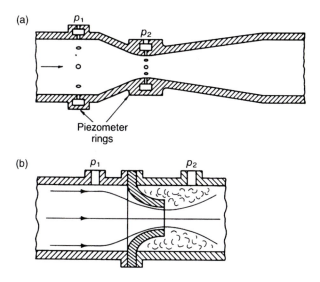

Figure 9.17 (a) Venturi and (b) nozzle flow meters.

Venturi and nozzle

A venturi tube consists of a converging and diverging nozzle with a short untapered section between them called the throat, see Figure 9.17(a). The device is usually positioned along a straight length of ducting to reduce disturbance to the flow. The pressures upstream of and at the throat are measured and the difference is related to the volume flow rate by Bernoulli's equation. A nozzle, on the other hand, is fitted inside the duct and the pressure upstream and downstream of the nozzle is measured at the duct wall.

Assuming an incompressible flow, the flow rate measured by a venturi or a nozzle is given by:

$$Q = C_d A_t \sqrt{\{2(p_1 - p_2)/[\rho(1 - \beta^4)]\}} \qquad (9.47)$$

where q = flow rate ($m^3\,s^{-1}$); C_d = discharge coefficient; A_t = throat (minimum) area of venturi or nozzle (m^2); $\beta = d/D$ = ratio of throat diameter to duct diameter; ρ = fluid density ($kg\,m^{-3}$) and $p_1 - p_2$ = pressure difference across the device (Pa).

The discharge coefficient, C_d, which allows for losses due to fluid friction, is dependent upon the geometry of the venturi or nozzle and the Reynolds number, Re. Values of C_d for a venturi with a cone angle in the diverging section of 5–7° (optimum angle for minimum flow disturbance) are plotted in Figure 9.18(a) versus Re ($\rho v d/\mu$), and values of C_d for a nozzle are plotted versus β in Figure 9.18(b).

Orifice plate

An orifice plate is a very common flow-measuring device because of its simplicity and reliability. It consists of a plate with a sharp orifice beveled around the edges in the downstream direction. As the fluid passes through the orifice it contracts to an area, which is smaller than the orifice hole, due to the sharp edges, which is called the

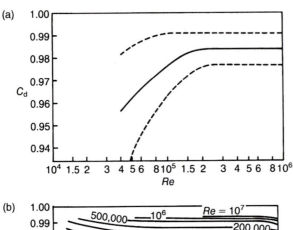

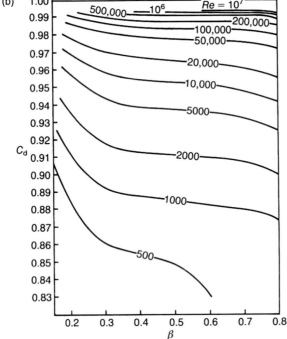

Figure 9.18 Discharge coefficients for (a) a venturi and (b) a nozzle [11].

'*vena contracta*'. This makes an orifice plate less susceptible to the value of Reynolds number as is the case for venturis and nozzles. The pressure difference across the orifice is measured at the duct wall. The recommended positions of the pressure tapings are normally at D and $x < D$ upstream and downstream of the orifice plate respectively in the case of duct wall tapings, D being the duct diameter [11] and [20]. The downstream position, x, is dependent on the ratio d/D as shown in Figure 9.19.

As for the venturi and nozzle the flow rate for an orifice plate is obtained from Bernoulli's equation and is given by:

$$Q = C_d A_o \sqrt{[(2/\rho)(p_1 - p_2)]} \tag{9.48}$$

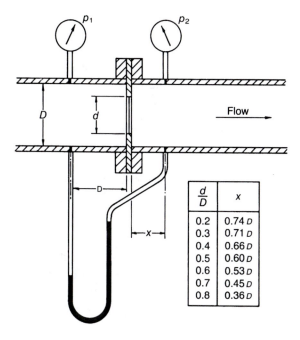

Figure 9.19 Orifice plate [11].

$\frac{d}{D}$	x
0.2	0.74 D
0.3	0.71 D
0.4	0.66 D
0.5	0.60 D
0.6	0.53 D
0.7	0.45 D
0.8	0.36 D

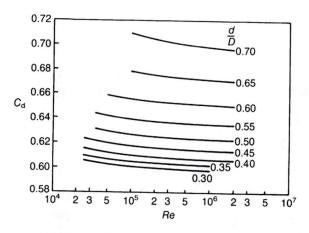

Figure 9.20 Discharge coefficients for an orifice plate with duct wall pressure tapings [11].

where C_d = discharge coefficient which includes the effect of flow contraction and energy losses and A_o = orifice area, $\pi d^2/4$ (m^2). The value of C_d is dependent on the ratio of orifice diameter to duct diameter, d/D, but less influenced by the Reynolds number, as shown in Figure 9.20. The pressure losses due to an orifice plate (insertion losses) are significantly higher than those for a venturi or a nozzle of the same diameter opening.

Flow-measuring grid

The flow-measuring devices described earlier require long straight lengths of ducting upstream of the device without obstructions, bends or other fittings. BS 1042 [20] gives values ranging between $10D$ and $80D$ depending on the type of fitting and the ratio of opening diameter to duct diameter, d/D. Such long straight sections of ducting are not always available in ventilation systems. Furthermore, calibration data for these devices are usually given for circular openings in circular ducts and they may not be sufficiently accurate when the devices are used for flow measurement in non-circular ducts. A flow grid, such as that shown in Figure 9.21, may be more suitable in these situations. This flow meter can be installed at virtually any location in a duct suitable for taking a velocity traverse by a pitot-static tube.

The grid shown in Figure 9.21 consists of a row of stainless steel tubes with closed ends, parallel to each other and forming an open 'fence' across the duct perpendicular to the duct axis. Some tubes have small holes facing upstream for measuring stagnation pressure and others have slots on the downstream side to record base or wake pressure. These tubes are usually arranged alternately across the duct so that the average of the two pressures is recorded across the duct section. All the upstream and downstream pressure tubes are connected to two common manifolds for recording the average upstream and downstream pressures across the grid. The difference between the two pressures, which is about 2.5 times greater than the velocity pressure in the duct, is proportional to the square of mean velocity in the duct, which can be converted to volume flow rate by calibration. Such devices can be fitted in both circular and rectangular ducts.

Figure 9.21 Flow grid for a rectangular duct. Courtesy of Airflow Development Ltd.

Laminar flow meter

Because of the square-law relationship between flow rate and differential pressure, the flow-measuring devices discussed so far have a limited turndown ratio on the flow, typically 4 : 1, i.e. the minimum measurable flow rate is a quarter of the maximum. A laminar flow meter is based on the linear-law relationship between flow rate and differential pressure as in laminar flow. This can produce a high turndown ratio of up to 100 : 1. To achieve laminar flow, the fluid is made to pass through a bundle of long, small bore tubes or between a pack of closely spaced parallel flat plates. The flow rate is related to the pressure differential along a known length of the meter by calibration through a linear relationship. For a laminar flow in a pipe the flow rate, Q, is given by the Hagen–Poiseuille equation:

$$Q = \frac{\pi d^4}{128 \mu L} \Delta p \tag{9.49}$$

where d = pipe diameter (m); L = pipe length (m); Δp = pressure drop along pipe (Pa); μ = dynamic viscosity of fluid (Pa s). Equation (9.49) applies up to a Reynolds number, $\rho v d / \mu$, of 2000 but most laminar flow meters are designed for a much lower Re.

Because this device relies on achieving a laminar flow, it is only suitable for measuring small air flow rates. The disadvantages of this device are that it is usually bulky and expensive in comparison with other flow meters. The advantages include the ability to use it for measuring average flow rates in a pulsating flow or flow that reverses direction.

Vortex flow meter

When a fluid flows past a long strut (bluff body) a wake forms downstream of the strut. Above a certain Reynolds number, vortices are formed and shed alternately at either side of the strut. These vortices travel downstream of the strut at a velocity, which is proportional to the velocity of the flow, forming a vortex street usually referred to as the '*Karman vortex street*'. The vortex-shedding frequency is proportional to the flow velocity, and hence the flow rate in the duct, when the strut is fitted across a duct. The frequency of vortices may be measured using various types of sensors, e.g. hot-film anemometer, pressure transducer or an ultrasonic wave sensor.

The strut can be of circular, square or triangular cross-section. For a circular strut of very large aspect ratio in a free uniform stream, the vortex-shedding frequency is given by the expression:

$$St = 0.198(1-(19.7/Re)) \tag{9.50}$$

where St = the Strouhal number = fd/v; Re = the Reynolds number = $\rho v d / \mu$; f = frequency of vortex shedding (Hz); d = cylinder diameter (m); v = stream velocity (m s^{-1}); μ = viscosity of fluid (Pa s) and ρ = fluid density (kg m^{-3}). Equation (9.50) applies to $250 < Re < 2 \times 10^5$.

Another type of flow-metering device founded on the vortex-shedding principle is the fibre-optic vortex flow meter. It is based on the oscillation of a light-transmitting fibre-optic wire arising from the vortex-shedding frequency and the resulting light refraction as the wire slightly bends. The wire is supplied with light from a light-emitting diode and the frequency of oscillation, which corresponds to the vortex-shedding frequency, is measured by a photodiode detector at the fibre exit. Fibre-optic wires of diameter 0.19–1.00 mm have been used. By calibration the instrument produces a frequency linearly proportional to flow rate. Barton and Saoudi [21] describe an instrument for water flow however this device is also suitable for air flow measurement.

9.7.2 Flow rates through air terminal devices

Velocity traverse in ducts

The velocity distribution in a duct is not uniform due to the presence of a boundary layer on the duct wall and the disturbance to the flow caused by duct branches and fittings. If the average velocity across the duct leading to the air terminal device is known, the volume flow rate through the duct and hence the device is calculated from the product of this velocity and the cross-sectional area of the duct. To determine the average velocity across the duct, measurements of velocity at a number of locations are required using one of the instruments described in Section 9.6.1. If a vane anemometer is used this should be of the small impeller type so that it can be inserted in the duct through a relatively small hole. When a pitot-static tube is used, the head must be aligned with the flow direction with the stagnation hole facing the oncoming flow. Details of measuring locations across ducts of various sections and sizes can be obtained from a BSRIA Application Guide [22] and Svensson [23]. The measuring section should be least $5D$ and $6D_h$ (D is duct diameter and D_h is hydraulic diameter) for a circular and a rectangular duct respectively, downstream of a branch, bend or obstruction such as a damper, and two diameters upstream of the air terminal device.

The diameter of the velocity-measuring device should be as small as possible to reduce the effect of blockage which could produce an overestimation of the duct velocity. When a pitot-static tube is used the tube diameter should not exceed 1/30th of the duct diameter so that the blockage ratio (projected area of tube/duct area) is kept to within 4%.

Face velocity method

The volume flow rate through a supply or an extract air terminal device can be determined from the measurement of the average velocity normal to the face of the device and its face area. For supply and extract grilles, extract louvers, wall-displacement ventilation units and perforated ceiling panels a rotating vane anemometer of 100 mm diameter is a suitable instrument for the measurement of face velocity. For slot diffusers (supply or extract) a small vane anemometer of a diameter smaller than the slot width or a thermistor anemometer can be used. If a vane anemometer is used then care should be exercised to ensure the correct direction of air flow through the anemometer head.

In the case of grilles, louvers, displacement ventilation units and ceiling panels, the face of the device should ideally be divided into squares of 150 mm and the face velocity at each square be measured with a vane anemometer of about 100 mm diameter to obtain an average face velocity. For larger devices the size of squares may be increased but should not exceed 300 mm. When velocity measurements are taken on a grille with deflecting vanes it is necessary to ensure that the vanes are set at right angles to the grille face.

The volume flow rate is obtained from the product of the average face velocity and the total area of the grille, louver, displacement unit or ceiling panel. In the case of a slot diffuser the flow rate per metre is determined from the product of air velocity at the slot and the effective slot width. The velocity along the slot is often uneven and a number of measurements may be necessary to obtain an average slot velocity. If a large variation in slot velocity is found or reverse flow is present, then another method of measuring the flow rate should be used such as the duct traverse or hood methods.

Hood method

The previous methods of flow measurement are not suitable for ceiling diffusers, grilles and diffusers with an uneven face velocity, or for swirl diffusers. In such cases a hood is most appropriate. The purpose of a hood is to confine air discharge or extraction through the device and direct it to an aperture of known cross-sectional area where the velocity can be measured. A vane anemometer can be used for measuring the air velocity through the hood but other instruments such as a thermistor anemometer may also be used. There are two basic hood types: the simple stub-duct and the venturi tube, see Figure 9.22.

The stub-duct is a short length of duct of a diameter, which is sufficiently large to mask the air terminal device or to cover a suitable area of a perforated ceiling. Aluminium sheet or light gauge-steel are suitable hood materials but large-diameter steel hoods are difficult to handle. The velocity is measured across the hood at various locations to obtain the average velocity through it. For this purpose, a vane anemometer is often used to cover a large area of the hood opening.

In a venturi hood the air flow is directed to a converging–diverging cone and the velocity is measured at the throat using a vane anemometer or in the case of the one shown in Figure 9.22(b) a thermistor anemometer. The advantage of this type of hood is that only one velocity reading is required to determine the flow rate and a linear relation between flow rate and velocity can be obtained by calibration. The

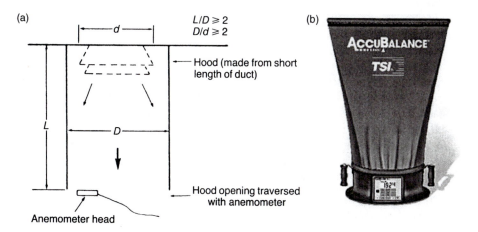

Figure 9.22 Flow measuring hoods. (a) Stub-duct hood and (b) venturi hood. (b) is courtesy of TSI Inc.

(a)

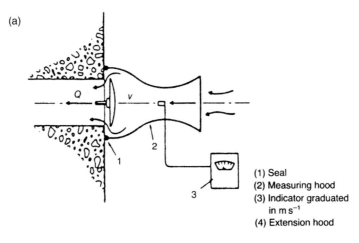

(1) Seal
(2) Measuring hood
(3) Indicator graduated
    in m s⁻¹
(4) Extension hood

(b)

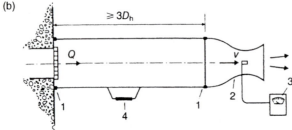

Figure 9.23 Venturi hoods for measuring the air flow rates through air terminal devices. (a) For extract terminals and (b) for supply terminals.

disadvantage is that a higher resistance is imposed on the air terminal device than the stub-duct hood and correction for this is normally required. However, the resistance to flow may be reduced by using a hood with a diverging section where some static pressure recovery can take place. A venturi hood can be used for measuring flow through supply and extract air terminals as shown in Figure 9.23(a) and (b). In the air supply case, a straight section of length at least $3D_h$ is required as a settling region for the flow.

All hood-type flow-measuring instruments influence the air flow rate through the air terminal device as certain resistance is imposed on the flow by the presence of the instrument. The effect of the instrument is more significant in the case of a venturi hood. However, allowance for the reduction in flow rate through the device can be made by measuring the static pressure difference across the device with and without the hood. The following equation is then used to predict the actual flow rate through the device in the absence of hood:

$$Q = Q_h\sqrt{(\Delta p/\Delta p_h)} \tag{9.51}$$

where Q = actual air flow rate through the air terminal device in the absence of hood ($m^3\ s^{-1}$); Δp = pressure difference across the device in the absence of hood (Pa);

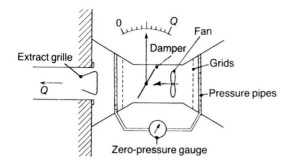

Figure 9.24 A venturi hood with a pressure drop compensating fan and a damper [24].

Q_h = air flow rate measured by the hood ($m^3 s^{-1}$) and Δp_h = pressure difference across the device with the hood attached to it (Pa).

The problem of hood resistance can be overcome by compensating for the pressure drop over the hood such that the static pressure on the duct side of the air terminal device remains the same with or without the hood over the device. The pressure compensation is made by inserting a fan in the hood. To increase the range of flow rate over which the hood can be used; either a variable-speed fan is employed or a fixed-speed fan and a flow-regulating damper are used. A hood using a fixed-speed fan and a damper is shown in Figure 9.24 [24]. Because a full compensation of the pressure loss within the hood can be made by the fan the hood design is normally improved by putting grids and flow straighteners in both the converging and diverging sections of the hood. These act as eddy breakers as well as flow straighteners and therefore long straight sections will not be necessary. As shown in Figure 9.24 the pressure difference across the venturi is measured using two static pressure taps and a pressure gauge. When a flow measurement is made the position of the damper (or fan speed) is adjusted until a zero pressure difference is recorded by the gauge, i.e. the pressure drop in the hood is compensated. By calibration the position of the damper (or the fan speed) can be related to the flow rate through the air terminal device.

Bag method

A rolled-up bag made of plastic material of thickness 0.02–0.04 mm and mounted on a frame, as shown in Figure 9.25, can be used to measure the air volume flow rate from a supply air terminal device. Knowing the volume of the fully inflated bag the air flow rate can be determined by recording the time that elapses until the bag is filled with air, using:

$$Q = V/t \tag{9.52}$$

where V = volume of fully inflated bag (m^3) and t = filling time (s).

To avoid pressurization of the bag the pressure in the bag is measured using a small tube, one end of which is inserted in the bag and the other end connected to a micromanometer. When taking a flow measurement, the inlet to the bag is placed over the device so that it is completely covered and the time required to fill the bag to

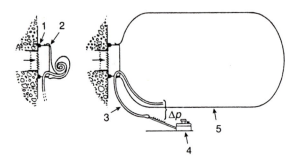

Figure 9.25 Measurement of air supply rate using a bag. (1) Seal; (2) frame for holding plastic bag; (3) pressure measuring tube; (4) micromanometer and (5) plastic bag of material thickness 0.02–0.04 mm.

a pressure of 3 Pa is noted. If the time is under 10 s then a larger bag should be used, otherwise the measurement should be repeated two or three times and the average value taken.

9.7.3   *Balancing of air supply*

To ensure that each zone of a ventilated building receives the design flow rate, balancing of the duct system is necessary during commissioning or regulation of the installation. The proportional balancing technique is the simplest and most effective way of regulating an air distribution system, for once a damper has been set it need not be altered. The volume of air delivered by each terminal device represents a certain ratio of the total flow through the duct system supplying the device. If the terminal damper setting is unaltered this ratio remains the same whatever the flow rate in the duct. To achieve the correct flow rate through each terminal the flow rate in the duct system must be adjusted to the required value using the duct damper, and the terminal dampers adjusted to produce the desired ratios of the total duct flow rate. This requires the measurement of volume flow rate in the duct using one of the methods described earlier. A step-by-step procedure of air flow balancing is found in a BSRIA Application Guide [22].

9.8   Measurement of thermal comfort

In Chapter 1 it was shown that thermal comfort depends on two personal parameters (activity and clothing) and four environmental parameters (air temperature, mean radiant temperature, air velocity and water vapour pressure). From a knowledge of the personal parameters and by measuring the environmental parameters, assessment of the thermal comfort of a person occupying a known environment can be made using one of the thermal comfort criteria described in Chapter 1. For this purpose equations or charts can be used to obtain the level of comfort or discomfort. Methods of measuring the four environmental parameters have been described earlier in this chapter.

   For thermal comfort research it is advantageous to know not only the level of comfort but also all the influencing parameters so that a better understanding of the effect of these parameters on comfort can be achieved. However, for applications where

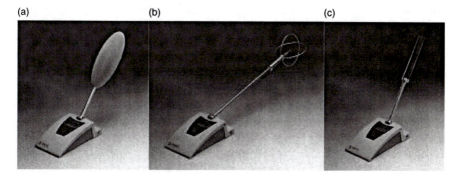

Figure 9.26 Thermal comfort measuring system. Courtesy of Dantec Dynamics A/S. (a) Temperature probe; (b) velocity probe and (c) humidity probe.

the primary objective is the assessment of the thermal environment as a whole, such as field studies, an instrument that is capable of integrating all the parameters and presenting a thermal comfort index without having to refer to equations, tables or charts is more preferable. A number of such instruments have been devised since the early part of last century and some of these are described by McIntyre [5] and in the ASHRAE Handbook [11]. The latest of these is the thermal comfort meter which was developed by Madsen [25], which is available commercially.

The thermal comfort measuring system shown in Figure 9.26, measures the combined effect of the four environmental parameters (the two temperatures, air velocity and humidity) and uses preset values of activity and clothing to calculate and display the comfort indices proposed by Fanger (described in Chapter 1). The indices displayed include: predicted mean vote (PMV), predicted percentage of dissatisfied (PPD) and the operative temperature. Measurement of the three environmental parameters is facilitated by an ellipsoidal electronic comfort transducer of length 165 mm and maximum diameter 55 mm, which is designed to simulate the heat exchange from a human body. The transducer contains a surface temperature sensor and a surface heating element whose power is controlled automatically to produce a transducer surface temperature similar to that of a thermally neutral person clothed as preset on the instrument panel. The rate of power supply needed to maintain this temperature (clothing temperature) gives a measure of the thermal environment. The shape of the comfort transducer is chosen to obtain the same ratio of horizontally and vertically projected radiant surface areas as for a person. To simulate a standing person, the transducer is mounted with its axis vertical and to simulate a seated person the transducer axis is tilted by 30° to the vertical. The dimensions of the comfort transducer have been chosen so that the ratio of radiant to convective heat loss matches that of a person, i.e.:

$$\frac{h_r}{h'_c} = f_{eff}\left(\frac{h_r}{h_c}\right) \tag{9.53}$$

where h_r = radiant heat transfer coefficient (W m^2 K^{-1}); h'_c = convective heat transfer coefficient of the transducer (W m^2 K^{-1}); h_c = convective heat transfer coefficient of a clothed person (W m^2 K^{-1}) and f_{eff} = effective radiation area factor of a clothed person.

The radiant heat transfer coefficients of both a person and the sensor are the same, but the convective heat transfer coefficients are different because they are a function of size, shape and temperature. Because the radiation factor of a person, f_{eff}, is less than unity (≈ 0.7), the convective heat transfer coefficient of the transducer, h'_c, must be greater than that of a person, h_c, so that equation (9.53) can be satisfied. The convective heat transfer coefficient is inversely proportional to the size of object and the dimensions of the transducer are chosen such that equation (9.53) is satisfied. The colour of the transducer has an emissivity for long- and short-wavelength radiation similar to light-coloured clothing or skin. The comfort meter can be used for evaluating a thermal environment that is not too different from the comfort requirement. The instrument is not suitable for extreme environments where heat stress can occur. In this case a wet-bulb and globe temperature (WBGT) heat stress monitor is used. Such an instrument is described in ISO 7243 [26].

9.9   Measurement of air pollutants

In Chapter 2 a number of common pollutants present in indoor air were described and their exposure limit values specified. The most common pollutants are: particulates or aerosols, radon, odours, CO_2 and volatile organic compounds (VOCs). Methods of measuring such indoor air pollutants are described in this section.

9.9.1   *Particulate measurement*

Indoor air could contain a number of particulates or aerosols generated by internal sources such as cigarette smoking, combustion appliances, chemical agents, mineral and synthetic fibres, biological materials and outdoor sources such as pollen and traffic aerosols. The sources of particulates and their health hazard can be identified by collecting samples and conducting chemical, physical and biological analyses of them. The primary method of particulate collection for subsequent analysis is by filtration from an air system, although electrostatic and thermal precipitators are also used. There are three measures of particulate concentration: number concentration, as number of particles above a given size per unit volume; mass concentration, as mass of particles per unit volume; and projected area, as stained area per unit volume. The first two methods are the most widely used for measuring particulates in indoor air.

A direct method of particle counting or number concentration measurement can be carried out by sampling particles on a membrane filter and examining the filter with an optical microscope which employs stage counts if the sample includes a range of particle sizes. The particle size limit using a light-field microscope is about $0.9\,\mu m$ but this can be reduced to $0.1\,\mu m$ by employing dark-field microscope techniques. With electron microscopes particle sizes below $0.1\mu m$ can easily be detected. The collection of particles from room air is carried out using a diaphragm-type pump in which a motor drives a diaphragm back and forth to pump the air. These have built-in flow metering (rotameter) and flow control devices. The number concentration is calculated by dividing the total number of particles collected by the volume of air sampled.

Indirect methods using the light-scattering principle for continuous particle counting and sizing that are much quicker and easier are now available. These are known as 'Aerosol Particle-Size Spectrometers' (APSs). An APS can be used to detect particles in the range $0.3\text{–}20\,\mu m$. Ultra-fine particles can be detected using a 'Condensation

Particle Counter' (CPC), which can measure particles down to 0.003 µm. This type of particle counter is also known by the name 'Scanning Mobility Particle Sizer' (SMPS). The technique employs vapour condensation (e.g. alcohol) on the particles to increase their size and allow detection by light scattering. This technique however, cannot be used to determine the particle size because all particles have almost equal size as a result of vapour condensation.

Particle mass concentration can also be determined using a direct method by capturing the particles on a filter made from a membrane or glass fibre. Knowing the volume flow rate of contaminated air through the filter and the difference in filter mass before and after sampling, the particle density can be calculated. This is known 'total' mass sampling. This is a simple, low-cost, accurate and widely used method for determining mass concentration. Other methods of particle capture such as electrostatic or thermal precipitators can also be used to measure particle density. Indirect methods such as laser light scattering can also be used for measuring particle size and density for a specific, selectable range of particle-size in real-time. This technique is extremely sensitive and has the advantages of minimal disturbance of the aerosol movement during sampling as well as instantaneous monitoring.

To determine the projected area for the purpose of the staining ability of particles, the sampling of particles is obtained on a filter paper and the light transmitted or scattered by this is compared with a standard unstained filter. Further information on particle measurement, instrumentation and calculation procedures can be found in [27].

9.9.2 Radon measurement

Because radon (Rn) and its progeny are recognized as health hazards a number of techniques have been developed for measuring their concentrations. For indoor air quality monitoring purposes ^{222}Rn and its daughters are of primary interest. These measurements are usually carried out to determine Rn concentration and Rn daughter concentration in air expressed as working levels (WL). Any variation in the sample collection time may be used in the measurements. Similarly, because of the rapid decay of Rn daughters, integration of concentration may be carried out over a short period or extended periods of days, weeks or months.

Two basic methods are used to determine Rn concentration. One relies on the removal of Rn daughters from the sample using filters or electrostatic precipitators and the other ensures that the daughter products are in equilibrium with Rn in a known ratio. The concentration of Rn is measured using an ionization chamber, a zinc sulphide (ZnS (Ag)) scintillator or a thermo-luminescence dosimeter detector to integrate the Rn levels. Most Rn daughter measuring techniques utilize a filter to collect the Rn decay products from an air sample for a subsequent analysis of their decay in a laboratory but instant measuring instruments are also becoming available. Radon daughter concentrations are usually presented as WL as defined in Chapter 2. Zns counters or thermo-luminescence detectors are used to determine the alpha particle activity for different integration periods which is then converted into WL. A more detailed review of Rn and progeny measuring techniques can be found in [28].

9.9.3 Odour measurement

Odour results from human bioeffluent, smoking, cooking, waste and other processes. It is usually associated with discomfort although certain odours can affect health over

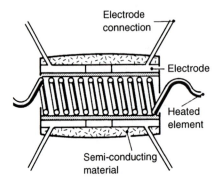

Figure 9.27 Air quality sensor. Courtesy of Staefa Control Systems.

long exposures. Cooking and smoking can produce gases such as nitrogen oxides (NO_x), CO and CO_2, and cleaning and similar activities can release organic vapours. In addition building materials, furnishings, HVAC plants, etc. can generate a variety of gases and vapours such as ammonia (NH_3), chlorine (Cl), formaldehyde (HCHO), VOCs and others. A detailed analysis of odour constituents is a major task involving a special instrument for each gas or vapour or group of gases and vapours, thus requiring many expensive instruments. Some of the instruments used for odour measurement are described in [29]. Such detailed analysis may sometimes be commissioned for assessing the air quality in industrial buildings where harmful gases or vapours are released from certain industrial processes, but is seldom required in commercial or residential buildings. Instead, air quality sensors may be used for controlling outdoor air supply rates to the building [30], such as the sensor shown in Figure 9.27. This sensor consists of a heating element inside a semi-conducting tube of zinc dioxide, which forms the electrical contact between two electrodes. The semi-conducting material is porous and has a large surface area, which adsorbs oxidizable gases. As a result of gas adsorption electrons are released and the electrical conductivity of the semiconductor increases, and this is accompanied by an increase in the current through the electrodes. Conversely, if the concentration of gases in the air is low, gases are diffused from the semiconductor and the electrical conductivity, and hence current, decreases. This sensor responds to a change in gas concentration within a few seconds, with varying degrees of sensitivity to many different gases and vapours, such as hydrogen, carbon monoxide, hydrocarbons, alcohols, esters, benzine, formaldehyde and others, as well as water vapour. However, it does not respond to CO_2 and therefore a special CO_2 sensor using infrared or gas chromatography is required for measuring CO_2 concentrations if necessary. This type of sensor has been found to assess the indoor air quality better than a CO_2 sensor [30].

Recognizing the difficulty in carrying out a complete chemical analysis of odour, and also the fact that a human nose is extremely sensitive to low concentrations of gases and vapours, Fanger and his co-workers [31, 32] developed an odour sniffing instrument which has come to be known as the 'olfmeter'. The instrument, shown in Figure 9.28, consists of a 3-litre glass jar covered with a plastic cap which contains two holes. In one hole a small battery-driven fan sucks the air from the jar and delivers it to a cone and the second hole serves as an air inlet. Inside the jar is placed the substance for which

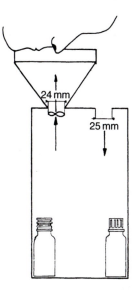

Figure 9.28 Odour sniffing instrument, the 'olfmeter' [31].

the odour concentration is required and the nostrils are placed in the centre of the cone while the chin is placed on the edge of the cone. The cone acts as an odour diffuser where vapour from the substance in the jar mixes with air sucked through the jar.

The olfmeter is a device for comparing the concentration of different odour sources, but before it can be used personnel have to be trained to perceive certain 'milestone' concentrations. The reference odour used is acetone (a substance present in blood and urine) at concentration levels of 1, 5, 10 and 20 decipol (see Chapter 1 for definition). It is claimed that trained personnel can perceive acetone concentration to a mean deviation of 2 decipol. These reference concentrations are used to perceive odour concentrations from other sources as well as the air quality of room air. The olfmeter has also been used to study pollution sources in ventilation systems by evaluating the air quality upstream and downstream of each component in an HVAC system [32]. For interest, the study showed that the average increase in air pollution in eight HVAC plants as the air passes through each plant (excluding ductwork) was 0.8 decipol with the air filters contributing an average of 0.4 decipol.

9.9.4 Carbon dioxide

Carbon dioxide (CO_2) gas is present at different concentrations indoors with typical outdoor concentrations of 350–400 p.p.m. It is often used as a surrogate indictor of indoor air quality (see also Chapter 1) and CO_2 sensors are often used in controlling fresh air supply rates to a building. In addition, CO_2 is used as a tracer gas for measuring ventilation rates. Hence, CO_2 sensors are widely used in indoor air and ventilation system measurements and control. There are many different types of CO_2 sensors in use that are based on photoelectric, acoustic, electrochemical or chemical principles of measurements. The most common instruments are the non-dispersive infrared (NIDR) sensor and the photoacoustic (PA) sensor.

The NIDR measuring technique utilizes the strong absorption of infrared energy by a gas at a certain wavelength when excited by an infrared light. For CO_2 this occurs at a wave length of 4.2 μm and NDIR cells are designed to facilitate a rapid CO_2 sample diffusion to facilitate instantaneous recording of CO_2 concentrations. The cell consists of a gas permeable membrane, an infrared source and detector which, by suitable calibration, give the CO_2 concentration in the air sample. The accuracy of this type of sensors is, however, influenced by water condensation, temperature, air pressure, dust, dirt and mechanical shock. Frequent calibration is therefore necessary using an inert gas, e.g. nitrogen, for the lower (zero) limit and a calibration gas mixture of known CO_2 concentration for the upper limit.

The PA CO_2 sensor consists of either an open or a closed cell supplied with pulsed infrared light filtered by a rotating sector disc (chopper) to permit light at the characteristic absorption wavelength of 4.2 μm to enter the cell. The light energy absorbed by the CO_2 gas heats the cell, which then produces a pressure wave that is sensed by either a piezo-electric transducer or a microphone. The open cell sensor has a permeable membrane through which air is naturally diffused whereas the closed cell sensor uses two openings through which the air sample is pumped in and out of the chamber for analysis. The accuracy of the open-cell CO_2 sensor is susceptible to vibrations and is not usually available as a portable unit. Both types also require frequent calibration to correct for electronic drift, pressure changes and other environmental factors that influence accuracy.

9.9.5   *Indoor air quality*

Apart from the well-known indoor air pollutants like odour, CO_2 and aerosols, there are many other gases and vapours present with different concentration. To measure the concentration of individual pollutants is impractical for most purposes and instead indoor air quality sensors have been developed to measure the total concentration of VOCs and other indoor pollutants using a single instrument. The principles of operation of these sensors are similar to those sensors used for measuring CO_2 concentrations. They normally consist of a sensing element and an output signal conditioner. The Indoor Air Quality (IAQ) sensor responds to the aggregate effect of the pollutants in the air. Although some of the commercial IAQ sensors are inexpensive and can give real-time readout, these devices have limited accuracy and their calibration is not a simple exercise.

9.10   Tracer gas measurements

The use of tracer gases in ventilation studies and the common methods used for measuring air infiltration and ventilation rates is described in Chapter 3. The types and properties of suitable tracer gases are described in Section 3.4.2. Here, the instruments that are used for the measurement of tracer concentration in the air will be briefly described.

There are basically two techniques in use for measuring tracer gas concentrations: the indirect method in which case an air sample is collected from the measuring site for analysis in a laboratory; and the direct method in which case the tracer concentration is measured on site. In the first case, the sample can be collected using plastic bags or bottles, or the gas adsorbed using charcoal, silica gel or porous polymers. The gas sample collected using a bag or bottle can be directly analysed whereas the adsorbed gas in a sample will require flushing out by heating the sample or using a suitable solvent. There are two main categories of gas analysis equipment: infrared absorption spectroscopy (IRAS) and gas chromatography (GC). These two systems will be described briefly here but the interested reader should refer to [33] for more details of these and other systems in use.

9.10.1 Infrared absorption spectroscopy

This is based on the principle that when infrared light passes through a gas, some of the energy of a certain part of the spectrum (certain wavelength) is absorbed while the energy for other parts of the spectrum is transmitted without being absorbed. Each gas has maximum infrared energy absorption characteristics at a certain wavelength, which corresponds to the natural frequency of the vibrations of the gas molecules. Infrared spectrometers are either of the *dispersive* or *non-dispersive* type. In the dispersive type, the spectrometer is manually tuned to the frequency of the spectrum that is specific to the gas to be measured, whereas in the non-dispersive type the full band of infrared radiation spectrum is used. The dispersive instrument requires the use of optical filters to obtain a narrow bandwidth of infrared energy required for the gas to be measured. The non-dispersive spectrometer is of the type described in Section 9.9.4 for measuring CO_2 gas concentration.

As in the case of the PA CO_2 sensor explained in Section 9.9.4, the infrared spectrometer uses a pressure transducer or microphone to detect the changes in pressure due to the heating of the gas being analysed. In the non-dispersive spectrometer two cells are used, one containing the gas sample and another containing air, assumed free from tracer gas, as a reference. The signals from the two chambers are compared to give the concentration of the gas. To distinguish between the infrared signal used in the measurement from other infrared signals that may also be present, the light is interrupted using a rotating sector disc (chopper) to generate a fluctuating signal. Figure 9.29 shows a non-dispersive infrared gas spectrometer.

9.10.2 Gas chromatography

Chromatography is a technique for the separation of closely related compounds. The main components of a GC are a sample injection system, a column and a detector,

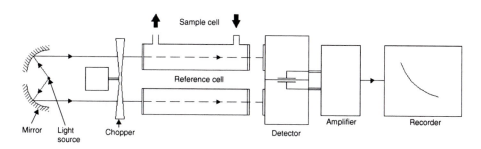

Figure 9.29 Non-dispersive type infrared gas spectrometer.

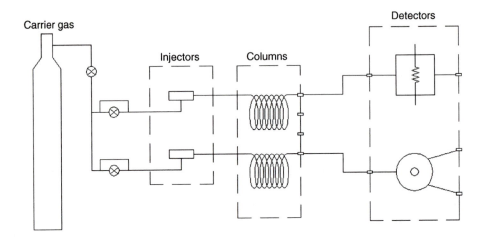

Figure 9.30 Basic outline of a GC.

see Figure 9.30. A GC, is one in which the 'mobile' phase is a gas or vapour. There are two basic types of GC, gas adsorption type and gas–liquid partition type. In gas–liquid chromatography, the 'mobile' phase is a gas and the 'stationary' phase is a liquid distributed on an inert solid. In gas–solid chromatography, the mobile phase is a gas and the stationary phase is an active solid such as alumina, activated charcoal, silica gel or porous polymer.

The sample injection system must be capable of introducing the sample unchanged instantly into the head of the column. The column is a rigid container made of metal, glass or other inert material, which contains the stationary phase and may have various shapes depending on the space available. The stationary phase is essentially a non-volatile liquid, which is capable of dissolving the sample component and then releasing it. The detector is an apparatus that measures the compositional changes in the mobile phase, which can be of numerous types, such as electron capture, photo-ionization, thermal conductivity, etc.

A small sample of air (≈ 0.5 c.c.) is carried into a separation column by a carrier gas (helium, hydrogen or nitrogen). The carrier gas is the mobile phase that is used to move the sample through the column. The gas component is detected when it reaches the end of the column, which generates an electrical signal as a function of time. Multi-gas samples may be analysed by using a multi-column GC.

Table 9.2 gives the minimum detection limits of commonly used tracer gases for the two types of analyzers discussed.

9.11 Air flow visualization

Flow visualization in a room or an enclosure is important in ventilation studies not only to provide a qualitative assessment of the flow pattern but in some cases it is used to yield quantitative information of the air motion. Because of the low air velocities involved, flow visualization methods in rooms are normally based on the reflection (scattering) of light by small solid or liquid particles introduced in the air stream, or on

Table 9.2 Detection limits of a selection of tracer gases

Tracer gas	Molecular mass	Density at NPT[a] ($kg\,m^{-3}$)	Maximum concentration (p.p.m.)		Minimum concentration (p.p.m.)		Comments
			For density	For safety	IR	GC	
Carbon dioxide (CO_2)	44	1.86	640	5000	0.4	—	Background concentration varies with occupancy
Nitrous oxide (N_2O)	44	1.86	640	25	0.2	—	Anaesthetic gas. Can form explosive mixture. Widely used as tracer
Sulphur hexafluoride (SF_6)	146	6.18	83	1000	0.01	1 part per 10^{13} air	Detection affected by other halogenated compounds in air. Decomposes into toxic components at 550°C. Widely used as tracer

Note
a Corresponding to 101.325 kPa and 15°C.

the refraction of light as a result of a difference in density of filaments or particles in the stream caused by heating the air, or injection of transparent gas of different refractive index from that of air. When solid or liquid particles are used for light scattering, it is important that they be of sufficiently small inertia to follow the local direction of the flow yet sufficiently large to reflect light for observation and photography.

9.11.1 Light-scattering methods

Smoke particles

Injection of smoke in the air is probably the most common method of visualizing the air motion in a room. The smoke particles, which can be liquid or solid, when illuminated cause scattering of light and their movement can then be traced and photographed. Because of low room air velocities, it is essential that the smoke used for visualization has a density close to that of air otherwise the effect of buoyancy could produce a false impression of the air motion. In addition, the smoke should be non-flammable, non-toxic, non-irritant and non-contaminating. Such a smoke can be produced by vaporization of suitable oil (e.g. Shell Ondina 17) in a portable smoke generator to which a long hose is attached to allow the smoke to cool before it enters the air stream. For local observations of air motion a smoke 'puff' produced chemically in a small smoke tube may be adequate.

Although smoke is a simple and cheap method of flow visualization, when used in an enclosure such as a room it quickly fills the space and requires a long time to disperse. This makes it suitable for only short visualization intervals. A typical visualization of air movement in an experimental test room at different time intervals is shown in Figure 9.31.

Localized flow visualization can be performed using a smoke filament and inter-mittent lighting. Photographic records taken at each lighting interval can produce qualitative and quantitative movement of the smoke filament with time, from which velocities may be obtained. This technique has some advantages over the usual smoke method whereby more detailed analysis of the air movement can be made and also measurements can be carried out over longer periods of time without filling the room with smoke because the smoke production rate is much smaller. The disadvantage is that a correction to the smoke velocity due to buoyancy is needed because of the high temperature of the smoke filament. Homma [34] used smoke produced by vaporizing a mixture of kerosene and paraffin using an electrically heated wire to study the convective currents around sedentary and standing person. Strings of smoke of length 100 and 300 mm were traced using computer-controlled stroboscopic lighting and photographic exposure. From these traces, velocities were calculated after correcting for the effect of buoyancy on the upward movement of the filament.

The effect of buoyancy associated with smoke visualization may be overcome by scattering laser light using liquid droplets (e.g. mist of liquid paraffin) or magnesium carbonate particles, which have strong light reflection characteristics. Enai *et al.* [35] used an argon laser and a scanning galvanometer to produce a two-dimensional light screen to observe the air movement patterns in a ventilated room. A 50 Hz laser shutter system was used to record the traces of reflected light on to a photographic film.

In another application using light reflected from liquid particles, Saunders and Albright [36] were able to determine the spatial distribution of contaminants in

Figure 9.31 Smoke visualization of impinging jet ventilation. Courtesy Dr T. Karimipanah.

a two-dimensional model enclosure. Assuming that glycerol aerosols injected in the model room are carried by the air motion in the same way as particles of contaminants, they used a narrow beam of white light to illuminate the length of the model. They further assumed that the intensity of light reflected by the glycerol particles was linearly proportional to the tracer particle concentration. By recording the reflected light intensity on a video camera and applying computerized image processing techniques, Saunders and Albright were able to obtain contours of constant relative light intensity which represented contaminant concentration contours.

Particle streak tracking and particle image velocimetry

'Particle streak tracking' (PST) and 'particle image velocimetry' (PIV) are used in many fluid flow applications to measure the velocity of tracer particles that follow the flow path. Recent advancements in image processing techniques and computer hardware have facilitated the application of these methods for flow visualization (qualitative) as well as velocity measurement (quantitative) in three-dimensional flow. In both cases a light source is projected across the flow field, which is seeded with particles, and the movement of particles is analysed. Usually, the light source is laser light and to cover a large area in a room, a laser sheet is used. The PST method is based on the measurements of an individual particle in the flow area of interest whereas the particle image velocimetry is based on analysing the movement of a number of particles with respect to one another. Thus the PIV method requires that the flow filed is seeded with a much larger number of particles than the PST method.

Kaga and Yoshikawa [37] used particle tracking and image processing techniques to trace the movement of smoke particles in three-dimensional ventilated models of enclosures. They traced the image of a particle in consecutive frames by recording the particle movement on a video recorder and identifying the same particle in different frames (known time intervals) to calculate the particle velocity. The photographed particle therefore appears as a streak. Scholzen and Moser [38] used a laser light sheet and three photographic cameras with synchronous triggering to measure the three-dimensional flow in a room seeded with helium-filled bubbles. Using three cameras solved some of the problems related to directional uncertainties of the streaks positions, however they found that synchronization of the cameras was critical particularly for velocities greater than about $0.5 \, \mathrm{m \, s^{-1}}$. Müller and Renz [39] used a low-cost light sheet system based on a commercial xenon laser lamp to perform two-dimensional measurements in three-dimensional room flows. The airflow was visualized by seeding small helium bubbles and analysing their movement using an image processing computer program. Linden *et al.* [40] used a PST system (they called particle streak velocimetry) to measure the velocity in an isothermal jet.

Because a PST method involves a complex optical and laser system for three-dimensional flow visualization, it is not suitable for field measurements and mainly used in laboratory measurements. PIV is a more suitable system for more general applications and is also available commercially. It allows measurement of particle velocity vectors at many points in the flow field. This system consists of a pulsed light sheet (usually laser) with appropriate optics to illuminate the flow that contains seeding particles. The time between two light pulses is varied according to the magnitude of the velocity being measured. The light scattered by the particles is recorded on a video camera for processing. Optical access to the flow is generally required at two locations, one for the light sheet and the other for the camera. The image fields are digitized (if not originally in digital format) and analysed on a computer to obtain the velocity vector images. The principle behind the PIV system is the measurement of small displacements, Δx and Δy, of the particles images in a given time step Δt (time between two light pulses) so that the velocity components, u and v (i.e. $\Delta x / \Delta t$ and $\Delta y / \Delta t$) are obtained. For accurate measurements, the pulse time Δt is chosen to be as small as possible for any given velocity field. The images of the particles are then digitized and processed using cross-correlation image processing techniques. Each calculated velocity vector represents the average displacement of the particle in a particular 'interrogation' cell. The global velocity field can be obtained by moving the interrogation cell across the image. Figure 9.32 shows sections across a three-dimensional flow field obtained using a volume mapping PIV system.

9.11.2 Light refraction methods

The refractive index, n, of an optical medium is related to the density, ρ, by the expression:

$$n = c\rho + 1$$

where c is a constant for the medium. The variation in air density due to pressure, temperature or both produces changes in the refraction of a light beam through an

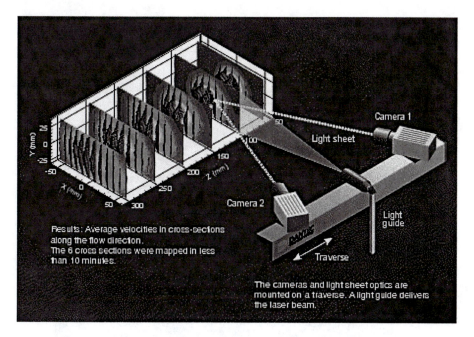

Camera 1

Light sheet

Camera 2

Light guide

Results: Average velocities in cross-sections along the flow direction.
The 6 cross sections were mapped in less than 10 minutes.

Traverse

The cameras and light sheet optics are mounted on a traverse. A light guide delivers the laser beam.

Figure 9.32 Flow field obtained using PIV. Courtesy of Dantec Dynamic A/S.

air stream, which may be used to form visual images of the flow. The same principle can also be applied to trace the movement of particles of a gas of different density injected in the air stream. There are three methods in use for the measurement of refractive index, or its derivatives, in air by observing the behaviour of a beam of light directed at the flow region of interest. These are the shadowgraph method, the Schlieren method and the Mach-Zehnder interferometer method. The principles of each method are explained in Pankhurst and Holder [41]. These methods are standard qualitative and quantitative flow visualization techniques in aerodynamic research particularly involving compressible flows. Generally however, they are not widespread in the visualization of large flow regions, such as in room air movement studies, because all density (i.e. refractive index) changes in the light path are integrated and this restricts their use to visualizing localized flow patterns only.

An interesting development of the Schlieren method for large-scale flow visualization in two- or three-dimensions has been carried by [42–44]. The system uses a light source with beamsplitter, a retroreflective 'source' grid made of parallel black and white strips, a focusing field lens, a cutoff grid, an image lens and an image plane, see Figure 9.33. The source grid has to be considerably larger than the test area (typically about twice the size) since the light convergence into the image lens placed ahead of the image plane. The cutoff grid is a photographic image of the source grid. The lens focuses the image from the source grid onto a sharply focused cutoff image, which is then adjusted to produce the desired cutoff. This lens also focuses the image from the test area onto a second focal (image) plane beyond the cutoff image. The image plane location is moved forward and backward to obtain a sharp focus of the image in the test plane.

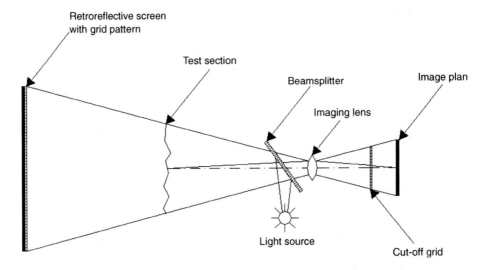

Figure 9.33 Retroreflective large-field Schlieren visualization system [43].

Figure 9.34 Large-field retroreflective Schlieren image [44].

This image can be photographed or analysed using image processing techniques to give the magnitude of velocity across the test plane. Figure 9.34 is an image produced using the large-scale Schlieren system described here. Modifications of this system are also available [42, 43].

References

1. Taylor, J. R. (1997) *An Introduction to Error Analysis – The Study of Uncertainties in Physical Measurements*, 2nd edn, University Science Books, California.
2. Vernon, H. M. (1932) The measurement of radiant heat in relation to human comfort. *J. md. Hyg.*, **14**, 95–111.
3. ISO Standard 7726 (2001) Ergonomics of the thermal environment – instruments for measuring physical quantities, International Standards Organisation, Geneva.
4. Humphreys, M. A. (1977) The optimum diameter for a globe thermometer for use indoors. *Ann. Occup. Hyg.*, **20**, 135–40.
5. McIntyre, D. A. (1980) *Indoor Climate*, Applied Science Publishers, London.
6. Funk, J. P. (1959) Improved polyethylene shielded net radiometer. *J. Sci. Instrum.*, **36**, 267–70.
7. Humphreys, M. A. (1974) Environmental temperature and comfort. *Build. Serv. Eng.*, **42**, 77–81.
8. Hellon, R. F. and Crockford, G. W. (1959) Improvements to the globe thermometer. *J. Appl. Physiol.*, **14**, 649–50.
9. Essing, G. and Steinhaus, I. (1982) Untersuchung und Verbesserung des Zeitverhaltens des Globe Thermometers. (Investigation and improvement of the adjustment time of the globe thermometer.) *Klima-Kälte-Heiz.*, part 4, **3**, 103–5.
10. Hurley, C. W. and Hasegawa, S. (1985) Humidity sensors for HVAC applications, *Proc. Int. Symp. Recent Advances in Control and Operation of Building HVAC Systems*, Trondheim, Norway.
11. ASHRAE Handbook of Fundamentals (2001) Ch. 14: Measurement and Instruments, American Society of Heating, Refrigeration and Air-Conditioning Engineers, Atlanta, GA.
12. Ower, E. and Pankhurst, R. C. (1966) *The Measurement of Air Flow*, Pergamon, Oxford.
13. Doebelin, E. O. (1990) *Measurement Systems: Application and Design*, McGraw-Hill, New York.
14. Finkelstein, W., Moog, W. and Fitzner, K. (1975) Measurement of room air velocities within air conditioned buildings, *Heat. Vent. Eng.*, May and June, 1975.
15. Hinze, J. O. (1975) *Turbulence*, McGraw-Hill, New York.
16. Melikov, A. K. (1998) Accuracy requirements and limitations for low velocity measurements, *Proc. Air Distribution in Rooms, Roomvent '98*, Vol. 2, pp. 577–84, Stockholm.
17. ASHRAE Standard-55 (1992) Thermal Environment Conditions for Human Occupancy, American Society of Heating, Refrigeration and Air-Conditioning Engineers, Atlanta, USA.
18. Welling, I., Koskela, H. and Hautalampi, T. (1993) Comparison of velocity transducers in laminar and turbulent flows, *Proc. Indoor Air '93*, Helsinki, Vol. 5, pp. 271–5.
19. Wasiolek, P. T., Whicker, J. J., Gong, H. and Rogers, J. (1999) Room airflow studies using sonic anemometry, *Indoor Air*, **9**, 125–33.
20. BS EN ISO 1042 (1997) Measurement of fluid flow by means of pressure differential devices. Part 1: Orifice plates, nozzles and Venturi tubes inserted in circular-section conduits running full, British Standards Institution, London.
21. Barton, J. S. and Saoudi, M. (1985) A fibre optic vortex flow meter. *J. Phys. E: Sci. Instrum.*, **19**, 64–6.
22. *BSRIA Application Guide* 1/75 (1975) Manual for regulating air conditioning installations, Building Services Research and Information Association, Bracknell, UK.
23. Svensson, A. (1983) Methods for measurement of air flow rates in ventilation systems, Swedish Institute of Building Research, English translation of Rep. No. T32, 1982, Gälve Sweden.
24. Phaff, H. (1988) Measurement of low air flow using a simple pressure compensating meter. *Air Infiltration Rev.*, **9** (4), 2–4.
25. Madsen, T. L. (1976) Thermal comfort measurements. *ASHRAE Trans.*, **82** (1), 60.

26. ISO 7243 (1989) Hot environments – estimation of the heat stress on working man, based on the WBGT-index (wet bulb globe temperature), International Standard Organisation, Geneva.
27. Hinds, W. C. (1999) *Aerosol Technology – Properties, Behavior, and Measurement of Airborne Particles*, 2nd edn, Wiley, New York.
28. Budnitz, R. J. (1974) Radon-222 and its daughters – a review of instrumentation for occupational and environmental monitoring. *Health Phys.*, 26, 145–63.
29. Walsh, P. J., Dudney, C. S. and Copenhaver, E. D. (1986) *Indoor Air Quality*, CRC Press, Boca Raton, FL.
30. Geerts, J. *et al.* (1987) Air quality control – measurements and experiences. *Energy-saving Control Applications (2)*, Staefa Control System AG, pp. 3–8.
31. Bluyssen, P. (1989) Olfbar. *Air Infiltration Rev.*, 10(2), 2–4.
32. Pejtersen, J. *et al.* (1989) Air pollution sources in ventilation systems, *Proc. 2nd World Congr. on Heating, Ventilation, Refrigeration and Air Conditioning – CLIMA 2000*, Vol. III, pp. 139–44.
33. Lodge, J. P. (ed.) (1989) *Methods of Air Sampling and Analysis*, 3rd edn, Lewis Publishers, Inc., Michigan, USA.
34. Homma, H. (1987) Free convection caused by metabolic heat around human body, *Proc. Int. Conf. on Air Distribution in Ventilated Spaces – Roomvent '87*, session 2a, Stockholm, June 1987.
35. Enai, M. *et al.* (1987) A feasibility study on open cooling – the characteristics of buoyant ventilation through high side openings, *Proc. Int. Conf. on Air Distribution in Ventilated Spaces – Roomvent '87*, session 2a, Stockholm, June 1987.
36. Saunders, D. D. and Albright, L. D. (1989) A quantitative air mixing visualisation technique for two-dimensional flow using aerosol tracers and digital imaging analysis, *Building Systems: Room Air and Air Contaminant Distribution* (ed. L. L. Christianson), ASHRAE, Atlanta, GA, pp. 84–7.
37. Kaga, A. and Yoshikawa, A. (1989) Velocity distribution measurement through digital processing of visualised flow images, *Building Systems: Room Air and Air Contaminant Distribution* (ed. L. L. Christianson), ASHRAE, Atlanta, GA, pp. 91–6.
38. Scholzen, F. and Moser, A. (1996) Three-dimensional particle streak velocimetry for room air flows with automatic stereo-photogrammetric image processing, *Proc. 5th Int. Conf. on Air Distribution in Rooms – Roomvent '96*, Vol. 1, pp. 555–62, Yokohama, Japan.
39. Müller, D. and Renz, U. (1998) A low cost particle streak tracking system (PST) and a new approach to three dimensional airflow velocity measurements, *Proc. 6th Int. Conf. on Air Distribution in Rooms (Roomvent '98)*, Stockholm, Vol. 2, pp. 593–600.
40. Linden, E., Todde, V. and Sandberg, M. (1998) Indoor low speed air jet flow: 3-dimensional particle streak velocimetry, *Proc. 6th Int. Conf. on Air Distribution in Rooms (Roomvent '98)*, Stockholm, Vol. 2, pp. 569–76.
41. Pankhurst, R. C. and Holder, D. E. (1952) *Wind-Tunnel Techniques*, Pitman, London.
42. Weinstein, L. (1993) Large-field high-brightness focusing Schlieren system, *AIAA J.*, 31, 1250–5.
43. Weinstein, L. (1998) Large field Schlieren visualization – from wind tunnels to flight, *Proc. VSJ-SPIE '98*, p. 13, Yokohama, Japan.
44. Settles, G. S. (2001) *Schlieren and Shadowgraph Techniques; Visualizing Phenomena in Transparent Media*, Springer-Verlag, Heidelberg, ISBN 3540661557.

Appendix A – Air infiltration calculation software

A.1   Air infiltration development algorithm (AIDA)

AIDA is a basic air infiltration calculation algorithm which was developed by the Air Infiltration and Ventilation Centre (AIVC) [A1, A2]. It uses a basic air infiltration and ventilation procedure intended for the calculation of air change rates in a single-zone enclosure. It also determines flow rates for any number of openings and calculates wind and stack pressures. The concepts for AIDA are outlined in chapter 12 of reference [A2]. The program is written in DOS BASIC and as its name suggests, it is a development algorithm that may be adapted to include additional flow paths, mechanical ventilation and other features. The program calculates the wind and stack pressure using equations (3.5) and (3.13) respectively at each flow opening and then solves the flow equation (3.39) iteratively. The solution of this equation is a combination of 'bisection' and 'addition'. Convergence is achieved when a flow balance within a defined limit is reached, which is usually fast.

A Windows® version of AIDA may be obtained from the following website:

http://freespace.virgin.net/vent.air/Software/software.htm

In the DOS version data entry is made in the following order:

- building volume (m^3)
- number of flow paths
- for each flow path:

  - height of flow path (m)
  - flow coefficient, k_i ($m^{-3}\,s^{-1}\,Pa^{-1}$)
  - flow exponent, n
  - wind pressure coefficient, C_p

- outdoor temperature, T_o (°C)
- internal temperature, T_i (°C)
- wind speed at building height, v ($m\,s^{-1}$)

Climate data is used to provide the values for the last three items. A listing of the BASIC code is given in Table A.1.

```
10 REM SET N
15 CLS
20 PRINT "Welcome to AIDA"
30 PRINT "Air Infiltration Development Algorithm"
40 PRINT "AIVC 1989"
50 DIM H(10),C(10),N(10),P(10),T(10),W(10),S(10),F(10)
55  PRINT:PRINT:PRINT
60 D=1.29 : REM Air Density at 0 Deg C
70 PRINT "Enter Building Data:"
80 INPUT "Building Volume (m3) = ";V
85  PRINT:PRINT:PRINT
90 PRINT "Enter Flow Path Data:"
100 INPUT "Number of Flow Paths = ";L
110 FOR J=1 TO L
115 PRINT:PRINT:PRINT
120    PRINT "Height (m)(Path";J;")    = ";: INPUT H(J)
130    PRINT "Flow Coef (Path";J;")    = ";: INPUT C(J)
140    PRINT "Flow Exp  (Path";J;")    = ";: INPUT N(J)
150    PRINT "Pres Coef (Path";J;")    = ";: INPUT P(J)
160 NEXT J
165 PRINT:PRINT:PRINT
170 PRINT "Enter Climatic Data:"
175 PRINT:PRINT:PRINT
180 INPUT "Ext Temp (Deg C)       =" ;E
190 INPUT "Int Temp (Deg C)       =" ;I
200 INPUT "Wind Spd(Bldg Ht)(m/s)=" ;U
210 REM Pressure Calculation
220 FOR J=1 TO L
230    REM Wind Pressure Calculation
240    W(J)=.5*D*P(J)*U*U
250    REM Stack Pressure Calculation
260    S(J)=-3455*H(J)*(1/(E+273)-1/(I+273))
270    REM Total Pressure
280    T(J)=W(J)+S(J)
290 NEXT J
300 REM Calculate Infiltration
310 CLS:PRINT:PRINT:PRINT
320 R=-100
330 X=50
340 Y=0
350 B=0
360 R=R+X
370 FOR J=1 TO L
390    O=T(J)-R
400    IF O=0 THEN F(J)=0: GOTO 430
410    F(J)=C(J)*(ABS(O)^N(J))*O/ABS(O)
```

```
420     B=B+F(J)
430 NEXT J
440 IF B<0 THEN R=R-X: X=X/2: GOTO 350
450 IF B<0.0001 THEN GOTO 470
460 GOTO 350
470 Q=0
480 FOR J=1 TO L
490     IF F(J)>0 THEN Q=Q+F(J)
495     PRINT "Infiltration through path (m3/s) ";J;"= ";F(J)
500 NEXT J
505 PRINT:PRINT:PRINT
510 REM Printing total infiltration
520 PRINT "Total infiltration rate (m3/s) = ";Q
530 A=Q*3600/V
540 PRINT "Air change rate (ach)          = ";A
545 PRINT:PRINT:PRINT
550 GOTO 170
```

Example A.1 To illustrate the application of AIDA, the air infiltration rate through a building with 3 flow paths is calculated and the input data and output results are given in Tables A.2 and A.3 respectively.

Table A.2 Input data

Building volume	$= 200\,\text{m}^3$		
Number of flow paths = 3	*Path 1*	*Path 2*	*Path 3*
Height of path (m)	1	3	6
Flow coefficient ($\text{m}^{-3}\,\text{s}^{-1}\,\text{Pa}^{-1}$)	0.05	0.03	0.04
Flow exponent	0.6	0.5	0.7
Wind pressure coefficient	0.4	−0.2	−0.45
Climate data	*Outdoor temp.* (°C)	*Inside temp.* (°C)	*Wind speed* ($\text{m}\,\text{s}^{-1}$)
Run 1	−1	20	0
Run 2	5	16	2
Run 3	22	22	4

Table A.3 Results

Infiltration rate ($\text{m}^3\,\text{s}^{-1}$)	Air change rate (h^{-1})	Path 1 flow rate ($\text{m}^3\,\text{s}^{-1}$)	Path 2 flow rate ($\text{m}^3\,\text{s}^{-1}$)	Path 3 flow rate ($\text{m}^3\,\text{s}^{-1}$)
0.798	1.436	0.0729	0.0069	−0.0798
0.0828	1.491	0.0828	−0.0125	−0.0703
0.1322	2.380	0.1322	−0.0320	−0.1002

A.2   Lawrence Berkeley Laboratory (LBL) model

The principles behind this single-zone model are described in Section 3.3.2 and given in more detail in references [A3, A4]. The equations given in Section 3.3.2 may be incorporated in a spreadsheet program to perform calculations using this model. The LBL website given below offers an on-line analysis program based on this model: http://epb.lbl.gov/ventilation/program.html

The user is required to answer a few questions about the building, its configuration, location, air tightness, etc. and the program calculates the air infiltration rate as well as the energy consumption. Both air infiltration and mechanical ventilation are considered and the program has dynamic defaults based on field measurements.

A.3   AIOLOS model

AIOLOS is a multi-zone air flow model which was developed under the ALTENER Energy Programme of the European Commission, DG XVII. The AIOLOS project was created for the development of passive ventilation in buildings. It is based upon the principles outlined in chapter 3 of reference [A5]. The calculations can be run for a short (few days) or extended (up to a year) time periods. The climate data can be statistically analysed using the AIOLOS climate pre-processor option. This useful feature offers the user the possibility to using climatic data for the region in which the building is located. It also incorporates a single-zone thermal model for calculating the cooling potential of natural ventilation techniques used in the building.

It is a Windows® based program with a main menu consisting of:

```
File  Climate  Edit  View  Calculations  Results  Sensitivity
Optimization  Thermal model  Cp Calc  Help
```

These functions and the use of the program with examples and typical output are explained in chapter 8 of Allard's book [A5]. The software code which runs under Windows® operating system is included in the CD-Rom attached to the book.

A.4   COMIS model

The Conjunction of Multizone Infiltration Specialists (COMIS) started as a one-year joint international research effort to develop a multi-zone infiltration model at the Lawrence Berkley Laboratory, USA during 1988–89. The task was to develop a detailed multi-zone programme taking in consideration crack flow, mechanical ventilation, single-sided ventilation and transport mechanisms through large openings. The one-year time period that was allocated to the development of the code was only used to select a suitable algorithm that was then coded under an International Energy Agency (IEA) research project. The code developed was evaluated in 1992 by tracer gas measurements, wind-tunnel data, inter-model comparison, and comparison with analytical solutions.

One of the major tasks of COMIS, and indeed all air infiltration algorithms, was to find a method for determining the wind pressure distribution for a building. This was accomplished in COMIS by developing a method based on a parametric study to determine the distribution of pressure coefficient, C_p, around a building. The algorithm is capable of modelling the flow through cracks and small openings, flow through large openings and also the flow due to mechanical ventilation. The mechanical ventilation part in the program is described in terms of the pressure losses in the duct components, air filters, cooling and heating coils, dampers, etc. using pressure loss coefficients for each component. The fan performance is expressed as a polynomial in terms of flow rate and pressure by the least-square method.

The solution of the non-linear flow equations between the pressure nodes is achieved using a modified version of the well-known Newton–Raphson method. The modification was made to avoid convergence problems associated with power functions which are particularly severe when the exponent in the crack flow equation is close to 0.5. The COMIS code is written in FORTRAN and was originally developed under DOS® environment but it is now also available under Windows® environment. A description of the COMIS model and its application is given in the COMIS 3.0 User Guide [A6]. A copy of the software can be downloaded from the website: http://eetd.lbl.gov/Software.html

A.5   CONTAMW model

CONTAMW [A7] is a multi-zone indoor air quality and ventilation analysis computer program for predicting:

- Air flow due to infiltration, exfiltration, and from room-to-room in buildings with natural (wind and stack effects) and mechanical ventilation.
- Contaminant dispersal and concentration calculations as a result of the predicted air flows.
- Personal exposure and risk assessment of occupants due to airborne contaminants.

The model was developed at the National Institute of Standards and Technology, USA [A7]. The computer programs **contamwz.exe** and **contamppz.exe** (CONTAMW Post Processor) which are needed for these analyses and the support data for CONTAMW may be downloaded from the website: http://bfrl.nist.gov/IAQanalysis. The program runs under Windows® operating system and Windows® graphical user interface.

References

A1 Liddament, M. (1989) AIDA – An air infiltration development algorithm. *Air Infiltration Rev.*, **11** (1), 10–12.

A2 Liddament, M. (1996) A guide to energy efficient ventilation, *Air Infiltration and Ventilation Centre*, International Network for Information on Ventilation, Brussels, Belgium (www.aivc.org).

A3 Sherman, M. H. and Grimsrud, D. T. (1980) Infiltration-pressurization correlation: Simplified physical modeling. *ASHRAE Trans.*, **86** (2), 778–807.

A4 Sherman, M. H. and Grimsrud, D. T. (1980) Measurement of infiltration using fan pressurization and weather data. *Proc. 1st AIC Conf. on Air Infiltration Instrumentation and Measuring Techniques*, 6–8 October 1980, Windsor, England.

A5 Allard, F. (ed.) (1997) *Natural Ventilation in Buildings: A Design Handbook*, James and James (Science Publishers), London.

A6 Feustel, H. E. and Smith, B. (eds) (1997) *COMIS 3.0 User Guide*, Lawrence Berkeley Laboratory, Berkeley, California, USA.

A7 Dols, W. S., Walton, G. N. and Denton, K. R. (2000) *CONTAMW 1.0 User Manual*, National Institute of Standards and Technology, Gaithersburg, MD, USA.

Appendix B – CFD codes

There are now a number of research and commercial CFD codes that are suitable for ventilation applications. Most commercial codes however have not been specifically developed for ventilation applications and may lack radiation, thermal comfort and indoor air quality assessment models. These codes tend to be very general and require considerable effort and tuition to benefit from their often advanced features, with few exceptions. Most of the commercial codes are available in trial versions and some are available as sharewares in limited editions with little or no charge. It is important however, to find out the important features of these codes before indulging into time-consuming exercise of discovering their limits. Many of the solvers used in research and commercial codes use FORTRAN 77 or 90 computer languages.

A brief description of some of the well-known CFD codes are given here.

VORTEX VORTEX© (Ventilation Of Rooms with Turbulence and Energy eXchange) is a 3D cartesian grid, finite volume code that uses the standard $k-\epsilon$ turbulence model and has been developed for application in buildings. Salient features are:

- User-friendly pre- and post-processing interfacing with default values
- Indoor pollution dispersion routine
- Surface-to-surface radiation routine
- Full thermal comfort analysis routine based on the PMV and PPD indices
- Age-of-air calculation routine
- Unsteady fire and smoke routines

A free trial version of 1000 cells with user manual is available free of charge from: www.rdg.ac.uk/~kcsawbi

FLOVENT FLOVENT© is a CFD software designed to calculate air flow, heat transfer and contamination distribution for the design and optimization of ventilation systems. It uses Cartesian grid system and has been widely used for simulating air flow in buildings. FLOVENT is available from Flowmerics Ltd, on: www.flowmerics.com

PHOENICS PHOENICS© (Parabolic Hyperbolic Or Elliptic Integration Code Series) is a general-purpose CFD package that predicts the fluid flow in and around engines, process equipment, buildings, human beings, rivers, lakes, oceans, etc. It has two-phase flow capabilities with associated changes in chemical or physical composition. The full version and limited editions of the software as shareware are available from Concentrated Heat and Mass Ltd, on: www.cham.co.uk

STAR-CD   STAR-CD© is a general-purpose CFD code suitable for steady-state and transient problems. It has flexible meshing capabilities for different cell shapes that make it suitable for adaptive meshing and solving problems with time-varying geometries by employing a grid marching technique. It offers alternative solvers including a parallelized algebraic multigrid option, a version optimized for high-speed compressible flows, free-surface multiphase and multiphysics capabilities. The code is available from CD Adapco Group on: www.cd-adapco.com

CFX   CFX© uses rectangular grid elements to avoid the problems associated with 'body-fitted coordinates' gridding. It employs an explicit numerical method to overcome the difficulties of under-relaxation associated with implicit methods for non-linear and coupled equations. The software is a coupled multigrid and unmatching grid solver with options for different turbulence models. It is capable of solving multi-phase flow, combustion and radiation problems. The current version CFX-5 is available from AEA Technology Ltd, on: www.software.aeat.com

FLUENT   FLUENT© uses the finite volume method with unstructured grid that can also handle moving and deforming meshes. It is capable of solving heat transfer with phase change problems, multi-phase flows, reacting flows, transient flows and has been extensively used in ventilation applications. FLUENT is available from Fluent Inc. on: www.fluent.com

CFD shareware codes   There are a number of shareware codes that are available on the website: www.icemcfd.com/cfd/CFD_codes_s.html

Index